Modeling and Simulation of Aerospace Vehicle Dynamics

Modeling and Simulation of Aerospace Vehicle Dynamics

Peter H. Zipfel
University of Florida
Gainesville, Florida

EDUCATION SERIES
J. S. Przemieniecki
Series Editor-in-Chief
Air Force Institute of Technology
Wright–Patterson Air Force Base, Ohio

Published by
American Institute of Aeronautics and Astronautics, Inc.
1801 Alexander Bell Drive, Reston, VA 20191-4344

American Institute of Aeronautics and Astronautics, Inc., Reston, Virginia

2   3   4   5

Library of Congress Cataloging-in-Publication Data

Zipfel, Peter H.
    Modeling and simulation of aerospace vehicle dynamics / Peter H. Zipfel.
        p.   cm.—(AIAA education series)
    Includes bibliographical references and index.
    1. Aerodynamics–Mathematics.   2. Airplanes–Mathematical models.
    3. Space vehicles–Dynamics–Mathematical model.   I. Title.   II. Series.
    TL573.Z64 2000   629.132′3′0151–dc21   00-046444
    ISBN 1-56347-456-5 (alk. paper)

"To Him who sits on the throne and to the Lamb
be praise and honor and glory and power,
for ever and ever!"
Revelation 5:13b (NIV)

Foreword

Modeling and Simulation of Aerospace Vehicle Dynamics by Peter H. Zipfel is an excellent introduction to the important subject of computer modeling and simulation of dynamics of aerospace vehicles that in recent years has evolved into a major discipline. This new discipline is used not only in the design process but also in the development and improvement of performance and operation of civil and military aircraft and missiles. The text is divided into two parts: Part 1 Modeling of Flight Dynamics and Part 2 Simulation of Aerospace Vehicles. Part 1 discusses the theoretical concepts that provide mathematical foundation for the simulation of aerospace systems. This includes frames of reference and coordinate systems, kinematics of translation and rotation, translational and attitude dynamics, as well as perturbation techniques used for modeling. In Part 2 the author describes in great detail the various types of simulations for aerospace vehicles for three-, five-, and six-degree-of-freedom systems, including real-time simulators.

Many of the AIAA Education Series texts include now either CDs or diskettes for computer programs, problem exercises, and any additional information. This author has introduced a novel approach of providing an Internet service for distributing such materials directly. This additional material for the present text can be obtained from the CADAC Web site, which can be accessed through the AIAA home page (www.aiaa.org) by selecting Market Pulse and then Web Links, where CADAC is listed. The advantage of this approach is obvious: it proves to be an easy avenue for disseminating any future new or updated materials (e.g., new classroom problems for the basic text). In writing this text, the author drew on his many years of experience as an educator at the University of Florida and as a research scientist with the U.S. Army and Air Force. This experience allowed him to produce an outstanding teaching text and a practical reference book on modeling and simulation of aerospace vehicles.

The AIAA Education Series of textbooks and monographs, inaugurated in 1984, embraces a broad spectrum of theory and application of different disciplines in aeronautics and astronautics, including aerospace design practice. The series also includes texts on defense science, engineering, and management. The books serve both as teaching texts for students and reference materials for practicing engineers, scientists, and managers. The complete list of textbooks published in the series (over 60 titles) can be found on the end pages of this volume.

J. S. Przemieniecki
Editor-in-Chief
AIAA Education Series

Table of Contents

Part 2   Simulation of Aerospace Vehicles

Preface

The time has come to give an account of modeling and simulation to aerospace students and professionals. What has languished in notebooks, papers, and reports should be made available to a wider audience. With modeling and simulation (M&S) penetrating technical disciplines at every level, engineers must understand its role and be able to exploit its strength. If you aspire to acquire a working knowledge of modeling and simulation of aerospace vehicle dynamics, this book is for you. It approaches modeling of flight dynamics in a novel way, covers many types of aerospace vehicles, and gives you hands-on experience with simulations.

The genesis of this text goes back to the years when the term M&S was still unknown. The challenges then were as great as today. Every new generation of computers was pressed into service as soon as it came on line. With analog computers, we could solve linear differential equations. Later, digital computers empowered us to master also nonlinear differential equations. Concurrently, flight dynamics evolved from Etkin's linearized equations to today's dominance of nonlinear equations of motion.

As computers became more powerful, the tasks grew more complex. The fidelity of models increased, the number of vehicles multiplied, and coordinate systems abounded. In the late 1960s, as I worked on my dissertation, it became clear that these complex models called for compact computer coding. Matrices are the conduit, and tensors are the theoretical underpinning. Thus evolved the invariant modeling of flight dynamics, my contribution to M&S.

In the late 1970s, I began to teach this approach at the University of Florida. What was first called "Advanced Flight Mechanics I and II" became in the 1990s "Modeling and Simulation of Aerospace Vehicles." In the meantime, as I worked for the U.S. Army and Air Force, I had the opportunity to apply these techniques to rockets, missiles, aircraft, and spacecraft.

Thus matured the two tracks of this book: invariant modeling of flight dynamics and computer simulations of aerospace vehicles—theory and praxis. The first part lays out the mathematical foundation of modeling with Cartesian tensors, matrices, and coordinate systems. Replacing the ordinary time derivative with the rotational time derivative and using the Euler transformation of frames enables the formulation of the equations of motions in tensor form, invariant under time-dependent coordinate transformations. Newton's law yields the translational equations, and Euler's law produces the attitude equations. Perturbation equations and aerodynamic derivatives complete the modeling of flight dynamics.

The second part applies these concepts to aerospace vehicles. Simple three degree-of-freedom (three-DoF) trajectory simulations are built for hypersonic aircraft, rockets, and single-stage-to-orbit vehicles. Adding two attitude degrees of freedom forms the five-degree-of-freedom (five-DoF) simulations. Cruise missiles, air intercept missiles, and aircraft simulations are introduced with flight controllers and guidance and navigation systems, culminating with six-degree-of-freedom (six-DoF) simulations of hypersonic aircraft, and missiles. Their components are modeled in greater detail. Aerodynamics, autopilots, actuators, inertial navigation systems, and seekers are matched with the full translational and attitude equations of motion. Real-time flight simulators and a glimpse at wargames round out the second part.

The aerospace vehicles discussed in this book find their actualization in the computer code stored on the CADAC Web site, which can be accessed through the AIAA home page (www.aiaa.org). Under the category Market Pulse, select Web Links to find CADAC among a list of Aerospace-related links. There you can locate eight complete simulations and four data sets for your own projects. The download is free of charge. I chose the Internet over the CD-ROM media because software is ever changing. A CD-ROM is stale and would be outdated at the time of publication, whereas the Web site is being updated periodically.

You can use the book in a formal class environment, or, with proper motivation, for self-study; some of you experts may just keep it as a reference manual. The following table gives suggestions for a one- or two-semester course. Chapters 2–7 can serve as a comprehensive study of flight dynamics with the complete nonlinear and linearized equations of motion. It could be followed by a second semester of immersion into flight vehicle simulations, using Chapters 8–11. If the students already have a solid foundation in flight dynamics, one semester could be devoted to just flight simulations, preceded by some familiarization with the notation.

Frequently, I use a third option. During a three-credit-hour course, I cover the essentials of modeling, Chapters 2–6, and introduce the students to simple simulations in a two-hour lab that meets every other week. The CADAC Primer, Appendix B, jumpstarts the computer orientation. Once Newton's law has been discussed, the students are prepared to work one of the projects of Chapters 8 or 9. For those who want to pursue six-DoF simulations in independent study, I assign Chapter 10 and one of its projects. A similar path can be chosen for self-study.

Suggestions for a one-/two-semester course

Chapter	1st semester	2nd semester	One semester with lab
1) Overview		Introductory reading	
Part 1			
2) Mathematical concepts			
3) Coordinate systems	Course in flight dynamics	—	Lecture
4) Kinematics			
5) Translational equations			
6) Attitude equations			
7) Perturbation equations			Optional study
Part 2			
8) Three DoF simulations		Training in flight-vehicle simulations	Lab
9) Five DoF simulations	—		
10) Six DoF simulations			Independent study
11) Real-time simulations			Optional reading

The problems at the end of each chapter are more than just exercises. Most of them relate to applications found in aerospace simulations. Within each chapter they increase in difficulty while also keeping pace with the development of the material. Some of them, labeled "Projects," are quite time consuming. Particularly the problems of Chapters 8–10 are better suited as semester projects. They challenge you to work with actual computer code and explore new designs. I trust the troika of instructional text, realistic problems, and prototype simulations delivers to you a complete learning environment.

I teach the course to aerospace (AE), operations research (OR), and electrical engineering (EE) students at the graduate level. Once in a while even a physicist may attend. The AE students come prepared with the prerequisite of a stability and control course, and the EE students, majoring in control systems, succeed also if they are willing to study the plant-to-be-controlled through some additional reading. Even physicists manage to make honorable grades. Part 1 can also be taught at the advanced undergraduate level, after the students have had an introductory course in dynamics. Part 2 requires some specialized knowledge in subsystem technologies. Particularly, Chapter 10 assumes familiarity with aerodynamics, classical and modern control, and stochastic effects. If you are a practicing engineer in the aerospace industry, you should be able to master the book even without a tutor.

I am indebted to my teachers Hermann Stuemke and Bertrand Fang who stirred in me the enthusiasm for flight mechanics and modeling techniques with tensors and matrices. My students are always an inspiration to me with their probing questions. Hopefully, they will find all the answers here. I must name four of them for their diligent review of the manuscript: Becky Hundley, Phil Webb, Chris Dennison, and Vy Nguyen. They rose to the challenge to spill red ink over the professor's work for the promise of better grades. Pat Sforza, my director, at the Research Center, was always ready with encouragement and useful suggestions. I thank him and Al Baker, my faithful colleague over two decades, for their cover-to-cover review of the manuscript. I extend also my thanks to Lynn Deibler, who reviewed the sections on radars and electro-optical sensors and made sure that I would not mistreat the radar range equation.

The members of my family were my cheerleaders. My daughter Heidi baked a bountiful supply of German Lebkuchen as "brain food" and my daughter Erika provided the champagne for our celebrations. Giving his dad some sorely needed advice, Jacob refereed the usage of the English language so that it would not come across like a German translation. Above all, my wife Barbara sustained me with her humor, despite my neglecting our nightly chess game.

Peter H. Zipfel
August 2000

Nomenclature

a_B^A = acceleration (first-order tensor) of point B wrt frame A

$D^A(*)$ = rotational time derivative of a vector or tensor $*$ wrt frame A

$\mathrm{d}(*)/\mathrm{d}t$ = ordinary time derivative

E = identity tensor

I_C^B = inertia tensor (second-order tensor) of body B referred to point C

l_C^{BA} = angular momentum (first-order tensor) of body B wrt frame A referred to point C

m^B = mass (zeroth-order tensor or scalar) of body B

P = projection tensor (second-order tensor)

p_B^A = linear momentum (first-order tensor) of particle B wrt frame A

R^{BA} = rotation tensor (second-order tensor) of frame B wrt frame A

s_{BA} = displacement vector (first-order tensor) of point B with respect to (wrt) point A

$\bar{s}_{AB}, [\bar{*}]^A$ = transposed vector or matrix $*$

T^{BA} = kinetic energy (scalar) of body B wrt frame A

v_B^A = velocity vector (first-order tensor) of point B wrt frame A

ω^{BA} = angular velocity vector (first-order tensor) of frame B wrt frame A

$[*]^A$ = vector or tensor $*$ expressed in the A coordinate system

Note: Scalars are represented by lower-case characters, vectors by bold lower-case characters, and tensors by bold upper-case characters. A subscript signifies a point and a superscript signifies a frame.

List of Acronyms

AGL	advanced guidance law
AMRAAM	advanced medium range air-to-air missile
ARDC	Air Research and Development Command
ASRAAM	advanced short-range air-to-air missile
BOGAG	bunch of guys and gals
BVR	beyond visual range
c.g.	center of gravity
c.m.	center of mass
CAD	computer aided design
CADAC	Computer Aided Design of Aerospace Concepts
CEO	chief executive officer
CEP	circular error probable
CIC	close-in combat
CRT	cathode ray tube
DEP	deflection error probable
DIS	distributed interactive simulation
DoF	degrees of freedom
EO	electro-optical
GPS	global positioning system
HIL	hardware-in-the-loop
HLA	high level architecture
IMU	inertial measuring unit
INS	inertial navigation system
IP	initial point
IR	infrared
ISO	International Standards Organization
LAR	launch acceptable region
LOA	line-of-attack
LOS	line-of-sight
M&S	modeling and simulation
MC	Monte Carlo
MOI	moment of inertia
MRAAM	medium range air-to-air missile
MRE	mean error probable
NASP	National Aerospace Plane
NOAA	National Oceanographic and Atmospheric Agency
PN	proportional navigation
RCS	reaction control system
RF	radio frequency
S/N	signal to noise ratio

SRAAM	short-range air-to-air missile
SSTO	single stage to orbit
TF/OA	terrain following/obstacle avoidance
TM	transformation matrix
TVC	thrust vector control
wrt	with respect to
WVR	within visual range

1
Overview

Imagine engineers without computers! It is true the great aeronautical discoveries were made without millions of transistors in pursuit of an optimal design. Heinkel used beer coasters to sketch out his famous airplanes. However, without digital computers solving the navigation equations, Neil Armstrong would not have set foot on the moon. It was in that decade, the 1960s, that I replaced my slide rule first by analog and then by digital computers. Certainly, I have no desire to return to the "good old days."

With the blessing of computers came also the curse to feed the beasts. They are insatiable, devouring innumerable lines of code. Who feeds them? Engineers do. Today, we design a big airplane like the Boeing 777 without a scrap of paper. Yes, we develop and use computer tools lavishly, but also try to keep our identity as visionaries of air and space travel.

In the following chapters I help you to model and simulate your visions. We presume that the design already exists and is defined by its subsystems, like aerodynamics, propulsion, guidance and control. You will learn how to formulate the dynamic behavior of your vehicle in a concise mathematical form and how to convert this model into computer code. You will write your own simulations in CADAC, a PC-based set of dynamic modeling tools. With its graphic charts you can promote your design among your peers.

We will use tensors to model vehicle dynamics, independent of coordinate systems. The simplest form of Cartesian tensors will suffice. They will serve us better than the vector formulation of so-called vector mechanics. The tensor's invariance under time-dependent coordinate transformations is a crucial characteristic in a dynamic environment that features a plethora of coordinate systems.

For programming we convert the tensor model into matrices by introducing suitable coordinate systems. Modern computers love to chew on matrices. Even the latest version of the venerable FORTRAN language features intrinsic matrix functions and instructions for parallel processing. So let us abandon the old habit of scalar coding and replace it by compact matrix expressions.

The poor man does flight testing on computers. Call it virtual testing, testing in cyberspace, or just plain computer runs. Instead of hardware, you build simulations and fly them without the whole world witnessing your new creation hitting the dirt.

Come join me in this adventure of modeling and simulation of aerospace vehicles. The ride will not be easy. There are some mathematical hairpins in the road we have to negotiate together. Once at the top, you can simulate all of the visions before you and only the sky will be the limit. Yet, what is even more important, you will have become a better engineer.

1.1   Virtual Engineering

Engineers are practical people. That is the original meaning of *virtual*; "being something in practice." The Encarta World English Dictionary[1] expands its meaning to "simulated by a computer for reasons of economics, convenience or performance." We deduct that *virtual engineering* is computer-based engineering for the sake of increased productivity.

The computer has replaced slide rules, spirules, drawing boards, mockups, and sometimes even brassboards and breadboards. *Virtual prototyping* has become the Holy Grail. Engineers are challenged to design, build, and test a prototype without ever bending metal. Will we ever reach this goal of so-called simulation-based acquisition? Let the future be the judge.

Modeling and simulation are important elements of virtual engineering. They do not replace creativity, but enable the engineer to define the design and explore its performance. With our emphasis on dynamic systems, modeling means the following to us: formulating dynamic processes in mathematical language. The foundation is physics, the blocks are the vehicle components, and calculus is the mortar that joins them together. The simulation is the finished structure, programmed for computer and ready for execution.

With modeling completed and the simulation validated, we have a powerful tool to carry out these important tasks:

1) Developing performance requirements—A variety of concepts are simulated to match up technologies with requirements and to define preliminary performance specifications.

2) Guiding and validating designs—Before metal is cut, designs are tested and validated by simulation.

3) Test support—Test trajectories and footprints are precalculated, and test results are correlated with simulations.

4) Reducing test cost—A simulation, validated by flight test, is used to investigate other points in the flight envelope.

5) Investigating inaccessible environments—Simulations are the only method to check out vehicles that fly through the Martian atmosphere or land on Venus.

6) Pilot and operator training—Thousands of flight simulators help train military and civilian pilots.

7) Practicing dangerous procedures—System failures, abort procedures, and extreme flight conditions can be tried safely on simulators.

8) Gaining insight into flight dynamics—Dynamic variables can be traced through the simulation, and limiting constraints can be identified.

9) Integrating components—Understanding how subsystems interact to form a functioning vehicle.

10) Entertainment—It is fun to fly simulators.

The history of modeling and simulation spans less than a lifetime. The first flight simulator was built by Link in the 1930s. It was a mechanical device with a simple cockpit, tilting with the pilot's stick input. The instructor used it to teach the fledgling student the three aircraft attitude motions: yawing, pitching, and rolling. When the first analog computers were introduced in the early 1960s, the linearized equations of motion of an aircraft could be solved electronically. I used a British-built Solartron computer while working on my Master's thesis on "Stability Augmentation of Helicopters" at the Helicopter Institute in Germany.

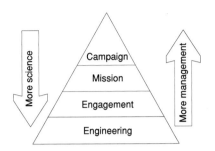

Fig. 1.1 Hierarchy of modeling and simulation.

During that time, the first digital computers came on line, but were still incapable of solving differential equations. In the 1970s analog computers were combined with digital computers. These hybrid computers were able to simulate the nonlinear vehicle motions and any other subsystem of interest. The high-frequency motions, like body-bending, rate, and acceleration control loops, were calculated by the analog circuitry, whereas the nonlinear equations of motion were solved by the digital components. Hybrid computers dominated the simulation industry for two decades. Today, advances in digital computing have made hybrid computers obsolete.

The ever-increasing computer power is harnessed at all levels of design, testing, and management. A hierarchy of modeling and simulation (M&S) has congealed at four distinctive levels of activities: engineering, engagement, mission, and campaign (see Fig. 1.1). Though the names have military connotations, they also apply to civilian enterprises. A good exposition can be found in the book *Applied Modeling and Simulation*.[2] I will just give a brief description of the four levels.

Engineering M&S provides the tools for design tradeoff at the subsystem and system level. It supports the development of design specifications, as well as test and performance evaluations. Physical laws shape the models. For instance, Newton's and Euler's laws generate the equations of motions, the radar range equation establishes the acquisition range, and the Navier–Stokes equation predicts aerodynamic forces. Engineering M&S establishes measures of performance for subsystem and systems. The majority of this book is devoted to its advancement.

Engagement M&S determines the effectiveness of systems. As they interact, reliability, survivability, vulnerability, and lethality are established. The scenarios are limited to one-on-one or few-on-few entities. For example, air combat simulators provide military measures of effectiveness, and air traffic simulators establish optimal approach patterns at airports. Engagement M&S is based on engineering M&S, but sacrifices fidelity to accommodate complexity. Chapter 11 covers flight simulators and in particular missile implementations in combat simulators.

Mission M&S investigates how operational goals are achieved. It broadens the scope to a greater number of players, both cooperative and adversarial. Some examples are the following: How can an airline beat the competition on the transatlantic route, or how can a carrier battle group defuse the tensions in the Persian Gulf? As various scenarios play out, measures of operational effectiveness are used to determine the best course of action.

Campaign M&S engages decision makers in broad-scale conflicts. Battle commanders practice winning strategies, and chief executive officers (CEOs) prepare for the next company takeover. Fidelity of individual models gives way to the emphasis on interplay amongst the myriad of elements, occupying large playing fields. With the emphasis on the outcome of the conflict, measures of outcome are derived to help congeal the best strategies. The ultimate campaign M&S is the war game. In Chapter 11 you have the opportunity to sample the art of wargaming.

The foundation of M&S is the engineering simulation, which establishes the performance of individual systems, based on scientific principles. As we climb the pyramid, the interplay of systems becomes more important. Synergism, tactics, and strategy exploit their performance for success. Scientific models yield to management principles, inert objects to human decision making.

With M&S penetrating so many technical and managerial disciplines, the paramount question becomes, can we trust the results. Is the simulation *verified, validated*, and *accredited*? Was the simulation built correctly according to specifications, was the right simulation built to do the job, and is it the rightfully accepted simulation for the study. These requirements are difficult to satisfy and often the time tried-and-true model wins by default.

Instead of roaming the esoteric heights of military campaigns, we will spend most of our time building the foundation of engineering models and simulations. Scientific principles will guide our venture, and high fidelity will characterize our simulations. M&S methods are like designing a model airplane, then building and flying it. You draw up specifications, lay out schematics, build the structure, and exercise the finished product.

M&S are demanding activities. Theoretical proficiency is paired with practical engineering skills. Because we are dealing with aerospace vehicles, I lay a solid foundation of flight dynamics, not shying away from some difficult modeling tasks. Chapters 2–7 are devoted to it under the umbrella of Part 1, Modeling of Flight Dynamics. Part 2, Simulation of Aerospace Vehicles, combines the dynamic equations with other engineering disciplines like aerodynamics, guidance, and control to fashion the simulations. Eight sample simulations should challenge you along the way.

As a virtual engineer you embrace the theoretical, practical, and programming challenges of M&S. Whether you are a novice or a seasoned veteran, I hope you will benefit from the following chapters. They are written to deepen your understanding of modeling of flight dynamics and to induce you to build sophisticated simulations of aerospace vehicles.

1.2   Modeling of Flight Dynamics

Flight dynamics is the study of vehicle motions through air or space. Unlike cars and trains, these motions are in three dimensions, unconstrained by road or rail. Flight dynamics is rooted in classical mechanics. Newton's and Euler's laws are quite adequate to calculate their motions. Relativistic effects are relegated to miniscule perturbations.

An aerospace vehicle experiences *six* degrees of freedom. Three *translational* degrees describe the motion of the center of mass (c.m.), also called the trajectory, and three *attitude* degrees orient the vehicle. If the c.m. of the vehicle is used as reference point, the translational and attitude motions can be described separately. Tracking a missile means recording the position coordinates of its c.m. Maintaining attitude of an aircraft requires the pilot to watch carefully the attitude indicator without reference to the aircraft's position.

Newton's second law governs the translational degrees of freedom and Euler's law controls the attitude dynamics. Both must be referenced to an inertial reference frame, which includes not just the linear and angular momenta but also their time derivatives. As long as the coordinate system is inertial, the equations are simple, but if body coordinates are introduced additional terms appear that make the adjustments for the time-dependent coordinate transformations.

My goal is to model flight dynamics in a form that is invariant under time-dependent coordinate transformations. To that end, these additional terms must be suppressed. A time operator, the rotational time derivative, will accomplish this feat. With it we can formulate the equations of motion in an invariant tensor form, independent of coordinate systems.

To clarify that approach, let me use Newton's second law as presented in any physics book. With p the linear momentum vector and f the external force vector, the time rate of change of the linear momentum equals the external force

$$\frac{d\boldsymbol{p}}{dt} = \boldsymbol{f}$$

Implied is that the time derivative is taken with respect to the inertial reference frame I. If we want to change the reference frame to the vehicle's body frame B, Newton's law must be written

$$\left.\frac{d\boldsymbol{p}}{dt}\right|_{B} + \boldsymbol{\omega} \times \boldsymbol{p} = \boldsymbol{f} \qquad (1.1)$$

with $\boldsymbol{\omega}$ the angular velocity of the body relative to the inertial frame. For programming, we have to coordinate the two equations. Because of the time derivatives, we express the first equation in inertial coordinates and the second one in body coordinates. Brackets and superscripts I or B indicate the coordinated vectors

$$\left[\frac{dp}{dt}\right]^{I} = [f]^{I}$$

$$\left[\frac{dp}{dt}\right]^{B} + [\Omega]^{B}[p]^{B} = [f]^{B}$$

where $[\Omega]^{B}$ is the skew-symmetric form of $\boldsymbol{\omega}$, expressed in body coordinates. The time derivative is *not* a tensor concept because it changes its form as the inertial coordinates are replaced by the body coordinates. It is not invariant under the transformation matrix $[T]^{BI}$ of the body coordinates with respect to the inertial coordinates, i.e., the right and left sides of the transformation are dissimilar:

$$\left[\frac{dp}{dt}\right]^{B} + [\Omega]^{B}[p]^{B} = [T]^{BI}\left[\frac{dp}{dt}\right]^{I}$$

If we introduce the rotational time derivative D^I relative to frame I, Newton's law has the same form in both coordinate systems,

$$[D^I p]^I = [f]^I$$

$$[D^I p]^B = [f]^B$$

and the rotational time derivative transforms like a first-order tensor:

$$[D^I p]^B = [T]^{BI}[D^I p]^I$$

With $[T]^{BI}$ representing any, even time-dependent, coordinate transformations, Newton's law can be expressed in the invariant tensor form

$$D^I p = f \qquad (1.2)$$

valid in any coordinate system. This tensorial formulation is the key to the invariant modeling of flight dynamics. It will allow us to derive the mathematical model first without consideration of coordinate systems. After having made desired changes, we pick the appropriate coordinate systems and code the component form.

The motto "from tensor modeling to matrix coding" will guide us through kinematics and dynamics to the simulation of aerospace vehicles. This approach has served me well over 30 years. I hope that you will also benefit from it by the diligent study of the following chapters.

The *second chapter*, "Mathematical Concepts in Modeling," lays the foundation through classical mechanics, a branch of physics. The axioms of mechanics and the principle of material indifference provide the sure footing for the modeling tasks.

With the hypothesis that points and frames are sufficient to model dynamic problems, I build a nomenclature that is self-defining. For instance, the displacement of missile M from the tracking radar R is modeled by the displacement vector s_{MR} of the two points, whereas the angular velocity of body frame B with respect to the Earth E is given by the angular velocity vector ω^{BE}. You will encounter other symbols that use points and frames, like linear velocity, angular momentum, moment of inertia, etc.

I permit only physical variables that are invariant under time-dependent coordinate transformations, that is, true tensor concepts. A construct like a radius vector has no place in our toolbox. Coordinate systems are abstract entities relating the components of a vector to Euclidean space. They have measure and direction, but no common origin. With these provisos we build our models with Cartesian tensors, as physical concepts, independent of coordinate systems.

With these tools we assail geometrical problems, like the near collision of two airplanes, both flying along straight lines; the miss distance of a missile impacting a plane; the imaging of an object on a focal plane array; and others. Problems at the end of the chapter invite you to practice your skills.

The *third chapter*, "Frames and Coordinate Systems," distinguishes carefully between the two concepts. Frames are models of physical objects consisting of mutually fixed points, but coordinate systems have no physical reality. They are, as already characterized in Chapter 1, mathematical abstracts. We make use of the nice properties of the transformation matrices between Cartesian coordinate systems. They are orthogonal, and therefore their inverse is the transpose. As the direction cosine matrix, they play an important part in flight mechanics.

No engineering discipline other than flight mechanics has to deal with so many coordinate systems. We will work with most of them: heliocentric, inertial, Earth, geographic, body, wind, and flight-path coordinate systems. We distinguish between round rotating Earth and flat Earth. In Chapter 10, I shall also introduce the oblate Earth and the geodetic coordinate system.

This chapter wraps up the modeling of geometrical problems. Do not underestimate their importance. In a typical aerospace simulation you may find that one-third to one-half of the effort is expended to get the geometry right. The next chapter leads us to the kinematics of flight vehicles.

The *fourth chapter*, "Kinematics of Translation and Rotation," introduces time and models the motions of vehicles without consideration of forces. We describe the translation of bodies by the displacement vector and their attitude by the rotation tensor. Their time derivatives are linear and angular velocities. It is here that I introduce the rotational time derivative, both for vectors and tensors. As already emphasized before, the rotational time derivative enables us to model flight dynamics by equations that are invariant under time-dependent coordinate transformations.

To shift reference frames, from inertial to Earth for instance, Euler's transformation is introduced. It is the generalization of the familiar form, shown in Eq. (1.1). Many derivations rely on it, particularly the formulation of the translational and attitude equations of motion. Shifting from the inertial to the Earth frame incurs such apparent forces as the Coriolis and centrifugal forces.

Finally in this chapter we solve the fundamental kinematic problem of flight dynamics, namely, given the body rates of the vehicle, determine the attitude angles. We take three approaches. The Euler method integrates the Euler angles directly with the penalty of singularities in the differential equations. Avoiding this disadvantage, the direction cosine and quaternion methods both solve linear differential equations. They are the preferred approach today because their higher computational load is no detraction any longer.

The *fifth chapter*, "Translational Dynamics," introduces Newton's second law for modeling the translational dynamics of aerospace vehicles. It is, together with Chapter 6, the heart of flight dynamics. Starting with the linear momentum, I formulate Newton's second law first for particles and then for rigid bodies. The earlier teaser on the invariancy of Newton's law will be fully developed. With Euler's transformation I derive the Coriolis and Grubin transformations for shifts in reference frames and reference points, respectively. You will also get the first taste of simulations from the derivation of the translational equations for three-, five-, and six-degree-of-freedom (DoF) models.

The *sixth chapter*, "Attitude Dynamics," formulates the attitude equations of motions based on Euler's law. Conventional wisdom says that the attitude equations are a consequence of Newton's law, but I will give evidence that Leonhard Euler developed them independently.

This chapter will challenge your mechanistic mind more than the rest of the book. I introduce the moment of inertia tensor with its axial and cross product of inertia. The moment of inertia ellipsoid gives a geometrical picture of the principal axes. As the linear momentum is at the center of Newton's law, so is the angular momentum the heartbeat of Euler's law. I start with particles and then expand the angular momentum to rigid bodies and eventually to clustered bodies. Euler's law states that the inertial time rate of change of the angular momentum equals the externally applied moments. Again, we use the rotational time derivative to present Euler's equation in tensor form, invariant under time-dependent coordinate transformations.

Now we are in a position to formulate the equations of motion of an aerospace vehicle and of a conventional spinning top. Of course, our emphasis is on free flight and on the significance of the c.m. of the vehicle. If the c.m. is used as reference point, Euler's equation simplifies greatly and becomes dynamically uncoupled from the translational equation. With l as the angular momentum and m the externally applied moment, we can formulate Euler's equation and combine it with Newton's Eq. (1.2) for the fundamental equations of flight dynamics:

$$D^I p = f, \quad D^I l = m \tag{1.3}$$

All modeling in flight dynamics begins with these equations. They are the backbone of six-DoF simulations.

The ultimate challenge is the formulation of the dynamics of clustered bodies. With the theorems and proofs you should be able to derive the equations of motion of a shuttle releasing a satellite, the swiveling nozzle of a missile, or an aircraft with rotating propellers.

Finally, I will introduce you to the mysterious world of gyrodynamics. The unexpected response of gyroscopes, their precession and nutation modes can easily be explained by Euler's law. With the energy theorem we derive two integrals of motion, the conservation of energy and angular momentum, which are pivotal for satellite dynamics.

The *seventh chapter*, "Perturbation Equations," completes the assortment of modeling techniques. Although perturbation equations are rarely used for full-up simulations, they are important for stability investigations and control system design. Even here I emphasize the invariant formulation of perturbations, which leads to component perturbations and the general perturbation equations of flight vehicles for unsteady reference flight.

The perturbations of aerodynamic forces and moments are given close attention. Taking advantage of the configurational symmetry of airplanes and missiles, vanishing derivatives of the Taylor series are sifted out and techniques presented for including higher-order derivatives.

As applications, we derive the roll, pitch, and yaw transfer functions for the autopilot designs of Chapter 10. More sophisticated examples are the perturbation equations of aircraft during pull-up, and of missiles executing high g maneuvers. These are illustrations of perturbation equations of unsteady reference flight, including nonlinear aerodynamic coupling effects.

Part 1 concludes here. It is a comprehensive treatment of Newtonian dynamics, sufficient for any modeling task in flight dynamics. The physical nature of the phenomena is emphasized by the invariant tensor formulation. Yet eventually, we have to feed our computers with instructions and numbers. That practical step is the subject of Part 2.

1.3 Simulation of Aerospace Vehicles

Having mastered the skills of modeling, you are prepared to face the challenge of simulation. The venture is not of a theoretical nature but one of encyclopedic knowledge of the subsystems that compose a flight vehicle. Who can claim to be an expert in aerodynamics, propulsion, navigation, guidance, and control all together? To be a good simulation engineer, however, you must be at least acquainted with all of these disciplines. In Part 2, I will expose you to these topics at increasing levels of sophistication. As we proceed from three- to six-DoF simulations, the prerequisites increase. You may have to do some background reading to keep up with the pace. Yet, let me also caution you that my treatment of subsystems is incomplete and that you must foster good relationships with experts in these fields to gain access to more detailed models.

Seldom will you be called to develop a simulation ex nihilo. Somebody has trodden that path before, and you should not hesitate to follow in his footsteps. At least pick up the outer shell, consisting of executive and input/output handling. A good graphics and postprocessing capability is also important. Then you can fill in the subsystem models and build your own vehicle simulation. But scrutinize the borrowed code carefully. Once you deliver your product, then you will be responsible for the entire simulation.

There are quite a few simulation environments from which you can choose. Appendix C gives you a selection. They are distinguishable by their programming language. All mature simulations are based on FORTRAN, with many years of verification and validation behind them. A new crop of symbolic simulations are available that use interactive graphics for modeling and a code generator to produce executable C code. That spawned another trend to build simulations in C^{++} directly, in adherence to its global penetration as a programming language.

The examples of Part 2 use prototype simulations from the CADAC (Computer Aided Design of Aerospace Concepts) environment. CADAC consists of the CADAC Studio and the CADAC Simulations. The Studio, written in Visual Basic, analyzes and plots the output and provides utility function for debugging. The CADAC Simulations encompass three-, five-, and six-DoF models. They are written in FORTRAN 77 with some common extensions. (A new $CADAC^{++}$ simulation environment is being developed exploiting encapsulation, inheritance, and polymorphism of the C^{++} language.) You can download these simulations, free of charge, from the CADAC Web site, and consult Appendix B for a quick overview. To access CADAC, go to the AIAA home page (www.aiaa.org), and under Market Pulse select Web Links where CADAC is listed.

Table 1.1 lists the prototype simulations. They encompass a broad selection of models from three to six DoF, from flat to elliptical Earth, from drag polars to full aerodynamic tables, from rocket to ramjet propulsion, and from simple to complex flight control systems. The number of lines of code gives you an idea of the size of the subroutines that model the subsystems of the vehicles.

Because practice makes perfect, you should attempt to carry out the projects at the end of Chapters 8–10. The required data are on the CADAC Web site. As you exercise your modeling skills, you add to you repertoire the simulations listed in Table 1.2: SSTO3 highlights the importance of trajectory shaping; AGM5 is an adaptation of the AIM5 simulation for the air-to-ground role; FALCON5 combines trimmed FALCON6 aerodynamics with the navigation aids of CRUISE5; and AGM6 is a detailed air-to-ground missile.

Table 1.1   Prototype simulations based on the CADAC architecture

DoF	Name	Type	Earth model	Lines of code
Three	GHAME3	NASA hypersonic vehicle	Spherical and rotating	1153
	ROCKET3	Three–stage-to-orbit rocket	Spherical and rotating	1048
Five	AIM5	Air intercept missile	Flat	1598
	SRAAM5	Short range air-to-air missile	Flat	5029
	CRUISE5	Subsonic cruise missile	Flat	5367
Six	SRAAM6	Short range air-to-air missile	Flat	5812
	FALCON6	F-16 aircraft	Flat	1339
	GHAME6	NASA hypersonic vehicle	Elliptical and rotating	4726

All of these simulations support the discussion of subsystem modeling, although the derivations in Chapters 8–10 are self-contained and apply to any simulation environment. We shall revisit the equations of motion, cover many aerodynamic modeling schemes, discuss all types of propulsion, design autopilots, and provide navigation and guidance aids where needed. Each chapter is devoted to one particular type of simulation.

The *eighth chapter*, "Three-Degree-of-Freedom Simulation," models point–mass trajectories. The three translational degrees of freedom of the c.m. of the vehicle are derived from Newton's second law for spherical rotating Earth and expressed in two formats. The Cartesian equations use the inertial position and velocity components as state variables, whereas the polar equations employ geographic speed, azimuth, and flight-path angles.

Here I introduce the environmental conditions, which are applicable to all simulations. The three most important standard atmospheres, ARDC 1959, ISO 1962, and US 1976, are compared. The analytical ISO 1962 model wins the popularity contest for simple endo-atmospheric simulations. Newton's law of attraction provides the gravitational acceleration. The term *gravity acceleration* is introduced for the apparent acceleration that objects are subjected near the Earth.

Aerodynamics is kept simple. Parabolic drag polars combined with linear lift slopes describe the lift and drag forces of aircraft and missile airframes. They

Table 1.2   Simulations you can build

DoF	Name	Type	Earth model	Project
Three	SSTO3	Single-stage-to-orbit vehicle	Spherical and rotating	Chapter 8
Five	AGM5	Air-to-ground missile	Flat	Chapter 9
	FALCON5	F-16 aircraft	Flat	
Six	AGM6	Air-to-ground missile	Flat	Chapter 10

are expressed in coordinates of the load factor plane. We touch on all types of propulsion systems: rocket, turbojet, ramjet, scramjet, and combined cycle engines. Although simple in nature, the propulsion models are used in many simulations, from three to six degrees of freedom.

The *ninth chapter*, "Five-Degree-of-Freedom Simulation," combines the three translational degrees of freedom with two attitude motions, either pitch/yaw or pitch/bank. We make use of a simplification that uses the autopilot transfer functions to model the attitude angles. This feature, i.e., supplementing nonlinear translational equations with linearized attitude equations, is called a pseudo-five-DoF simulation. As the examples show, it finds wide applications with aircraft and missiles.

These pseudo-five-DoF equations of motion are derived for spherical Earth and specialized for flat Earth. Because the Euler equations are not solved, the body rates are derived from the incidence rates of the autopilot and the flight-path angle rates of the translational equations. They are needed for the rate gyros of the inertial navigation systems (INS) and the rate feedback of gimbaled seekers.

Subsystems are the building blocks of simulations. I cover them at various levels of detail, either in Chapter 8, here, or in Chapter 10. Some of the treatment, especially aerodynamics and autopilots, is tailored to the type of simulation. However, the sections on propulsion, guidance, and sensors are universally applicable. Table 1.3 lists the features available to you.

A detailed description of the AIM5 simulation concludes the chapter. It exemplifies a typical pseudo-five-DoF simulation. As you follow my presentation, you will discover how the angle of attack, as output of the autopilot, is used in the aerodynamic table look-up. The guidance loop, wrapped around the control loop, exhibits the key elements: a kinematic seeker, proportional navigation, and miss distance calculations. If you want to work a simple, but complete missile simulation, the AIM5 model is the place to start.

The *tenth chapter*, "Six-Degree-of-Freedom Simulation," explores the sophisticated realm of complete dynamic modeling. The three attitude degrees of freedom,

Table 1.3   Subsystem features discussed in Chapter 9

Subsystem	Features	Section
Aerodynamics	Trimmed tables for aircraft and missiles	9.2.1
Propulsion	Turbojet, Mach hold controller	9.2.2
Autopilot	Acceleration controller, pitch/yaw and pitch/bank	9.2.3
	Altitude hold autopilot	
Guidance	Proportional navigation	9.2.4
	Line guidance	
Sensor	Kinematic seeker	9.2.5
	Dynamic seeker	
	Radars	
	Imaging infrared sensors	

Table 1.4   Subsystem features discussed in Chapter 10

Subsystem	Features	Section
Aerodynamics	Models for aircraft, hyper-sonic vehicles and missiles	10.2.1
Autopilot	Rate damping loop Roll position tracker Heading controller Acceleration autopilot Altitude hold autopilot Flight-path angle controller	10.2.2
Actuator	Aerodynamic control Thrust vector control	10.2.3
Inertial navigation	Space stabilized error model Local level error model	10.2.4
Guidance	Compensated proportional navigation Advanced guidance law	10.2.5
Sensor	IIR gimbaled seeker	10.2.6

governed by Euler's law, join Newton's translational equations. Creating a six-DoF simulation is the ambition of every virtual engineer.

We ease into the topic with the derivation of the equations of motion for flat Earth and its expansions to spinning missiles and Magnus rotors. Afterward, we accept the challenge and consider the Earth to be an ellipsoid. An excursion to geodesy will expose you to the geodetic coordinate system and the second-order model of gravitational attraction. All will culminate with the six-DoF equations of motion for elliptical rotating Earth, complemented by the methods of quaternion and direction cosine for attitude determination.

The description of subsystems is continued from Chapter 9 and summarized in Table 1.4. Whereas aerodynamics, autopilots, and actuators are partial to six-DoF simulations, the remaining three topics of inertial navigation guidance and seeker apply also to five-DoF models. The best way to master these diverse subjects is by experimenting with simulations. You will find all features modeled at least in one of the simulations SRAAM6, FALCON6, or GHAME6.

Monte Carlo analysis is the prerogative of six-DoF simulations. Their high fidelity, including nonlinearities and random effects, can only be exploited by a large number of sample runs, followed by statistical postprocessing. The methodology of accuracy analysis is discussed for univariate and bivariate distributions, with particular emphasis on miss-distance calculations.

Wind and turbulence is another field reserved for six-DoF models. With the standard NASA wind profile over Wallops Islands and the classic Dryden turbulence model, you can investigate environmental effects on your vehicle design. Because of the stochastic nature of the phenomena, the Monte Carlo approach will yield the most realistic assessment.

The *eleventh chapter*, "Real-Time Applications," gives you a taste of exploring the higher levels of the pyramid of Fig. 1.1. After having spent 10 chapters building the solid foundation of engineering simulations, you can lift your head and strive for piloted engagement simulations, hardware-in-the-loop facilities (HIL), or even participate in war games.

Flight simulators model the dynamic behavior of aerospace vehicles with human involvement. I discuss simple workstation and sophisticated cockpit simulators with their motion, vision, and acoustic environments. They find many uses, from control law development, flight-test analysis to pilot training.

When flight simulators are linked together, role playing can be staged. Blue fighters engage red aircraft, and blue and red missiles fly through the air. I will survey close-in air-to-air combat with its tactics and standardized maneuvers. Particularly, I will discuss the need for high-fidelity missile models and the proper use of five- and six-DoF simulations. To simplify the validation process, a real-time conversion process is described that prepares a complete CADAC model for the flight simulator.

A HIL facility combines hardware with software and executes in real time without humans-in-the-loop. Although expensive to build, it is indispensable for flight hardware integration and checkout. Our discussion will be brief, highlighting the main elements of flight table, target simulator, and main processor. Some of the elements of HIL simulators like aerodynamics, propulsion, and the equations of motion have to be implemented on the processor. Yet seekers, guidance and control systems can be hardware or software based; it just depends on the maturity of the development program.

Finally, let the games begin! Wargaming is an old art that has experienced a renaissance of unprecedented scope. The U.S. Armed Forces try to outdo each other at their annual games: Army After Next, Global (Navy), and Global Engagement (Air Force). You will kibitz a typical scenario and see how war games are built, conducted, and evaluated. But it will hardly make you a commanding general.

We will be content building the foundational engineering simulations on which engagement, mission, and campaign models rest. This book is intended to be your guide for modeling flight dynamics and simulating aerospace vehicles, providing you with virtually everything you need to become a better virtual engineer.

References

[1] *Encarta World English Dictionary*, Microsoft Encarta, St. Martin's Press, 1999.

[2] Cloud, D. J., and Rainey, L. B. (eds.), *Applied Modeling and Simulation: An Integrated Approach to Development and Operation*, Space Technology Series, McGraw-Hill, New York, 1998.

Part 1
Modeling of Flight Dynamics

2
Mathematical Concepts in Modeling

Modeling is a broad term with many meanings. Would it not be more exciting if this were a book about fashion models and a collection of pretty pictures? Well, a *model* is something uncommon or unreal. It is the copy of an object. The objects that I will focus on are inert, but nevertheless exciting. We are dealing with aircraft, spacecraft, and missiles. However, instead of building scaled replicas of these vehicles, we construct mathematical models of their dynamic behavior. Launching models is always more fun than just having them sitting on your shelf. I will teach you how to make them soar on your computer. But first we have to lay the foundation. Classical mechanics, a branch of physics, will be our cornerstone. Digging deep into the past, I found an interesting axiomatic treatment of the principles of mechanics. It will serve us well when we lay out the canon of modeling. Particularly useful is the principle of material indifference, which we will employ for several proofs.

The mathematical language we use consists of tensors and matrices. That may get you excited, but calm down—the bare essentials of Cartesian tensors will suffice. We will talk about frames, coordinate systems, transformation matrices, and so on, in a systematic order. If you are rusty in matrix algebra, brush up with Appendix A.

Of course, all theory is only as good as it is able to solve practical problems; at least that is the opinion of most engineers. I subscribe to that philosophy also and will show you in this chapter just how well tensors model geometrical problems. Throughout this book they will be our companions. Our motto is "from tensor modeling to matrix coding." Thus, expand your mind and go back to explore the future!

2.1 Classical Mechanics

At the turn of the last century, physicists thought that all of the laws of the physical universe were known. Over three centuries, Galileo, Newton, Bernoulli, D'Alembert, Euler, and Lagrange built the structure of the branch of physics that we call *mechanics*. Today, after another century of breathtaking progress in the physical sciences, we fondly remember that fully developed branch as *classical mechanics*. Although physicists have turned their back on it, engineers have explored it through many adventures, from first flight to a visit to the moon.

2.1.1 Elements of Classical Mechanics

So confident were the researchers that Hamel would write in the 1920s in the famous *Handbuch der Physik*[1] an axiomatic treatment of mechanics—an axiom is a statement that is generally accepted as self-evident truth. I follow Hamel's lead and delineate the basic elements of classical mechanics:

1) *Material body*: A body is a three-dimensional, differentiable manifold whose elements are called *particles*. It possesses a nonnegative scalar measure that is called the *mass distribution* of the body. In particular, a body is called *rigid* if the distances between every pair of its particles are time invariant.

2) *Force*: The force describes the action of the outside world on a body and the interactions between different parts of the body. We distinguish between volume forces and surface forces.

3) *Euclidean space-time*: The interaction of the forces with the material body occurs in space and time and is called an *event*. Events in classical mechanics occur in Euclidean space-time. The Euclidean space exhibits a metric that abides, for infinitesimal displacements ds, the law of Pythagoras over the three-dimensional space $\{x_1, x_2, x_3\}$:

$$ds^2 = dx_1^2 + dx_2^2 + dx_3^2 = \sum_{i=1}^{3} dx_i^2$$

The concept of a *particle*, so important in classical mechanics, defines a mathematical point with volume and mass attached to it. We could also call it an atom or molecule, but prefer the mathematical notion to the physical meaning. By accumulating particles we form material bodies with volume and mass. If the particles do not move relative to each other, we have the all-important concept of a rigid body.

Without forces, the body would, according to Newton's first law, persist at rest or continue its rectilinear motion. However, we shall have plenty of opportunity to model forces. There are aerodynamic and propulsive forces acting on the outside of the body as surface forces. We will deal with gravitational effects, which belong to the volume forces, acting on all particles, and not only on those at the surface.

In classical mechanics space and time are entirely different entities. Space has three dimensions with positive and negative extensions, but time is a uniformly increasing measure. For us, this so-called Galilean space-time model will suffice. However, we should remember that in 1905, just after the turn of the century, Albert Einstein revitalized physics with his Special Theory of Relativity, where time becomes just a fourth dimension.

Einstein did not abolish Newton's laws, but expanded the knowledge of space and time. He relegated Newton to a sphere where velocities are much less than the speed of light. However, that sphere encompasses all motions on and near the Earth. Even planetary travel is adequately represented by Newtonian dynamics, consigning relativistic effects to small perturbations.

2.1.2 Axioms of Classical Mechanics

Classical mechanics is the investigation of the interactions of material bodies and forces in Euclidean space-time. According to Hamel it is governed by four axioms[1]:

1) Time and space are *homogeneous*. There exists no preferred instant of time or special location in space.

2) Space is *isotropic*. There exists no preferred direction in space.

3) Every effect must have its cause by which it is uniquely determined. This is also called the *causality principle*.

4) No particular length, velocity, or mass is singled out.

Homogeneity of space is not natural to us. We think we are at the center of the universe and everything else turns around us. Yet we are just one reference frame. Every person can make the same claim. Homogeneity expresses the fact that all reference frames are equally valid, and therefore there is no preferred location in space. Does the sun revolve around Earth or Earth around the sun? Either statement is valid. It is just a matter of reference.

Homogeneity of time means that there exists no preferred instant of time. In the western world the Julian calendar begins with the birth of Christ, but other civilizations have their own calendars with different starting times. These are just arbitrary man-made beginnings. However, because time is a uniformly increasing measure, it must have had a beginning. That instant, when time was created, is distinct, but we do not know when it occurred. All other times have equal stature.

Space is not only homogeneous, but also isotropic, meaning that all directions in space have equal significance. On Earth we fly by the compass, which indicates magnetic north. But Mars probes navigate in an inertial, sun-centered frame, which is unrelated to terrestrial north. These are man's preferences. Space itself has no preferred direction.

We all have experienced the causality principle in our lives. I cut my finger (cause), and blood drips (effect). The pilot increases the throttle, the engine increases thrust, and the aircraft gains speed or altitude. There are two effects possible, speed and altitude, but each is uniquely determined by the thrust increase. All laws of classical mechanics abide by this causality principle.

The fourth axiom is a source of distress for all of those scientists who have tried for centuries to define the length of a meter. Eventually they agreed to make two marks on a bar of platinum and store it at the Bureau International des Poids et Mesures near Paris at a temperature of 0°C. There you also find the kilogram, well preserved for those who cherish precision. Yet, classical mechanics does not recognize any of these human endeavors.

Modern physics brakes with tradition and violates at least one of these axioms. In relativistic mechanics space is inhomogeneous and nonisotropic (Riemannian space); quantum mechanics does not recognize the causality principle; and the theory of relativity singles out the speed of light.

2.1.3 Principle of Material Indifference

Material bodies consist of matter whose behavior is modeled by *constitutive equations*. Because it is impossible to capture all of the nuances, special ideal materials are devised that approximate the phenomena. Their behavior is governed by constitutive equations.

When I searched the literature for basic modeling principles of material bodies, I found a very useful account by Noll[2] on the invariancy of constitutive equations. It was enshrined later in the new edition of the *Handbuch der Physik*, jointly authored by Truesdell and Noll.[3] These constitutive equations satisfy three principles:

1) *Coordinate invariance:* Constitutive equations are independent of coordinate systems.

2) *Dimensional invariance:* Constitutive equations are independent of the unit system employed.

3) *Material indifference:* Constitutive equations are independent of the observer. Or expressed in other words, the constitutive equations of materials are invariant under spatial rigid rotations and translations.

Material interactions do not depend on the coordinate system used for their numerical evaluations. As an example, the airflow over an aircraft wing and the resulting pressure distribution exist a priori, without specification of a coordinate system. You could record it in aircraft coordinates or, via telemetry, in ground coordinates. In both cases you would calculate the same lift. Or consider the thrust vector of a turbojet engine. It could be measured in aircraft or engine coordinates. The resultant force is still the same.

Does it matter whether you use metric or English strain gauges to record the thrust? You will get different numbers, but certainly the aircraft responds to the thrust unfettered by human schemes of measuring units. Physical phenomena transcend the artificiality of units.

The principle of material indifference, or, more precisely, the principle of material *frame*-indifference, as Truesdell and Noll[3] call it, is tantamount to the general theory of material behavior. It asserts "that the response of a material is the same for all observers."[3] Let the captain delight in the bulge of the sails or a dockside bystander conclude that a stiff easterly blows. Their emotions may be different, but, nevertheless, the bulge has not budged.

You may be part of an international calibration team. You take that norm-sphere and measure its drag in the wind tunnel at the University of Florida and then travel to Stuttgart, Germany, and repeat your test. If the measurements differ, you would not explain the discrepancy by the fact that the facilities are separated by 4000 miles and tilted by 67 deg with respect to each other (different longitude and latitude); rather, you would look for physical differences in the tunnels.

The Principle of Material Indifference (PMI) is the cornerstone of mathematical modeling of dynamic systems. It will enable us to formulate the equations of motions of aerospace vehicles in an invariant form and serve us to prove several theorems.

2.1.4   Building Blocks of Mathematical Modeling

With the general principles of classical mechanics under our belt, we employ a mathematical language that allows us to formulate dynamic problems concisely and to solve them readily with computers. We make use of two fundamental mathematical notions: *Points* are mathematical models of a physical object whose spatial extension is irrelevant. *Frames* are unbounded continuous sets of points over the Euclidean three-space whose distances are time invariant and which possess a subset of at least three noncollinear points.

Points and frames, although mathematical concepts, are regarded as idealized physical objects that exist independently of observers and coordinate systems. A point designates the location of a particle, but it is not a particle in itself. It does not have any mass or volume associated with it. For instance, a point marks the c.m. of a satellite; but for modeling the dynamics of the trajectory, the satellite's mass has to join the point to become a particle. Only then can Newton's second law be applied.

Combining at least three noncollinear points, mutually at rest, creates a frame. The best known frames are the *frames of reference*. Any frame can serve as a frame of reference. We will encounter inertial frames, Earth frames, body frames, and others. A frame can fix the position of a rigid body, but it is not a rigid body in itself. Only a collection of particles, mutually at rest, forms a rigid body. It is essential for you to remember that both, points and frames, are physical objects, albeit idealized.

Points and frames are the building blocks for modeling aerospace vehicle dynamics. I will show by example that they are the only two concepts needed to formulate any problem in flight dynamics. Surprised? Follow me and you be the judge and jury.

We need a mathematical shorthand notation to describe points and frames and their interactions in space and time. *Tensors* in their simple Cartesian form will serve us splendidly. They exist independently of observers and coordinate systems, and their physical content is invariant under coordinate transformations.

Coordinate systems are required for measurements and numerical problem solving. They establish the relationship between tensors and algebraic numbers and are a purely mathematical concept. Be careful however! Truesdell[3] warns, "In particular, frame of reference should not be regarded as a synonym for coordinate system." They are two different entities. Frames model physical objects, while coordinate systems embed numbers, called *coordinates*.

These coordinates are ordered numbers, arranged as matrices. *Matrices* are algebraic arrays that present the coordinates of tensors in a form that is convenient for algebraic manipulations. You will build simulations mostly from matrices. Computers love to chew on these arrays.

The modeling chain is now complete. The mathematical modeling of aerospace vehicles is a three-step process: 1) formulation of vehicle dynamics in invariant tensor form, 2) introduction of coordinate systems for component presentation, and 3) formulation of problems in matrices for computer programming and numerical solutions.

First, you have to think about the physics of the problem. What laws govern the motions of the vehicle? What are the parameters and variables that interact with each other? Which elements are modeled by points and which by frames? Then introduce tensors for the physical quantities and model the dynamics in an invariant form, independent of coordinate systems. Manipulate the equations until they divulge the variables that you want to simulate.

As a physicist you would be finished, but as an engineer your toil has just begun. You have to select the proper coordinate systems for numerical examination. What coordinate systems underlie the aerodynamic and thrust data? In what coordinates are the moments of inertia given? Does the customer want the trajectory output in inertial coordinates or in longitude and latitude? There are many questions that you have to address and translate into the mathematical framework of coordinate systems.

Eventually all equations are coordinated and linked by coordinate transformations. The tensors have become matrices and are ready for programming. Any of the modern computer languages enable programming of matrices directly or at least permit you to create appropriate objects or subroutines. Finally, building the simulation should be straightforward, although very time consuming.

2.1.5 Notation

Now we come to a nettlesome issue. What notation is best suited for modeling of aerospace vehicle dynamics? It should be concise, self-defining, and adaptable to tensors and matrices. By "self-defining" I mean that the symbol expresses all characteristics of the physical quantity. For intricate quantities it may require several sub- and superscripts.

Surveying the field, I go back to my vector mechanics book. There, as an example, velocity vectors are portrayed by symbols like v, $\vec{v}$, $\mathbf{v}$, or $\underline{v}$. An advanced physics book will most likely use the subscripted tensor notation, emphasizing the transformation properties of tensors. The velocity vector is written as v_i; $i = 1, 2, 3$ over the Euclidean three-space, and the transformation between coordinates is

$$v_i = t_{ij}v_j; \quad j = 1, 2, 3; \quad i = 1, 2, 3$$

with the summation convention over the dummy index j implied, meaning

$$v_i = \sum_{j=1}^{3} t_{ij}v_j; \quad i = 1, 2, 3$$

Draper Laboratory at the Massachusetts Institute of Technology has modified this convention, favoring the form

$$v^i = t^i_j v^j; \quad j = 1, 2, 3; \quad i = 1, 2, 3$$

as a vector transformation.

Our need is driven by our modeling approach, i.e., from invariant tensors to programmable matrices. Vector mechanics emphasizes the symbolic, coordinate-independent notation, whereas the tensor notation focuses on the components. We adopt the best of both worlds. Bolded lower-case letters are used for vectors (first-order tensors) and bolded upper-case letters for tensors (second-order tensors). For scalars (zeroth-order tensor) we use regular fonts. These are the only three types of variables that occur in the Euclidean space of Newtonian mechanics.

The sub- and superscript positions immediately after the main symbol are reserved for further specification of the physical quantity. Here we make use of our postulate that points, and frames suffice to describe any physical phenomena in flight dynamics. We fix indelibly the following convention: subscripts for points and superscripts for frames. For both we use capital letters. Some examples should crystallize this practice.

The displacement vector of point A with respect to point B is the vector s_{AB}; the velocity vector of point B with respect to the inertial frame I is modeled by v_B^I; and the angular velocity vector of frame B with respect to frame I is annotated by ω^{BI}. All three are first-order tensors. The moment of inertia tensor I_C^B of body (frame) B referred to the reference point C is a second-order tensor. If there are two sub- or two superscripts, they are always read from left to right, joined by the phrase "with respect to" (wrt).

For expressing the tensors in coordinate systems, we could use the subscript notation of tensor algebra or the sub/superscript formulation of the Massachusetts Institute of Technology. However, our sub- and superscript locations would become overloaded. I prefer to emphasize the fact that the tensor has become a matrix (through coordination) by using square brackets with the particular coordinate system identified by a raised capital letter. Let us expand on the four examples.

To express the displacement vector s_{AB} in Earth coordinates E, we write $[s_{AB}]^E$; the velocity vector v_B^I becomes $[v_B^I]^E$; and the angular velocity vector ω^{BI}, stated in inertial coordinates, is $[\omega^{BI}]^I$. All three are 3×1 column matrices. The moment of inertia tensor I_C^B, expressed in body coordinates B, is the 3×3 matrix $[I_C^B]^B$. Usually the bolding of the symbols will be omitted once the variable is enclosed in brackets, and we will write plainly $[s_{AB}]^E$, $[v_B^I]^E$, $[\omega^{BI}]^I$, and $[I_C^B]^B$.

The nomenclature at the front of this volume summarizes most of the variables that you will encounter throughout the book. I will adhere to these symbols closely, only changing the sub- and superscripts. Let me just point out a few things. All variables are considered tensors either of zeroth-, first-, or second order, but I will use mostly the term *vector* for the first-order tensor. The transpose is indicated by an overbar. We will distinguish carefully between an ordinary and rotational time derivative.

The advantage of the nomenclature lies in the clear distinction between coordinate-independent (invariant) tensor notation and the coordinate-dependent bracketed matrix formulation. General tensor algebra, with its sub- and superscript notation, emphasizes many types of tensors, e.g., covariant, contravariant tensors, Kronecker delta, and permutation symbol. The dummy indices and contraction (summation) play an important part. This mathematical language was created for the sophisticated world of general relativity embedded in Riemannian space. Our world is still Newtonian and Euclidean. Simple Cartesian tensors are completely adequate. Therefore, I forego the tensorial sub- and superscript notation in favor of the matrix brackets and am able to readily distinguish between the many coordinate systems of flight mechanics.

2.2 Tensor Elements

We attribute tensor calculus to the Italian mathematicians Ricci and Levi-Civita,[4] who provided the modeling language for Einstein to formulate his famous General Theory of Relativity.[5] More recently, tensor calculus is also penetrating the applied and engineering sciences. Some of the references that shaped my research are the three volumes by Duschek and Hochrainer,[6] which emphasize the coordinate invariancy of physical quantities; the book by Wrede,[7] with its concept of the rotational time derivative; and the engineering text by Betten.[8]

The world of the engineer is simple, as long as he remains in the solar system and travels at a fraction of the speed of light. His space is Euclidean and has three dimensions. Newtonian mechanics is adequate to describe the dynamic phenomena. In flight mechanics we can even further simplify the Euclidean metric to finite differences Δ, the so-called *Cartesian metric*

$$\Delta s^2 = \Delta x_1^2 + \Delta x_2^2 + \Delta x_3^2 = \sum_{i=1}^{3} \Delta x_i^2$$

The elements Δx_i are mutually orthogonal, and the metric expresses the Pythagorean theorem of how to calculate the finite distance Δs. In this world tensors are called *Cartesian tensors*. As we will see, they are particularly simple to use and completely adequate for our modeling tasks.

The elements of Cartesian tensor calculus are few. I will summarize them for you, discuss products of tensors, and wrap it up with some examples. Keep an open mind! I will break with some traditional concepts of vector mechanics in favor of a modern treatment of modeling of aerospace vehicles. Before we discuss Cartesian tensors however, we need to define coordinates and coordinate systems.

2.2.1 Coordinate Systems

Coordinates are ordered algebraic numbers called *triples* or *n-tuples*.

Coordinate systems are abstract entities that establish the one-to-one correspondence between the elements of the Euclidean three-space and the coordinates.

Cartesian coordinate systems are coordinate systems in the Euclidean space for which the Cartesian metric $\Delta s^2 = \sum_i \Delta x_i^2$ holds.

Coordinate axes are the geometrical images of mathematical scales of algebraic numbers.

Coordinate transformation is a relabeling of each element in Euclidean space with new coordinates according to a certain algorithm. A coordinate system is said to be *associated* with a frame if the coordinates of the frame points are time invariant. All coordinate systems embedded in one frame form a class $\aleph$. All classes over all frames form the entity of the *allowable* coordinate systems.

These definitions necessitate some explanations. Coordinates are arranged as numbered elements of matrices, e.g., the coordinates of the velocity vector v_B^I, expressed in the Earth coordinate system $]^E$, are

$$\left[v_B^I \right]^E = \begin{bmatrix} v_1 \\ v_2 \\ v_3 \end{bmatrix}$$

The triple occupies three ordered positions in the column matrix. The moment of inertia tensor, expressed in the body coordinate system $]^B$, exhibits the 9-tuple of ordered elements

$$\left[I_C^B \right]^B = \begin{bmatrix} I_{11} & I_{12} & I_{13} \\ I_{21} & I_{22} & I_{23} \\ I_{31} & I_{32} & I_{33} \end{bmatrix}$$

By the way, not every matrix and its elements constitute the coordinates of a tensor. There must exist a one-to-one correspondence between the three-dimensional

Euclidean space and the coordinates. For instance, the three velocity coordinates are related to the three orthogonal directions of Euclidean space by

$$\begin{bmatrix} v_1 \\ v_2 \\ v_3 \end{bmatrix} \Leftrightarrow \begin{bmatrix} \text{first direction} \\ \text{second direction} \\ \text{third direction} \end{bmatrix}$$

The moment of inertia tensor, on the other hand, has two directions associated with each element.

Because we are dealing with physical quantities, their numerical coordinates imply certain units of measure. The same units are embedded in every coordinate, e.g., v_1, v_2, and v_3 all have the units of meters per second. This requirement to give measure to the coordinates leads to the geometrical concept of coordinate axes. They can be envisioned as rulers, etched with the unit measures, and given a positive direction.

At this point we pause and compare coordinate systems with frames of reference. We defined a frame as a physical entity, consisting of points without relative movement. On the other hand, coordinate systems are mathematical abstracts without physical existence. This distinction is essential. Let me again quote Truesdell,[3] "It is necessary to distinguish sharply between changes of frame and transformation of coordinate systems." This separation will enable us to model the dynamics of flight vehicles in a coordinate-independent form, using points and frames, and defer the coordination and numerical evaluation until the building of the simulation.

Let us explore this conversion process. Given frame A and two of its points A_1 and A_2 (see Fig. 2.1), the displacement vector of point A_1 wrt A_2 is $s_{A_1 A_2}$. This vector is a well-defined quantity without reference to a coordinate system. Now we create a Cartesian coordinate system that establishes one-to-one relationships between the three-dimensional Euclidean space and the coordinates of the displacement vector. Designating it by $]^A$, we have one particular matrix realization

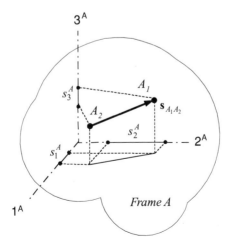

Fig. 2.1 Frame A and coordinate system $]^A$.

of the displacement vector

$$\left[s_{A_1 A_2} \right]^A = \begin{bmatrix} s_1^A \\ s_2^A \\ s_3^A \end{bmatrix}$$

The coordinates are shown in Fig. 2.1, superimposed on the coordinate axes. We label the axes in the $1 - 2 - 3$ sequence with the name of the coordinate system as superscript. If the coordinates do not change in time, the coordinate system $]^A$ is said to be *associated* with frame A. There are many, actually an infinite number of coordinate systems that have the same characteristic. They form a class $\aleph$, the so-called associated coordinate systems with frame A.

Moreover, there are other coordinate systems. Picture a spear A, whose center-line is modeled by the displacement vector $s_{A_1 A_2}$, with point A_1 marking the tip and A_2 the tail. We already discussed the coordinating in the associated coordinate systems of its frame A. But suppose, you as observer, modeled by frame B, watch the spear in flight. In a coordinate system $]^B$ associated with your frame, the centerline would have the coordinates

$$\left[s_{A_1 A_2} \right]^B = \begin{bmatrix} s_1^B \\ s_2^B \\ s_3^B \end{bmatrix}$$

However, the coordinates are now changing in time. Your frame has a whole class $\aleph$ of such coordinate systems, just like the frame of the spear. There could be many frames (persons) present. All of these classes of coordinate systems form an entity, called the *allowable* coordinate systems.

Converting from one coordinate system to another is a relabeling process:

$$\begin{bmatrix} s_1^B \\ s_2^B \\ s_3^B \end{bmatrix} \xleftarrow{\text{RELABELING}} \begin{bmatrix} s_1^A \\ s_2^A \\ s_3^A \end{bmatrix}$$

Because our coordinate systems are Cartesian, the relabeling algorithm is the multiplication of a 3×3 matrix with the coordinates of the vector

$$\begin{bmatrix} s_1^B \\ s_2^B \\ s_3^B \end{bmatrix} = \begin{bmatrix} t_{11} & t_{12} & t_{13} \\ t_{21} & t_{22} & t_{23} \\ t_{31} & t_{32} & t_{33} \end{bmatrix} \begin{bmatrix} s_1^A \\ s_2^A \\ s_3^A \end{bmatrix} \tag{2.1}$$

We symbolize the transformation matrix by $[T]^{BA}$, meaning that it establishes the $]^B$ coordinates wrt the $]^A$ coordinates. Notice my strict adherence to the double-index convention, reading from left to right, $B \rightarrow A$, and linking them with the words *with respect to*. Equation (2.1) is abbreviated by

$$\left[s_{A_1 A_2} \right]^B = [T]^{BA} \left[s_{A_1 A_2} \right]^A \tag{2.2}$$

The substance of the spear has not changed. We just have expressed the coordinates of its centerline in two different coordinate systems. If the coordinate systems are associated with frames that change attitude relative to each other, the elements of $[T(t)]^{BA}$ are, as in our spear example, a function of time. Only if they are part of the same frame or other fixed frames are the elements constant.

By convention and convenience we use only *right-handed* Cartesian coordinate systems. (This terminology refers to the motion of the right hand, symbolically rotating the 1-axes into the 2-axis—shortest distance—while the index finger points in the positive direction of the 3-axis.) We use them exclusively because they have the pleasant feature of their determinants always being positive one and their inverse equaling the transpose.

The discussion of Euclidean space would be incomplete without mentioning other coordinate systems that satisfy the Euclidean metric. Best known among them are the cylindrical and spherical coordinates. They are also orthogonal in the infinitesimal small sense of the Euclidean metric. However, only Cartesian coordinates satisfy the finite orthogonality of the Cartesian metric within the Euclidean space. With the definition of Cartesian coordinate systems in place, we can finally turn to the definition of tensors.

2.2.2 Cartesian Tensors

A **first-order tensor** (vector) x is the aggregate of ordered triples, any two of which satisfy the transformation law:

$$[x]^B = [T]^{BA}[x]^A \qquad (2.3)$$

where $]^A$ and $]^B$ are any allowable coordinate system.

A **second-order tensor** (tensor) X is the aggregate of ordered 9-tuples, any two of which satisfy the transformation law:

$$[X]^B = [T]^{BA}[X]^A[\bar{T}]^{BA} \qquad (2.4)$$

where $]^A$ and $]^B$ are any allowable coordinate system.

By aggregate I mean the collection of all possible transformations among all allowable coordinate systems. That class could contain as many as ∞^2 elements: an infinite number for every frame over infinite numbers of frames.

How useful is this definition with infinite components? We can take two viewpoints:

1) Physical point of view—Tensors describe properties of intrinsic geometrical or physical objects, i.e., objects that do not depend on the form of presentation (coordinate system).

2) Mathematical point of view—Because tensors are the total aggregate of all ordered n-tuples, they are defined in all coordinate systems and thus are not tied to any particular one.

As we model aerospace systems, we deal with the physical world. Therefore, I adopt the physical point of view in this book and interpret these definitions in an intrinsic or invariant sense. The physical quantities exist a priori, with or without coordinates. The introduction of coordinates is just an expedience for numerical evaluation.

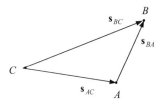

Fig. 2.2   Vector triangle.

A *scalar* is a particularly simple tensor. It remains the same in any allowable coordinate system, a useful characteristic that will serve us well in many proofs. We define a scalar as a physical quantity without any directional content, like mass, density, or pressure. Even particles and points belong in this category.

Some examples should clarify these definitions. Let us begin with the concept of a point and ask the question how do we determine its position? A point can have any position in space, or better, a point has no position unless related to another point, an intuitive postulate that follows from the Galilean relativity principle. The statement "I stand on the moon" uses the moon as reference. By just asserting "I stand," nothing meaningful has been conveyed. Suppose I am modeled by point B and the footprint of Neil Armstrong is point A. Then my displacement from A is given by the vector s_{BA}, and the displacement of the footprint with respect to me is s_{AB}. Both are related by $s_{BA} = -s_{AB}$. Notice the important Rule 1: Subscript reversal changes the sign. If Michael Collins, at point C, wants to determine my position s_{BC} and he knows the displacement of the footprint from him s_{AC}, and my displacement from the footprint s_{BA}, he can add the vectors to obtain the result (Fig. 2.2):

$$s_{BC} = s_{BA} + s_{AC} \qquad (2.5)$$

Rule 2 becomes evident: Vector addition is by subscript contraction. The subscripts A are deleted to get the BC sequence

$$BC \leftarrow B \underbrace{A + A}_{\text{contraction}} C$$

Notice also the location of the arrowheads in Fig. 2.2. They are always at the first point of the subscript. The displacement of point B wrt point C is s_{BC} with the arrowhead at point B.

To describe this scenario, there was no need to refer to any coordinate systems. We modeled the three entities by points and related them by invariant displacement vectors. For numerical evaluation, however, we have to choose coordinate systems. Suppose Collins wants to use a coordinate system $]^C$ associated with his frame of reference. He expresses all terms in Eq. (2.5) by $]^C$ coordinates

$$[s_{BC}]^C = [s_{BA}]^C + [s_{AC}]^C \qquad (2.6)$$

However, my displacement from the footprint is given in a coordinate system $]^B$, associated with my reference frame B. He therefore transforms my coordinates from $]^B$ to $]^C$ using the transformation matrix $[T]^{CB}$ according to Eq. (2.3):

$$[s_{BA}]^C = [T]^{CB}[s_{BA}]^B$$

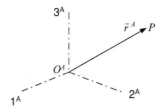

Fig. 2.3   Radius vector.

and substitutes the known coordinates into Eq. (2.6):

$$[s_{BC}]^C = [T]^{CB}[s_{BA}]^B + [s_{AC}]^C$$

If Collins knows the transformation matrix $[T]^{CB}$, he can compute my position in his coordinates.

Suppose his calculation furnishes $[\overline{s_{BC}}]^C = [30,100 \ -1,500 \ \ 2,800]$ km (I use the transposed to conserve space). To interpret the result, he has to know the direction and positive sense of the coordinate axes. If 1^C is forward, 2^C right, and 3^C down, then the numbers would indicate to him that I am 30,100 km in front of him; 1,500 km to the left; and 2,800 km below. Please observe an important fact: no mention of a coordinate system origin was made. The numbers make perfect sense without Collins being at the origin. You may be shocked, but here is Rule 3: Coordinate systems have no origins.

Without origins, we have to dispose of radius vectors. They are used in vector mechanics to locate points in coordinate systems. But by doing so, the origins are made reference points. Because coordinate systems are purely mathematical entities, we cannot intermingle such physical reference points.

I will show that radius vectors are not vectors in the sense of the first-order tensor definition Eq. (2.3) and therefore cannot be part of our toolbox. Displacement vectors, on the other hand, are legitimate tensors.

Figure 2.3 depicts the radius vector $\vec{r}^A$ emanating from the origin O^A of the coordinate system A to the point P. Two radius vectors associated with two coordinate systems A and B are related by the vector addition $\vec{r}^B = \vec{r}^A + \vec{l}^B$, where $\vec{l}^B$ is the radius vector of the origin O^A, referenced in the B coordinate system (see Fig. 2.4). With $[T]^{BA}$ the transformation matrix of coordinates B wrt A, the radius vector $\vec{r}^A$ in coordinate system A transforms to the radius vector $\vec{r}^B$ in coordinate

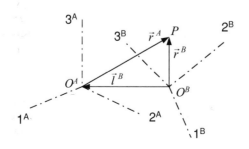

Fig. 2.4   Radius vector addition.

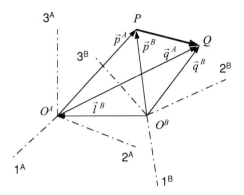

Fig. 2.5   Displacement vector.

system B, as follows:

$$\begin{bmatrix} r_1^B \\ r_2^B \\ r_3^B \end{bmatrix} = \begin{bmatrix} t_{11} & t_{12} & t_{13} \\ t_{21} & t_{22} & t_{23} \\ t_{31} & t_{32} & t_{33} \end{bmatrix} \begin{bmatrix} r_1^A \\ r_2^A \\ r_3^A \end{bmatrix} + \begin{bmatrix} l_1^B \\ l_2^B \\ l_3^B \end{bmatrix}$$

and the abbreviated matrix notation

$$[r^B]^B = [T]^{BA}[r^A]^A + [l]^B \tag{2.7}$$

Compare this transformation with the transformation that defines a tensor Eq. (2.3). Because of the additional term $[l]^B$, the radius vector $\vec{r}^A$ does not transform into $\vec{r}^B$ like a first-order tensor and therefore is not a tensor.

Conversely, let us prove that displacement vectors are tensors. Introduce the vector s_{QP} as the displacement of point Q wrt point P, as shown in Fig. 2.5. It is related to the radius vector of coordinate system A by

$$[s_{QP}]^A = [q^A]^A - [p^A]^A \tag{2.8}$$

and to those of coordinate system B by

$$[s_{QP}]^B = [q^B]^B - [p^B]^B \tag{2.9}$$

The radius vectors $[q^B]^B$ and $[p^B]^B$ transform according to Eq. (2.7):

$$[q^B]^B = [T]^{BA}[q^A]^A + [l]^B$$

$$[p^B]^B = [T]^{BA}[p^A]^A + [l]^B$$

Substituting these into Eq. (2.9) and with Eq. (2.8) yields

$$[s_{QP}]^B = [T]^{BA}\big([q^A]^A - [p^A]^A\big) = [T]^{BA}[s_{QP}]^A$$

Indeed, the displacement vector s_{QP} transforms like a first-order tensor

$$[s_{QP}]^B = [T]^{BA}[s_{QP}]^A$$

and therefore is a tensor.

I hope that you appreciate now the difference between radius and displacement vectors. Because we want to use only invariant tensor concepts, I have to give Rule 4: Radius vectors are not used. Before we move on to geometrical applications, let us summarize the four rules.

Rule 1: Subscript reversal of displacement vectors changes their sign.
Rule 2: Vector addition of displacement vectors must be consistent with subscript contraction.
Rule 3: Coordinate systems have no origins.
Rule 4: Radius vectors are not used.

2.2.3 Tensor Algebra

To work with tensors, we need to know their properties of addition and multiplication. With your a priori knowledge of vectors and matrices, there will be little new ground to cover. Basically, tensors are manipulated like vectors and matrices. We only need to spend some time discussing the vector and dyadic products that take on a different flavor.

First-order tensors (vectors) can be added and subtracted, abiding by the *commutative* rule

$$s_{BC} = s_{BA} + s_{AC} = s_{AC} + s_{BA} \qquad (2.10)$$

and the *associative* rule

$$(s_{BC} + s_{CA}) + s_{AB} = s_{BC} + (s_{CA} + s_{AB}) \qquad (2.11)$$

However, I do not recommend the exchange of the terms on the right-hand side of Eq. (2.10), because it will upset the rule of contraction of subscripts. Notice in Eq. (2.11) I arranged the order of the subscripts to form the null vector s_{BB}, which is characterized by zero displacement. You can verify this fact by contraction of the subscripts.

Multiplication of a vector s_{BC} with scalars α and β is *commutative*, *associative*, and *distributive*:

$$\alpha(\beta s_{BC}) = \beta(\alpha s_{BC}) = (\alpha\beta)s_{BC}$$
$$(\alpha + \beta)s_{BC} = \alpha s_{BC} + \beta s_{BC} \qquad (2.12)$$

Next I will deal with the multiplication of two vectors. There are three possibilities, distinguishable by the results. The outcome could be a scalar, vector, or tensor. Therefore, we call them scalar, vector, or dyadic products. The word *dyadic* is borrowed from vector mechanics, which calls stress and inertia tensors dyadics. I will use the symbols x and y for the two vectors, but also write them in the bracketed form $[x]$ and $[y]$, to emphasize their column matrix property. If they do not carry a superscript to mark the coordinate system, they abide in any allowable coordinate system and are therefore first-order tensor symbols just like x and y.

2.2.4 Scalar Product

If the transpose of vector x is multiplied with the vector y, we obtain a scalar

$$\bar{x}\,y = \text{scalar} \tag{2.13}$$

This is the scalar product of matrix algebra. For any coordinate system $]^A$ the scalar product is

$$[\bar{x}]^A[y]^A = \begin{bmatrix} x_1^A & x_2^A & x_3^A \end{bmatrix} \begin{bmatrix} y_1^A \\ y_1^A \\ y_1^A \end{bmatrix} = x_1^A y_1^A + x_2^A y_2^A + x_3^A y_3^A$$

If multiplied by itself, the resulting scalar is the square of the vector's length. For any $]^A$

$$[\bar{x}]^A[x]^A = \left(x_1^A\right)^2 + \left(x_2^A\right)^2 + \left(x_3^A\right)^2 \equiv |x|^2$$

Because the scalar $|x|^2$ is invariant under coordinate transformations, this statement holds for all coordinate systems

$$\bar{x}\,x = |x|^2 \tag{2.14}$$

What is the form of the scalar of Eq. (2.13)? Consider the vector triangle of Fig. 2.6. From the law of cosines, the squares of the length of the three vectors are related by

$$|z|^2 = |x|^2 + |y|^2 - 2|x||y|\cos\theta$$

On the other hand, $|z|^2$ is also obtained from the scalar product of the vector subtraction $z = x - y$

$$|z|^2 = \bar{z}z = (\bar{x} - \bar{y})(x - y)$$

$$= \bar{x}x + \bar{y}y - \bar{y}x - \bar{x}y$$

$$= |x|^2 + |y|^2 - 2\bar{x}y$$

Comparing the last equation with the law of cosines furnishes the value for the

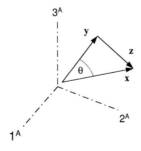

Fig. 2.6   Vector triangle.

scalar. The scalar product is therefore

$$\bar{x}y = |x||y|\cos\theta \tag{2.15}$$

The scalar is $|x|$ times the projection of y on x and assumes the sign of $\cos\theta$. In Eq. (2.15) we can exchange the symbols without changing the result; therefore

$$\bar{x}y = \bar{y}x$$

However, beware of moving the transpose sign. The result is $\bar{x}y \neq x\bar{y}$ (why?).

Example 2.1   Angle Between Two Beams

Problem.   A traffic control radar R tracks two aircraft A and B and records their displacements in geographic coordinates $]^G$ as $[\overline{s_{AR}}]^G = [20 \ 10 \ -9]$ and $[\overline{s_{BR}}]^G = [30 \ -20 \ -9]$ km. What is the angle between the two radar beams?

Solution.   Solve Eq. (2.15) for θ

$$\theta = \arccos\left(\frac{[\overline{s_{AR}}]^G [s_{BR}]^G}{|s_{AR}||s_{BR}|}\right)$$

substitute the matrices and multiply out

$$\theta = \arccos\left(\frac{[20 \quad 10 \quad -9]\begin{bmatrix}30\\-20\\-9\end{bmatrix}}{\sqrt{20^2 + 10^2 + (-9)^2}\sqrt{30^2 + (-20)^2 + (-9)^2}}\right)$$

$$= \arccos\left(\frac{481}{895.75}\right) = 57.5 \text{ deg}$$

2.2.5   *Vector Product*

Let us ask another question: What kind of multiplication of two vectors x, y yields a vector z? From matrix algebra we know that a 3×1 vector is only obtained by the multiplication of a 3×3 matrix with a 3×1 vector

$$[X][y] = [z] \tag{2.16}$$

We want to explore the form of $[X]$. Vector algebra gives us three conditions for a vector product, as visualized by Fig. 2.7:

1) z normal to x and y.
2) $|z|$ = area of x and y parallelogram.
3) right-handed sequence: $x \rightarrow y \rightarrow z$.

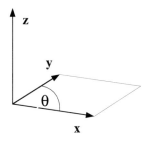

Fig. 2.7   Vector product.

These conditions will lead to the conclusion that $[X]$ is the skew-symmetric form of $[x]$:

$$[x]^A = \begin{bmatrix} x_1 \\ x_2 \\ x_3 \end{bmatrix}^A \Leftrightarrow [X]^A = \begin{bmatrix} 0 & -x_3 & x_2 \\ x_3 & 0 & -x_1 \\ -x_2 & x_1 & 0 \end{bmatrix}^A$$

for any allowable coordinate system $]^A$. Therefore, the vector product in tensor algebra is expressed by

$$Xy = z \tag{2.17}$$

with the understanding that X is the skew-symmetric tensor of x.

Proof: If X is skew-symmetric, it satisfies the three conditions of the vector product:

1) The first condition consists of two orthogonality conditions.

a) For z to be normal to x, their scalar product must be zero: $\bar{z}x = 0$; with Eq. (2.17) $\bar{y}\bar{X}x = 0$, which is satisfied if $\bar{X}x = 0$. This is indeed the case as demonstrated by the matrix multiplication (for any $]^A$)

$$[\bar{X}]^A[x]^A = \begin{bmatrix} 0 & x_3 & -x_2 \\ -x_3 & 0 & x_1 \\ x_2 & -x_1 & 0 \end{bmatrix}^A \begin{bmatrix} x_1 \\ x_2 \\ x_3 \end{bmatrix}^A = \begin{bmatrix} x_3 x_2 - x_2 x_3 \\ -x_3 x_1 + x_1 x_3 \\ x_2 x_1 - x_1 x_2 \end{bmatrix}^A = \begin{bmatrix} 0 \\ 0 \\ 0 \end{bmatrix}^A$$

b) For z to be normal to y, their scalar product must be zero: $\bar{z}y = 0$; with Eq. (2.17) $\bar{y}\bar{X}y = 0$, which is satisfied if $\bar{X}$ is skew-symmetric

$$[\bar{y}]^A[\bar{X}]^A[y]^A = [y_1 \quad y_2 \quad y_3]^A \begin{bmatrix} 0 & x_3 & -x_2 \\ -x_3 & 0 & x_1 \\ x_2 & -x_1 & 0 \end{bmatrix}^A \begin{bmatrix} y_1 \\ y_2 \\ y_3 \end{bmatrix}^A$$

$$= [y_1 \quad y_2 \quad y_3]^A \begin{bmatrix} x_3 y_2 - x_2 y_3 \\ -x_3 y_1 + x_1 y_3 \\ x_2 y_1 - x_1 y_2 \end{bmatrix}^A = 0$$

The matrix multiplication holds for any allowable coordinate system. Therefore, if $\bar{X}$ is skew-symmetric, so is X, and the first condition is satisfied.

2) The second condition requires that the length of vector $|z|$ = area of the parallelogram subtended by the vectors x and y. Let us express Eq. (2.17), with X as a skew-symmetric matrix, in an arbitrary coordinate system $]^A$.

$$\begin{bmatrix} z_1 \\ z_2 \\ z_3 \end{bmatrix}^A = \begin{bmatrix} 0 & -x_3 & x_2 \\ x_3 & 0 & -x_1 \\ -x_2 & x_1 & 0 \end{bmatrix} \begin{bmatrix} y_1 \\ y_2 \\ y_3 \end{bmatrix}^A = \begin{bmatrix} -x_3y_2 + x_2y_3 \\ x_3y_1 - x_1y_3 \\ -x_2y_1 + x_1y_2 \end{bmatrix}^A$$

The scalar product of z with itself is the square of its absolute value

$$[\bar{z}]^A[z]^A = |z|^2 = z_1^2 + z_2^2 + z_3^2$$

$$= (x_2y_3 - x_3y_2)^2 + (x_3y_1 - x_1y_3)^2 + (x_1y_2 - x_2y_1)^2$$

$$= (x_1^2 + x_2^2 + x_3^2)(y_1^2 + y_2^2 + y_3^2) - (x_1y_1 + x_2y_2 + x_3y_3)^2$$

The last line can be written as

$$|z|^2 = |x|^2|y|^2 - ([\bar{x}][y])^2 = |x|^2|y|^2 - |x|^2|y|^2 \cos^2\theta = |x|^2|y|^2(1 - \cos^2\theta)$$

Replacing $(1 - \cos^2\theta) = \sin^2\theta$ and taking the square root yields

$$|z| = |x||y| \sin\theta$$

which is the area of the parallelogram between vectors x and y. By choosing for X the skew-symmetric form, we have indeed satisfied the second condition of vector multiplication.

3) The third condition states the right handedness of the vector product. Let us introduce a right-handed Cartesian coordinate system 1^A, 2^A, and 3^A. For the particular situation in Fig. 2.8, the vector product assumes the form

$$[z]^A = [X]^A[y]^A = \begin{bmatrix} 0 & 0 & x_2 \\ 0 & 0 & -x_1 \\ -x_2 & x_1 & 0 \end{bmatrix}^A \begin{bmatrix} 0 \\ y_2 \\ 0 \end{bmatrix}^A = \begin{bmatrix} 0 \\ 0 \\ x_1y_2 \end{bmatrix}^A$$

If x_1 and y_2 are positive, so is z_3. Therefore, the skew-symmetric form of $[X]$ satisfies the right-handedness condition as demonstrated by this specific example.

We have indeed confirmed that the vector product of two vectors x and y consists of the multiplication of the skew-symmetric form of vector x with vector y. In coordinate form we execute a matrix multiplication.

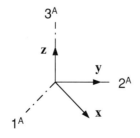

Fig. 2.8   Right handedness.

I introduced vector multiplication in a simplified form, adhering to vector mechanics rather than to tensor algebra. The right-handed convention eliminates the need for third-order tensors, which are really required to define the vector product properly. For the theoretically inclined among you, I provide the tensor definition of a vector product. Let x_i, y_j, and z_k be three vectors in Euclidean space, and ε_{ijk} the third-order permutation tensor, then the vector product is defined (dummy indices summation implied):

$$z_i = \varepsilon_{ijk} x_j y_k$$

It is valid for any type of coordinate system compatible with the Euclidean metric. By agreeing to use only right-handed Cartesian coordinate systems and the right-hand rule of vector products, we can use the simpler Eq. (2.17) as a definition of the vector product. You will see that this version is adequate for modeling all situations related to aerospace vehicles.

Example 2.2   Area Calculation

Problem.   A farmer's son inherits a rhombus-shaped field (Fig. 2.9). The barn is located on the corner B. With a global positioning system (GPS) set he records the coordinates of B and the two adjacent corners C and D. Then he converts the data to two vectors in geographic coordinates $[\bar{s}_{CB}]^G = [0.5\ 2\ 0]$ and $[\bar{s}_{DB}]^G = [2\ 0.5\ 0]$ km. How many square kilometers of land did he inherit?

Solution.   Apply Eq. (2.17) and take the absolute value of the cross product to obtain the area A:

$$A = \left| [S_{CB}]^G [s_{DB}]^G \right|$$

where $[S_{CB}]^G$ is the skew-symmetric form of $[s_{CB}]^G$.

$$A = \left| \begin{bmatrix} 0 & 0 & 2 \\ 0 & 0 & -0.5 \\ -2 & 0.5 & 0 \end{bmatrix} \begin{bmatrix} 2 \\ 0.5 \\ 0 \end{bmatrix} \right| = \left| \begin{bmatrix} 0 \\ 0 \\ -3.75 \end{bmatrix} \right| = 3.75 \text{ km}^2$$

Notice, there was no need for a coordinate system origin.

2.2.5.1   Vector triple product.   For any three vectors x, y, and z with X and Y, the skew-symmetric forms of x and y, the vector triple product, is defined as

$$XYz = y\bar{x}z - z\bar{x}y \tag{2.18}$$

which are to be interpreted as matrix multiplication

$$[X][Y][z] = [y][\bar{x}][z] - [z][\bar{x}][y]$$

The left-hand side involves two vector products $[w] = [Y][z]$ and $[X][w]$. The right-hand side is the subtraction of the vector $[z]$, multiplied by scalar $[\bar{x}][y]$ from the vector $[y]$, multiplied by scalar $[\bar{x}][z]$.

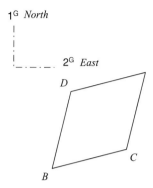

Fig. 2.9 Rhombus-shaped field.

2.2.5.2 Scalar triple product. For any three vectors x, y, and z with X, the skew-symmetric form of x, the scalar triple product, is defined as

$$V = (\overline{Xy})z = \bar{y}\bar{X}z \tag{2.19}$$

implying the matrix multiplication

$$V = [\overline{X}][y][z] = [\bar{y}]\bar{X}[z]$$

V is the result of the scalar product of two vectors $[w] = [X][y]$ and $[z]$ and equals the volume contained in the parallelepiped formed by x, y, and z.

2.2.6 Dyadic Product

The third possibility is the multiplication of vector x with the transpose of vector y to obtain tensor Z,

$$x\bar{y} = Z \tag{2.20}$$

which represents, for any allowable coordinate system $]^A$, the matrix multiplication

$$[x]^A[\bar{y}]^A = [Z]^A \Leftrightarrow \begin{bmatrix} x_1 \\ x_2 \\ x_3 \end{bmatrix}^A [y_1 \quad y_2 \quad y_3]^A = \begin{bmatrix} x_1y_1 & x_1y_2 & x_1y_3 \\ x_2y_1 & x_2y_2 & x_2y_3 \\ x_3y_1 & x_3y_2 & x_3y_3 \end{bmatrix}^A$$

We borrowed the name *dyadic product* from vector mechanics, which avoids tensors by decomposing them into vector form and calling that hybrid construct a *dyad*. We have no need of dyads themselves, having committed ourselves to modeling flight mechanics by tensors only.

If both vectors x and y are one and the same unit vector u, then the dyadic product produces the projection tensor P. Given any vector t, its projection on the u direction is the vector r, resulting from

$$r = Pt \quad \text{with} \quad P = u\bar{u} \tag{2.21}$$

You can easily convince yourself of that fact by substituting the definition of P into Eq. (2.21) $r = u\bar{u}t$ and recognizing that the scalar product $\bar{u}t$ gives the length and u the direction of vector r. In any coordinate system, say $]^A$, the projection tensor has the form

$$[P]^A = [u]^A[\bar{u}]^A = \begin{bmatrix} u_1^2 & u_1u_2 & u_1u_3 \\ u_2u_1 & u_2^2 & u_2u_3 \\ u_3u_1 & u_3u_2 & u_3^2 \end{bmatrix}^A$$

It is a symmetric matrix because the scalar product of coordinates is commutative.

Example 2.3   Thrust Vector Projection

Problem.   The direction of the centerline of a missile is given by the unit vector $[\bar{u}]^G = [0.2 \ 0.3 \ -0.9327]$ in geographic coordinates. To make a course correction, the gimbaled rocket motor turns the thrust vector away from the centerline to a position given by $[\bar{t}]^G = [7.6 \ 12.8 \ -36]$ kN. Determine the thrust vector along the centerline of the missile $[r]^G$ in geographic coordinates. (See Fig. 2.10.)

Solution.   We just substitute the values of the example into Eq. (2.21) and calculate the centerline thrust vector

$$[r]^G = [u]^G[\bar{u}]^G[t]^G = \begin{bmatrix} 0.2 \\ 0.3 \\ -0.9327 \end{bmatrix} [0.2 \ 0.3 \ -0.9327] \begin{bmatrix} 7.6 \\ 12.8 \\ -36 \end{bmatrix} = \begin{bmatrix} 7.787 \\ 11.681 \\ -36.317 \end{bmatrix} kN$$

The matrix product can be executed any two ways. Either first calculate the projection tensor and multiply it with the thrust vector, or evaluate the scalar product first and then multiply it with the unit vector. Both options lead to the same result.

We have come to the vista overlooking the foundations of mathematical modeling of aerospace vehicles. Classical mechanics is the environment, and Cartesian tensors are the building blocks. We reviewed the axioms of mechanics as well as the Principle of Material Indifference and postulated the sufficiency of points and frames as modeling elements. Simple Cartesian tensors are our language with emphasis on the invariant, coordinate-free formulation of physical phenomena. There are many more overlooks still ahead. However, we pause and apply our newfound skills to some important geometrical modeling tasks.

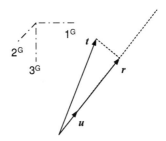

Fig. 2.10   Vector projection.

2.3 Modeling of Geometry

As you build your aerospace vehicle simulations, you will be surprised at how much time you spend just getting the geometrical situation right. You have to keep track of vehicles in various coordinate systems. They may be moving along lines, and their proximity to certain planes may be of interest. Some of the vectors must be projected into new directions and others into planes. We may even have to deal with reflection and rotational symmetries.

Geometrical models will be with us throughout this book. In this section we build on the elements of tensor algebra and formulate such mundane things as lines, planes, and projections. Yet you will be surprised how different the invariant formulations are from the customary treatment.

2.3.1 Displacement of Points

Let us recap: The location of a point is meaningless unless it is referred to a reference point. So, for instance, the location of the center of mass (c.m.) of a missile B must be related to a tracking station, the launch point, or the target coordinates. The term *displacement* implies that mutual relationship. If the reference point is R, then the displacement vector of the missile is s_{BR}. Its time dependency is expressed by $s_{BR}(t)$. The displacement of a point is an invariant tensor concept, valid in any allowable coordinate system. For computational attainment this first-order tensor must be expressed in a coordinate system to be processed numerically. For instance, the missile may be measured in the tracking station coordinates $]^R$. Then the displacement vector's coordinates are

$$[s_{BR}(t)]^R = \begin{bmatrix} s_1(t) \\ s_2(t) \\ s_3(t) \end{bmatrix}$$

Example 2.4   Helical Displacement

Problem.   A missile makes an evasive circular maneuver toward a tracking radar R. Its closing speed is v, revolving at the angular velocity ω on a helix of radius r. Suppose the 1^R direction of the radar coordinate axes is parallel to the helix centerline. Formulate the displacement vector of the missile wrt to the radar in radar coordinates.

Solution.   You should be able to verify the result with the help of Fig. 2.11.

$$[s_{BR}(t)]^R = \begin{bmatrix} -vt \\ r\cos(\omega t) \\ r\sin(\omega t) \end{bmatrix}$$

For emphasis I point out that the coordinate axes are completely defined by direction and sense. They are free floating without origin. Drawing them from a common point, as I have done in Fig. 2.11, is convenient but unnecessary.

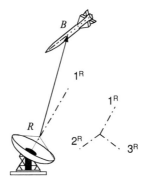

Fig. 2.11 Missile displacement.

2.3.2 Straight Line

Straight lines arise as models of straight trajectories, star sightings, surveying of landmarks, or just a person walking down the aisle of an aircraft. They are considered of infinite length, but contain a displacement vector, whose endpoint moves along the line.

We let the point B slide along the straight line, starting from an initial point B_0, while maintaining R as a fixed reference point (see Fig. 2.12). The sliding process is generated by a scalar parameter u, $\in$: $-\infty \leq u \leq +\infty$ that lengthens or shortens the vector $s_{BB_0} = u s_{UB_0}$ on the line. The vector s_{UB_0} establishes the direction of the line and could be a vector of unit length.

Definition: A straight line, with direction s_{UB_0} and anchored at $s_{B_0 R}$, is defined by the sliding of point B, referred to point R

$$s_{BR} = u s_{UB_0} + s_{B_0 R} \tag{2.22}$$

The line is a one-dimensional manifold with parameter u.

It takes three extra points to describe the sliding of point B: the reference point R and the two points U and B_0, which establish the direction of the line. If the reference point should be on the line, B_0 could assume its function.

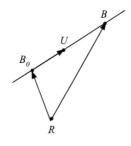

Fig. 2.12 Line geometry.

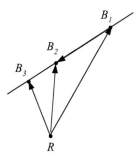

Fig. 2.13   Flight line.

Example 2.5   Straight Line Trajectory

Problem.   A surveillance radar R took two fixes of an incoming attack fighter: $[\overline{s_{B_1 R}}]^G = [30 \ \ 10 \ \ -8]$ and $[\overline{s_{B_2 R}}]^G = [20 \ \ -5 \ \ -6]$ km (see Fig. 2.13). Timing the two fixes gave an elapsed time of $\Delta t = 50$ s.

1) Determine the average speed of the aircraft.
2) Where will the aircraft be ($[\overline{s_{B_3 R}}]^G$) after $\Delta \tau = 20$ s has elapsed beyond B_2, assuming it continues its steady and straight flight?

Solution.   1) The speed of the aircraft V is calculated from the distance $|s_{B_2 B_1}|$ divided by time

$$\left[\overline{s_{B_2 B_1}}\right]^G = \left[\overline{s_{B_2 R}}\right]^G - \left[\overline{s_{B_1 R}}\right]^G = [-10 \ \ -15 \ \ 2]$$

$$V = \frac{\left|s_{B_2 B_1}\right|}{\Delta t} = \frac{18.1}{50} = 0.3628 \text{ km/s} = 362.8 \text{ m/s}$$

2) Because the aircraft flies along a straight line, its displacement after 20 s from the radar station is according to Eq. (2.22)

$$\left[s_{B_3 R}\right]^G = u\left[s_{B_2 B_1}\right]^G + \left[s_{B_1 R}\right]^G$$

where $[s_{B_2 B_1}]^G$ points in the direction of flight. The parameter u is calculated from the time ratios

$$u = \frac{\Delta \tau + \Delta t}{\Delta t} = 1.4$$

then the radar will pick up the aircraft after 20 s at

$$\left[\overline{s_{B_3 R}}\right]^G = 1.4 \times [-10 \ \ -15 \ \ 2] + [30 \ \ 10 \ \ -8] = [16 \ \ -11 \ \ -5.2] \text{ km}$$

The parameter u in this example is actually a time ratio, which occurs quite often in these types of problems.

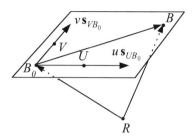

Fig. 2.14   Plane.

2.3.3 Plane

Whereas lines are one-dimensional, planes are two-dimensional manifolds. In Euclidean space we can safely speak about straight lines and flat planes, corroborating our experience. Our simulation may have to define such elements like ground planes, imaging planes, or target planes. Let us see how to model them in an invariant, coordinate-independent form (see Fig. 2.14).

Let the point B sweep over a whole plane, starting at B_0 and maintaining R as the reference point. The movement over the plane is generated by two parameters $u, \in: -\infty \le u \le +\infty$ and $v, \in: -\infty \le v \le +\infty$ that modify the two directional vectors s_{UB_0} and s_{VB_0}.

Definition: A plane, subtended by s_{UB_0} and s_{VB_0}, is defined by the sweeping motion of the displacement vector of point B, referred to point R

$$s_{BR} = u s_{UB_0} + v s_{VB_0} + s_{B_0 R} \tag{2.23}$$

The plane is a two-dimensional manifold with parameters u and v.

The embedded vectors $[s_{UB_0}]$ and $[s_{VB_0}]$ stretch out the level of the plane. They do not have to be mutually orthogonal nor be unit vectors. We needed four extra points to describe the sweeping motion of B. The three points B_0, U, and V establish the orientation of the plane, and R serves as reference point. If R should be in the plane, B_0 could assume its function.

Example 2.6   Helicopter Landing Aid

Problem.   Figure 2.15 shows a helicopter H preparing to land on an oddly shaped landing patch of a swaying ship. The pilot uses his landing aid, which displays his projection B on the landing surface, for touchdown near the center. This instrument has trackers that measure the three corners B_0, U, and V relative to the vehicle and provides them in geographic coordinates with the aid of an onboard INS. Develop the equations that are used to establish the orientation of the platform and the point B in geographic coordinates.

Solution.   The platform is described by B sweeping out the plane just as Eq. (2.23) indicates

$$s_{BH} = u s_{UB_0} + v s_{VB_0} + s_{B_0 H}$$

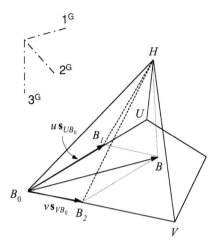

Fig. 2.15   Helicopter landing patch.

where $s_{B_0 H}$ is measured directly and the other two vectors are obtained from the additional measurements s_{UH} and s_{VH}

$$s_{UB_0} = s_{UH} + s_{HB_0}; \quad s_{VB_0} = s_{VH} + s_{HB_0}$$

where $s_{HB_0} = -s_{B_0 H}$. What remains to be determined are u and v. We can calculate them with the help of the scalar product and trigonometry. Let us do it for u, and, by analogy, the solution for v follows. Referring to Fig. 2.16 and Eq. (2.15), we derive

$$\cos \alpha = \frac{\bar{s}_{HB_0} s_{UB_0}}{|s_{HB_0}| \, |s_{UB_0}|} = \frac{u \, |s_{UB_0}|}{|s_{HB_0}|}$$

and solving for u

$$u = \frac{\bar{s}_{HB_0} s_{UB_0}}{|s_{UB_0}|^2}$$

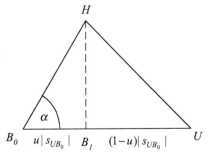

Fig. 2.16   Calculation of u.

Similarly,

$$v = \frac{\bar{s}_{HB_0} s_{VB_0}}{|s_{VB_0}|^2}$$

The problem was solved entirely in symbolic form. The processor of the instrument carries the calculations out in the same geographic coordinates that the measurements $[s_{B_0 H}]^G$, $[s_{UH}]^G$, and $[s_{VH}]^G$ are given:

$$[s_{UB_0}]^G = [s_{UH}]^G + [s_{HB_0}]^G; \quad [s_{VB_0}]^G = [s_{VH}]^G + [s_{HB_0}]^G$$

$$u = \frac{[\overline{s_{HB_0}}]^G [s_{UB_0}]^G}{|s_{UB_0}|^2}; \quad v = \frac{[\overline{s_{HB_0}}]^G [s_{VB_0}]^G}{|s_{VB_0}|^2}$$

and the projection of the helicopter on the landing platform is calculated from

$$[s_{BH}]^G = u[s_{UB_0}]^G + v[s_{VB_0}]^G + [s_{B_0 H}]^G$$

or in the platform plane

$$[s_{BB_0}]^G = [s_{BH}]^G - [s_{B_0 H}]^G$$

Review briefly our two-step approach. We first derived the solution in vector form without reference to coordinate systems or their origins, followed by the coordinate form for programming. You can practice computing a sample numerical solution by solving Problem 2.2.

2.3.4   Normal Form of a Plane

The definition of a plane by Eq. (2.23) may be somewhat intimidating for its complexity. There exists an alternate, simpler formulation, which is also quite useful for modeling planes. It is called the *normal form* because the unit normal vector defines its orientation.

Let the unit normal vector of a plane be given by u (see Fig. 2.17). Premultiplying Eq. (2.23) by the transpose of u generates scalar products on both sides of the equation

$$\bar{u}s_{BR} = u\bar{u}s_{UB_0} + v\bar{u}s_{VB_0} + \bar{u}s_{B_0 R}$$

Because u is orthogonal to s_{UB_0} and s_{VB_0}, the first two terms on the right-hand side

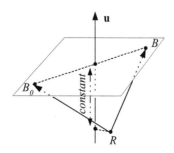

Fig. 2.17   Normal form of plane.

are zero; and, with s_{B_0R} being constant (see Fig. 2.17), we have the condition for B to sweep out the plane: $\bar{u}s_{BR} = $ const.

Definition: A plane, with unit normal direction u, is defined by the sweeping point B, referred to point R, such that the scalar product of the two vectors is constant

$$\bar{u}s_{BR} = \text{const} \tag{2.24}$$

In your pursuit of modeling planes, you may wonder which definition is appropriate for your situation. Here are the telltales: If the plane is given by points or embedded vectors, use Eq. (2.23), otherwise apply Eq. (2.24) with the unit normal vector.

Example 2.7 Angle Distance Measuring

Problem. An aircraft R with an INS and distance-measuring equipment on-board is to determine the distances to points B_1 and B_2, given its own known displacement vector s_{B_0R} (see Fig. 2.18). All three points lie on the surface with unit vertical vector u, established by the INS. The sensor of the distance-measuring equipment, however, can only measure the angles β_1 and β_2 between the local vertical and the direction to the points. Derive the equations that calculate the two distances $|s_{B_1R}|$ and $|s_{B_2R}|$.

Solution. Because all three points lie on the surface, we have from Eq. (2.24) the relationships

$$\bar{u}s_{B_0R} = \bar{u}s_{B_1R} = \bar{u}s_{B_2R} = \text{const}$$

The first term is given and establishes the value of the constant. The scalar product of the remaining terms can be expressed in terms of the length of the vectors and their angles from the vertical

$$|s_{B_1R}|\cos\beta_1 = |s_{B_2R}|\cos\beta_2 = \text{const}$$

From which we obtain the desired lengths

$$|s_{B_1R}| = \frac{\text{const}}{\cos\beta_1}; \quad |s_{B_2R}| = \frac{\text{const}}{\cos\beta_2}$$

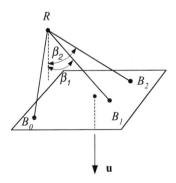

Fig. 2.18 Distance measuring.

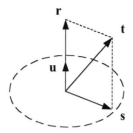

Fig. 2.19   Plane projection.

This is a typical application for the normal form of a plane. The unit vector of a plane is a free vector in accordance with the characteristics of a plane as a two-dimensional manifold. The free parallel displacement of the vector u in two dimensions corresponds to the two parameters u and v of the original definition, (Sec. 2.3.3).

2.3.5 Plane Projection Tensor

With the definitions of planes in plain view, we can address the task of projecting vectors into planes. Your simulation may require the velocity vector of an aircraft be projected on the ground or the silhouette of a missile be imaged on a charge-coupled device (CCD) planar array. For all of these situations, the plane projection tensor will be a useful tool.

We met the line projection tensor P earlier. It is formed by the dyadic product of the unit vector u, $P = u\,\bar{u}$. According to Eq. (2.21), P produces the vector r from t by projecting it onto u: $r = Pt$. Now, u does not only establish the direction of a line, but also the unit normal of a plane (see Fig. 2.19).

The challenge is to find the projection tensor N that projects vector t onto the plane given by u. The projected vector is labeled s. From the vector triangle we derive

$$s = t - r$$

and substituting Eq. (2.21) for r

$$s = t - Pt = (E - P)t = (E - u\bar{u})t$$

We define the plane projection tensor with E the unit tensor and u normal of the plane

$$N = E - u\bar{u} \tag{2.25}$$

Just like the line projection tensor, the plane projection tensor is symmetric.

Example 2.8   Focal Plane Imaging

Problem.   An aircraft is imaged on a focal plane array. To simulate that process, we need to develop the equations that project the aircraft's silhouette on the focal plane. We keep it simple by modeling the perspective of the aircraft with the displacement vectors of the tip, stern, right wing tip, and left wing tip wrt the geometrical center C, t_{B_1C}, t_{B_2C}, t_{B_3C}, and t_{B_4C}. The displacement of the aircraft center C wrt the focal plane center F is given by t_{CF} and the orientation of the planar array by the unit normal vector u. Separation distance and optics reduce the scale of the projections on the focal plane by a factor f. Determine the aircraft attitude vectors s_{B_1C}, s_{B_2C}, s_{B_3C}, and s_{B_4C} and the displacement vector s_{CF} in the focal plane. (To practice, make a sketch.)

Solution. Subjecting the displacement vectors to the plane projection tensor $N = E - u\bar{u}$ and reducing the magnitude by f produces the image

$$s_{B_1C} = fNt_{B_1C}, \quad s_{B_2C} = fNt_{B_2C}, \quad s_{B_3C} = fNt_{B_3C}, \quad s_{B_4C} = fNt_{B_4C}$$

and the displacement of the aircraft from the focal plane center

$$s_{CF} = fNt_{CF}$$

For building the simulation, the vectors have to be converted to matrices. Most likely, the aircraft data are in geographic coordinates $]^G$, and the image should be portrayed in focal plane coordinates $]^F$. Therefore, a transformation between the two coordinate systems $[T]^{GF}$ will enter the formulation.

2.3.6 Reflection Tensor

A plane of symmetry has the characteristics of a mirror. The left side repeats the right side. Any aircraft exhibits this reflectional symmetry—what ever happened to the oblique wing? Even right-hand maneuvers can be reflected into left-hand maneuvers just by geometrical manipulation. The tensor that makes this happen is called the reflection tensor M.

What are the characteristics of this tensor M that reflects the vector t into t' through the mirror plane with unit vector u (see Fig. 2.20)? First, project t onto u with the projection tensor $P = u\bar{u}$ to get $r = Pt$ and then derive t' from the vector triangle

$$t' = t - 2r = t - 2Pt = (E - 2P)t$$

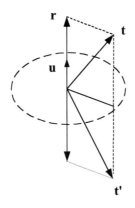

Fig. 2.20   Reflection.

The reflection tensor of the mirror plane with unit normal u and unit tensor E is therefore

$$M = E - 2u\bar{u} \qquad (2.26)$$

It is not only symmetric but also orthogonal, as we can demonstrate by

$$M\bar{M} = (E - 2u\bar{u})(E - 2u\bar{u}) = E - 4u\bar{u} + 4u\bar{u}\,u\bar{u} = E$$

The reflection tensor plays an important role in Chapter 7, where we will sort out the existence of higher-order aerodynamic derivatives of airframes exhibiting reflectional symmetry. If the mirror plane is oriented in body coordinates $]^B$ such that its unit normal has the coordinates $[\bar{u}]^B = [0 \ 1 \ 0]$ (right wing of aircraft), then the reflection tensor has the coordinates

$$[M]^B = [E]^B - 2[u]^B[\bar{u}]^B = \begin{bmatrix} 1 & 0 & 0 \\ 0 & 1 & 0 \\ 0 & 0 & 1 \end{bmatrix} - 2\begin{bmatrix} 0 & 0 & 0 \\ 0 & 1 & 0 \\ 0 & 0 & 0 \end{bmatrix} = \begin{bmatrix} 1 & 0 & 0 \\ 0 & -1 & 0 \\ 0 & 0 & 1 \end{bmatrix}$$

We conclude that the reflection tensor changes the sign of the second coordinate, but keeps the other two coordinates unchanged.

Example 2.9   Application of Reflection Tensor

Problem.   An aircraft (see Fig. 2.21), with a canted twin tail, executes a push-down maneuver. What is the resultant force of both control surfaces f if the force on one surface is f_1? Derive the equations in invariant form and then introduce body coordinates $]^B$.

Solution.   We use the reflection tensor to determine the force on the other control surface and add both together

$$f = f_1 + f_2 = f_1 + Mf_1 = (E + M)f_1 = 2(E - P)f_1$$

With the unit normal of the mirror plane being u

$$f = 2(E - u\bar{u})f_1$$

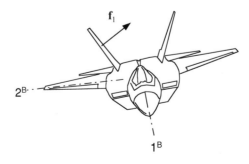

Fig. 2.21   Control forces.

This is the desired result in symbolic tensor notation. Introducing body coordinates for the unit normal $[\bar{u}]^B = [0 \ 1 \ 0]$, and for the force on the right control surface $[\bar{f}_1]^B = [0 \ f_{12} \ f_{13}]$ yields the result

$$[f]^B = 2 \begin{bmatrix} 1 & 0 & 0 \\ 0 & 0 & 0 \\ 0 & 0 & 1 \end{bmatrix} \begin{bmatrix} 0 \\ f_{12} \\ f_{13} \end{bmatrix} = 2 \begin{bmatrix} 0 \\ 0 \\ f_{13} \end{bmatrix}$$

We find that the horizontal force components cancel, and the vertical component is doubled by the second surface.

2.4 Summary

If you are reading this, you have persevered until this chapter's end. We are indebted to the physicists of the 18th and 19th centuries for the foundations of classical mechanics. Its axiomatic treatment puts our modeling tasks on a sure footing. Simple Cartesian tensors in Euclidean three-space are the symbolic language, and their realization as matrices by coordinate systems are the fodder for computers. I hypothesized that points and frames suffice to model flight mechanics—a statement that still needs verification. Some of the cherished traditions of vector mechanics had to be abandoned. Coordinate systems have no origins, and the radius vector has no place in our tool chest.

Then, with the help of tensor algebra we assembled some basic operations. The scalar, vector, and dyadic products are essential for general modeling tasks, and they are applied to specific geometric problems. Some of them we readied for the toolbox: straight line, plane, normal form of plane, plane projection tensor, and reflection tensor.

Needless to say, but worth emphasizing, we just got started! There is so much more for you in store. Although we already introduced frames and coordinate systems, we need to dig deeper. I shall attempt to clearly delineate their distinctly separate purposes in the next chapter.

References

[1]Hamel, G., "Die Axiome der Physik," *Handbuch der Physik*, Band 5, Springer-Verlag, Berlin, 1929, Chap. 1.

[2]Noll, W., "On the Continuity of the Solid and Fluid States," *Journal of Rational Mechanical Analysis*, Vol. 4, No. 1, 1955, p. 17.

[3]Truesdell, C., and Noll, W., "The Nonlinear Field Theories in Mechanics," *Handbuch der Physik*, Vol. III/3, edited by S. Fluegge, Springer-Verlag, Berlin, 1965, pp. 36, 41, and 42.

[4]Ricci, G., and Levi-Civita, T., "Methodes de Calcul Differentiel Absolu et Leurs Applications," *Mathematische Annalen*, Vol. 54, 1901.

[5]Einstein, A., "Die Grundlagen der Allgemeinen Relativitaetstheorie," *Annalen der Physik*, Vol. 4, No. 49, 1916, pp. 769–822.

[6]Duschek, A., and Hochrainer, A., *Tensorrechnung in Analytischer Darstellung*, Vols. I, II, and III, Springer-Verlag, Berlin, 1968, 1970, and 1965.

[7]Wrede, R. C., *Introduction to Vector and Tensor Analysis*, Wiley, New York, 1963.

[8]Betten, J., *Elementare Tensorrechnung fuer Ingenieure*, Vieweg, Brunswick, Germany, 1977.

Problems

2.1   Scalar triple product. Show that the scalar triple product $V = (\overline{X}y)z = \overline{y}\overline{X}z$ is the volume contained within the three vectors x, y, and z by using scalar and vector products. For the body coordinate system $]^B$ the vector coordinates are $[\overline{x}]^B = [8\ -2\ -3]$, $[\overline{y}]^B = [3\ 14\ 1]$, and $[\overline{z}]^B = [3\ -2\ 10]$ m. What is the value of V?

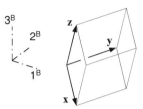

2.2   Helicopter landing aid (Example 2.6). Refer to Fig. 2.15, the helicopter landing patch, to visualize the following numerical measurements: $[\overline{s_{B_0H}}]^G = [-22\ -10\ 121]$, $[\overline{s_{VH}}]^G = [-12\ 16\ 124]$, and $[\overline{s_{UH}}]^G = [9\ -5\ 116]$. Calculate the displacement vector of projection B wrt the helicopter H, $[s_{BH}]^G$ with the equations derived in the example. (Numbers are in meters.)

2.3   Parallel lines. Determine the equation of a line $[s_{PR}]^R = u[s_{UP_0}]^R + [s_{P_0R}]^R$ that intersects the point P_0, given by its displacement vector $[\overline{s_{P_0R}}]^R = [1\ 2\ 3]$. This line is parallel to another line $[\overline{s_{QR}}]^R = v[1\ 1\ 1] + [1\ 0\ 0]$. The coordinate system $]^R$ is arbitrary (*Solution*: $[\overline{s_{PR}}]^R = [u+1\ \ u+2\ \ u+3]$).

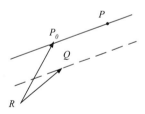

2.4   Intersecting lines. Determine the point of intersection I, $[s_{IR}]^R$, of the two straight lines with the same reference point $[\overline{s_{PR}}]^R = u[1\ 1\ 1] + [0\ 1\ 0]$ and $[\overline{s_{QR}}]^R = v[0\ 1\ 1] + [1\ 1\ 0]$. (*Hint*: Two lines intersect if there exist u and v such that $[s_{PR}]^R = [s_{QR}]^R$.) Complement the sketch by a three-dimensional scaled figure. The coordinate system $]^R$ is arbitrary (*Solution*: $[\overline{s_{IR}}]^R = [1\ 2\ 1]$).

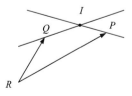

2.5   Closest distance to a line. Determine the point P^*, s_{P^*R}, on the straight line $s_{PR} = us_{UP_0} + s_{P_0R}$ that has the shortest distance from R. (*Hint*: Find u^* such that $\bar{s}_{PR}\, s_{PR}$ is minimized.) Derive the general equation first, then apply it to the example $[\overline{s_{PR}}]^R = u[0\ 2\ 1] + [0\ 0\ 1]$. Make a three-dimensional sketch. The coordinate system $]^R$ is arbitrary (*Solution*: $[\overline{s_{P^*R}}]^R = [0\ -2/5\ 4/5]$).

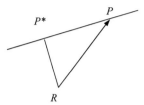

2.6 Closest distance to a plane. Determine the point P^* of the plane $s_{PR} = u s_{UP_0} + v s_{VP_0} + s_{P_0R}$ that has the shortest distance from R. *Procedure*: Let u^* and v^* be the parameters of the point P^* in the plane that is closest to R and show that

$$u^* = \frac{\bar{s}_{VP_0} s_{UP_0}\bar{s}_{P_0R} s_{VP_0} - \bar{s}_{VP_0} s_{VP_0}\bar{s}_{P_0R} s_{UP_0}}{\bar{s}_{UP_0} s_{UP_0}\bar{s}_{VP_0} s_{VP_0} - \left(\bar{s}_{VP_0} s_{UP_0}\right)^2}$$

$$v^* = \frac{\bar{s}_{UP_0} s_{VP_0}\bar{s}_{P_0R} s_{UP_0} - \bar{s}_{UP_0} s_{UP_0}\bar{s}_{P_0R} s_{VP_0}}{\bar{s}_{UP_0} s_{UP_0}\bar{s}_{VP_0} s_{VP_0} - \left(\bar{s}_{VP_0} s_{UP_0}\right)^2}$$

by minimizing $\bar{s}_{PR} s_{PR}$ with respect to u and v.

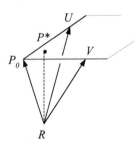

2.7 Projected angle on focal plane array. A building is imaged on a focal plane array. Two unit vectors t_1 and t_2 model the sides of the building, emanating from one of the rectangular corners. The normal unit vector of the focal plane array is u.

(a) Derive the tensor equation that allows you to calculate the angle ϕ between the two sides as seen on the focal plane array.

(b) Calculate ϕ from the numerical values:

$$[\bar{t}_1]^G = [1\quad 0\quad 0]$$

$$[\bar{t}_2]^G = [0\quad 1\quad 0]$$

$$[\bar{u}]^G = [0.648\quad 0.3\quad 0.7]$$

(*Solution*: $\phi = \arccos(\frac{\bar{t}_1 \bar{N} N t_2}{|N t_1 \| N t_2|})$, where $N = E - u\bar{u}$).

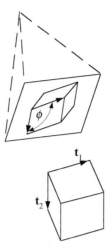

2.8   Miss vector in target plane.   Given the target plane P and its associated coordinate system $]^P$ with its 1^P, 2^P– axes parallel to the surface. The point target T is contained in this plane. The missile's last two calculated positions are at $\bar{B}$ and B. Derive the miss vector $[s_{B^*T}]^P$ with B^* the penetration point. By interpolating the straight line $[s_{B\bar{B}}]$, you should obtain

$$[s_{B^*T}]^P = \frac{(s_{T\bar{B}})_3^P}{(s_{B\bar{B}})_3^P}[s_{B\bar{B}}]^P - [s_{T\bar{B}}]^P$$

Furthermore, derive the intersection time t_{B^*} with the target plane, if the time $t_{\bar{B}}$ and the integration interval Δt are given (assume constant velocity):

$$t_{B^*} = \frac{(s_{T\bar{B}})_3^P}{(s_{B\bar{B}})_3^P}\Delta t + t_{\bar{B}}$$

You will need these two equations for any simulation that models the intercept of a plane. Air-to-ground and air-to-ship missiles are typical examples. For most of these applications, the linear interpolation of the intercept point is adequate. You will find them implemented in the CADAC simulation CRUISE5, Subroutine G4.

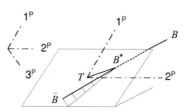

2.9 Closest approach of two lines. Determine the displacement vector $s_{B^*T^*}$ of the closest approach of a missile c.m. B and a target T by linear interpolation of the last two integration steps. Assume that both velocities are constant during the integration interval Δt. First, derive the time of closest approach $t^* = t_{\bar{B}} + \tau$ by minimizing

$$\underset{\tau}{\mathrm{Min}}\{\bar{s}_{B^*T^*}\bar{s}_{B^*T^*}\}$$

subject to the elapsed time τ,

$$t^* = t_{\bar{B}} - \Delta t \frac{(s_{B\bar{B}} - s_{T\bar{T}})(s_{\bar{B}E} - s_{\bar{T}E})}{(s_{B\bar{B}} - s_{T\bar{T}})(s_{B\bar{B}} - s_{T\bar{T}})}$$

then derive the miss vector of closest approach:

$$s_{B^*T^*} = \frac{\tau}{\Delta t}(s_{B\bar{B}} - s_{T\bar{T}}) + s_{\bar{B}E} - s_{\bar{T}E}$$

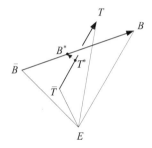

These equations are indispensable for an air-to-air intercept simulation. They calculate the terminal miss distance, which determines the probability of kill, for a particular fuse and warhead. You will find them implemented in the CADAC simulations AIM5, SRAAM5, and SRAAM6 in their respective Subroutines G4.

2.10 Tilt angle of maneuver plane—project. An aircraft executes a coordinated banking maneuver (zero sideslip angle) with bank angle ϕ_{BL} and generates the normal load factor acceleration a_N by the angle of attack α. Its orientation is given by the base vectors t_1, t_2, and t_3 and its velocity of the c.m. T wrt to Earth E by v_T^E. The gravitational acceleration g, acting on the aircraft, points in the direction of the unit local vertical l_3 and lies in the vertical plane v_T^E, l_3. The load factor plane, subtended by a_N and the base vectors t_1 and t_3, differs from the maneuver plane by the gravitational acceleration g.

(a) Derive the equation for the tilt angle ϕ_M in tensor form between the maneuver plane and the vertical plane. (*Hint:* v_T^E is contained in both planes). Write down all of the steps.

(b) Develop the matrix equations to calculate the tilt angle. Consider as given the bank angle ϕ_{BL}, the normal load factor n, mass m, reference area S (wing loading $w = mg/S$), normal force derivative C_{N_α}, aircraft velocity V, heading angle ψ_{VL}, and flight-path angle θ_{VL}. Write down all matrix equations that must be programmed in (d).

(c) Calculate, before you do (d), the tilt angle ϕ_M for the following numerical values:

$$\phi_{BL} = 40 \text{ deg}; \quad V = 250 \text{ m/s}$$

$$n = 3; \quad \psi_{VL} = 0$$

$$w = 3249 \text{ N/m}^2; \quad \theta_{VL} = 10 \text{ deg}$$

$$C_{N_\alpha} = 0.0523 \text{ deg}^{-1}$$

Sketch the attitude of the aircraft from the rear looking forward and indicate the approximate magnitude of the angles ϕ_M, ϕ_{BL} (*Solution:* $\phi_M = 84.1$ deg).

(d) Write a computer program that automates the calculation of the tilt angle. Consider as input: V, ψ_{VL}, θ_{VL}, n, ϕ_{BL}, and the other variables as parameters. Use the matrix utilities UTL.FOR of CADAC as much as possible. As output, record the tilt angle, angle of attack, and the direction cosine matrix $[T]^{BL}$. First run the test case of (c) for verification, then change the following values from the baseline for Case 2: $\psi_{VL} = 90$ deg; and Case 3: $n = 7$; and Case 4: $V = 350$ m/s. Provide the source code and the numerical results of all five cases.

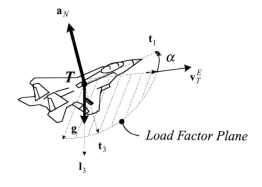

3
Frames and Coordinate Systems

Chapter 2 introduced frames and coordinate systems. *Frames* are models of physical references, whereas *coordinate systems* establish the association with Euclidean space. Both entities are important elements of aerospace vehicle dynamics. Frequently, they are presumed to be the same, but I will maintain a careful distinction throughout this book, heeding Truesdell's warning, quoted earlier, that frames should not be regarded as a synonym for coordinate systems.

This chapter will expand our understanding of these concepts. Important frames of reference, such as inertial, Earth, and body frames will be introduced. The triad of base vectors will emerge as a keystone to define their location and orientation. It will bridge the chasm between frames and coordinate systems with the so-called preferred coordinate systems.

Coordinate systems are the spider web of simulations, providing structure, direction, and focus. They structure the Euclidean space into application-specific associations, establish sense of direction, and focus on numerical solutions— the sustenance of any simulation. However, they can also lead to bafflement and mystification and may ensnare the careless user. (I once developed an air combat simulation that dealt with 24 different coordinate systems.) Clarity of definition is essential. We shall deal with a host of systems: inertial, Earth, geographic, local-level, perifocal, velocity, body, gimbal, relative wind, stability, aeroballistic, and some more—hopefully all of the coordinate systems you will ever need.

3.1 Frames

Recall the definition from Chapter 2: A frame is an unbounded continuous set of points over the Euclidean three-space with invariant distances and which possesses, as a subset, at least three noncollinear points. The inertial frame is such an unbounded set of points. We will give it a precise definition shortly. The Earth, although bounded, has also a frame associated with it. Theoretically, the Earth frame extends beyond the confines of the geoid, but when we refer to the points of the Earth's frame, we remain on the Earth. A similar approach is taken with the body frame. Strictly speaking, the body must be rigid to be modeled by a frame. For elastic modeling it is common practice to divide the body into finite, rigid elements, each of which is represented by a frame.

We must be able to identify at least three noncollinear points of a frame. Otherwise, the frame may not occupy the three-dimensional manifold of space. In particular, we may pick the three points such that, if connected with a fourth base point, they establish a triad that defines the position of the frame completely.

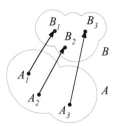

Fig. 3.1   Frame relationship.

3.1.1   Frame Positioning

Let A and B be two frames containing three noncollinear points A_1, A_2, A_3 and B_1, B_2, B_3, respectively (see Fig. 3.1). The position of frame B wrt frame A is determined by three displacement vectors $s_{B_i A_i}, i = 1, 2, 3$. Six of the nine vector coordinates are independent, i.e., a frame has six DoF. Are you surprised to encounter six rather than three degrees of freedom in the Euclidean three-space? Frames have the property of location and orientation, whereas space has only the attribute of location. That explains the difference. I will define these frame properties in more detail.

3.1.1.1   Location of a frame.
Let A and B be two points of frames A and B, respectively (see Fig. 3.2). The displacement vector s_{BA} determines the location of frame B relative to frame A. We call the two points *base points*. A displacement vector can only relate to one point in each frame. It fixes the location of the base point, but leaves the frame the freedom of orientation. Yet remember that the frame points are mutually unchanging. Two more noncollinear and nonplanar points would define the complete location and orientation of the frame B wrt frame A. However, we chose three additional points such that they form the endpoints of an orthonormal vector triad and thus define the orientation of a frame.

3.1.1.2   Orientation of a frame.
Build a triad from a set of three orthonormal base vectors (see Fig. 3.3):

$$a_1, a_2, a_3 \quad \text{where} \quad \bar{a}_i a_j = \begin{cases} 0 & \text{for } i \neq j \\ 1 & \text{for } i = j \end{cases}; \quad i = 1, 2, 3; \quad j = 1, 2, 3$$

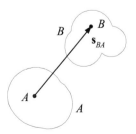

Fig. 3.2   Location.

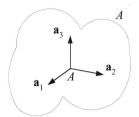

Fig. 3.3 Orientation.

with magnitudes one, which connect the base point A with three other points of a frame A. The orientation of a frame is given by this triad.

Just like location requires a reference point, so orientation needs a reference frame. A displacement vector models location, whereas orientation, as we shall see in Sec. 4.1, is portrayed by the rotation tensor. Both, location and orientation determine the position of a frame. A triad, with its base point and three vectors, is sufficient to define a frame.

You may interject that in Fig. 3.3 you recognize the familiar coordinate axes system. But not so! A triad is a physical entity, consisting of points and vectors, and is not a coordinate system. However, now we can finally bridge the gap. Among the coordinate systems associated with each frame, there is a particular one that coordinates the base vectors in the simple form $[a_1]^A = [1 \ 0 \ 0]$, $[a_2]^A = [0 \ 1 \ 0]$, and $[a_3]^A = [0 \ 0 \ 1]$. It is called the *preferred* coordinate system and is the most important one among all the possible coordinate systems associated with a frame.

3.1.2 Reference Frames

Without reference points and frames we could not model positions or motions of aerospace vehicles. Reference frames in particular have captured the interest of astronomers over millenniums. Just recall the argument whether the Earth, the sun or the stars should constitute the primary reference. In recent times, the launching of satellites and interplanetary travel of spacecraft have made it a matter of practical importance to clearly understand reference frames and coordinate systems. Vallado[1] gives an up-to-date account in his book *Fundamentals of Astrodynamics and Applications.*

We will concentrate on those frames that are of primary importance for the modeling of aerospace vehicles. The Sun-centered (heliocentric) frame serves all planetary space travel, whereas the Earth-centered frame suffices for Earth satellite trajectory work. Both are considered *inertial* frames in the Newtonian sense; the choice depends on the application. For Earth-bound flights the rotation of the Earth can often be neglected. Under these circumstances the Earth frame itself becomes the inertial reference. Even the vehicle's body can become a reference frame for rotating turbine blades, propellers, or gimbaled seeker heads. We shall define these frames now in detail using the concept of base point and base vectors.

3.1.2.1 Heliocentric frame.
For all of humanity the sun is the major frame of reference. It separates day and night, the seasons, and the years. It is the main source of energy, and its gravitational pull keeps the Earth and the planets on their

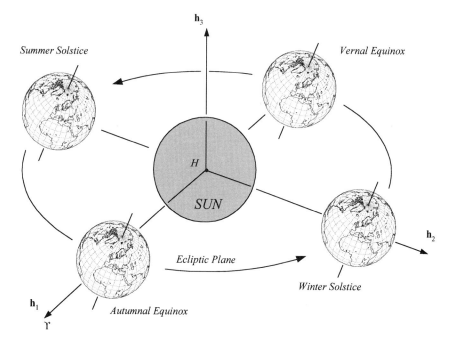

Fig. 3.4   Heliocentric frame of reference.

elliptical paths. As the Earth revolves around the sun in one year, it encircles the ecliptic plane. The astronomers speak of the mean ecliptic, which averages some minor periodic fluctuations. Because the Earth is tilted about 23.5-deg from the ecliptic, we experience the four seasons. Figure 3.4 visualizes the yearly cycle for you.

The sun is not solid but gaseous and possibly liquid. In the midst of all of the explosions, there are no reference points mutually at rest; therefore, we have to create them artificially. We pick the gravitational center as the base point H. The third base vector h_3 is normal to the ecliptic, pointing upward, based on the Earth's orbital direction and the sense of the vector abiding by the right-hand rule. We need at least one more base vector to fully define the triad (the third one follows from the orthonormal condition). The choice is the first base vector h_1, which points to the First Point of Aries; or at least where the star constellation was during Christ's lifetime. Today it points in the direction of the constellation Pisces.

How is that direction defined? At the first day of spring (vernal equinox), position yourself at the center of the Earth. As you look out, you see the intersection of the equator and the ecliptic lining up with the sun. If you had been born 2000 years ago, you would have seen the constellation Aries beyond the sun—an impossible feat with human eyes. The sign indicating Aries is a modified Greek Υ, resembling a ram.

Summarizing, the heliocentric frame is modeled by the base point H and the three base vectors h_1, h_2, and h_3, with h_1 pointing to Υ, h_3 being the normal of the ecliptic plane, and h_2 completing the triad. If you ever get to travel to Mars, you

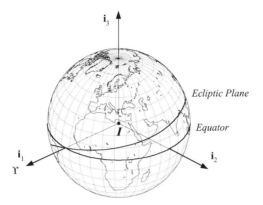

Fig. 3.5 Inertial reference frame.

would become very familiar with this heliocentric frame. However, back at planet Earth, even as astronaut of the International Space Station, you would more likely be dealing with the geocentric-inertial reference frame.

3.1.2.2   Geocentric-inertial (J2000) frame

Now let us concentrate on the Earth as we view it in Fig. 3.5. The most useful inertial frame is collocated with the center of the Earth, but its orientation remains fixed in the ecliptic. This is a good example of why we have to distinguish between location and orientation of a frame. The location of the frame is given by the displacement vector s_{IH} of the center of the Earth I wrt the center of the sun H. Its orientation is described by the base vectors i_1, i_2, and i_3.

To define the vector i_1, we have to carry out a *Gedanken* experiment ("Gedanken" is German for "thought"). Imagine a nonrotating shell around the Earth with the etched trace of the equator. The ecliptic, projected on the shell, is the path of the sun during one year. It will intersect the equator at two points. Of particular interest is the point when the sun crosses the equator in spring. This point is called the *vernal equinox*, already introduced in Fig. 3.4. We align with this direction the first base point i_1. It points at the constellation Aries and therefore is fixed with the stars. The remaining two base vectors are easily defined. The Earth's axis of rotation serves as direction and sense for i_3, and i_2 completes the right-handed triad.

Unfortunately, the equatorial plane, just as the ecliptic and the axis of rotation, moves very slightly over time. We can only speak of a truly inertial frame if we refer to its position at a particular epoch. Therefore, the astronomers defined the J2000 System that is based on the Fundamental Katalog, FK5.[1] This system is the best realization of an ideal inertial frame. It is the foundation for all near-Earth modeling and simulation. We will therefore refer to it just as the *inertial frame*.

3.1.2.3   Earth frame.

Look around you and take notice of the Earth (see Fig. 3.6). Its particles form the Earth's frame. The base point E is at the Earth's center, and the triad consists of the base vectors e_1, e_2, and e_3. One meridian of the Earth assumes particular significance. It is the prime meridian that traces through the Royal Observatory at Greenwich, a suburb of London. Its intersection with the

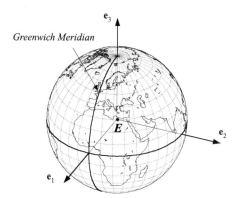

Fig. 3.6   Earth frame.

equator establishes the penetration point of the first base vector e_1. The Earth's axis of rotation serves as the direction and sense for the third base vector e_3, and e_2 completes the triad.

The prime meridian serves as origin for measuring longitude (positive in an easterly direction), whereas latitude is surveyed from the equatorial plane (positive in northerly direction). These reference lines are fixed on the Earth and ideal for locating sites on the Earth's surface.

3.1.2.4 Body frame.
Aircraft, missiles, and spacecraft are of primary concern for us. If they can be considered rigid bodies, they are represented by a frame, the so-called *body frame*. Although this frame is usually not used as a reference, it is nevertheless important for modeling location and orientation of vehicles under study. Its base point B coincides with the c.m., and the base vectors b_1, b_2, and b_3 are aligned with the principal axes of the moment of inertia tensor.

Sometimes other directions are more relevant. For instance, the b_1 vector for an aircraft is more likely aligned with some salient geometrical feature, like the tip of the nose or the zero-lift line at a particular Mach number (see Fig. 3.7). The aircraft designer has even adopted the terminology of his colleague, the ship architect, and calls it the waterline. However, for all vehicles b_2 is parallel to the second principal moment of inertia axis.

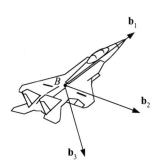

Fig. 3.7   Body frame.

Table 3.1 Summary of frames

Frame	Base point	Base vectors	First direction	Third direction
Heliocentric	H center of sun	h_1, h_2, h_3	h_1 Aries	h_3 normal of ecliptic
Inertial	I center of Earth	i_1, i_2, i_3	i_1 vernal equinox	i_3 Earth's spin axis
Earth	E center of Earth	e_1, e_2, e_3	e_1 Greenwich	e_3 Earth's spin axis
Body	B center of mass	b_1, b_2, b_3	b_1 nose	b_3 down

For aircraft, the positive sense of the b_1 base vectors is out the nose, b_2 out of the right wing, and b_3 down, completing the triad. For missiles with rotational symmetry, any direction for b_2 and b_3 is a principal axis. Therefore, control fins or geometrical marks may be used to fix them.

If elastic phenomena need be modeled, a body frame is still required as a reference for the bending and vibrating modes. This frame could be defined as coinciding with the vehicle under no-load conditions; or, in the case of wing flutter, the fuselage alone could be chosen. Sometimes the whole vehicle is divided into many rigid body subframes, and one of them is chosen as the primary reference.

3.1.2.5 Summary.
In this section I have introduced four important frames. Starting first with the heliocentric frame of reference for interplanetary travel, then zeroing in on the geocentric inertial frame for orbital trajectories, we eventually come down to Earth to define the Earth frame, which serves as reference for most atmospheric flight. In addition, the body frame is of particular significance. It models the position and orientation of the vehicle, which we want to simulate.

Table 3.1 summarizes the four frames. Notice the significance of the ecliptic and the equatorial planes. Their normal unit vectors define the direction of three base vectors (the Earth's spin axis is normal to the equator). You may be puzzled by the fact that the center of the Earth is the base point for both the inertial and the Earth frames. Indeed they coincide. Of all the points of the Earth frame, the center is the only point shared with the inertial frame.

You probably cannot wait any longer to meet the more familiar coordinate systems. Their time has come. I only hope that you keep in mind their fundamental difference with frames.

3.2 Coordinate Systems

The significance of coordinate systems is their ability to enable numerical calculations of symbolic equations. After all, modeling of aerospace vehicles finds its fulfillment in simulations, and computers can only chew on numbers and not on symbolic letters.

As you build your simulations, you will require many types of coordinate systems. Let me list those that are the most important ones: heliocentric, inertial, Earth, perifocal, geographic, local level, velocity, body, stability, aeroballistic, and relative wind coordinate systems. Others arise as applications require it: gimbal, sensors, nozzle, target coordinate systems, etc.

With such a confounding multitude it is understandable that order had to be established by standardization. In Germany, the LN Standard 9300[2] has been in use for many years. The U.S. has lagged behind. In the past I had to rely on a sole U.S. Navy document[3] for aeroballistic modeling. In 1992 AIAA published, in collaboration with the American National Standards Institute, the *Recommended Practice for Atmospheric and Space Vehicle Coordinate Systems.*[4] Of course, over the years, many textbooks have served as references as well: Etkin,[5] Bate et al.,[6] Britting[7]; and more recently, Pamadi,[8] Vallado,[1] and Chatfield.[9]

As we make the transition from frames to coordinate systems, the preferred coordinate system, defined earlier, will play an important role. If a triad has been defined for a frame, it is most convenient to pick from the infinite number of associated systems one that lines up with the base vectors.

Just like frames refer to each other—establishing their relative position—coordinate systems are related by coordinate transformations. Let us briefly review (see Chapter 2): coordinates are ordered algebraic numbers that are related to the Euclidean space by coordinate systems and relabeled by coordinate transformations. We employ only right-handed Cartesian coordinate systems. Before we detail the most important coordinate systems and their transformations, we will discuss the properties of coordinate transformation matrices.

3.2.1   Coordinate Transformation Matrix

Coordinate systems are related by coordinate transformation matrices that relabel the coordinates of a tensor. We will dissect them and examine their elements more closely. Two representations will help us to better understand their composition.

3.2.1.1   Base vector representation.
Introduce x, a first-order tensor, and any two allowable coordinate systems $]^A$, $]^B$ with their transformation matrix $[T]^{BA}$. From the definition of a first-order tensor, Eq. (2.3), we have

$$[x]^B = [T]^{BA}[x]^A \tag{3.1}$$

Furthermore, introduce the triad a_1, a_2, and a_3 (see Fig. 3.8), associated with the frame A. Decompose x into the vectors of the triad

$$x = x_1^A a_1 + x_2^A a_2 + x_3^A a_3$$

$x_i^A, i = 1, 2, 3$ are the coordinates, and $x_i^A a_i, i = 1, 2, 3$ are the components of the vector. Now express x in $]^B$ coordinates:

$$[x]^B = x_1^A [a_1]^B + x_2^A [a_2]^B + x_3^A [a_3]^B \tag{3.2}$$

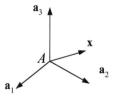

Fig. 3.8   Triad A.

This equation can be written as a scalar product

$$[x]^B = \begin{bmatrix} [a_1]^B & [a_2]^B & [a_3]^B \end{bmatrix} \begin{bmatrix} x_1^A \\ x_2^A \\ x_3^A \end{bmatrix} \tag{3.3}$$

If we select $]^A$ as the *preferred* from all of the associated coordinates systems, then the vector x is coordinated directly as

$$[x]^A = \begin{bmatrix} x_1^A \\ x_2^A \\ x_3^A \end{bmatrix}$$

and therefore Eq. (3.3) becomes

$$[x]^B = \begin{bmatrix} [a_1]^B & [a_2]^B & [a_3]^B \end{bmatrix}[x]^A$$

Comparing this equation with Eq. (3.1), we find a representation for the coordinate transformation matrix

$$[T]^{BA} = \begin{bmatrix} [a_1]^B & [a_2]^B & [a_3]^B \end{bmatrix} \tag{3.4}$$

Its columns consist of the frame A base vectors, coordinated in $]^B$, and predicated on using the preferred coordinate system for frame A.

We can reverse the derivation and decompose vector x into the triad b_1, b_2, and b_3 of frame B

$$x = x_1^B b_1 + x_2^B b_2 + x_3^B b_3$$

and coordinate it in the $]^A$ system

$$[x]^A = x_1^B [b_1]^A + x_2^B [b_2]^A + x_3^B [b_3]^A = \begin{bmatrix} [b_1]^A & [b_2]^A & [b_3]^A \end{bmatrix}[x]^B$$

Solving for $[x]^B$ and comparing with Eq. (3.1) yields another representation for the transformation matrix:

$$[T]^{BA} = \begin{bmatrix} [\bar{b}_1]^A \\ [\bar{b}_2]^A \\ [\bar{b}_3]^A \end{bmatrix} \tag{3.5}$$

The rows are the base vectors of frame B transposed as row vectors and coordinated in the $]^A$ system. Both presentations can be summarized in the following schematic with t_{ij} the elements of $[T]^{BA}$:

$$\begin{array}{c} [a_1]^B [a_2]^B [a_3]^B \\ \downarrow \quad \downarrow \quad \downarrow \\ \begin{matrix} [\bar{b}_1]^A \rightarrow \\ [\bar{b}_2]^A \rightarrow \\ [\bar{b}_3]^A \rightarrow \end{matrix} \begin{bmatrix} t_{11} & t_{12} & t_{13} \\ t_{21} & t_{22} & t_{23} \\ t_{31} & t_{32} & t_{33} \end{bmatrix} \end{array}$$

In summary, the coordinate transformation matrix consists of base vectors coordinated in their preferred coordinate systems. Frequently, I will abbreviate the expression "coordinate transformation matrix" simply by TM.

3.2.1.2 Direction cosine transformation matrix.
There is another interpretation of the elements of TM. If we write out the coordinates of Eq. (3.4),

$$[T]^{BA} = \begin{bmatrix} a_{11}^B & a_{12}^B & a_{13}^B \\ a_{21}^B & a_{22}^B & a_{23}^B \\ a_{31}^B & a_{32}^B & a_{33}^B \end{bmatrix}$$

we can interpret each element as the cosine of the angle between two base vectors. The first subscript indicates the b triad and the second subscript the a triad. Let us check it out using the a_{11}^B term (see Fig. 3.9). To calculate the cosine of the angle between the base vectors b_1 and a_1, take their scalar product and express them in the $]^B$ coordinate system

$$\cos \angle(b_1, a_1) = [\bar{b}_1]^B [a_1]^B = [1 \quad 0 \quad 0] \begin{bmatrix} a_{11}^B \\ a_{21}^B \\ a_{31}^B \end{bmatrix}^B = a_{11}^B$$

Indeed, a_{11}^B is the cosine of the angle between the two base vectors. You can also start with Eq. (3.5) and reach the same conclusion. In general, for any b_i and a_k we can calculate the cosine of the angle according to the formula

$$\cos \angle(b_i, a_k) = a_{ik}^B = t_{ik}; \quad i = 1, 2, 3; \quad k = 1, 2, 3 \tag{3.6}$$

Because the elements are the cosines between two directions, the TM is also called the *direction cosine matrix*, a term quite frequently assigned to the TM of the body wrt the geographic coordinates.

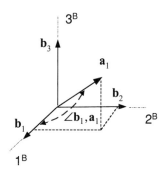

Fig. 3.9   Direction cosines.

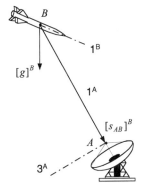

Fig. 3.10   Tracking coordinate system.

Example 3.1   Tracking Coordinate System

Problem.   Figure 3.10 shows a missile B being tracked by a search radar A. The onboard computer must determine the tracking radar's coordinate system and the TM $[T]^{BA}$ of the missile's coordinates wrt those of the tracking radar. Available are the gravity vector $[g]^B$ and the displacement vector of the radar wrt the missile $[s_{AB}]^B$, both measured in body coordinates. Some additional information is known about the coordinate system of the tracking radar. Its 1^A axis is parallel and opposite of the displacement vector $[s_{AB}]^B$, and the 2^A axis is horizontal (into the page). What are the equations that the missile processor has to execute?

Solution.   We start by formulating the radar's base vectors in missile axes. Then it is just a matter of using Eq. (3.4) to calculate the transformation matrix. The first base vector is the unit vector of the negative displacement vector

$$[a_1]^B = -\frac{[s_{AB}]^B}{|s_{AB}|}$$

The second base vector is obtained from the vector product

$$[a_2]^B = \frac{[S_{AB}]^B[g]^B}{|s_{AB}||g|}$$

where the upper case $[S_{AB}]^B$ indicates the skew-symmetric matrix of the lower case displacement vector $[s_{AB}]^B$. The third base vector completes the triad

$$[a_3]^B = [A_1]^B[a_2]^B = -\frac{[S_{AB}]^B[S_{AB}]^B[g]^B}{|s_{AB}|^2|g|}$$

We program the following matrix with three columns for the missile computer:

$$[T]^{BA} = \left[-\frac{[s_{AB}]^B}{|s_{AB}|} \quad \frac{[S_{AB}]^B[g]^B}{|s_{AB}||g|} \quad -\frac{[S_{AB}]^B[S_{AB}]^B[g]^B}{|s_{AB}|^2|g|} \right]$$

As the missile continues on its trajectory, the $[s_{AB}]^B$ and $[g]^B$ coordinates change (why does $[g]^B$ also change?), and the calculations have to be continually updated.

3.2.1.3 Properties of transformation matrices.

As we model and simulate aerospace vehicles, we will be manipulating many coordinate transformations. It is, therefore, important for you to have a good understanding of their properties.

Property 1

Transformation matrices are orthogonal.

Proof: Use the fact that the scalar product is the same in two coordinate systems. Introduce any vector x and any two allowable coordinate systems $]^A$ and $]^B$. Formulate the scalar product in both coordinate systems and then use Eq. (3.1):

$$[\bar{x}]^A[x]^A = [\bar{x}]^B[x]^B = [\bar{x}]^A[\bar{T}]^{BA}[T]^{BA}[x]^A$$

The outer equations form

$$[\bar{x}]^A\big([\bar{T}]^{BA}[T]^{BA} - [E]\big)[x]^A = 0$$

Because $[x]^A$ is arbitrary,

$$[\bar{T}]^{BA}[T]^{BA} = [E] \tag{3.7}$$

which expresses the orthogonality condition

$$[\bar{T}]^{BA} = ([T]^{BA})^{-1} \qquad \text{QED}$$

Property 2

The determinant of the TM is ± 1.

Proof: Take the determinant of Eq. (3.7):

$$|[\bar{T}]^{BA}[T]^{BA}| = |[E]|$$

$$|[\bar{T}]^{BA}||[T]^{BA}| = |[E]| = 1$$

$$|[T]^{BA}|^2 = 1$$

$$|[T]^{BA}| = \pm 1 \qquad \text{QED}$$

To maintain the right handedness, we choose $|[T]^{BA}| = +1$ only.

Property 3

Taking the transpose of the TM corresponds to changing the sequence of transformation, i.e.,

$$[\bar{T}]^{BA} = [T]^{AB}$$

Proof: Premultiply Eq. (3.1) by $[\bar{T}]^{BA}$,

$$[\bar{T}]^{BA}[x]^B = [\bar{T}]^{BA}[T]^{BA}[x]^A = [x]^A$$

$$[x]^A = [\bar{T}]^{BA}[x]^B$$

Because in Eq. (3.1) $]^A$ and $]^B$ are arbitrary, we can exchange their positions:

$$[x]^A = [T]^{AB}[x]^B$$

Comparing the last two equations yields $[\bar{T}]^{BA} = [T]^{AB}$ QED.

Property 4

Let $]^A$, $]^B$, and $]^C$ be any allowable coordinate systems, then $[T]^{CA} = [T]^{CB}[T]^{BA}$, i.e., consecutive transformations are contracted by canceling adjacent superscripts.

Proof: Let us apply Eq. (3.1) three times:

$$[x]^B = [T]^{BA}[x]^A \tag{3.8}$$

$$[x]^C = [T]^{CB}[x]^B \tag{3.9}$$

$$[x]^C = [T]^{CA}[x]^A \tag{3.10}$$

Substituting Eq. (3.8) into Eq. (3.9),

$$[x]^C = [T]^{CB}[T]^{BA}[x]^A$$

and comparing with Eq. (3.10), we conclude that

$$[T]^{CA} = [T]^{CB}[T]^{BA}$$ QED

The sequence of combining the TMs is important. Because these are matrix multiplications, they do not commute.

Property 5

Transformation matrices are *not* tensors.

Proof: If $[T]^{BA}$ were a tensor, it would have to be invariant under any coordinate transformation. Introduce a TM between any two allowable coordinate systems, say $[T]^{CD}$. According to Sec. 2.2.2, an entity like $[T]^{BA}$ is a second-order tensor if it maintains its characteristic under the transformation

$$[T]^{BA} = [T]^{CD}[T]^{BA}[\bar{T}]^{CD}$$

Clearly, the right and the left sides are completely different types of matrices. We cannot contract the superscripts of the right-hand side to conform to the left side. Therefore, $[T]^{BA}$ is not a tensor.

You will avoid errors in modeling if you remember three rules (Rules 1–4 are in Chapter 2).

Rule 5: $[T]^{BA}$ is always read as the TM of coordinate system $]^B$ with respect to (wrt) coordinate system $]^A$.

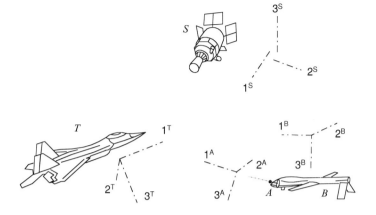

Fig. 3.11   UAV B tracking target T with antenna A and uplinking to satellite S.

Rule 6: The transpose of the TM reverses the order of transformation $[\bar{T}]^{BA} = [T]^{AB}$.

Rule 7: In transformation sequences, adjacent superscripts must be the same.

Example 3.2   Multiple Coordinate Transformations

Problem.   The antenna A of an unmanned aerial vehicle B images the target T and sends the information to a satellite tracker S (see Fig. 3.11). The uplinked information consists of the following TMs: $[T]^{TA}$, $[T]^{BA}$, and $[T]^{BS}$. How does the satellite processor calculate the TM of the target coordinates $]^T$ wrt its own coordinate system $]^S$?

Solution.   String the TMs together according to Rules 6 and 7:

$$[T]^{TS} = [T]^{TA}[\bar{T}]^{BA}[T]^{BS}$$

Notice that the sequence of adjacent superscripts is maintained if the transposed TM is replaced by Rule 6:

$$[T]^{TS} = [T]^{TA}[T]^{AB}[T]^{BS}$$

These strings of TMs are quite common in full-up simulations. Naturally, you do not multiply them term-by-term, but leave the manipulations to the computer.

You may have noticed in Fig. 3.11 that the coordinate axes of the satellite do not meet. I offset them intentionally, to emphasize that coordinate systems have no origin and do not have to emanate from a common point. They are completely defined by direction and sense.

With deeper insight into the structure of TMs, we are prepared for the multitude of special coordinate systems and their associations. I will introduce the most important ones next, but defer some to later sections as they naturally occur in the context of specific applications.

3.2.2 Coordinate Systems and Their Transformations

Coordinate systems are pervasive in computer simulations of aerospace vehicles. In contrast to frames with their base point and base vectors, coordinate systems have no physical substance. They are just mathematical schemes of relabeling the coordinates of tensors. However, we have seen, if base vectors are expressed in preferred coordinate systems, they take on a particularly simple form. This relationship invites us to display geometrically the direction and positive sense of the coordinate axes. Remember, however, that the coordinate axes do not have to emanate from the base point, although it will be convenient for us to do so most of the time.

We are already acquainted with the triads of the heliocentric, inertial, Earth, and body frames. Over these triads we superimpose the preferred coordinate systems with the same names. The axes are labeled 1, 2, 3, rather than x, y, z, with a capital superscript indicating the associated frame. We let the coordinate axes pierce the unit sphere and connect the piercing points to create surface triangles like orange peels. After some practice it will not be necessary to draw the axes any longer. The sketches of the orange peels will suffice to help you visualize the coordinate systems. At least that's what happened to me. Professor Stuemke, University of Stuttgart and formerly Peenemuende, was an orange lover and demanded from his students to think of coordinate systems as orange peels. Having dealt over the past 40 years with many coordinate systems, I am thankful that he did not waver in his devotion.

Rather than introducing individual coordinate systems, I pair them up and show mutual relationships that lead to coordinate transformations. We begin with the two most important reference systems.

3.2.2.1 Heliocentric and inertial coordinate systems.
Go back and review Figs. 3.4 and 3.5. We choose the preferred coordinate system of the inertial frame triad i_1, i_2, and i_3 (see Fig. 3.12). The 1^I inertial axis is aligned with the

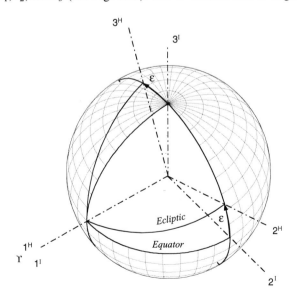

Fig. 3.12   Heliocentric and inertial coordinate systems.

vernal equinox, and the 3^I axis with the north pole. The 1^I and 2^I axes lie in the equator. The heliocentric coordinate axes are tilted by the obliquity of the ecliptic ε. Let us build the TM $[T]^{HI}$ of the heliocentric wrt the inertial coordinate systems with the help of Eq. (3.5)

$$
\begin{bmatrix} [\bar{h}_1]^I \\ [\bar{h}_2]^I \\ [\bar{h}_3]^I \end{bmatrix} = [T]^{HI}
$$

where the $[h_i]^I$, $i = 1, 2, 3$ are the base vectors of the heliocentric triad coordinated in inertial axes. By inspection of Fig. 3.12, you should be able to verify

$$
\begin{bmatrix} [\bar{h}_1]^I \\ [\bar{h}_2]^I \\ [\bar{h}_3]^I \end{bmatrix} = \begin{bmatrix} 1 & 0 & 0 \\ 0 & \cos \varepsilon & \sin \varepsilon \\ 0 & -\sin \varepsilon & \cos \varepsilon \end{bmatrix} = [T]^{HI} \tag{3.11}
$$

Notice the pattern of the TM. The 1 appears in the first column and row indicating that the transformation is taking place about the 1 direction without change in coordinates. The diagonals are the cosine of the transformation angle, a fact verified from the direction cosine matrix Eq. (3.6). The remaining off-diagonal elements are the sine of the angle, again verifiable by Eq. (3.6). You only have to decide where to put the negative sign. A simple rule says that the negative sign appears before that sine function, which is above the row containing the 1. If that row is on top, as in our example, imagine continuing rows in the sequence 3-2-1.

An alternate rule focuses on the positive sign. Inspect Fig. 3.12. The new axis that lies between the original axes indicates the row with the positive sine function. In our case 2^H lies between 2^I and 3^I; therefore, the second row of the matrix carries the positive sine.

As promised, I peel off the connections between the piercing points and drawn them in Fig. 3.13. To help you in the transition, I still show the coordinate axes. With practice you can soon do without them.

By earlier agreement we use only right-handed coordinate transformations. Transforming the $]^I$ system about the 1^I axis, we bring the 2^I axis through the angle ε to establish the 2^H axis by a right-handed motion with the index finger

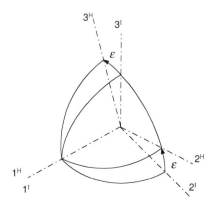

Fig. 3.13 Orange peels of heliocentric wrt inertial coordinate systems.

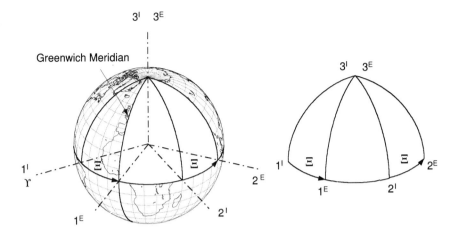

Fig. 3.14 Earth and inertial coordinate systems.

pointing in the 1^I direction. This transformation $[T]^{HI}$ is symbolically expressed as $]^H \overset{\varepsilon}{\longleftarrow}]^I$, verbalizing transformation of coordinate system H wrt I through the angle ε. Sometimes I may also use the expression, "transforming from I to H through the angle ε," although the first form is preferred. Taking the transpose of the TM $[\bar{T}]^{HI}$ changes the sequence of transformation to $[T]^{IH}$, through the negative angle $-\varepsilon$, or symbolically $]^I \overset{-\varepsilon}{\longleftarrow}]^H$.

3.2.2.2 Earth coordinate system. The Earth coordinate system is the preferred coordinate system of the Earth's frame triad e_1, e_2, and e_3 (see Fig. 3.6). Its 1^E axis pierces through the unit sphere at the intersection of the equator with the Greenwich meridian. The 3^E axis overlays the Earth's spin axis, and the 2^E axis completes the right-handed coordinate system (see Fig. 3.14).

To relate the Earth coordinates to the inertial coordinates, we have to heed the Earth's rotation. Every 24 h the Earth presents the same face to the sun. This is called the *solar day*. However, a full rotation of the Earth relative to the stars, the so-called *sidereal day*, is actually shorter by about 4 min. The lengthening of the solar day is caused by the progression of the Earth on the ecliptic during 24 h. To present the same face to the sun, the Earth has to rotate further. We are interested in the sidereal time that has elapsed since the 1^E axis coincided with the 1^I axis. Rather than using time, however, we use an angular measurement, equating 360° to one sidereal day. The angle between 1^E and 1^I is called the hour angle Ξ and establishes the Greenwich meridian relative to the meridian of the vernal equinox.

The transformation matrix $[T]^{EI}$ of the Earth coordinates wrt the inertial coordinates is obtained by inspection:

$$[T]^{EI} = \begin{bmatrix} \cos \Xi & \sin \Xi & 0 \\ -\sin \Xi & \cos \Xi & 0 \\ 0 & 0 & 1 \end{bmatrix} \qquad (3.12)$$

The rules, just given, should enable you to do the same. Figure 3.14 displays also the frugal orange peel schematic, which in its simplicity conveys all of the important information of the picture on the left.

3.2.2.3 Geographic coordinate system.
As you become comfortable with the Earth system, you want to navigate on the surface of the Earth. A grid, blanketing the Earth's surface, determines any point you want to reach. It consists of lines of longitude and latitude. Longitude is divided into $\pm180°$ with the positive direction starting at the Greenwich meridian in an easterly direction. Latitude is measured from the equator, positive to the north from 0 to $90°$ and negative south.

Inside of simulations it is better to work with radians. Longitude can extend from 0 to 2π or $\pm\pi$ and latitude between $\pm\pi/2$. Yet, to be kind to the customer, you can allow the data to be input in degrees, minutes, and seconds, and in turn you convert your output to the same units. The unit of arc minutes takes on a particular significance on a great circle like the longitude meridians or the equator because the *nautical mile* is defined as the arc length of 1 min. Ergo, the circumference of the Earth on the equator is $60 \times 360 = 21,600$ n miles.

At a specific point on the surface of the Earth, with its longitude l and latitude λ, the geographic coordinate system $]^G$ is defined as follows: The 1^G axis points north, the 3^G axis points at the center of the Earth, and the 2^G axis, pointing east, completes the right-handed coordinate system.

To relate the geographic coordinate system to the Earth system requires a few steps (refer to Fig. 3.15). The first transformation is from $]^E$ to an intermediate system $]^X$ with the longitude angle l, symbolically written as $]^X \xleftarrow{l}]^E$, followed by another intermediate system $]^Y$, obtained through the compliment of the latitude angle $90° - \lambda$, or symbolically $]^Y \xleftarrow{90°-\lambda}]^X$.

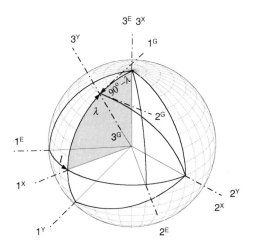

Fig. 3.15   Geographic wrt Earth coordinates.

Before we formulate the complete transformation, let us determine the TM of these two steps:

$$[T]^{YE} = [T(90° − λ)]^{YX}[T(l)]^{XE}$$

$$[T]^{YE} = \begin{bmatrix} \sin λ & 0 & −\cos λ \\ 0 & 1 & 0 \\ \cos λ & 0 & \sin λ \end{bmatrix} \begin{bmatrix} \cos l & \sin l & 0 \\ −\sin l & \cos l & 0 \\ 0 & 0 & 1 \end{bmatrix}$$

$$= \begin{bmatrix} \sin λ \cos l & \sin λ \sin l & −\cos λ \\ −\sin l & \cos l & 0 \\ \cos λ \cos l & \cos λ \sin l & \sin λ \end{bmatrix}$$

Notice the first transformation $[T]^{XE}$ is about the 3^E axis, followed by the transformation $[T]^{YX}$ about the 2^X axis.

To reach the geographic axes, we have to make a 180-deg somersault. The 1^G and 3^G axes take the opposite direction of the 1^Y and 3^Y axes, respectively, while the 2^G axis maintains the same sense as 2^Y. How do we determine this TM? We go back to Eq. (3.5) and visualize the base vectors associated with the preferred coordinate system $]^G$ being expressed in the $]^Y$ system. From this perspective we obtain the somersault transformation

$$[T]^{GY} = \begin{bmatrix} [\bar{g}_1]^Y \\ [\bar{g}_2]^Y \\ [\bar{g}_3]^Y \end{bmatrix} = \begin{bmatrix} −1 & 0 & 0 \\ 0 & 1 & 0 \\ 0 & 0 & −1 \end{bmatrix}$$

Stringing all of the transformations together,

$$[T]^{GE} = [T(180°)]^{GY}[T(90° − λ)]^{YX}[T(l)]^{XE}$$

yields the important TM of geographic wrt Earth coordinates:

$$[T]^{GE} = \begin{bmatrix} −\sin λ \cos l & −\sin λ \sin l & \cos λ \\ −\sin l & \cos l & 0 \\ −\cos λ \cos l & −\cos λ \sin l & −\sin λ \end{bmatrix} \qquad (3.13)$$

As a vehicle moves across the Earth, its longitude and latitude coordinates change and so does the TM $[T]^{GE}$. This phenomenon has led to the expression, "the geographic coordinate system is attached to the vehicle and its origin moves with it," a perspective that attributes physical substance to coordinate systems.

We are taking the "road less traveled" and follow an interpretation that is consistent with our premise that coordinate systems are purely mathematical entities. If you inspect Eq. (3.13), all you see are longitude and latitude angles. The TM, therefore, does not depend on an origin moving with the vehicle. We need only the longitude and latitude angles of the vehicle. As it moves over the Earth, the directions of the coordinate axes will change. However, the altitude of the vehicle is irrelevant. Therefore, you do not have to keep track of a coordinate origin travelling with the vehicle. Coordinate systems have no origins!

The 1^G and 2^G axes are tangential to the Earth at the longitude and latitude of the vehicle. At least that is true for our current assumption that a spherical Earth is adequate for our modeling tasks. Later, in Sec. 10.1.2, for sophisticated six-DoF simulations we will consider the Earth to be a spheroid. There we will introduce the *geodetic* coordinate $]^D$, which is tangential to the spheroid at the vehicle's latitude and longitude. Its 3^D axis does not point at the center of the Earth anymore. This direction will be maintained by the 3^G axis as before, and for that reason the geographic system will be renamed in that section as the *geocentric system*.

3.2.2.4 Body coordinate system.

The preferred body coordinate system is aligned with the body triad of Fig. 3.7. The 1^B axis points through the nose of the vehicle and lies with the downward-pointing 3^B axis in the plane of symmetry. The 2^B axis, out the right wing, completes the coordinate system.

A prominent transformation matrix in flight mechanics is the TM of body coordinates wrt geographic coordinates $[T]^{BG}$. It is composed of three transformations by the so-called Euler angles: yaw, pitch, and roll or ψ, θ, and ϕ. Two intermediate systems $]^X$ and $]^Y$ are needed to complete the chain:

$$[T]^{BG} = [T(\phi)]^{BY}[T(\theta)]^{YX}[T(\psi)]^{XG}$$

As Fig. 3.16 shows, the unit sphere is getting more cluttered, and you should be grateful for the orange peel rendering, showing only the essentials without coordinate axes jumbling up the picture.

Let us start the chain reaction with the geographic $]^G$ to the first intermediate $]^X$ system through the yaw angle ψ. The transformation occurs about the 3^G axis.

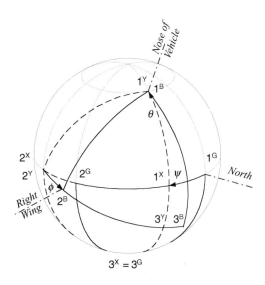

Fig. 3.16 Body wrt geographic coordinate systems.

Therefore we have the pattern

$$[T]^{XG} = \begin{bmatrix} \cos\psi & \sin\psi & 0 \\ -\sin\psi & \cos\psi & 0 \\ 0 & 0 & 1 \end{bmatrix}$$

The second transformation is about the 2^X axis through the pitch angle θ to the second intermediate system $]^Y$:

$$[T]^{YX} = \begin{bmatrix} \cos\theta & 0 & -\sin\theta \\ 0 & 1 & 0 \\ \sin\theta & 0 & \cos\theta \end{bmatrix}$$

and the third transformation leads us to the body axes through the roll angle ϕ about the 1^Y axis

$$[T]^{BY} = \begin{bmatrix} 1 & 0 & 0 \\ 0 & \cos\phi & \sin\phi \\ 0 & -\sin\phi & \cos\phi \end{bmatrix}$$

Now it is just a matter of multiplying the three matrices to obtain the Euler transformation matrix:

$$[T]^{BG} = \begin{bmatrix} \cos\psi\cos\theta & \sin\psi\cos\theta & -\sin\theta \\ \cos\psi\sin\theta\sin\phi - \sin\psi\cos\phi & \sin\psi\sin\theta\sin\phi + \cos\psi\cos\phi & \cos\theta\sin\phi \\ \cos\psi\sin\theta\cos\phi + \sin\psi\sin\phi & \sin\psi\sin\theta\cos\phi - \cos\psi\sin\phi & \cos\theta\cos\phi \end{bmatrix}$$

$$(3.14)$$

Quite frequently this transformation is also called the direction cosine matrix, as I already mentioned in Sec. 3.2.1.

In your simulation this TM may be given, and you are required to calculate the Euler angles. By inspection of Eq. (3.14), you deduce from the element t_{13} of the matrix that the pitch angle is

$$\theta = \arcsin(-t_{13}) \qquad (3.15)$$

and from the elements t_{11} and t_{12} the yaw angle

$$\psi = \arctan\left(\frac{t_{12}}{t_{11}}\right) \qquad (3.16)$$

and lastly from the elements t_{23} and t_{33} the roll angle

$$\phi = \arctan\left(\frac{t_{23}}{t_{33}}\right) \qquad (3.17)$$

If you change the sequence of the matrices, you do not get the same transformation matrix because matrix multiplications do not commute. However, because matrix multiplications are associative, you can change the order of the multiplications without changing the result. Convince yourself of these facts by using the matrices of this example.

3.2.2.5 Wind coordinate systems.

Now the atmosphere thickens, and we take note of the air flowing over the vehicle. The air mass may be at rest or moving wrt the Earth. We assume, however, that it is monolithic, i.e., the air molecules remain mutually fixed. This characteristic qualifies the air to be modeled as a frame A.

As the vehicle moves through the air mass, it experiences a *relative wind* over its body, which gives rise to aerodynamic forces. We introduce the wind coordinate system $]^W$. Only the 1^W axis is defined unambiguously. It is parallel and in the direction of the velocity vector v_B^A of the c.m. of the vehicle B wrt the air A. The type of vehicle determines the other two axes. We distinguish between aircraft and missiles. An aircraft's planar symmetry gives rise to the Cartesian incidence angles: angle of attack α and sideslip angle β, whereas missiles with rotational symmetry are frequently modeled by polar aeroballistic angles—total angle of attack α' and aerodynamic roll angle ϕ'. We shall treat the TMs of wind wrt body coordinates for aircraft and missiles separately.

Cartesian incidence angles for aircraft. For aircraft the TM of wind wrt body coordinates $[T]^{WB}$ consists of two transformations with the interim stability coordinate system $]^S$. Unfortunately by convention, one cannot reach $]^W$ from $]^B$ by two positive transformations. The sequence is rather $]^W \xleftarrow{\beta}]^S \xleftarrow{-\alpha}]^B$ with a negative alpha transformation (see Fig. 3.17). The stability system takes on a particular significance because both TMs $[T(\alpha)]^{BS}$ and $[T(\beta)]^{WS}$ are from its perspective reached by positive angles. Therefore, we first derive these individual transformations and then combine them to form $[T]^{WB}$.

The stability coordinate system is defined as follows. The 1^S axis is parallel and in the direction of the projection of the velocity vector v_B^A on the symmetry plane 1^B, 3^B, and the 2^S axis stays with 2^B. The TM $[T]^{WB}$ is about this 2^S axis by the angle of attack α

$$[T]^{BS} = \begin{bmatrix} \cos\alpha & 0 & -\sin\alpha \\ 0 & 1 & 0 \\ \sin\alpha & 0 & \cos\alpha \end{bmatrix}$$

The second TM $[T]^{WS}$ connects with the wind axis through the sideslip angle β

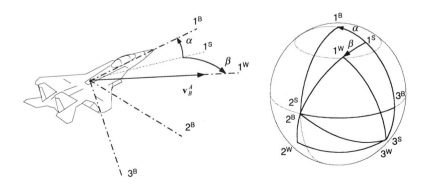

Fig. 3.17   Wind wrt body coordinate systems.

transformed about the 3^S axis

$$[T]^{WS} = \begin{bmatrix} \cos\beta & \sin\beta & 0 \\ -\sin\beta & \cos\beta & 0 \\ 0 & 0 & 1 \end{bmatrix}$$

To arrive at our final destination, we multiply the two TMs but have to transpose first $[T]^{BS}$, which has the effect of reversing the direction of the transformation, i.e., $[\bar{T}(\alpha)]^{BS} = [T(-\alpha)]^{BS}$

$$[T]^{WB} = [T]^{WS}[\bar{T}]^{BS}$$

$$[T]^{WB} = \begin{bmatrix} \cos\alpha\cos\beta & \sin\beta & \sin\alpha\cos\beta \\ -\cos\alpha\sin\beta & \cos\beta & -\sin\alpha\sin\beta \\ -\sin\alpha & 0 & \cos\alpha \end{bmatrix} \qquad (3.18)$$

This TM distinguishes itself by keeping the 3^W axis in the aircraft's plane of symmetry and aligning it with the 3^S axis of the stability coordinate system.

Polar aeroballistic incidence angles for missiles. For missiles with rotational symmetry, the *load factor plane* is more important than a body symmetry plane. It contains the total incidence angle α' that gives rise to the aerodynamic force. As the stability axes subtend the aircraft symmetry plane, so does the aeroballistic coordinate system $]^R$ line up with the load factor plane (see Fig. 3.18). In particular, the 1^R and 3^R axes lie in the load factor plane, with 1^R coinciding with 1^B. To change from the aeroballistic coordinates to the body coordinates, the aerodynamic roll angle ϕ' determines the transformation $]^B \xleftarrow{\phi'}]^R$ about the 1^R axis.

Because the wind coordinates for missiles are different than for aircraft, we rename them *aeroballistic wind coordinates* with the label $]^A$. Their 1^A axis is defined just like the 1^W axis for the aircraft, namely it is parallel and in the direction of the relative velocity vector v^A_B; but its 3^A axis lies in the load factor plane, and the 2^A axis remains in the 2^B, 3^B plane. The transformation of the aeroballistic

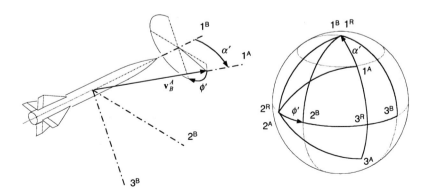

Fig. 3.18 Aeroballistic wind wrt body coordinate systems.

coordinates wrt the aeroballistic wind coordinates $[T]^{RA}$ is by the total angle of attack α', $]^R \xleftarrow{\alpha'}]^A$, about the 2^A axis. From the orange peels of Fig. 3.18, we deduct the TM of the body wrt aeroballistic coordinates

$$[T]^{BR} = \begin{bmatrix} 1 & 0 & 0 \\ 0 & \cos\phi' & \sin\phi' \\ 0 & -\sin\phi' & \cos\phi' \end{bmatrix}$$

and the TM of the aeroballistic wrt the wind coordinates

$$[T]^{RA} = \begin{bmatrix} \cos\alpha' & 0 & -\sin\alpha' \\ 0 & 1 & 0 \\ \sin\alpha' & 0 & \cos\alpha' \end{bmatrix}$$

Our goal is the TM of the aeroballistic wind wrt the body coordinates $[T]^{AB}$. We get there in two steps. First, combine the two transformations $[T]^{BA} = [T]^{BR}[T]^{RA}$, and then take the transpose

$$[T]^{AB} = [\bar{T}]^{BA} = [\bar{T}]^{RA}[\bar{T}]^{BR}$$

$$[T]^{AB} = \begin{bmatrix} \cos\alpha' & \sin\alpha'\sin\phi' & \sin\alpha'\cos\phi' \\ 0 & \cos\phi' & -\sin\phi' \\ -\sin\alpha' & \cos\alpha'\sin\phi' & \cos\alpha'\cos\phi' \end{bmatrix} \qquad (3.19)$$

We have accomplished our task of deriving the TMs of the relative wind wrt the body coordinates for aircraft and missiles. If you compare Eqs. (3.18) and (3.19), you verify by inspection that they are not the same. Their transformation angles are dissimilar—Cartesian vs polar—and the wind axes are indeed defined differently.

From these TMs we can derive the definitions for the incidence angles in terms of the velocity components in body coordinates $[\bar{v}_B^A]^B = [u\ v\ w]$ and relative wind axes for aircraft $[\bar{v}_B^A]^W = [V\ 0\ 0]$ and missile $[\bar{v}_B^A]^A = [V\ 0\ 0]$, where $V = \sqrt{u^2 + v^2 + w^2}$. The angle of attack α follows from the application of Eq. (3.18):

$$\left[v_B^A\right]^W = [T]^{WB}\left[v_B^A\right]^B$$

$$\begin{bmatrix} V \\ 0 \\ 0 \end{bmatrix} = \begin{bmatrix} \cos\alpha\cos\beta & \sin\beta & \sin\alpha\cos\beta \\ -\cos\alpha\sin\beta & \cos\beta & -\sin\alpha\sin\beta \\ -\sin\alpha & 0 & \cos\alpha \end{bmatrix} \begin{bmatrix} u \\ v \\ w \end{bmatrix}$$

From the last line we derive

$$\alpha = \arctan\left(\frac{w}{u}\right) \qquad (3.20)$$

To obtain a similar relationship for the sideslip angle β, we use the two top lines

$$V = u\cos\alpha\cos\beta + v\sin\beta + w\sin\alpha\cos\beta$$

$$0 = -u\cos\alpha\sin\beta + v\cos\beta - w\sin\alpha\sin\beta$$

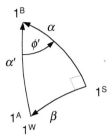

Fig. 3.19 Spherical triangle.

Multiplying the first equation by $\sin \beta$ and the second by $\cos \beta$ and adding both together yields the definition of β

$$\beta = \arcsin\left(\frac{v}{V}\right) \tag{3.21}$$

A similar procedure provides the definitions of the polar incidence angles of missiles:

$$\alpha' = \arccos\left(\frac{u}{V}\right) \tag{3.22}$$

$$\phi' = \arctan\left(\frac{v}{w}\right) \tag{3.23}$$

Equations (3.21) and (3.22) are undefined for $V = 0$, a case of no interest to us, but $v = w = 0$ can happen in unperturbed flight, causing ϕ' to be undefined according to Eq. (3.23).

The relationship between the Cartesian and polar incidence angles can be derived from the spherical triangle that nestles around the 1^B axis. Superimposing the angles from Figs. 3.17 and 3.18 enables us to draw Fig. 3.19. It is a right spherical triangle that engages all four incidence angles. As an exercise, you should be able to derive

$$\alpha' = \arccos(\cos \alpha \cos \beta)$$

$$\phi' = \arctan\left(\frac{\tan \beta}{\sin \alpha}\right) \tag{3.24}$$

$$\alpha = \arctan(\cos \phi' \tan \alpha')$$

$$\beta = \arcsin(\sin \phi' \sin \alpha')$$

Note again that ϕ' is undefined if both α and β are zero.

3.2.2.6 Flight-path coordinate system. After having related the velocity vector of the vehicle wrt the air mass v_B^A to the body axes, we define the *flight-path coordinate system* $]^V$ that relates the velocity vector of the vehicle wrt Earth v_B^E to the geographic system. (If the air is at rest, $A = E$, and the two velocity vectors are one and the same.) The 1^V axis is parallel and in the direction of v_B^E, and the

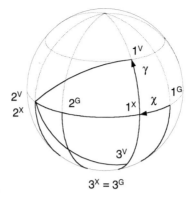

Fig. 3.20   Flight path wrt geographic coordinates.

2^V axis remains in the horizontal plane subtended by 1^G and 2^G (see Fig. 3.20). Two angles relate the flight path coordinates to the geographic system. The heading angle χ is measured from north to the projection of v_B^E into the local tangent plane and the flight-path angle γ takes us vertically up to v_B^E. The TM consists of the two individual transformations $[T]^{VG} = [T(\gamma)]^{VX}[T(\chi)]^{XG}$

$$[T]^{VG} = \begin{bmatrix} \cos\gamma & 0 & -\sin\gamma \\ 0 & 1 & 0 \\ \sin\gamma & 0 & \cos\gamma \end{bmatrix} \begin{bmatrix} \cos\chi & \sin\chi & 0 \\ -\sin\chi & \cos\chi & 0 \\ 0 & 0 & 1 \end{bmatrix}$$

and multiplied

$$[T]^{VG} = \begin{bmatrix} \cos\gamma\cos\chi & \cos\gamma\sin\chi & -\sin\gamma \\ -\sin\chi & \cos\chi & 0 \\ \sin\gamma\cos\chi & \sin\gamma\sin\chi & \cos\gamma \end{bmatrix} \tag{3.25}$$

From this TM we can derive the definitions for χ and γ. Let the velocity components in the geographic coordinates be $[\bar{v}_B^E]^G = [u_G \; v_G \; w_G]$ and in flight path coordinates $[\bar{v}_B^E]^V = [V \; 0 \; 0]$. The TM of Eq. (3.25) provides the relationship $[v_B^E]^V = [T]^{VG}[v_B^E]^G$

$$\begin{bmatrix} V \\ 0 \\ 0 \end{bmatrix} = \begin{bmatrix} \cos\gamma\cos\chi & \cos\gamma\sin\chi & -\sin\gamma \\ -\sin\chi & \cos\chi & 0 \\ \sin\gamma\cos\chi & \sin\gamma\sin\chi & \cos\gamma \end{bmatrix} \begin{bmatrix} u_G \\ v_G \\ w_G \end{bmatrix}$$

From the second line we glean

$$\chi = \arctan\left(\frac{v_G}{u_G}\right) \tag{3.26}$$

and you should also show that

$$\gamma = \arctan\left(\frac{-w_G}{\sqrt{u_G^2 + v_G^2}}\right) \tag{3.27}$$

When you program these equations, you have to be careful with the arctan function because it is multivalued. Particularly, the heading angle can take on values from 0 to 360 deg. It is best to use the ATAN2 intrinsic routine that most computer languages provide.

3.2.2.7 Local-level coordinate system.
We conclude our exposition of coordinate systems with a special case of geographic coordinates, suitable for many aircraft and missile simulations. If the vehicle flies in the atmosphere with speeds less than Mach 5 (below hypersonic velocity), the Earth can be presumed an inertial reference frame. Furthermore, if the particular location on the globe is irrelevant to the simulation, any local tangent plane can serve as a geographic coordinate system, independent of the longitude and latitude designations. This special geographic coordinate system is called the *local-level coordinate system*. It maintains its fixed, level orientation, usually that of the launch point, although the vehicle is traversing the ground. Envision the longitude and latitude grid unfurled into this local-level plane. The vehicle's trajectory is calculated relative to this plane, and altitude and ground distance are accurately portrayed. If you wanted to plot the trajectory on the globe, you could drape the ground track and altitude over the Earth's curvature and assign longitude and latitude coordinates.

The local-level coordinate system $]^L$ embeds its 1^L and 2^L axes into the horizontal plane and points the 3^L axis downward. The direction of 1^L is arbitrary, but, by convention, it is said to point north and the 2^L axis to point east. For this reason it is sometimes also called the north-east-down (NED) coordinate system.

For those simulations that abide by these assumptions, you can replace the geographic by the local level coordinate system. The TMs derived in this chapter still maintain their validity. The body wrt geographic TM $[T]^{BG}$ becomes $[T]^{BL}$ with the same Euler angles ψ, θ, and ϕ

$$[T]^{BL} = \begin{bmatrix} \cos\psi\cos\theta & \sin\psi\cos\theta & -\sin\theta \\ \cos\psi\sin\theta\sin\phi - \sin\psi\cos\phi & \sin\psi\sin\theta\sin\phi + \cos\psi\cos\phi & \cos\theta\sin\phi \\ \cos\psi\sin\theta\cos\phi + \sin\psi\sin\phi & \sin\psi\sin\theta\cos\phi - \cos\psi\sin\phi & \cos\theta\cos\phi \end{bmatrix}$$

$$\tag{3.28}$$

and the velocity wrt geographic TM $[T]^{VG}$ is replaced by $[T]^{VL}$ with the path angles χ and γ

$$[T]^{VL} = \begin{bmatrix} \cos\gamma\cos\chi & \cos\gamma\sin\chi & -\sin\gamma \\ -\sin\chi & \cos\chi & 0 \\ \sin\gamma\cos\chi & \sin\gamma\sin\chi & \cos\gamma \end{bmatrix} \tag{3.29}$$

Distinguish carefully that $]^L$ is associated with frame E but is not $]^G$.

Have you kept up with the number of coordinate systems? Without the intermediate systems $]^X$, and $]^Y$ I count a total of nine. That does not include the geodetic system, which will be introduced in Sec. 10.1.2, or the perifocal system,

Table 3.2 Summary of coordinate systems and their transformation angles

System	Directions		Angles	
Heliocentric H	1^H—Aries	3^H—normal of ecliptic		——
Inertial I	1^I—vernal equinox	3^I—Earth's spin axis	Obliquity of the ecliptic ε Hour angle Ξ	——
Earth E	1^E—Greenwich	3^E—Earth's spin axis	Longitude l	——
Geographic G	1^G—north	3^G—Earth's center	Latitude λ Yaw ψ	——
Body B	1^B—nose	3^B—down	Pitch θ Roll ϕ	——
Wind aircraft W	1^W—v_B^A	3^W—symmetry plane	Angle of attack α Sideslip angle β Total angle of attack α'	——
Wind missile A	1^A—v_B^A	3^A—load factor plane	Aero roll angle ϕ'	Heading angle χ
Flight path V	1^V—v_B^E	2^V—horizontal plane		Flight-path angle γ

the subject of an exercise. As we discuss practical implementations, a few more will make their appearance in conjunction with seeker gimbals, variable nozzles, and INSs. For now, however, Table 3.2 summarizes the coordinate systems and transformation angles of this section.

3.2.2.8 Summary. Surely by now you are thoroughly familiar with frames and coordinate systems. I hope I have convinced you that they are different entities and not synonymous. Just remember that frames are physical, whereas coordinate systems are mathematical models.

We discussed the representation of a frame by base point and triad. The location of its base vector and the orientation of the triad determine the position of the frame. I defined the important heliocentric, inertial, Earth, and body frames. Then we moved over to the mathematical ward and dissected coordinate transformation matrices. We found them to consist of base vectors expressed in preferred coordinate systems or direction cosines of enclosed angles. I will spare you the drudgery of repeating the coordinate systems and their transformations, but point you again to Table 3.1 for a summary.

This wraps up the geometrical part of our exposition. With our tool chest filled we can model lines, planes, bodies, and reference frames, and place them into the Euclidean space with coordinate systems. However, so far time has eluded us. Now we must bring time into play and embark for the shores of kinematics, the study of motions in space and time.

References

[1] Vallado, D. A., *Fundamentals of Astrodynamics and Applications*, Space Technology Series, McGraw–Hill, New York, 1997.

[2] Luftfahrt, Norm, LN 9300, *Flugmechanik*, Beuth Vertriebe GmbH, Koeln, Germany, 1970.

[3] Wright, J., "A Compilation of Aerodynamic Nomenclature and Axes Systems," U.S. Naval Ordnance Lab., NOLR 1241, White Oak, MD, Aug. 1962.

[4] American National Standard, *Recommended Practice for Atmospheric and Space Vehicle Coordinate Systems*, AIAA and ANSI, Washington, DC, 1992.

[5] Etkin, B., *Dynamics of Atmospheric Flight*, John Wiley, New York, 1972.

[6] Bate, R. R., Mueller, D. D., and White, J. E., *Fundamentals of Astrodynamics*, Dover, New York, 1971.

[7] Britting, K. R., *Inertial Navigation Systems Analysis*, Wiley-Interscience, New York, 1971.

[8] Pamadi, B. N., *Performance, Stability, Dynamics, and Control Applications*, AIAA Education Series, AIAA, Reston, VA, 1998.

[9] Chatfield, A. B., *Fundamentals of High Accuracy Inertial Navigation*, Progress in Astronautics and Aeronautics, AIAA, Reston, VA, 1997.

Problems

3.1 Position of an aircraft relative to a tracking radar. The position of an aircraft A with c.m. A is to be defined wrt to a tracking radar R and its antenna, represented by point R. The radar is referenced to north. How do you model the aircraft's and the radar's location and orientation. Be specific in your definitions. Make a sketch.

3.2 Conversion of satellite velocity. The velocity of a satellite S wrt Earth E is measured in geographic coordinates $[v_S^E]^G$. What are its coordinates in the heliocentric system $]^H$? Use the TMs that were introduced in Sec. 3.2.2.

3.3 Aerodynamic force component conversion. The aerodynamic force f is commonly coordinated in the stability system $]^S$ by its components lift L, drag D, and side force Y, $[\bar{f}]^S = [-D \; Y \; -L]$. In six-DoF simulations, however, f is frequently expressed in body coordinates $[\bar{f}]^B = [X \; Y \; Z]$. Derive the conversion transformation between the two component forms.

3.4 Transformation matrix between satellite and inertial coordinates. The normal satellite coordinate system $]^S$ (see Ref. 1, p. 46) is used to express drag forces on a satellite B. The 1^S axis is parallel to the satellites inertial velocity v_B^I, the 2^S axis normal to the orbital plane, and the 3^S axis in the general direction of the satellite's displacement vector from the center of the Earth s_{BI}. Given $[v_B^I]^I$ and $[s_{BI}]^I$ in inertial coordinates, express the TM $[T]^{SI}$ in terms of these two vectors.

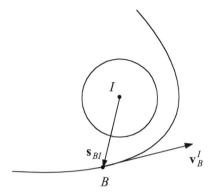

3.5   Angle of missile from north.   The TM of the body wrt geographic coordinates $[T]^{BG}$ is given by Eq. (3.14). How do you determine the angle between the centerline of the missile and the north direction?

3.6   Euler angles of gyro dynamics.   The elements of the direction cosine matrix, Eq. (3.14), contain the trigonometric functions of the Euler angles. More precisely they should be called the Euler angles of flight mechanics. In the study of the dynamics of gyroscopes, a different set of Euler angles is frequently used. Although the same symbols ψ, θ, and ϕ are adopted, the TM of body coordinates wrt inertial coordinates uses the sequence $3 > 1 > 3$ and not $3 > 2 > 1$ as in the case of flight mechanics. Make an orange peel diagram and derive the TM of gyro dynamics $[T]^{BI}$ with the sequence $]^B \xleftarrow{\phi}]^Y \xleftarrow{\theta}]^X \xleftarrow{\psi}]^I$. Compare the two transformations. Are there any similarities?

3.7   Sequence of transformation is all important.   The standard Euler transformation of flight mechanics $[T]^{BG}$ is sequenced $]^B \xleftarrow{\phi}]^Y \xleftarrow{\theta}]^X \xleftarrow{\psi}]^G$ and also called the $3 > 2 > 1$ transformation. Let us reverse the sequence to $1 > 2 > 3$ or $]^G \xleftarrow{\psi}]^X \xleftarrow{\theta}]^Y \xleftarrow{\phi}]^B$ and name it $[T]^{GB}$. Sketch the orange peel diagram and derive $[T]^{GB}$. Is $[T]^{GB}$ the transpose of $[T]^{BG}$? Why not? What is the sequence of transformation for the transpose of $[T]^{BG}$?

3.8   Perifocal coordinate system.   The trajectory of a satellite is best described in the perifocal coordinate system. Determine the transformation matrix $[T]^{PI}$ of the perifocal wrt inertial coordinates given in the figure. The sequence of individual transformations is $3 > 1 > 3$, or in symbolic form $]^P \xleftarrow{\omega}]^Y \xleftarrow{i}]^X \xleftarrow{\Omega}]^I$, with Ω the longitude of the ascending node, i the inclination, and ω the argument of the periapsis. These three angles are part of the six orbital elements that describe size, shape, and orientation of a satellite orbit. The remaining three are the semimajor axis and eccentricity of the elliptical orbit and the time of the periapsis (closest point to the Earth) passage.

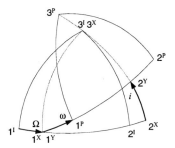

3.9 Seeker wrt vehicle transformation matrix.

An infrared seeker head of a missile has two gimbals. Its inner gimbal with the optics and the focal plane array executes pitching motions while its outer gimbal allows for rolling excursions. Two coordinate systems are of interest: the body coordinates $]^B$ with the 1^B axis parallel to the roll axis, pointing forward, and the head coordinates $]^H$ with the 2^H axis parallel to the pitch axis, pointing to the right. Determine the transformation matrix $[T]^{HB}$ of the head coordinates wrt the body coordinates using the roll angle ϕ_{HB} and pitch angle θ_{HB}. Sketch the orange peel diagram, clearly indicating the two angles.

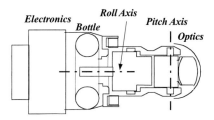

3.10 Antenna angles and transformation.

The main beam of a radar antenna is deflected by the azimuth a_z and elevation e_l angles from the centerline of the missile 1^B. The transformation sequence is 3 > 2 or symbolically $]^A \xleftarrow{e_l}]^X \xleftarrow{a_z}]^B$. Make an orange peel diagram and derive the TM $[T]^{AB}$. What are the coordinates of the antenna base vector $[a_1]^B$ in body coordinates? What are the angles between 1^A, 1^B; 2^A, 2^B; and 3^A, 3^B? (*Hint:* Use direction cosine matrix form.)

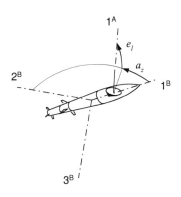

3.11 Initialization of flight-path angles. To initialize six-DoF simulations, it is most convenient to use the Euler angles ψ, θ, ϕ and the incidence angles α, β.

(a) Given these angles, derive the equations that determine the flight-path angles χ and γ.

(b) Introduce coordinate systems and express the equations in a form suitable for programming.

4
Kinematics of Translation and Rotation

After spending two chapters in the three-dimensional world of geometry, we are ready to launch into the fourth dimension, time. We will study *kinematics*, the branch of mechanics that deals with the motion of bodies without reference to force or mass. Later, in Chapters 5 and 6 we will add mass and force, apply Newton's and Euler's laws, and study the *dynamics* of aerospace vehicles.

If you watch the space shuttle take off at the Cape and track its altitude gain in time, you study its launch kinematics. However, if you are in Mission Control, responsible for ascent and orbit insertion, you concern yourself with the effect of mass, drag, and thrust and therefore are accountable for the dynamics of the space shuttle. Dynamics builds on kinematics. Hence we begin with kinematics.

I first introduce the *rotation tensor*, which defines the mutual orientation of two frames. It is the physical equivalent of the abstract coordinate transformation. Then I go right to the essence of the coordinate-independent formulation of kinematics and introduce the *rotational time derivative*. It will enable us in Chapters 5 and 6 to formulate Newton's and Euler's laws in an invariant form, valid in any allowable coordinate system. Afterward you are ready for the discussion of linear and angular motions of aircraft, missile, and spacecraft in greater detail. Finally, we wrap up the chapter with the fundamental problem in kinematics of flight, namely, how to calculate the attitude angles from the body's angular velocity.

Throughout these minutiae we shall remain true to our principle "from invariant modeling to matrix simulations." All of the forthcoming kinematic concepts are valid in any allowable coordinate system and thus are true tensor concepts. So welcome aboard, bring your tool chest, and I will fill it up with more goodies.

4.1 Rotation Tensor

Actually, we are not quite finished with geometry. In Sec. 2.1 I emphasized the importance of referencing points and frames to other points and frames. With the displacement vector s_{BA} we model the displacement of point B wrt point A. For frames we shall ascertain that the rotation tensor R^{BA} references the orientation of the frame B wrt the frame A.

As we study the properties of the rotation tensor, we establish the connection with coordinate transformations. Special rotations will give us more insight into the structure of the rotation tensor, and particularly, the small rotation tensor proves useful in perturbations like the inertial navigation system (INS) error model. Finally, a special rotation tensor, the tetragonal tensor, models the tetragonal symmetry of missiles, a feature we exploit in Sec. 7.3.1 for aerodynamic derivatives.

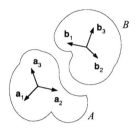

Fig. 4.1   Frames A and B and their triads.

4.1.1 Properties of the Rotation Tensor

The orientation of a frame A is modeled by its base triad consisting of the three orthonormal base vectors a_1, a_2, and a_3 (Sec. 3.1.1). Figure 4.1 shows the two frames A and B and their base triads. The orientation of frame B wrt frame A is established by the rotation tensor R^{BA}, which maps the a_i into the b_i base vectors:

$$b_i = R^{BA}a_i, \quad i = 1, 2, 3 \tag{4.1}$$

Our first concern is whether R^{BA} is a tensor. If we can show that it transforms like a tensor [see Eq. (2.4)], then it is a tensor.

Property 1

The rotation tensor R^{BA} of frame B wrt frame A is a tensor, i.e., for any two allowable coordinate systems $]^A$ and $]^B$ with their transformation matrix $[T]^{BA}$ it transforms like a second-order tensor

$$[R^{BA}]^B = [T]^{BA}[R^{BA}]^A[\bar{T}]^{BA}$$

Proof: Coordinate Eq. (4.1) by any two allowable coordinate systems, say $]^A$ and $]^B$,

$$[b_i]^A = [R^{BA}]^A[a_i]^A \tag{4.2}$$

and

$$[b_i]^B = [R^{BA}]^B[a_i]^B \tag{4.3}$$

Because the base vectors themselves are tensors, they transform like first-order tensors, Eq. (2.3):

$$[b_i]^B = [T]^{BA}[b_i]^A \tag{4.4}$$

and

$$[a_i]^B = [T]^{BA}[a_i]^A \tag{4.5}$$

Substitute Eq. (4.2) into Eq. (4.4) and replace $[a_i]^A$ by the transposed of Eq. (4.5)

$$[b_i]^B = [T]^{BA}[R^{BA}]^A[\bar{T}]^{BA}[a_i]^B$$

Comparing with Eq. (4.3), we deduct

$$[R^{BA}]^B = [T]^{BA}[R^{BA}]^A[\bar{T}]^{BA}$$

Because $]^A$ and $]^B$ can be any allowable coordinate systems, $\mathbf{R}^{BA}$ transforms like a second-order tensor and therefore is a tensor.

Property 2

Sequential rotations are obtained by multiplying the individual rotation tensors. For three frames A, B, and C and the rotation tensors $\mathbf{R}^{BA}$ and $\mathbf{R}^{CB}$, the rotation tensor $\mathbf{R}^{CA}$ of frame C wrt frame A is obtained from

$$\mathbf{R}^{CA} = \mathbf{R}^{CB}\mathbf{R}^{BA}$$

Note the contraction of the superscripts: the adjacent B's are deleted to form CA.

Proof: Let each frame A, B, and C be modeled by the triads $\mathbf{a}_i$, $\mathbf{b}_i$, and $\mathbf{c}_i$, $i = 1, 2, 3$ and related by

$$\mathbf{b}_i = \mathbf{R}^{BA}\mathbf{a}_i, \quad \mathbf{c}_i = \mathbf{R}^{CB}\mathbf{b}_i, \quad \mathbf{c}_i = \mathbf{R}^{CA}\mathbf{a}_i, \quad i = 1, 2, 3$$

Substituting the first into the second equality and comparing with the third one proves the property

$$\mathbf{c}_i = \mathbf{R}^{CB}\mathbf{b}_i = \mathbf{R}^{CB}\mathbf{R}^{BA}\mathbf{a}_i \Rightarrow \mathbf{R}^{CA} = \mathbf{R}^{CB}\mathbf{R}^{BA}$$

Property 3

Rotation tensors, coordinated in preferred coordinate systems, are related to their transformation matrices. For any two triads $\mathbf{a}_i$, $\mathbf{b}_i$, with the rotation tensor $\mathbf{R}^{BA}$, and the preferred coordinate systems $]^A$, $]^B$ with the transformation matrix $[T]^{BA}$, the following relationships hold:

$$[R^{BA}]^A = [R^{BA}]^B = [\bar{T}]^{BA} \tag{4.6}$$

Note first the surprising result that the rotation tensor has the same coordinates in both of its preferred coordinate systems. Furthermore, the rotation sequence is the reverse of the transformation sequel. This reversal becomes clear when we exchange the transpose sign of the TM for the reversal of the transformation order $[R^{BA}]^A = [R^{BA}]^B = [T]^{AB}$.

Proof: If the base vectors are coordinated in their respective preferred coordinate systems, they have the same coordinates:

$$[b_i]^B = [a_i]^A, \quad i = 1, 2, 3 \tag{4.7}$$

First substitute this equation into Eq. (4.2) and then replace $[b_i]^B$ by Eq. (4.4):

$$[b_i]^A = [R^{BA}]^A[a_i]^A = [R^{BA}]^A[b_i]^B = [R^{BA}]^A[T]^{BA}[b_i]^A$$

Because $[b_i]^A$ is arbitrary and certainly not zero, it follows that $[R^{BA}]^A[T]^{BA} = [E]$ and therefore $[R^{BA}]^A = [\bar{T}]^{BA}$. This completes the first part of the proof. Similarly, if we start with Eq. (4.3) and replace $[b_i]^B$ by Eq. (4.7) and then transpose Eq. (4.5) for substitution, we can prove the second relationship

$$[R^{BA}]^B[a_i]^B = [b_i]^B = [a_i]^A = [\bar{T}]^{BA}[a_i]^B \Rightarrow [R^{BA}]^B = [\bar{T}]^{BA}$$

Combining both results delivers the proof

$$[R^{BA}]^A = [R^{BA}]^B = [\bar{T}]^{BA}$$

Property 4

The rotation tensor is orthogonal.

Proof: It follows from Eq. (4.6) and the proof of Sec. 3.2, Property 1. Because the coordinate transformation is orthogonal, at least one matrix realization of the rotation tensor is orthogonal. But if one coordinate form is orthogonal, so are all and, therefore, the tensor is orthogonal.

Just as the determinant of the transformation matrix is ±1, so is that of the rotation tensor. Every such orthogonal linear transformation in Euclidean three-space preserves absolute values of vectors and angles between vectors. In addition, if the determinant is $+1$, it also preserves the relative orientation of vectors embedded in the frame. These are very useful properties, and therefore we limit ourselves to these rotations. The case with a "negative one" determinant is the reflection tensor, which we already encountered in Sec. 2.3.6.

Property 5

Transposing the rotation tensor reverses the direction of rotation $\bar{R}^{BA} = R^{AB}$.

Proof: Exchanging the b and a, for both lower- and upper-case letters in Eq. (4.1) provides

$$a_i = R^{AB}b_i, \quad i = 1, 2, 3$$

Substituting Eq. (4.1) yields

$$a_i = R^{AB}R^{BA}a_i, \quad i = 1, 2, 3$$

Because a_i is nonzero and the rotation tensor is orthogonal, $R^{AB}R^{BA} = E \Rightarrow \bar{R}^{BA} = R^{AB}$.

We have established the rotation tensor as an absolute model of the mutual orientation of two frames. The nomenclature R^{BA} expresses that relationship of frame B relative to frame A. You can read it as the orientation of frame B obtained from frame A, or just as the rotation of B wrt A.

No reference point is needed. Rotations are independent of points; they only engage frames. This independence becomes clear by an example. Suppose you stand on the east side of a runway and watch an airplane A take off and climb out at 10 deg. The airplane's 10-deg rotation wrt to the runway R is modeled by R^{AR}. On the next day you position yourself at the west end and watch the same airplane take off and climb out. The same R^{AR} will give its orientation, although you changed your reference point from E to W. To define the airplane's orientation, no reference points are needed.

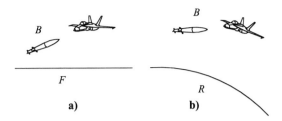

Fig. 4.2 Earth displays: a) flat and b) round.

Example 4.1 Conversion from Flat to Round Earth

Problem. You build a three-dimensional visualization of a long-range air intercept missile. Vectors of polygons model the missile shape. The simulation calculates the missile attitude R^{BF} wrt a flat Earth. You are required to display the missile orientation over a round Earth. How do you convert the vectors of the missile shape?

Solution. The missile frame (geometry) is related to the flat Earth by R^{BF} (see Fig. 4.2a). As the missile flies toward the intercept, the Earth's local level tilts wrt to the flat surface by R^{RF} (see Fig. 4.2b). The orientation of the missile wrt the local level is therefore

$$R^{BR} = R^{BF} \bar{R}^{RF}$$

Any geometrical vector of the missile, say t_i, is oriented wrt the flat Earth by $R^{BF}t_i$ and wrt the local level by $R^{BR}t_i$.

4.1.2 Special Rotations

Let us build up our confidence by constructing the general rotation tensor from special rotations. Beginning with planar rotations, we extend them to the third dimension and eventually obtain a general formulation that presents the rotation tensor in terms of its rotation axis and angle.

4.1.2.1 Planar rotations. In Fig. 4.3 I have plotted two unit vectors b', and c', embedded in the plane subtended by the 1^A, 2^A axes of the coordinate system $]^A$. Unit vector c' is obtained by rotating unit vector b' through the angle ψ. We determine the elements of the rotation tensor in the $]^A$ coordinate system. From Eq. (4.1) we deduct, considering the two vectors as base vectors,

$$[c']^A = [R]^A[b']^A \qquad (4.8)$$

To determine $[R]^A$, we calculate first the components of the two vectors from elementary trigonometric relationships:

$$\begin{bmatrix} c_1' \\ c_2' \\ c_3' \end{bmatrix}^A = \begin{bmatrix} \cos(\varepsilon + \psi) \\ \sin(\varepsilon + \psi) \\ 0 \end{bmatrix} = \begin{bmatrix} \cos\varepsilon\,\cos\psi - \sin\varepsilon\,\sin\psi \\ \sin\varepsilon\,\cos\psi + \cos\varepsilon\,\sin\psi \\ 0 \end{bmatrix}$$

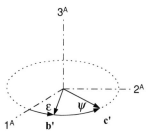

Fig. 4.3   Planar rotation.

and

$$\begin{bmatrix} b_1' \\ b_2' \\ b_3' \end{bmatrix}^A = \begin{bmatrix} \cos \varepsilon \\ \sin \varepsilon \\ 0 \end{bmatrix}$$

Substituting both relationships into Eq. (4.8) yields

$$\begin{bmatrix} \cos \varepsilon \ \cos \psi - \sin \varepsilon \ \sin \psi \\ \sin \varepsilon \ \cos \psi + \cos \varepsilon \ \sin \psi \\ 0 \end{bmatrix} = \begin{bmatrix} r_{11} & r_{12} & r_{13} \\ r_{21} & r_{22} & r_{23} \\ r_{31} & r_{32} & r_{33} \end{bmatrix} \begin{bmatrix} \cos \varepsilon \\ \sin \varepsilon \\ 0 \end{bmatrix}$$

Our task is to establish the elements of the rotation matrix r_{ij}. By inspection we deduce the first two columns:

$$[R]^A = \begin{bmatrix} \cos \psi & -\sin \psi & ? \\ \sin \psi & \cos \psi & ? \\ 0 & 0 & ? \end{bmatrix} \tag{4.9}$$

To determine the third column, we have to introduce the third dimension.

4.1.2.2 Nonplanar rotation.   We expand Fig. 4.3 to the third dimension and reinterpret b' and c' as the projections of the two vectors b and c of equal height (see Fig. 4.4). To determine the elements of the rotation tensor in the $]^A$ coordinate

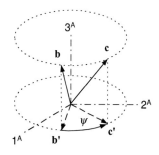

Fig. 4.4   Nonplanar rotation.

system, we first recognize that the 1^A, 2^A coordinate axes of the two vectors $[c]^A$ and $[b]^A$ are the same as those of the planar rotation example just shown. Second, to determine the remaining last column we apply the fact that the third component of $[c]^A$ and $[b]^A$ remains unchanged. Thus we supplement Eq. (4.9) and receive the three-dimensional rotation tensor

$$[R]^A = \begin{bmatrix} \cos\psi & -\sin\psi & 0 \\ \sin\psi & \cos\psi & 0 \\ 0 & 0 & 1 \end{bmatrix} \tag{4.10}$$

Clearly evident is the similarity with the coordinate transformation matrix, but note that the negative sign of the sine function is two rows above the 1 entry and not right above it, as it is the case with coordinate transformations. Be careful however, this derivation is only valid if the rotation is about the third direction. The generalization occurs in the next section.

4.1.2.3 General rotation.

According to a theorem accredited to Euler,[1] the general rotation of two frames can be expressed by an angle about a unit rotation vector. We set out to discover this form of the rotation tensor and ask, what are the elements of the rotation tensor in any allowable coordinate system $]^B$ expressed in terms of the angle of rotation ψ and the unit vector of the axis of rotation $[n]^B$?

Our itinerary starts with the special rotation tensor $[R]^A$ of Eq. (4.10) and the special axis of rotation $[\bar{n}]^A = [\bar{a}_3]^A = [0 \quad 0 \quad 1]$, followed by the transformation to any allowable coordinate system, say $]^B$, with the TM $[T]^{BA}$:

$$[R]^B = [T]^{BA}[R]^A[\bar{T}]^{BA}$$

$$[n]^B = [T]^{BA}[n]^A$$

We reach our goal by expressing $[R]^B$ such that it is a function of $[n]^B$ and ψ only. Let us begin with rewriting $[R]^A$ of Eq. (4.10):

$$[R]^A = \cos\psi\,[E]^A + (1-\cos\psi)\begin{bmatrix} 0 & 0 & 0 \\ 0 & 0 & 0 \\ 0 & 0 & 1 \end{bmatrix}$$

$$+ \sin\psi\left(\begin{bmatrix} 0 & 0 & 0 \\ 1 & 0 & 0 \\ 0 & 0 & 0 \end{bmatrix} - \overline{\begin{bmatrix} 0 & 0 & 0 \\ 1 & 0 & 0 \\ 0 & 0 & 0 \end{bmatrix}}\right)$$

Transforming to $[R]^B$,

$$[R]^B = \cos\psi\,[E]^B + (1-\cos\psi)[T]^{BA}\begin{bmatrix} 0 & 0 & 0 \\ 0 & 0 & 0 \\ 0 & 0 & 1 \end{bmatrix}[\bar{T}]^{BA}$$

$$+ \sin\psi\left([T]^{BA}\begin{bmatrix} 0 & 0 & 0 \\ 1 & 0 & 0 \\ 0 & 0 & 0 \end{bmatrix}[\bar{T}]^{BA} - [T]^{BA}\overline{\begin{bmatrix} 0 & 0 & 0 \\ 1 & 0 & 0 \\ 0 & 0 & 0 \end{bmatrix}}[\bar{T}]^{BA}\right)$$

Substituting Eq. (3.4) yields

$$[R]^B = \cos\psi[E]^B + (1 - \cos\psi)[a_3]^B[\bar{a}_3]^B + \sin\psi\left([a_2]^B[\bar{a}_1]^B - [a_1]^B[\bar{a}_2]^B\right) \tag{4.11}$$

Use the coordinates of the base vectors

$$[a_1]^B = \begin{bmatrix} a_{11}^B \\ a_{21}^B \\ a_{31}^B \end{bmatrix}, \quad [a_2]^B = \begin{bmatrix} a_{12}^B \\ a_{22}^B \\ a_{32}^B \end{bmatrix}, \quad [a_3]^B = \begin{bmatrix} a_{13}^B \\ a_{23}^B \\ a_{33}^B \end{bmatrix}$$

to express the last term bracketed in Eq. (4.11):

$$\left([a_2]^B[\bar{a}_1]^B - [a_1]^B[\bar{a}_2]^B\right)$$

$$= \begin{bmatrix} 0 & -a_{11}^B a_{22}^B + a_{12}^B a_{21}^B & a_{12}^B a_{31}^B - a_{11}^B a_{32}^B \\ a_{11}^B a_{22}^B - a_{12}^B a_{21}^B & 0 & -a_{21}^B a_{32}^B + a_{22}^B a_{31}^B \\ -a_{12}^B a_{31}^B + a_{11}^B a_{32}^B & a_{21}^B a_{32}^B - a_{22}^B a_{31}^B & 0 \end{bmatrix}$$

$$= \begin{bmatrix} 0 & -a_{33}^B & a_{23}^B \\ a_{33}^B & 0 & -a_{13}^B \\ -a_{23}^B & a_{13}^B & 0 \end{bmatrix}$$

where we used the triad property $[a_3]^B = [A_1]^B[a_2]^B$, with $[A_1]^B$ the skew-symmetric form of $[a_1]^B$. The last matrix has the skew-symmetric structure of $[a_3]^B$, which is also the unit vector of rotation $[A_3]^B = [N]^B$. Replacing in Eq. (4.11) $[a_3]^B, [A_3]^B$ by $[n]^B, [N]^B$ yields

$$[R]^B = \cos\psi[E]^B + (1 - \cos\psi)[n]^B[\bar{n}]^B + \sin\psi[N]^B$$

Because $]^B$ is any coordinate system, this equation holds for all allowable coordinate systems and, therefore, is the general tensor form of rotations

$$R = \cos\psi E + (1 - \cos\psi)n\bar{n} + \sin\psi N \tag{4.12}$$

So indeed, we confirmed Euler's theorem that the rotation tensor is completely defined by its angle of rotation ψ and the unit vector of rotation n.

Example 4.2   Boresight Error

Problem.   The seeker centerline s of a missile, carried on the right wing tip of an aircraft, is boresighted before takeoff to the aircraft's radar centerline r (see Fig. 4.5). The antenna unit vector expressed in aircraft coordinates is $[\bar{r}]^B = [1\ 0\ 0]$. In flight the wing tip twists upward by 3 deg about the wing box, which is swept back by 30 deg.

1) Calculate the elements of the rotation tensor $[R^{MA}]^B$ of the missile M wrt the aircraft A in flight.

2) Calculate the components of the seeker centerline in flight $[s]^B$ and give the bore-sight errors in the aircraft pitch and yaw planes in degrees.

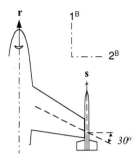

Fig. 4.5   Missile aircraft geometry.

Solution.   In general, the relationship between the missile and antenna center-line is

$$s = R^{MA}r \tag{4.13}$$

1) To calculate $[R^{MA}]^B$, we make use of Eq. (4.12) in body coordinates

$$[R^{MA}]^B = \cos\psi[E]^B + (1 - \cos\psi)[n]^B[\bar{n}]^B + \sin\psi[N]^B$$

With $\psi = 3$ deg,

$$[\bar{n}]^B = [-\sin 30\text{ deg} \quad \cos 30\text{ deg} \quad 0] = [-0.5 \quad 0.866 \quad 0]$$

$$[R^{MA}]^B = 0.99863[E] + 0.00137 \begin{bmatrix} 0.25 & -0.433 & 0 \\ -0.433 & 0.750 & 0 \\ 0 & 0 & 0 \end{bmatrix}$$

$$+ 0.05234 \begin{bmatrix} 0 & 0 & 0.866 \\ 0 & 0 & 0.5 \\ -0.866 & -0.5 & 0 \end{bmatrix}$$

$$= \begin{bmatrix} 0.99897 & -0.000593 & 0.045326 \\ -0.000593 & 0.99966 & 0.02617 \\ -0.045326 & -0.02617 & 0.99863 \end{bmatrix}$$

2) To calculate the missile centerline $[s]^B$ in flight from the antenna boresight $[r]^B$, we apply Eq. (4.13):

$$[s]^B = [R^{MA}]^B[r]^B \Rightarrow [\bar{s}]^B = [0.99897 \quad -0.000593 \quad -0.045326]$$

The second component is the displacement of the tip of the boresight vector in the 2^B direction, which corresponds to an in-turning about the 3^B axis (yaw) of -0.000593 rad or-0.034 deg. The third component moves the tip up and is therefore a positive pitch twist of 0.045326 rad or 2.6 deg.

Now we turn to a special rotation tensor with the angle $\psi = 90$ deg that describes the tetragonal symmetry of missiles.

4.1.2.4   Tetragonal tensor.   Missiles with four fins possess tetragonal symmetry, i.e., their external configuration duplicates after every 90-deg rotation about their symmetry axis. The rotation tensor that models these replications is called the *tetragonal tensor* R_{90}. We derive it directly from Eq. (4.12) by setting $\psi = 90$ deg:

$$R_{90} = n\,\bar{n} + N$$

The tetragonal tensor is composed of the projection tensor $P = n\bar{n}$ [(see Eq. (2.21)] and the skew-symmetric tensor of the unit rotation vector N. An example should clarify the operation.

Example 4.3   Missile with Fins 90 Deg Apart

Problem.   In Fig. 4.6, f_1 points to the root of fin #1. Use the tetragonal tensor to point to the root of the second fin f_2.

Solution.   With the base vector b as the unit vector of rotation, we have

$$f_2 = R_{90} f_1 = b\bar{b} f_1 + B f_1 \tag{4.14}$$

The vector f_2 is composed of the vector f_1 projected on b and the component that is the result of the vector product $B f_1$. With Fig. 4.6 you should be able to visualize this vector addition.

A numerical example can be of further assistance. Introduce the body coordinate system $]^B$. The vectors b and f_1 have the following coordinates in the body system: $[b]^B = [1\ 0\ 0]$ and $[\bar{f}_1]^B = [f_{11}\ f_{12}\ f_{13}]$. The tetragonal tensor becomes

$$[R_{90}]^B = \begin{bmatrix} 1 & 0 & 0 \\ 0 & 0 & 0 \\ 0 & 0 & 0 \end{bmatrix} + \begin{bmatrix} 0 & 0 & 0 \\ 0 & 0 & -1 \\ 0 & 1 & 0 \end{bmatrix} = \begin{bmatrix} 1 & 0 & 0 \\ 0 & 0 & -1 \\ 0 & 1 & 0 \end{bmatrix}$$

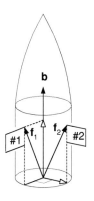

Fig. 4.6   Tetragonal symmetry.

Applying Eq. (4.14), the second fin has the following coordinates in body axis:

$$[f_2]^B = \begin{bmatrix} 1 & 0 & 0 \\ 0 & 0 & -1 \\ 0 & 1 & 0 \end{bmatrix} \begin{bmatrix} f_{11} \\ f_{12} \\ f_{13} \end{bmatrix} = \begin{bmatrix} f_{11} \\ -f_{13} \\ f_{12} \end{bmatrix}$$

This result is confirmed by Fig. 4.6. The 1 coordinate remains unchanged; the second and third coordinates are exchanged with a sign reversal.

4.1.3 Axis and Angle of Rotation

You have learned that a rotation tensor is orthogonal with the determinant value $+1$. Conversely, every orthogonal 3×3 tensor with determinant $+1$ represents a rigid right-handed rotation. Given such a rotation tensor, how do we calculate its rotation axis and rotation angle?

4.1.3.1 Determination of the axis of rotation.
Let n be the axis of the rotation tensor R. Because rotating the rotation vector about itself does not change its direction or length, $n = Rn$. The invariancy of n holds also if the sense of the rotation is reversed: $n = \bar{R}n$. Subtracting both equations yields $(R - \bar{R})n = 0$. For any allowable coordinate system, say $]^A$,

$$([R]^A - [\bar{R}]^A)[n]^A = [0]^A$$

Now, substituting the elements of the rotation tensor

$$[R]^A = \begin{bmatrix} r_{11} & r_{12} & r_{13} \\ r_{21} & r_{22} & r_{23} \\ r_{31} & r_{32} & r_{33} \end{bmatrix} \qquad (4.15)$$

we obtain

$$\begin{bmatrix} 0 & -(r_{21} - r_{12}) & (r_{13} - r_{31}) \\ (r_{21} - r_{12}) & 0 & -(r_{32} - r_{23}) \\ -(r_{13} - r_{31}) & (r_{32} - r_{23}) & 0 \end{bmatrix} \begin{bmatrix} n_1 \\ n_2 \\ n_3 \end{bmatrix} = \begin{bmatrix} 0 \\ 0 \\ 0 \end{bmatrix}$$

The left-hand side is a vector product. Therefore, for the nontrivial case, the vector equivalent of $([R]^A - [\bar{R}]^A)$ is parallel to $[n]^A$ with the scaling factor k:

$$\begin{bmatrix} n_1 \\ n_2 \\ n_3 \end{bmatrix} = k \begin{bmatrix} r_{32} - r_{23} \\ r_{13} - r_{31} \\ r_{21} - r_{12} \end{bmatrix} \qquad (4.16)$$

Choose k such that n is a unit vector, pointing in the direction given by the right-hand rule. As a scaling factor, k is the same for all allowable coordinate systems. To calculate it, we chose the most convenient coordinate system, that is, $]^B$, such that

$$[R]^B = \begin{bmatrix} \cos\psi & -\sin\psi & 0 \\ \sin\psi & \cos\psi & 0 \\ 0 & 0 & 1 \end{bmatrix} \quad \text{and} \quad [n]^B = \begin{bmatrix} 0 \\ 0 \\ 1 \end{bmatrix} \qquad (4.17)$$

Substituting into Eq. (4.16),

$$\begin{bmatrix} 0 \\ 0 \\ 1 \end{bmatrix} = k \begin{bmatrix} 0 \\ 0 \\ 2 \sin \psi \end{bmatrix} \Rightarrow k = \frac{1}{2 \sin \psi}$$

If ψ were known, k could be calculated, and Eq. (4.16) would provide the unit vector of rotation. Let us determine ψ.

4.1.3.2 Determination of the angle of rotation.
Determining the angle of rotation is simplified by the fact that the trace of a tensor is preserved under transformation. We can take the trace of Eq. (4.17) and set it equal to the diagonal elements of the general rotation tensor Eq. (4.15):

$$1 + 2 \cos \psi = r_{11} + r_{22} + r_{33}$$

Solving for ψ gives the desired equation

$$\psi = \arccos \left(\frac{1}{2} \sum_{i=1}^{3} r_{ii} - \frac{1}{2} \right)$$

In summary, for a rotation tensor $[R]^A$, coordinated in any allowable coordinate $]^A$, the angle and axis of rotation are given by the following equations:

$$\psi = \arccos \left(\frac{1}{2} \sum_{i=1}^{3} r_{ii} - \frac{1}{2} \right)$$

$$[n]^A = \frac{1}{2 \sin \psi} \begin{bmatrix} r_{32} - r_{23} \\ r_{13} - r_{31} \\ r_{21} - r_{12} \end{bmatrix}$$

$$(4.18)$$

If $[R]^A = [E]$, then ψ is zero, and $[n]^A$ is undefined.

Example 4.4   Satellite Positioning

Problem.   A satellite S is placed in a geosynchronous orbit. The onboard INS provides its initial orientation S_0 wrt the inertial frame I by the rotation tensor $R^{S_0 I}$ (see Fig. 4.7). To point the satellite's antenna at Earth E, it needs to be adjusted such that its new orientation wrt Earth conforms to the rotation tensor $R^{S_1 E}$. The most fuel-efficient way to torque the satellite is by direct rotation about an inertial axis. Provide the equations for the INS-based attitude controller.

Solution.   The rotation tensor to be achieved by the attitude controller (where R^{EI} is known from almanac tables) is

$$R^{S_1 I} = R^{S_1 E} R^{EI} \qquad (4.19)$$

Because the INS provides the main reference of the satellites, the unit vector of rotation should be declared in inertial coordinates. In preparation we express

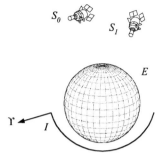

Fig. 4.7 Satellite, Earth, and inertial frames.

Eq. (4.19) in inertial coordinates $[R^{S_1 I}]^I = [R^{S_1 E}]^I [R^{EI}]^I$. Yet, most likely, the $R^{S_1 E}$ tensor is given in Earth coordinates $[R^{S_1 E}]^E$. Therefore, Eq. (4.19) is modified

$$[R^{S_1 I}]^I = [T]^{IE} [R^{S_1 E}]^E [\bar{T}]^{IE} [R^{EI}]^I \tag{4.20}$$

With $[R^{S_1 I}]^I$ thus given and its coordinates stated as

$$[R^{S_1 I}]^I = \begin{bmatrix} r_{11} & r_{12} & r_{13} \\ r_{21} & r_{22} & r_{23} \\ r_{31} & r_{32} & r_{33} \end{bmatrix}$$

we apply Eq. (4.18)

$$\psi = \arccos \left(\frac{1}{2} \sum_{i=1}^{3} r_{ii} - \frac{1}{2} \right)$$

$$[n]^I = \frac{1}{2 \sin \psi} \begin{bmatrix} r_{32} - r_{23} \\ r_{13} - r_{31} \\ r_{21} - r_{12} \end{bmatrix} \tag{4.21}$$

and thus Eqs. (4.20) and (4.21) are the equations for the attitude controller.

4.1.4 Small Rotations

Isn't it interesting that the orientation between two frames, although determined by unit vector and angle, requires a second-order rotation tensor for portrayal? If several rotations are combined, the sequence is not arbitrary because the multiplication of second-order tensors does not commute. Both attributes are simplified, however, if we deal with so-called small rotations. Small rotations reduce to vectors and can be combined in any order.

Engineers like to deal with small quantities. They are so much easier to model because the mathematics is simpler. Consider the stability equations of aircraft with their implicit small angle assumption and the error equations of INS systems. These small perturbations simplify the analysis, resulting in linear differential equations, the darlings of engineering analysis.

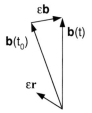

Fig. 4.8   Small rotation.

I will guide you from the general rotation tensor to the perturbation tensor of rotation, the small rotation vector, and finally, the small rotation tensor. All of these concepts are important and should be in your toolbox. We will start with the question: what is the form of the rotation tensor if a vector is rotated by an infinitesimal small amount?

Rotate vector b from its initial position $b(t_0)$ through rotation tensor R to its current position

$$b(t) = Rb(t_0) \tag{4.22}$$

The difference εb between the two vectors is, as shown in Fig. 4.8, $\varepsilon b = b(t) - b(t_0)$ and with Eq. (4.22)

$$\varepsilon b = (R - E)b(t_0) \tag{4.23}$$

Introducing the definition of the perturbation tensor of rotation

$$\varepsilon R = R - E \tag{4.24}$$

into Eq. (4.23) yields

$$\varepsilon b = \varepsilon R b(t_0) \tag{4.25}$$

The displacement vector εb is the result of the multiplication of a tensor εR with the initial vector. What is the property of this tensor?

Because R is orthogonal, $\bar{R}R = E$, and from Eq. (4.24)

$$E = (\varepsilon \bar{R} + \bar{E})(\varepsilon R + E) = E + \varepsilon R + \varepsilon \bar{R} + \varepsilon \bar{R} \varepsilon R$$

Now we assume that ε indicates a small quantity. Then the last term is small to the second-order wrt the other terms and can be neglected. Thus $\varepsilon R = -\varepsilon \bar{R}$, i.e., the perturbation tensor εR is skew-symmetric and thus can be represented by a vector. For any allowable coordinate system $]^A$ the perturbation tensor has the components

$$[\varepsilon R]^A = \begin{bmatrix} \varepsilon r_{11} & \varepsilon r_{12} & \varepsilon r_{13} \\ \varepsilon r_{21} & \varepsilon r_{22} & \varepsilon r_{23} \\ \varepsilon r_{31} & \varepsilon r_{32} & \varepsilon r_{33} \end{bmatrix}$$

but if it is skew-symmetric it has only three independent components,

$$[\varepsilon R]^A = \begin{bmatrix} 0 & \varepsilon r_{12} & \varepsilon r_{13} \\ -\varepsilon r_{12} & 0 & \varepsilon r_{23} \\ -\varepsilon r_{13} & \varepsilon r_{23} & 0 \end{bmatrix} \tag{4.26}$$

from which we obtain by contraction the small rotation vector

$$[\varepsilon r]^A \equiv \begin{bmatrix} \varepsilon r_1 \\ \varepsilon r_2 \\ \varepsilon r_3 \end{bmatrix} = \begin{bmatrix} -\varepsilon r_{23} \\ \varepsilon r_{13} \\ -\varepsilon r_{12} \end{bmatrix} \tag{4.27}$$

Finally, the tensor R for small rotations consists according to Eq. (4.24) of the unit tensor and the perturbation tensor Eq. (4.26). Expressed in any allowable coordinate system $]^A$, it becomes

$$[R]^A = \begin{bmatrix} 1 & \varepsilon r_{12} & \varepsilon r_{13} \\ -\varepsilon r_{12} & 1 & \varepsilon r_{23} \\ -\varepsilon r_{13} & -\varepsilon r_{23} & 1 \end{bmatrix} \tag{4.28}$$

By now you may be thoroughly perturbed by all of these perturbations. Let us recap. The rotation tensor of a small rotation between two frames, say B wrt A, R^{BA}, consists of the unit tensor E and the perturbation tensor of rotation εR^{BA}

$$R^{BA} = E + \varepsilon R^{BA}$$

where εR^{BA} can be reduced to a small rotation vector εr. To clarify these concepts, let us look at an example.

Example 4.5   INS Tilt

An INS maintains a level reference either by a gimbaled platform or a computed reference frame. Yet, mechanical or computational imperfections cause the reference to deviate from the true values. These errors, called *tilts*, are rather small and are measured in arc seconds.

The tilt angle θ (pitch) of an INS platform about the east direction gives rise to the rotation tensor expressed in geographic axes $]^G$

$$[R(\theta)]^G = \begin{bmatrix} \cos\theta & 0 & \sin\theta \\ 0 & 1 & 0 \\ -\sin\theta & 0 & \cos\theta \end{bmatrix}$$

Because the tilt is a very small angle $\varepsilon\theta$, we introduce the small angle assumption

$$[R(\varepsilon\theta)]^G = \begin{bmatrix} 1 & 0 & \varepsilon\theta \\ 0 & 1 & 0 \\ -\varepsilon\theta & 0 & 1 \end{bmatrix} \tag{4.29}$$

The skew-symmetric perturbation tensor of rotation is

$$[\varepsilon R(\varepsilon\theta)]^G = \begin{bmatrix} 0 & 0 & \varepsilon\theta \\ 0 & 0 & 0 \\ -\varepsilon\theta & 0 & 0 \end{bmatrix}$$

with the corresponding small rotation vector $[\overline{\varepsilon r(\varepsilon\theta)}]^B = [0 \ \ \varepsilon\theta \ \ 0]$. Similar small rotation tensors can be developed for the tilt ϕ about the north direction and the azimuth error ψ:

$$[R(\varepsilon\phi)]^G = \begin{bmatrix} 1 & 0 & 0 \\ 0 & 1 & -\varepsilon\phi \\ 0 & \varepsilon\phi & 1 \end{bmatrix}; \quad [R(\varepsilon\psi)]^G = \begin{bmatrix} 1 & -\varepsilon\psi & 0 \\ \varepsilon\psi & 1 & 0 \\ 0 & 0 & 1 \end{bmatrix}$$

The complete tilt tensor consists of the product

$$[R(\varepsilon\phi, \varepsilon\theta, \varepsilon\psi)]^G = [R(\varepsilon\phi)]^G [R(\varepsilon\theta)]^G [R(\varepsilon\psi)]^G \tag{4.30}$$

and the multiplications carried out

$$[R(\varepsilon\phi, \varepsilon\theta, \varepsilon\psi)]^G = \begin{bmatrix} 1 & -\varepsilon\psi & \varepsilon\theta \\ \varepsilon\psi & 1 & -\varepsilon\phi \\ -\varepsilon\theta & \varepsilon\phi & 1 \end{bmatrix} \tag{4.31}$$

Subtracting the unit tensor yields the perturbation tensor

$$[\varepsilon R(\varepsilon\phi, \varepsilon\theta, \varepsilon\psi)]^G = \begin{bmatrix} 0 & -\varepsilon\psi & \varepsilon\theta \\ \varepsilon\psi & 0 & -\varepsilon\phi \\ -\varepsilon\theta & \varepsilon\phi & 0 \end{bmatrix} \tag{4.32}$$

from which we obtain by analogy with Eqs. (4.26) and (4.27) the small rotation vector

$$[\overline{\varepsilon r(\varepsilon\phi, \varepsilon\theta, \varepsilon\psi)}]^G = [\varepsilon\phi \ \ \ \varepsilon\theta \ \ \ \varepsilon\psi] \tag{4.33}$$

Here you experienced all three notions: small rotation tensor, perturbation tensor of rotation, and small rotation vector. You can convince yourself of the commutative property of small rotations by changing the sequence of multiplication in Eq. (4.30) and each time arriving at the same result Eq. (4.31)

This insight into perturbations prepares you to follow the derivation of an INS error model in Sec. 10.2.4, which is used in six-DoF simulations. The tilt tensor, with its tilt vector, models three of the nine INS error states. Another application of importance is the derivation of the perturbation equations of air vehicles, Sec. 7.2. You will find again a small rotation tensor describing the deviation of the perturbed flight from the reference flight.

We have about exhausted the topic of rotation tensors. As the displacement vector models the location of point B wrt point A, so does the rotation tensor establish the orientation of frame B wrt frame A. Both together define the position of frame B wrt frame A (refer to Sec. 3.1.1). We established the tensor characteristic of rotations and the special circumstances under which they relate to transformation matrices. You should by now be convinced that both are different entities! We dealt with special rotations like the tetragonal symmetry tensor and small rotations. A correct understanding of these concepts will make it easier for you to grasp the following kinematic developments.

4.2   Kinematics of Changing Times

We are now prepared to invite time into our Euclidean three-dimensional world and develop the tools to study the kinematics of motions. Motions are modeled by including time dependency in the displacement vector and the rotation tensor. Hence, if s_{BA} is the displacement vector of point B wrt point A, $s_{BA}(t)$ describes the motion of point B wrt reference point A. Likewise, $R^{BA}(t)$ models the changing orientation of frame B wrt the reference frame A.

Other kinematic concepts are formed by the time rate of change of displacement and rotation. Linear velocity is the first and acceleration the second time derivative of displacement. Angular velocity is obtained by the time derivative of rotation. For instance, the linear velocity of point B wrt frame A is produced by $v_B^A = (d/dt)s_{AB}$. Yet these new entities require close scrutiny if we want them to be tensors. Remember, physical models, like velocity, maintain their tensor property only if their structure remains invariant under coordinate transformations. As long as the coordinate transformations are time invariant, the time derivative does not change the tensor characteristic of its operand, but our modeling strategy is more ambitious. We require that the time derivatives of tensors are tensors in themselves, even under time-dependent coordinate transformations.

Pick up any textbook on mechanics and you find the following treatment of the time derivative of vector s transformed from A to B coordinate system:

$$\frac{ds}{dt}\bigg|_A = \frac{ds}{dt}\bigg|_B + \omega \times s \qquad (4.34)$$

where ω is the angular velocity between B and A. Unfortunately, the right side has an additional term, and thus the time derivative destroyed the tensor property of the derivative of vector s.

Let us translate Eq. (4.34) into our nomenclature. With any two allowable coordinate systems $]^A$ and $]^B$ the vector s transforms like a first-order tensor:

$$[s]^A = [T]^{AB}[s]^B$$

Taking the time derivative and applying the chain rule

$$\left[\frac{ds}{dt}\right]^A = [T]^{AB}\left[\frac{ds}{dt}\right]^B + \left[\frac{dT}{dt}\right]^{AB}[s]^B$$

$$= [T]^{AB}\left(\left[\frac{ds}{dt}\right]^B + [\bar{T}]^{AB}\left[\frac{dT}{dt}\right]^{AB}[s]^B\right)$$

and exchanging the sequence of transformation in the last term by transposition yields

$$\left[\frac{ds}{dt}\right]^A = [T]^{AB}\left(\left[\frac{ds}{dt}\right]^B + [T]^{BA}\overline{\left[\frac{dT}{dt}\right]}^{BA}[s]^B\right) \qquad (4.35)$$

Comparison with Eq. (4.34) shows that the right and left sides are related by the TM $[T]^{AB}$ and that the ω vector corresponds to the term $[T]^{BA}\overline{[dT/dt]}^{BA}$.

From Eq. (4.35) it is clear that the time derivative of s does not transform like a first-order tensor. If, however, we could define a time operator that would give the right and left side the same form, the tensor property would be maintained. Does such an operator exist? Indeed it does, and it is called the *rotational time derivative*.

Back in 1968, pursuing my doctoral dissertation, I found in Wrede's book[2] on vector and tensor analysis just the concept I needed. He defines a rotational time derivative, whose operation on tensors preserves their tensor property. It is couched in hard-to-understand tensor lingo; but applied to right-handed Cartesian coordinate systems, the concept is easy to grasp. Later I discovered that Wrede's work was preceded by Wundheiler's (Warsaw, Poland) research, in 1932.[3]

4.2.1 Rotational Time Derivative

The challenge is to find the time operator that preserves the form on both sides of Eq. (4.35). We venture to define the terms in parentheses as that operator and see what happens to the left side.

Definition: The rotational time derivative of a first-order tensor x wrt any frame A, $D^A x$, and expressed in any allowable coordinate system $]^B$ is defined by

$$[D^A x]^B \equiv \left[\frac{dx}{dt}\right]^B + [T]^{BA} \overline{\left[\frac{dT}{dt}\right]}^{BA} [x]^B \qquad (4.36)$$

Notice in $[D^A x]^B$ the appearance of frame A and coordinate system $]^B$. Both are arbitrary, but the coordinate system $]^A$ of $[T]^{BA}$ on the right-hand side has to be associated with frame A.

Let us break the suspense and apply the rotational derivative to the left side of Eq. (4.35). Instead of $]^B$ the coordinate system is now $]^A$:

$$[D^A s]^A = \left[\frac{ds}{dt}\right]^A + [T]^{AA} \overline{\left[\frac{dT}{dt}\right]}^{AA} [s]^A = \left[\frac{ds}{dt}\right]^A$$

which is, with the time derivative of the unit matrix being zero, exactly in the desired form. Therefore, we can write Eq. (4.35) as

$$[D^A s]^A = [T]^{AB}[D^A s]^B$$

That looks to me like a tensor transformation. Although the $]^B$ coordinate system is arbitrary, the $]^A$ system is unique by its association with frame A. For a true tensor form any allowable coordinate system must be admitted. Therefore, consider an arbitrary allowable coordinate system $]^D$ to replace $]^A$ on the left side of Eq. (4.35):

$$\left[\frac{ds}{dt}\right]^D + [T]^{DA} \overline{\left[\frac{dT}{dt}\right]}^{DA} [s]^D = [T]^{DB}\left(\left[\frac{ds}{dt}\right]^B + [T]^{BA} \overline{\left[\frac{dT}{dt}\right]}^{BA} [s]^B\right)$$

By definition of the rotational derivative, we obtain the true tensor transformation

$$[D^A s]^D = [T]^{DB}[D^A s]^B$$

Therefore, the rotational derivative of a vector transforms like a tensor, and $D^A s$ is a tensor.

A comparable rationale defines the rotational derivative for second-order tensors. I will just state the results. The details can be found in my dissertation.[4]

Definition: The rotational time derivative of a second-order tensor X wrt any frame A, $D^A X$, and expressed in any allowable coordinate system $]^B$, is defined as

$$[D^A X]^B \equiv \left[\frac{dX}{dt}\right]^B + [T]^{BA}\left[\frac{dT}{dt}\right]^{\overline{BA}}[X]^B + [X]^B\left[\frac{dT}{dt}\right]^{BA}[\bar{T}]^{BA} \qquad (4.37)$$

We have two terms pre- or postmultiplying the tensor with the term $[T]^{BA}\overline{[dT/dt]}^{BA}$ and its transpose.

The rotational derivative of tensors transforms like a second-order tensor. For any two allowable coordinate systems $]^B$ and $]^D$,

$$[D^A X]^D = [T]^{DB}[D^A X]^B [\bar{T}]^{DB}$$

Therefore if X is a second-order tensor, so is $D^A X$.

The rotational time derivative has some nice properties that will make it a joy to work with. For detailed proofs I refer you again to my dissertation.[4]

Property 1

The rotational derivative wrt any frame A is a linear operator.
1) For any constant scalar k and vector x

$$D^A(kx) = k D^A x \qquad (4.38)$$

2) For any two vectors x and y

$$D^A(x + y) = D^A x + D^A y \qquad (4.39)$$

Property 2

The rotational time derivative abides by the chain rule. For any tensor Y and vector x

$$D^A(Yx) = (D^A Y)x + Y D^A x$$

Note that you must maintain the order of the tensor multiplication. The parentheses on the right side are redundant because, by convention, the derivative operates only on the adjacent variable.

The rotational derivative strengthens the modeling process of aerospace vehicles. It enables the formulation of Newton's and Euler's laws as invariants under time-dependent coordinate transformations. As you will see in Chapters 5 and 6, we remain true to our principle from tensor formulation to matrix coding. With the instrument of an invariant time operator in hand, we can create linear and angular velocities and their accelerations.

4.2.2   Linear Velocity and Acceleration

At several occasions we have encountered already the linear velocity v_B^A of point B wrt frame A. Now we have to assign it a definition (refer to Fig. 4.9). Given

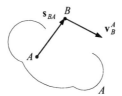

Fig. 4.9   Linear velocity.

is the frame A and the displacement vector s_{BA}, where B is any point and A is a point of frame A. The linear velocity of point B wrt frame A is obtained from the rotational time derivative wrt frame A, applied to the displacement vector s_{BA}:

$$v_B^A = D^A s_{BA} \tag{4.40}$$

The notation of the linear velocity vector does not carry over the identity of point A, but just refers to the frame A. This simplification is justified by the fact that any point of frame A can serve as reference point.

Proof: Show that the point A in Eq. (4.40) can be any point of frame A. Let $A1$ and $A2$ be any two points of frame A. Then $s_{BA1} = s_{BA2} + s_{A2A1}$. Take the rotational time derivative on both sides and use the second part of Property 1:

$$D^A s_{BA1} = D^A s_{BA2} + D^A s_{A2A1}$$

However, $D^A s_{A2A1} = \mathbf{0}$ because the displacement vector s_{A2A1} is fixed in frame A. Therefore,

$$D^A s_{BA1} = D^A s_{BA2} = v_B^A$$

The fact that $D^A s_{A2A1} = \mathbf{0}$ can be verified from Eq. (4.36). Pick $]^A$ as a coordinate system associated with frame A. Then

$$\left[D^A s_{A2A1} \right]^A = \left[\frac{ds_{A2A1}}{dt} \right]^A + [T]^{AA} \left[\overline{\frac{dT}{dt}} \right]^{AA} [S_{A2A1}]^A.$$

Both terms are zero. The first term is an ordinary time derivative of a vector with constant length, and the second term takes the time derivative of a unit matrix. If a tensor is zero in one coordinate system, it is zero in all coordinate systems.

Example 4.6   Differential Velocity

Problem.   A missile with c.m. M is attacking a target, point T (see Fig. 4.10). Both are tracked by the point antenna R of the radar frame R. What is the velocity of the target wrt the missile as observed from the radar?

Solution.   The displacement vector of the target wrt to the missile is

$$s_{TM} = s_{TR} - s_{MR} \tag{4.41}$$

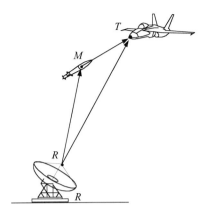

Fig. 4.10   Differential velocity.

We take the rotational derivative wrt the radar frame R and apply the linear operator property

$$D^R s_{TM} = D^R(s_{TR} - s_{MR}) = D^R s_{TR} - D^R s_{MR}$$

According to the linear velocity definition Eq. (4.40), the right-hand sides are $D^R s_{TR} = v_T^R$ and $D^R s_{MR} = v_M^R$. The term on the left however is different. None of its points belongs to the radar frame. Rather the two points relate the target to the missile. Its rotational derivative wrt to the radar frame forms the desired velocity. We name it the *differential velocity* of T wrt M as observed from frame R and write

$$v_{TM}^R = v_T^R - v_M^R \tag{4.42}$$

The concept of differential velocity is particularly important in missile engagements and will surface again in our discussion of proportional navigation.

In this example we encounter two types of linear velocities distinguishable by their reference points. If the reference point is part of the reference frame, we speak of the *relative* velocity, or just velocity (e.g., v_T^R); otherwise, if the reference point is not part of the reference frame, we refer to it as *differential* velocity (e.g., v_{TM}^R).

The *linear acceleration* a_B^A of a point B wrt the reference frame A is defined as the second rotational derivative of the displacement vector s_{BA}, or the first derivative of the velocity vector

$$a_B^A = D^A D^A s_{BA} = D^A v_B^A \tag{4.43}$$

If the reference point A is part of the reference frame A, any point of that frame can be used, and thus the nomenclature a_B^A does not explicitly identify the reference point A. (It may be beneficial at this point to remind you that without exception points are subscripts and frames are superscripts.) This type of acceleration, relative to reference frame A, is also called *relative* acceleration to distinguish it from the *differential* acceleration.

Referring to Fig. 4.10, the concept of differential acceleration is similar to differential velocity. If we take the second rotational derivative of the displacement vector Eq. (4.41), we obtain

$$D^R D^R s_{TM} = D^R D^R s_{TR} - D^R D^R s_{MR}$$

On the right-hand side the reference point R is part of the reference frame R, and therefore the two terms are the relative accelerations $D^R D^R s_{TR} = a_T^R$ and $D^R D^R s_{MR} = a_M^R$. The term on the left contains the reference points T and M that do not belong to frame R. Hence, we are dealing with the differential acceleration $D^R D^R s_{TM} = a_{TM}^R$, which is the difference of the two relative accelerations

$$a_{TM}^R = a_T^R - a_M^R \tag{4.44}$$

Linear velocities and accelerations are important building blocks in any simulations. You will have ample opportunity to use them as modeling tools of aerospace vehicles. Yet our toolbox is incomplete without the angular velocities and accelerations.

4.2.3   Angular Velocity

As the linear velocity is born of the displacement vector, so is the angular velocity derived from the rotation tensor by the rotational time derivative. We take a two-track approach. To pay homage to vector mechanics, I present the classical approach to be found in any mechanics text like Goldstein,[1] and the general development via the rotational time derivative.

4.2.3.1   Classical Approach.   Consider a vector b of constant magnitude rotating about its fixed base point B (see Fig. 4.11). At time t it is displaced from its original position at time t_0 by

$$\varepsilon b = b(t) - b(t_0)$$

The time rate of change of εb is the tip velocity v, formed from the limit as $\Delta t \to 0$:

$$v = \lim_{\Delta t \to 0} \frac{b(t) - b(t_0)}{\Delta t} = \frac{d(\varepsilon b)}{dt}$$

Substitute Eq. (4.25) into this equation:

$$v = \frac{d(\varepsilon R)}{dt} b(t_0) \tag{4.45}$$

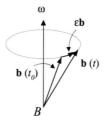

Fig. 4.11   Incremental vector.

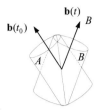

Fig. 4.12   Rotation.

Because εb is small, εR and consequently $[d(\varepsilon R)/dt]$ are skew-symmetric tensors (see Sec. 4.1.4). Therefore, Eq. (4.45) presents the vector product of the classical form $v = \omega \times b(t_0)$ with ω as the angular velocity.

4.2.3.2 General development.

Introduce frame A associated with the constant $b(t_0)$ and frame B associated with $b(t)$ (see Fig. 4.12). The two vectors are related by the rotation tensor R^{BA} of the frame B wrt frame A:

$$b(t) = R^{BA}b(t_0) \tag{4.46}$$

To determine the tip velocity of $b(t)$, take the rotational derivative wrt frame A and apply Eq. (4.46) twice:

$$v_B^A = D^A b(t) = D^A R^{BA}b(t_0) = D^A R^{BA}\overline{R^{BA}}b(t)$$

We define the angular velocity tensor of frame B wrt frame A as

$$\Omega^{BA} \equiv D^A R^{BA}\overline{R^{BA}} \tag{4.47}$$

and obtain the tip velocity

$$v_B^A = \Omega^{BA}b(t) \tag{4.48}$$

Comparing Eqs. (4.45) and (4.48) and assuming small displacements $b(t_0) \approx b(t)$, we can relate the time derivative of the perturbation tensor of rotation with the angular velocity tensor

$$\frac{d(\varepsilon R)}{dt} = \Omega^{BA} \tag{4.49}$$

and conclude that because εR is skew symmetric then so is Ω^{BA}. The angular velocity tensor possesses, therefore, the vector equivalent ω^{BA}. For any coordinate system, say $]^C$, we have the following correspondence:

$$[\Omega^{BA}]^C = \begin{bmatrix} 0 & -r & q \\ r & 0 & -p \\ -q & p & 0 \end{bmatrix} \Leftrightarrow [\omega^{BA}]^C = \begin{bmatrix} p \\ q \\ r \end{bmatrix} \tag{4.50}$$

Let us solidify the fact that the angular velocity tensor is skew symmetric.

Proof: Show that the angular velocity tensor is skew symmetric.
Start with the null tensor and develop the result:

$$0 = \frac{d}{dt} E = D^A(R^{BA}\overline{R^{BA}}) = D^A R^{BA}\overline{R^{BA}} + R^{BA}\overline{D^A R^{BA}} = \Omega^{BA} + \overline{\Omega^{BA}}$$

Therefore, $\overline{\Omega^{BA}} = -\Omega^{BA}$ and Ω^{BA} is skew symmetric.

The nomenclature of angular velocity abides by our standards. As indicated, only frames determine its definition, and the superscripts are read from left to right, e.g., Ω^{BA} is the angular velocity tensor of frame B wrt frame A. Section 4.2.4 discusses additive properties and other characteristics, which can only be proven after the introduction of the Euler transformation.

Example 4.7   Missile Separating from an Aircraft

Problem.   A missile B separates from an aircraft A and descends in the vertical plane. The rotation tensor in aircraft coordinates is given by

$$[R^{BA}(t)]^A = \begin{bmatrix} \cos\theta(t) & 0 & \sin\theta(t) \\ 0 & 1 & 0 \\ -\sin\theta(t) & 0 & \cos\theta(t) \end{bmatrix}$$

What is the angular velocity vector in aircraft coordinates?

Solution.   Use Eq. (4.47) expressed in aircraft coordinates $]^A$:

$$[\Omega^{BA}]^A = [D^A R^{BA}]^A [\overline{R^{BA}}]^A$$

Because of Eq. (4.37), $[D^A R^{BA}]^A = [dR^{BA}/dt]^A$, and thus

$$[\Omega^{BA}]^A = \left[\frac{dR^{BA}}{dt}\right]^A [\overline{R^{BA}(t)}]^A = \begin{bmatrix} -\dot{\theta}\sin\theta(t) & 0 & \dot{\theta}\cos\theta(t) \\ 0 & 1 & 0 \\ -\dot{\theta}\cos\theta(t) & 0 & -\dot{\theta}\sin\theta(t) \end{bmatrix}$$

$$\times \begin{bmatrix} \cos\theta(t) & 0 & -\sin\theta(t) \\ 0 & 1 & 0 \\ \sin\theta(t) & 0 & \cos\theta(t) \end{bmatrix} = \begin{bmatrix} 0 & 0 & \dot{\theta} \\ 0 & 0 & 0 \\ -\dot{\theta} & 0 & 0 \end{bmatrix}$$

and the angular velocity vector is $[\overline{\omega^{BA}}]^A = [0 \ \dot{\theta} \ 0]$.

We have added several new concepts to our collection while remaining true to our hypothesis that any phenomena in flight dynamics can be modeled solely by points and frames. Let us review. Displacement s_{BA} uses only points, whereas rotations R^{BA} use only frames. Linear velocities v_B^A and accelerations a_B^A mix it up, whereas angular velocities ω^{BA} employ only frames. Finally, the rotational time derivative refers to frames only. Changing its reference frames is the subject of the next section.

4.2.4 Euler Transformation

The rotational time derivative is the pass key to the invariant formulation of time-phased dynamic systems. As an operator that preserves the tensor characteristic, it depends on a reference frame. Sometimes it is desirable to change this reference frame, for instance from inertial to body frame. Euler's generalized transformation governs that change of frame.

We approach the derivation of this vital transformation fastidiously in increasing complexity. An ad hoc introduction points to the possible formulation, followed by a heuristic derivation based on the isotropy of space. The eggheads among you are referred to Appendix A of my dissertation,[4] which provides an analytical proof.

The classical Euler transformation is embodied in Eq. (4.34). It transforms the time derivative from frame A to frame B. Comparison with Eq. (4.35) leads to the conjecture that

$$[T]^{BA} \left[\frac{\mathrm{d}T}{\mathrm{d}t} \right]^{BA} = [\Omega^{BA}]^B$$

so that the last term of Eq. (4.35) reflects the vector product $[\Omega^{BA}]^B [s]^B$

$$\left[\frac{\mathrm{d}s}{\mathrm{d}t} \right]^A = [T]^{AB} \left(\left[\frac{\mathrm{d}s}{\mathrm{d}t} \right]^B + [\Omega^{BA}]^B [s]^B \right)$$

Introducing the rotational time derivatives for the two special cases $[\mathrm{d}s/\mathrm{d}t]^A = [D^A s]^A$ and $[\mathrm{d}s/\mathrm{d}t]^B = [D^B s]^B$, we can formulate

$$[D^A s]^A = [T]^{AB}([D^B s]^B + [\Omega^{BA}]^B [s]^B)$$

The key question is whether this transformation is a tensor concept. As written, it holds only for the associated coordinate systems $]^A$ and $]^B$. Fortuitously, it can be generalized for any allowable coordinate system. We call it Euler's general transformation, or just plain *Euler transformation*.

Theorem: Let A and B be two arbitrary frames related by the angular velocity tensor Ω^{BA}. Then, for any vector x the following transformation of the rotational time derivatives holds:

$$D^A x = D^B x + \Omega^{BA} x \tag{4.51}$$

Proof: The theorem is a direct consequence of the isotropic property of space. If R^{BA} is the rotation tensor of frame B wrt frame A, then, because of the isotropic property of space, the rotational derivative of x wrt frame A, $D^A x$ can also be evaluated as follows:

1) First rotate x through R^{BA} to obtain $R^{BA} x$.

2) Take the rotational time derivative wrt the rotated frame, now called B, D^B $(R^{BA} x)$.

3) Rotate the result back through R^{BA} into the original orientation. The three steps produce

$$D^A x = \overline{R^{BA}} D^B (R^{BA} x)$$

The chain rule applied to the right side yields

$$D^A x = D^B x + \overline{R^{BA}} D^B R^{BA} x \tag{4.52}$$

If we can show that $\overline{R^{BA}} D^B R^{BA} = \Omega^{BA}$, the theorem is proved. Let us interchange A and B and execute the same three steps again:

$$D^B x = D^A x + \overline{R^{AB}} D^A R^{AB} x$$

Adding the last two equations provides

$$\overline{R^{BA}} D^B R^{BA} = -\overline{R^{AB}} D^A R^{AB} \tag{4.53}$$

The right-hand side is the desired Ω^{BA} because the transpose changes the order of rotation

$$-\overline{R^{AB}} D^A R^{AB} = -R^{BA} \overline{D^A R^{BA}} = -\overline{D^A R^{BA} R^{BA}}$$

and recall the definition of the angular velocity tensor, Eq. (4.47), where

$$-\overline{D^A R^{BA} \overline{R^{BA}}} = -\overline{\Omega^{BA}} = \Omega^{BA}$$

Therefore, Eq. (4.53) is Ω^{BA} and indeed Eq. (4.52) proves the theorem:

$$D^A x = D^B x + \Omega^{BA} x$$

With the Euler transformation at our disposal, we can model many interesting kinematic phenomena. First, let us have another look at linear velocities and accelerations and then state and prove several properties of angular velocity and acceleration.

Example 4.8   Relative Velocities

Problem.   Refer back to Example 4.6 and Fig. 4.10. Let us assume that the radar lost track of the target, but still receives the missile's measured target velocity v_T^M via data link while tracking the missile's velocity v_M^R. What is the velocity of the target wrt the radar v_T^R, and how can it be calculated?

Solution.   All three quantities are relative velocities. It is tempting to add them vectorially $v_T^R = v_T^M + v_M^R$, but this is only possible if both the missile and the target were modeled by points only. However, the seeker is measuring the target velocity v_T^M wrt the missile reference frame M. Therefore, the angular velocity of the missile relative to the radar frame enters the calculation. I will derive the proper equation from the basic displacement triangle, employing the rotational time derivative and Euler transformation.

From Fig. 4.10,

$$s_{TR} = s_{TM} + s_{MR}$$

We impose the rotational time derivative to create the desired relative velocity $v_T^R = D^R s_{TR}$

$$D^R s_{TR} = D^R s_{TM} + D^R s_{MR} \tag{4.54}$$

The differential velocity $D^R s_{TM} = v_{TM}^R$ is unavailable. Instead, the missile seeker measures $v_T^M = D^M s_{TM}$. The kinship is provided by Euler's transformation

$$D^R s_{TM} = D^M s_{TM} + \Omega^{MR} s_{TM}$$

Substituting back into Eq. (4.54) yields an equation, exclusively with relative velocities:

$$v_T^R = v_T^M + \Omega^{MR} s_{TM} + v_M^R \tag{4.55}$$

To implement this equation in the radar processor, we need to know the coordinate system of the measured data. The target velocity is measured in missile coordinates $[v_T^M]^M$; the angular velocity tensor, recorded by the onboard INS, is beamed down in radar coordinates $[\Omega^{MR}]^R$ (the R frame is also the reference frame of the INS); and finally, missile position and velocity are measured by the radar in its own coordinates $[s_{MR}]^R$ and $[v_M^R]^R$. Because $]^R$ is prevalent, we coordinate Eq. (4.55) first in radar coordinates

$$[v_T^R]^R = [v_T^M]^R + [\Omega^{MR}]^R [s_{TM}]^R + [v_M^R]^R$$

followed by transforming the target velocity into missile coordinates

$$[v_T^R]^R = [T]^{RM}[v_T^M]^M + [\Omega^{MR}]^R [s_{TM}]^R + [v_M^R]^R \tag{4.56}$$

We have succeeded in deriving the component equation for implementation, provided the coordinate transformation $[T]^{RM}$ from the missile INS is also beamed down to the radar station.

Example 4.9   Relative Accelerations

Problem.   Continuing with the Example 4.8, we ask how to calculate the target's acceleration wrt the radar a_T^R when it is measured by the missile a_T^M?

Solution.   We start by taking the rotational time derivative wrt frame R of Eq. (4.55) to obtain the desired acceleration $a_T^R = D^R v_T^R$:

$$D^R v_T^R = D^R v_T^M + D^R(\Omega^{MR} s_{TM}) + D^R v_M^R \tag{4.57}$$

Let us discuss the right-hand terms one at a time.
1) The first term $D^R v_T^M$ relates to two different frames. To change it into the desired target/missile acceleration $a_T^M = D^M v_T^M$, we need to apply the Euler transformation

$$D^R v_T^M = D^M v_T^M + \Omega^{MR} v_T^M = a_T^M + \Omega^{MR} v_T^M$$

2) The second term $D^R(\Omega^{MR}s_{TM})$ is converted by the chain rule and the Euler transformation to obtain v_T^M:

$$D^R\left(\Omega^{MR}s_{TM}\right) = D^R\Omega^{MR}s_{TM} + \Omega^{MR}D^Rs_{TM}$$

$$= D^R\Omega^{MR}s_{TM} + \Omega^{MR}\left(D^Ms_{TM} + \Omega^{MR}s_{TM}\right)$$

$$= D^R\Omega^{MR}s_{TM} + \Omega^{MR}v_T^M + \Omega^{MR}\Omega^{MR}s_{TM}$$

The right side consists of the angular acceleration term, one-half of the Coriolis acceleration, and the centrifugal acceleration.

3) The third term $D^Rv_M^R$ is the easy one. It is the missile acceleration wrt the radar $D^Rv_M^R = a_M^R$.

Collecting terms, Eq. (4.57) expresses the target acceleration in known quantities:

$$a_T^R = a_T^M + a_M^R + 2\Omega^{MR}v_T^M + \Omega^{MR}\Omega^{MR}s_{TM} + D^R\Omega^{MR}s_{TM}$$

What a formidable equation to implement! If we only could model the missile by a simple point, then Ω^{MR} would not exist, and the target acceleration would simply be

$$a_T^R = a_T^M + a_M^R$$

This shortcut is made frequently. But do not forget to assess the neglected terms.

These two examples should give you the working knowledge for modeling linear velocity and acceleration problems. We now turn to angular velocities and accelerations. Refer back to Sec. 4.2.3 and recall the definition of the angular velocity tensor Eq. (4.47). Do not forget that it is a skew-symmetric tensor and therefore contracts to an angular velocity vector. With our freshly acquired Euler transformation we are able to prove several properties of angular velocity.

4.2.4.1 Properties of angular velocities.   The angular velocity tensor and vector have additive properties. Their superscripts reveal the rotational direction, and they relate to coordinate transformations in a special way.

Property 1

Angular velocities are *additive*. For example, if frame B revolves relative to frame A with Ω^{BA} and frame C wrt frame B with Ω^{CB} , then C revolves wrt A with the angular velocity tensor

$$\Omega^{CA} = \Omega^{CB} + \Omega^{BA} \tag{4.58}$$

The vector equivalent is

$$\omega^{CA} = \omega^{CB} + \omega^{BA} \tag{4.59}$$

The superscripts reflect the sequence of addition. By contracting adjacent letters, you get the result on the left-hand side.

Proof: Apply the Euler transformation twice to an arbitrary vector x

$$D^A x = D^B x + \Omega^{BA} x; \quad D^B x = D^C x + \Omega^{CB} x$$

and substitute the second into the first equation:

$$D^A x = D^C x + \Omega^{CB} x + \Omega^{BA} x = D^C x + (\Omega^{CB} + \Omega^{BA}) x$$

Compare this result with a third application of Euler's transformation

$$D^A x = D^C x + \Omega^{CA} x$$

and conclude: $\Omega^{CA} = \Omega^{CB} + \Omega^{BA}$.

Property 2

Reversing the sequence between two frames changes the sign of the angular velocity tensor:

$$\Omega^{BA} = -\Omega^{AB} \tag{4.60}$$

The vector equivalent is

$$\omega^{BA} = -\omega^{AB} \tag{4.61}$$

Proof: Replace C by A in Eq. (4.58) and, because $\Omega^{AA} = 0$, the relationship is proven. The angular velocity vector equivalent follows by contraction.

Property 3

The rotational time derivative of the angular velocity vector between two frames can be referred to either frame

$$D^A \omega^{BA} = D^B \omega^{BA} \tag{4.62}$$

Proof: Transform the rotational time derivative of ω^{BA} from frame A to B,

$$D^A \omega^{BA} = D^B \omega^{BA} + \Omega^{BA} \omega^{BA}$$

and recognize that the last term, being the vector product of the same vector, is zero.

Note that Eq. (4.62) also holds for regular time derivatives expressed in the associated coordinate systems $]^A$ and $]^B$:

$$\left[\frac{d\omega^{BA}}{dt} \right]^A = [T]^{AB} \left[\frac{d\omega^{BA}}{dt} \right]^B$$

Proof: From Eq. (4.62)

$$[D^A \omega^{BA}]^A = [T]^{AB} [D^B \omega^{BA}]^B$$

Introduce the definition of the rotational time derivative Eq. (4.36) on both sides:

$$\left[\frac{d\omega^{BA}}{dt}\right]^A + [T]^{AA}\underbrace{\overline{\left[\frac{dT}{dt}\right]}^{AA}[\omega^{BA}]^A}_{=0} = [T]^{AB}\left([\frac{d\omega^{BA}}{dt}]^B + [T]^{BB}\underbrace{\overline{\left[\frac{dT}{dt}\right]}^{BB}[\omega^{BA}]^B}_{=0}\right)$$

which proves the relationship.

Property 4

The coordinated angular velocity can be calculated from the coordinate transformations of the associated coordinate systems:

$$[\Omega^{BA}]^A = \overline{\left[\frac{dT}{dt}\right]}^{BA}[T]^{BA}; \quad [\Omega^{BA}]^B = [T]^{BA}\overline{\left[\frac{dT}{dt}\right]}^{BA} \qquad (4.63)$$

Proof: From the definition of the angular velocity tensor, Eq. (4.47) expressed in the associated coordinate system $]^A$ and because of Eq. (4.6) we form

$$[\Omega^{BA}]^A = [D^A R^{BA}]^A[\overline{R^{BA}}]^A = \left[\frac{dR^{BA}}{dt}\right]^A[\overline{R^{BA}}]^A = \overline{\left[\frac{dT}{dt}\right]}^{BA}[T]^{BA}$$

The second part is proven by transforming to the $]^B$ coordinate system

$$[\Omega^{BA}]^B = [T]^{BA}[\Omega^{BA}]^A[\bar{T}]^{BA} = [T]^{BA}\overline{\left[\frac{dT}{dt}\right]}^{BA}[T]^{BA}[\bar{T}]^{BA} = [T]^{BA}\overline{\left[\frac{dT}{dt}\right]}^{BA}$$

Example 4.10   Turbojet Spooling

Problem.   A pilot increases thrust as he pulls up the aircraft. The turbine's T angular velocity $[\omega^{TA}]^A$ wrt the airframe A is recorded in airframe coordinates $]^A$, and the aircraft pitch-up rate $[\omega^{AE}]^L$ wrt Earth E is measured by the onboard INS in local-level coordinates $]^L$. Both angular velocities are changing in time. Determine the angular acceleration of the turbine $[d\omega^{TE}/dt]^L$ wrt Earth in local-level coordinates. The INS provides also the direction cosine matrix $[T]^{AL}$.

Solution.   The solution follows the two-step approach: Solve the problem in tensor form, followed by coordination. We use the additive property of angular velocities then apply the rotational derivative wrt Earth and the Euler transformation to recover the turbine revolutions-per-minute measurement. Then we introduce the appropriate coordinate systems to present the turbine acceleration in local-level coordinates.

The angular velocity of the turbine wrt Earth is

$$\omega^{TE} = \omega^{TA} + \omega^{AE}$$

Take the rotational derivative wrt Earth

$$D^E\omega^{TE} = D^E\omega^{TA} + D^E\omega^{AE}$$

Transform the turbine derivative $D^E \omega^{TA}$ to the airframe A and obtain

$$D^E \omega^{TE} = D^A \omega^{TA} + \Omega^{AE} \omega^{TA} + D^E \omega^{AE}$$

Introduce local-level coordinates

$$[D^E \omega^{TE}]^L = [D^A \omega^{TA}]^L + [\Omega^{AE}]^L [\omega^{TA}]^L + [D^E \omega^{AE}]^L$$

Because the local-level coordinate system is associated with the Earth frame, the first and last rotational derivative simplify to the ordinary time derivative. The second rotational derivative, as well as the turbine speed, must be transformed to the airframe axis to get the desired result:

$$\left[\frac{d\omega^{TE}}{dt} \right]^L = [\bar{T}]^{AL} \left[\frac{d\omega^{TA}}{dt} \right]^A + [\Omega^{AE}]^L [\bar{T}]^{AL} [\omega^{TA}]^A + \left[\frac{d\omega^{AE}}{dt} \right]^L$$

As you see, you cannot just add together the angular accelerations. The turbine speed also couples with the aircraft angular rate as an additional acceleration term.

Sections 4.1 and 4.2 that you have just mastered are the trailblazers for invariant modeling. Particularly, the rotational time derivative and the Euler transformation preserve the tensor characteristics of kinematics, even under time-dependent co-ordinate transformations. To adopt them as your own, and apply them whenever the opportunity arises, should become your ambition. I have put them to good use for 30 years, and they spared me some major headaches.

4.3 Attitude Determination

A final subject of importance is the *fundamental problem of kinematics* of flight vehicles. In six-degree-of-freedom (DoF) simulations the application of Euler's law renders the differential equations of body rates. More precisely, the solution of the differential equations yields the angular velocity vector of the vehicle frame B wrt the inertial frame I, expressed in body coordinates $[\omega^{BI}]^B$. Specialists call them the p, q, r components of

$$[\overline{\omega^{BI}}]^B = [p \quad q \quad r]$$

The fundamental problem states as follows: Given the body rates $[\omega^{BI}]^B$, determine the orientation of the vehicle wrt the inertial frame. The orientation can be expressed in terms of the rotation tensor, Euler angles, or quaternion. Because orientation is removed from angular velocity by integration, the solution will embody differential equations. I shall treat each of the three possibilities individually.

4.3.1 Rotation Tensor Differential Equations

The first approach is based on the definition of the angular velocity tensor, Eq. (4.47) referenced to body frame B. If expressed in body coordinates, the rotational derivative reduces to the ordinary time derivative [see Eq. (4.37)]:

$$[\Omega^{IB}]^B = [D^B R^{IB}]^B [\overline{R^{IB}}]^B = \left[\frac{dR^{IB}}{dt} \right]^B [\overline{R^{IB}}]^B$$

Solving for the time derivative

$$\left[\frac{dR^{IB}}{dt}\right]^{B} = [\Omega^{IB}]^{B}[R^{IB}]^{B}$$

reversing the frame sequence to conform to the known body rates $[\Omega^{BI}]^{B}$

$$\left[\frac{dR^{BI}}{dt}\right]^{B} = [\overline{\Omega^{BI}}]^{B}[\overline{R^{BI}}]^{B} \tag{4.64}$$

and taking the transposed on both sides

$$\left[\frac{dR^{BI}}{dt}\right]^{B} = [R^{BI}]^{B}[\Omega^{BI}]^{B} \tag{4.65}$$

produces the differential equations for the nine elements of the rotation tensor.

In simulations, rather than calculating the rotation tensor, the focus is on the TM of body wrt inertial coordinates. The connection is provided by Eq. (4.6), $[T]^{BI} = [\overline{R^{BI}}]^{B}$, which we can apply directly to Eq. (4.64):

$$\left[\frac{dT}{dt}\right]^{BI} = [\overline{\Omega^{BI}}]^{B}[T]^{BI} \tag{4.66}$$

These are the famous *differential equations of the direction cosine matrix.* You will see them again in Sec. 10.1.2 when we formulate the six-DoF equations of motion over an elliptical Earth. In simpler simulations, when Earth E serves as an inertial frame, the local-level system is used. Therefore, we replace frame I by E:

$$\left[\frac{dT}{dt}\right]^{BL} = [\overline{\Omega^{BE}}]^{B}[T]^{BL} \tag{4.67}$$

where $[T]^{BL}$ relates to the Euler angles of Eq. (3.28). If the Euler angles are given at launch, use Eq. (3.28) to initialize the nine elements.

The nine differential equations are not independent. Because the TM is composed of three orthonormal base vectors, only six differential equations need be solved, and the remaining three elements can be calculated from the orthogonality conditions. Express the direction cosine matrix in the three base vectors of the geographic frame $[T]^{BG} = [[g_1]^{B} \ [g_2]^{B} \ [g_3]^{B}]$ and substitute them into the differential equation, Eq. (4.67),

$$\frac{d}{dt}[[g_1]^{B} \ [g_2]^{B} \ [g_3]^{B}] = [\overline{\Omega^{BE}}]^{B}[[g_1]^{B} \ [g_2]^{B} \ [g_3]^{B}]$$

Take the first two columns individually, add an orthogonality condition, and you have the set of six differential and three algebraic equations:

$$\frac{d}{dt}[g_1]^{B} = [\overline{\Omega^{BE}}]^{B}[g_1]^{B}$$

$$\frac{d}{dt}[g_2]^{B} = [\overline{\Omega^{BE}}]^{B}[g_2]^{B} \tag{4.68}$$

$$[g_3]^{B} = [G_1]^{B}[g_2]^{B}$$

To recover the Euler angles, use Eqs. (3.15–3.17) or an alternate form that is used in flight simulators:

$$\theta = \arcsin(-t_{13})$$

$$\psi = \arccos\left(\frac{t_{11}}{\cos\theta}\right)\operatorname{sign}(t_{12}) \qquad (4.69)$$

$$\phi = \arccos\left(\frac{t_{33}}{\cos\theta}\right)\operatorname{sign}(t_{23})$$

where the t_{ij} are the elements of the direction cosine matrix. Equation (4.69) delivers similar results as Eqs. (3.15–3.17). The sign function—FORTRAN intrinsic function SIGN(A,B)—controls the sign like the FORTRAN ATAN2(A,B) function in Eqs. (3.15–3.17). The range of the angles is $-\pi \le \psi \le +\pi$, $-\pi \le \phi \le +\pi$, and $-\pi/2 < \theta < +\pi/2$.

With these types of Euler angles adopted for flight mechanics, singularities occur at $\theta = \pm\pi/2$, where no distinction is possible between yaw and roll. These correspond to a vertical dive or climb, situations seldom encountered in airplanes, but important in missile launches. Resequencing the Euler angles can place the singularity at another less conspicuous attitude. Fortunately, this undesirable behavior occurs only in the output calculations and not in the differential equations themselves, where they would provoke much more trouble.

Integrating the differential equations on digital computers with finite word length and incremental time steps produces errors that corrupt the TM. To maintain orthogonality, a correction term is applied from integration step n to $n + 1$:

$$[T(n+1)]^{BI} = [T(n)]^{BI} + \tfrac{1}{2}\left([E] - [T(n)]^{BI}[\overline{T(n)}]^{BI}\right)[T(n)]^{BI}$$

As a sanity check, note that the correction term is zero for orthogonal TMs. For details however you have to wait until Sec. 10.1.2, which describes the implementation as part of a six-DoF simulation. The actual code can be found in the CADAC GHAME6 simulation.

The rotation tensor and direction cosine matrix solutions have superior features that lend themselves to computer applications. The differential equations are linear, well behaved, and without singularities. Above all, the important direction cosine matrix is directly computed. However, because Euler angles are so much easier to visualize, initialization and output have to be converted. To avoid this conversion, the Euler angles can be integrated directly.

4.3.2 Euler Angle Differential Equations

In days past, when computational efficiency was of prime concern, the direct integration of Euler angles was the preferred method. Let us investigate its merit and see why it has fallen from favor. We continue using the Earth as an inertial reference frame and the local-level coordinate system. Starting with the body rates $[\omega^{BE}]^B = [p \quad q \quad r]$, we develop the differential equations of the three Euler angles ψ, θ, and ϕ. A general solution can be derived from Eq. (4.66), but we use a simpler

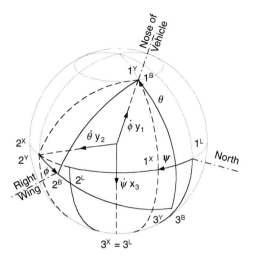

Fig. 4.13   Euler angular velocity equations.

derivation based on the property that angular rates can be added vectorially. Figure 4.13 highlights the three Euler rates, which make up the body rates

$$\omega^{BE} = \dot{\psi}x_3 + \dot{\theta}y_2 + \dot{\phi}y_1$$

Selecting body coordinates

$$[\omega^{BE}]^B = \dot{\psi}[x_3]^B + \dot{\theta}[y_2]^B + \dot{\phi}[y_1]^B$$

and expressing the base vectors in their preferred coordinate systems

$$[x_3]^B = [T]^{BX}[x_3]^X$$

$$[y_2]^B = [T]^{BY}[y_2]^Y$$

$$[y_1]^B = [T]^{BY}[y_1]^Y$$

yields the convenient expression of body rates

$$[\omega^{BE}]^B = \dot{\psi}[T]^{BX}[x_3]^X + [T]^{BY}\left(\dot{\theta}[y_2]^Y + \dot{\phi}[y_1]^Y\right)$$

With the TMs leading up to Eq. (3.14), we can coordinate the body rates

$$
[\omega^{BE}]^B =
\begin{bmatrix}
\cos\theta & 0 & -\sin\theta \\
\sin\phi\,\sin\theta & \cos\phi & \sin\phi\,\cos\theta \\
\cos\phi\,\sin\theta & -\sin\phi & \cos\phi\,\cos\theta
\end{bmatrix}
\begin{bmatrix}
0 \\
0 \\
\dot{\psi}
\end{bmatrix}
$$

$$
+
\begin{bmatrix}
1 & 0 & 0 \\
0 & \cos\phi & \sin\phi \\
0 & -\sin\phi & \cos\phi
\end{bmatrix}
\left(
\begin{bmatrix}
0 \\
\dot{\theta} \\
0
\end{bmatrix}
+
\begin{bmatrix}
\dot{\phi} \\
0 \\
0
\end{bmatrix}
\right)
$$

$$\begin{bmatrix} p \\ q \\ r \end{bmatrix} = \begin{bmatrix} -\dot{\psi}\,\sin\theta + \dot{\phi} \\ \dot{\psi}\,\sin\phi\,\cos\theta + \dot{\theta}\,\cos\phi \\ \dot{\psi}\,\cos\phi\,\cos\theta - \dot{\theta}\,\sin\phi \end{bmatrix} = \begin{bmatrix} 1 & 0 & -\sin\theta \\ 0 & \cos\phi & \sin\phi\,\cos\theta \\ 0 & -\sin\phi & \cos\phi\,\cos\theta \end{bmatrix} \begin{bmatrix} \dot{\phi} \\ \dot{\theta} \\ \dot{\psi} \end{bmatrix}$$

Solving for the Euler angular rates yields the desired differential equations:

$$\begin{bmatrix} \dot{\phi} \\ \dot{\theta} \\ \dot{\psi} \end{bmatrix} = \begin{bmatrix} 1 & \sin\phi\,\tan\theta & \cos\phi\,\tan\theta \\ 0 & \cos\phi & -\sin\phi \\ 0 & \sin\phi/\cos\theta & \cos\phi/\cos\theta \end{bmatrix} \begin{bmatrix} p \\ q \\ r \end{bmatrix} \qquad (4.70)$$

These three nonlinear differential equations, although compact and easily initialized, suffer from singularities at vertical climb and dive. Approaching these attitudes, the integration deteriorates and breaks down completely at the singularities. Only in older simulations will you still find these equations. They are used by the CADAC FALCON6 simulation, which you can download from the CADAC CD-ROM. With modern, high-performance computers the old requirement for computational efficiency has given way to accuracy and flexibility. In its train was swept in the ancient quaternion to slove the fundamental kinematic problem. We revive it for the third method.

4.3.3 Quaternion Differential Equations

Would you believe that the introduction of quaternions preceded vectors and tensors? Sir W. R. Hamilton published in 1843 his quaternion algebra,[5] which contained, albeit hidden, three-dimensional vectors.

Quaternions are vectors in four-dimensional space, therefore their distinctive name. A simpler version, complex vectors, are their cousins in two dimensions. In Fig. 4.14 we consider the number 1 on the real axis as the first base vector and, on the imaginary axis, i as the second base vector. The two-dimensional vector p can be expressed in the component form $p = 1p_0 + ip_1$. With the angle of rotation ε and absolute value $|p|$ known, we can also represent the complex variable in the polar form $p = |p|e^{i\varepsilon}$.

Now let us transition to four-dimensional space. We already encountered in Sec. 4.1.1 Euler's theorem of rotation, which states that four parameters specify an arbitrary rigid rotation. There are three components for the axis of rotation n in addition to the angle of rotation ε (see Fig. 4.15). Quaternions with unit norm represent such rotations in four-parameter space. Hamilton generalized the

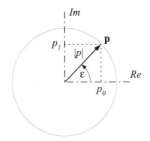

Fig. 4.14   Complex numbers.

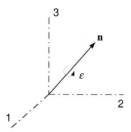

Fig. 4.15   Euler's four parameters.

two-dimensional complex number space by adding two more imaginary axes j and k. Embedded in this four-dimensional space is the rotation quaternion in the component form

$$q = q_0 + iq_1 + jq_2 + kq_3 \qquad (4.71)$$

where $ij = k$, $jk = i$, $ki = j$ and $i^2 = j^2 = k^2 = -1$ and its norm $q_0^2 + q_1^2 + q_2^2 + q_3^2 = 1$.

Hamilton created a complete quaternion algebra with vectors as a special case. It is therefore not surprising that many concepts, which we have introduced in three-space, have their counterparts in four-space. To develop the quaternion differential equations, I shall make use of several entities. The rotation quaternion $\{q\}$, just introduced, the rotation tensor quaternion of frame B wrt frame E $\{Q^{BE}\}$, and the angular velocity quaternion of frame B wrt frame E $\{\Omega^{BE}\}$. The use of braces identifies them as tensors in four-dimensional space, which, when coordinated, receive a superscript for the coordinate system.

4.3.3.1 Rotation quaternion.
The rotation quaternion has four coordinates q_0, q_1, q_2, and q_3 with direct relationship to the four Euler parameters n and ε. In any coordinate system, with braces designating four-dimensional and brackets three-dimensional Euclidean space

$$\{q\} = \begin{Bmatrix} q_0 \\ q_1 \\ q_2 \\ q_3 \end{Bmatrix} = \begin{Bmatrix} \cos(\varepsilon/2) \\ \sin(\varepsilon/2) \cdot n_1 \\ \sin(\varepsilon/2) \cdot n_2 \\ \sin(\varepsilon/2) \cdot n_3 \end{Bmatrix}, \quad \text{where} \quad [n] = \begin{bmatrix} n_1 \\ n_2 \\ n_3 \end{bmatrix} \qquad (4.72)$$

This relationship gives us a vivid picture of a rotation quaternion. The scalar component $q_0 = \cos(\varepsilon/2)$ contains the half-angle of rotation, and the vector part

$$[q] = \begin{bmatrix} q_1 \\ q_2 \\ q_3 \end{bmatrix} = \sin\left(\frac{\varepsilon}{2}\right) \begin{bmatrix} n_1 \\ n_2 \\ n_3 \end{bmatrix}$$

relates to the unit vector of rotation $[n]$. The mystery of four-dimensional space is explained if we consider the rotation quaternion consisting of a scalar part q_0 and

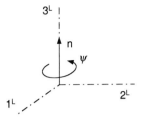

Fig. 4.16   Yaw rotation.

a three-dimensional vector $[q]$

$$\{q\} = \left\{ \frac{q_0}{[q]} \right\}$$
(4.73)

A simple example may help your intuition.

Example 4.11   Rotation Quaternion

Problem.   Body frame B is rotated wrt Earth frame E by the angle $\psi = 60$ deg about the vector of rotation $[\bar{n}]^L = [0\ 0\ 1]$ in local-level coordinates (see Fig. 4.16). Calculate the rotation quaternion $\{q^{BE}\}^L$.

Solution.   Introducing local-level coordinates into Eq. (4.73) and with Eq. (4.72) provides

$$\{q^{BE}\}^L = \left\{ \frac{q_0}{[q]^L} \right\}^L = \left\{ \frac{\cos(\psi/2)}{\sin(\psi/2)[n]^L} \right\} = \left\{ \begin{matrix} \cos(\psi/2) \\ \hline \sin(\psi/2)n_1 \\ \sin(\psi/2)n_2 \\ \sin(\psi/2)n_3 \end{matrix} \right\}$$

Using the rotation vector and the numerical values of the example yields the rotation quaternion

$$\{q^{BE}\}^L = \left\{ \begin{matrix} \cos(\psi/2) \\ \hline 0 \\ 0 \\ \sin(\psi/2)n_3 \end{matrix} \right\} = \left\{ \begin{matrix} 0.5\sqrt{3} \\ \hline 0 \\ 0 \\ 0.5 \end{matrix} \right\}$$

This is the rotation quaternion of the first Euler rotation about the angle yaw. The other two rotations about pitch and roll follow similar patterns.

4.3.3.2   Rotation tensor quaternion.   As a vector has a skew-symmetric tensor equivalent, so does the rotation quaternion have a tensor equivalent. It consists of the scalar part multiplied by the unit tensor quaternion and an additive skew-symmetric tensor quaternion

$$\{Q\} = q_0\{E\} + \left\{\begin{array}{c|c} 0 & [\bar{q}] \\ \hline -[q] & [Q] \end{array}\right\} = \left\{\begin{array}{cccc} q_0 & 0 & 0 & 0 \\ 0 & q_0 & 0 & 0 \\ 0 & 0 & q_0 & 0 \\ 0 & 0 & 0 & q_0 \end{array}\right\}$$

$$\underbrace{\phantom{\left\{\begin{array}{cccc}0&0&0&0\\0&0&0&0\end{array}\right\}}}_{\text{symmentric}}$$

$$+ \underbrace{\left\{\begin{array}{c|ccc} 0 & q_1 & q_2 & q_3 \\ \hline -q_1 & 0 & -q_3 & q_2 \\ -q_2 & q_3 & 0 & -q_1 \\ -q_3 & -q_2 & q_1 & 0 \end{array}\right\}}_{\text{skew-symmentric}} = \left\{\begin{array}{cccc} q_0 & q_1 & q_2 & q_3 \\ -q_1 & q_0 & -q_3 & q_2 \\ -q_2 & q_3 & q_0 & -q_1 \\ -q_3 & -q_2 & q_1 & q_0 \end{array}\right\}$$

We can absorb the unit tensor quaternion into one term:

$$\{Q\} = \left\{\begin{array}{c|c} q_0 & [\bar{q}] \\ \hline -[q] & q_0[E] + [Q] \end{array}\right\} \tag{4.74}$$

Notice that $[Q]$ is the familiar skew-symmetric tensor of vector $[q]$. Our preceding example, elevated to the rotation tensor quaternion, and again expressed in geographic coordinates becomes

$$\{Q^{BE}\}^L = \left\{\begin{array}{c|ccc} 0.5\sqrt{3} & 0 & 0 & 0.5 \\ \hline 0 & 0.5\sqrt{3} & -0.5 & 0 \\ 0 & 0.5 & 0.5\sqrt{3} & 0 \\ -0.5 & 0 & 0 & 0.5\sqrt{3} \end{array}\right\}$$

The rotation tensor quaternion $\{Q^{BE}\}$ is the equivalent of the rotation tensor R^{BE} in three-dimensional space and has similar properties, like orthogonality and rotation sequencing. It is also related to the angular velocity quaternion.

4.3.3.3 Angular velocity quaternion.
We define the angular velocity quaternion in a similar fashion as the angular velocity tensor Eq. (4.47). But instead of defining the rotational time derivative for quaternions, we content ourselves with the regular time derivative and pick the local-level coordinate system

$$\{\Omega^{BE}\}^L \equiv 2\left\{\frac{dQ^{BE}}{dt}\right\}^L \{\overline{Q^{BE}}\}^L \tag{4.75}$$

Compare this definition with that of the three-dimensional space

$$[\Omega^{BE}]^L = \left[\frac{dR^{BE}}{dt}\right]^L [\overline{R^{BE}}]^L$$

The factor two of the quaternion definition grabs your attention, but is understandable because quaternions operate with half angle of rotation. Because Eq. (4.75) involves rotation tensor quaternions, the angular velocity quaternion is also a tensor quaternion, yet without a scalar part

$$\{\Omega^{BE}\}^L = \left\{ \begin{array}{c|c} 0 & [\overline{\omega^{BE}}]^L \\ \hline -[\omega^{BE}]^L & [\Omega^{BE}]^L \end{array} \right\} \tag{4.76}$$

Not surprising, the vector part is the angular velocity vector of three-space.

We have assembled all required elements to proceed with the quaternial solution of the fundamental kinematic problem in flight simulations. Rearranging Eq. (4.75) will deliver the differential equations.

4.3.3.4 Differential equations.

The time derivative of the rotation quaternion is buried in Eq. (4.75), but we have to transform the angular velocity quaternion first to body coordinates—the body rates are given in body coordinates—and then solve for it. I will spare you the details, although the conversion is not difficult, because the rotation tensor quaternion $\{Q^{BE}\}^L$ has its dual in the quaternion transformation matrix $\{Q\}^{BL}$ related by $\{Q^{BE}\}^L = \{\bar{Q}\}^{BL}$. The result is

$$\{\dot{\overline{Q}}^{BE}\}^L = \frac{1}{2} \{\overline{\Omega^{BE}}\}^B \{\overline{Q}^{BE}\}^L$$

Partitioning the matrices

$$\left\{ \begin{array}{c|c} \dot{q}_0 & -[\dot{\bar{q}}] \\ \hline [\dot{q}] & \dot{q}_0[E] - [\dot{Q}] \end{array} \right\} = \frac{1}{2} \left\{ \begin{array}{c|c} 0 & -[\overline{\omega^{BE}}] \\ \hline [\omega^{BE}] & -[\Omega^{BE}] \end{array} \right\} \left\{ \begin{array}{c|c} q_0 & -[\bar{q}] \\ \hline [q] & q_0[E] - [Q] \end{array} \right\}$$

and equating the first partitioned columns

$$\left\{ \begin{array}{c} \dot{q}_0 \\ \hline [\dot{q}] \end{array} \right\} = \frac{1}{2} \left\{ \begin{array}{c|c} 0 & -[\overline{\omega^{BE}}] \\ \hline [\omega^{BE}] & -[\Omega^{BE}] \end{array} \right\} \left\{ \begin{array}{c} q_0 \\ \hline [q] \end{array} \right\}$$

we finally obtain, with $[\omega^{BE}]^B = [p \; q \; r]$, the desired differential equations

$$\begin{Bmatrix} \dot{q}_0 \\ \dot{q}_1 \\ \dot{q}_2 \\ \dot{q}_3 \end{Bmatrix} = \frac{1}{2} \begin{bmatrix} 0 & -p & -q & -r \\ p & 0 & r & -q \\ q & -r & 0 & p \\ r & q & -p & 0 \end{bmatrix} \begin{Bmatrix} q_0 \\ q_1 \\ q_2 \\ q_3 \end{Bmatrix} \tag{4.77}$$

These differential equations are a joy to implement, because they are linear, have no singularities, and number only four. Yet, the initialization with Euler angles is not quite straightforward.

For initialization, we need to express the quaternion components in terms of Euler angles because who wants to describe the launch attitude of a missile in quaternions. Using Eq. (4.71) with Eq. (4.72) to build the three Euler rotations and combining them $q(\psi)q(\theta)q(\phi)$ leads to the relationships

$$q_0 = \cos\left(\frac{\psi}{2}\right)\cos\left(\frac{\theta}{2}\right)\cos\left(\frac{\phi}{2}\right) + \sin\left(\frac{\psi}{2}\right)\sin\left(\frac{\theta}{2}\right)\sin\left(\frac{\phi}{2}\right)$$

$$q_1 = \cos\left(\frac{\psi}{2}\right)\cos\left(\frac{\theta}{2}\right)\sin\left(\frac{\phi}{2}\right) - \sin\left(\frac{\psi}{2}\right)\sin\left(\frac{\theta}{2}\right)\cos\left(\frac{\phi}{2}\right)$$

$$q_2 = \cos\left(\frac{\psi}{2}\right)\sin\left(\frac{\theta}{2}\right)\cos\left(\frac{\phi}{2}\right) + \sin\left(\frac{\psi}{2}\right)\cos\left(\frac{\theta}{2}\right)\sin\left(\frac{\phi}{2}\right)$$

$$q_3 = \sin\left(\frac{\psi}{2}\right)\cos\left(\frac{\theta}{2}\right)\cos\left(\frac{\phi}{2}\right) - \cos\left(\frac{\psi}{2}\right)\sin\left(\frac{\theta}{2}\right)\sin\left(\frac{\phi}{2}\right)$$

(4.78)

Computational implementation of Eq. (4.77) must maintain the unit norm of the rotation quaternion even in the presence of rounding errors. A proven method is the addition of the factor $k\lambda\{q\}$ to the right-hand side of Eq. (4.77) with $k\Delta t \le 1(\Delta t$ integration interval) and $\lambda = 1 - (q_0^2 + q_1^2 + q_2^2 + q_3^2)$ to maintain the unit norm.

The output of the differential equations must be converted to physical meaningful quantities. Those are primarily the Euler angles. Yet, rather than calculating them from the quaternion directly, we take the intermediate steps of rotation tensor and direction cosine matrix to obtain the Euler angles. By this approach we also make available the TM of body wrt local-level coordinates $[T]^{BL}$.

4.3.3.5 Rotation tensor.

We have already exploited the close ties between quaternions and tensors of three space. Just glance back at the relationship between the angular velocity quaternion and the angular velocity tensor. Not surprisingly, a similar association exists between the rotation quaternion and the rotation tensor. We could make the derivation from the general rotation quaternion tensor, but prefer a simpler route. I will state the result and prove that it leads to the rotation tensor Eq. (4.12).

Given the rotation quaternion $\{q\}$ in any allowable coordinate system with the scalar q_0 and the vector $[q]$, [see Eq. (4.73)], the rotational tensor $[R]$ in the same coordinate system is

$$[R] = q_0^2[E] - [\bar{q}][q][E] + 2[q][\bar{q}] + 2q_0[Q] \tag{4.79}$$

where $[Q]$ is the skew-symmetric version of $[q]$. To prove that this formulation leads to the classical form

$$[R] = \cos\varepsilon[E] + (1 - \cos\varepsilon)[n][\bar{n}] + \sin\varepsilon[N]$$

we substitute the rotation quaternion $q_0 = \cos(\varepsilon/2)$, $[q] = \sin(\varepsilon/2)[n]$ into Eq. (4.79)

$$[R] = \cos^2\frac{\varepsilon}{2}[E] - \sin^2\frac{\varepsilon}{2}(n_1^2 + n_2^2 + n_3^2)[E] + 2\sin^2\frac{\varepsilon}{2}[n][\bar{n}] + 2\cos\frac{\varepsilon}{2}\sin\frac{\varepsilon}{2}[N]$$

$$= \cos\varepsilon[E] + (1 - \cos\varepsilon)[n][\bar{n}] + \sin\varepsilon[N] \qquad \text{QED}$$

Therefore we can use Eq. (4.79) and relate the elements of the rotation tensor to the elements of the rotation quaternion. Substituting $[\bar{q}] = [q_1 \quad q_2 \quad q_3]$ into Eq. (4.79) yields

$$[R] = \begin{bmatrix} q_0^2 + q_1^2 - q_2^2 - q_3^2 & 2(q_1q_2 - q_0q_3) & 2(q_1q_3 + q_0q_2) \\ 2(q_1q_2 + q_0q_3) & q_0^2 - q_1^2 + q_2^2 - q_3^2 & 2(q_2q_3 - q_0q_1) \\ 2(q_1q_3 - q_0q_2) & 2(q_2q_3 + q_0q_1) & q_0^2 - q_1^2 - q_2^2 + q_3^2 \end{bmatrix} \quad (4.80)$$

Specifically, if the quaternion $\{q^{BE}\}^L$ models the rotation of body frame B wrt Earth frame E, expressed in local-level coordinates, then its rotational tensor is $[R^{BE}]^L$, and the direction cosine matrix

$$[T]^{BE} = [\overline{R^{BE}}]^L \quad (4.81)$$

The Euler angles can now be derived from $[T]^{BE}$ via Eq. (3.28):

$$\tan \psi = \frac{2(q_1q_2 + q_0q_3)}{q_0^2 + q_1^2 - q_2^2 - q_3^2}, \quad \sin \theta = -2(q_1q_3 - q_0q_2)$$

$$\tan \phi = \frac{2(q_2q_3 + q_0q_1)}{q_0^2 - q_1^2 - q_2^2 + q_3^2} \quad (4.82)$$

The first equation has singularities at $\psi = \pm 90$ deg, and the last equation at $\phi = \pm 90$ deg. But do not despair; because these are just output calculations, you can program around them, and the accuracy suffers only in the close vicinity of the singularities.

The quaternion methodology is complete. Given the Euler angles, initialize with Eq. (4.78) the quaternion differential Equation(4.77), calculate the direction cosine matrix from Eqs. (4.80) and (4.81). Obtain the Euler angles again from Eq. (4.82), as the vehicle flies to its destination. You can acquire the detailed implementation from the CADAC SRAAM6 code, an air-to-air simulation over the flat Earth.

4.3.3.6 Summary.

Here you have three methods for solving the fundamental kinematic problem of flight dynamics. You can use the archaic, compact method of Euler angles, the wasteful direction cosine matrix approach, or the elusive quaternion technique. You will need them only in full-up six-DoF simulations, where the solution of the Euler differential equations provides you with the body rates. In three- and five-DoF simulations the lack of attitude equations (Euler's equations) makes it infeasible and unnecessary to solve the kinematic differential equations—an advantage, if these simple simulations satisfy your need.

Table 4.1 summarizes the options. I favor the latter two methods. Quaternions are well suited for the near-Earth simulations that can use the Earth as inertial frame and employ the local-level coordinate system. In this case the quaternions and the direction cosine matrix are closely linked to the Euler angles. For round and oblate Earth simulations the inertial frame is used, and the $[T]^{BI}$ direction cosine matrix is best solved directly because its angles are not the traditional Euler angles.

Table 4.1   Three methods of solving the fundamental kinematic problem in simulations with $\overline{[\omega^{BE}]}^B = [p\ q\ r]$ from Euler's attitude equations

Features	Euler angles	Direction cosines	Quaternions
Differential equation	$\begin{bmatrix} \dot\phi \\ \dot\theta \\ \dot\psi \end{bmatrix} = [f(\phi, \theta, \psi)] \begin{bmatrix} p \\ q \\ r \end{bmatrix}$	$\left[\dfrac{dT}{dt}\right]^{BI} = [\overline{\Omega^{BI}}]^B [T]^{BI}$	$\overline{\left\{\dfrac{dQ^{BE}}{dt}\right\}}^L = \dfrac{1}{2}\{\Omega^{BE}\}^B \{Q^{BE}\}^L$
Initialization	Directly by ϕ_0, θ_0, ψ_0	$[T(t=0)]^{BI}$	$\{q(t=0)\} = f(\phi_0, \theta_0, \psi_0)$
TM	$[T]^{BL} = f(\phi, \theta, \psi)$	Directly calculated	$[T]^{BL} = [\overline{R^{BE}}]^L$
Euler angles	Directly calculated	From $[T]^{BI}$	$\begin{bmatrix} \phi \\ \theta \\ \psi \end{bmatrix} = f(q_0, q_1, q_2, q_3)$
Advantage	Three differential equations. Angular attitude calculated directly. Direct initialization	Transformation matrix calculated directly. Six linear differential equations.	Four linear differential equations. Simple orthogonality condition.
Disadvantage	Singularity at $\theta = \pm\pi/2$. Nonlinear differential equations. Transformation matrix not directly available.	Computationally ineffective. Euler angles not directly available. Initial calculations necessary.	Transformation matrix and Euler angles not directly available. Initial calculations necessary.

References

[1]Goldstein, H., *Classical Mechanics*, Addison Wesley Longman, Reading, MA, 1965, p. 118.

[2]Wrede, R. C., *Introduction to Vector and Tensor Analysis*, Wiley, New York, 1963, p. 169.

[3]Wundheiler, A., "Kovariante Ableitung und die Cesaroschen Unbeweglichkeitsbedingungen," *Mathematische Zeitschrift*, Vol. 36, 1932, pp. 104–109.

[4]Zipfel, P. H., "On Flight Dynamics of Magnus Rotors," Dept. of the Army, Technical Rept. 117, Defense Technical Information Center AD-716-345, Cameron Station, Alexandria, VA, Nov. 1970.

[5]Hamilton, W. R., "On a New Species of Imaginary Quantities Connected with a Theory of Quaternions," *Dublin Proceedings*, Vol. 2, No. 13, Nov. 1843, pp. 424–434.

Problems

4.1   Orthogonality of rotation tensors.   Property 4 states that the rotation tensor is orthogonal. The proof suggested using the orthogonality property of TMs. You are to develop a direct proof employing the rotation tensor.

4.2   Elevator rotation tensor.   The hinge line of the elevator E is parallel to the 2^B axis of an aircraft B. The deflection angle δe is measured positive when the trailing edge moves downward. Derive $[R^{EB}]^B$, the rotation tensor of the elevator wrt the airframe, in aircraft coordinates.

4.3   Tip speed of missile fin.   A Missile B executes rolling motions wrt the Earth frame E that are recorded by the onboard INS as $\phi(t) = \omega t$, with ω a constant. Derive the rotational tensor $[R^{BE}(\phi)]^B$ and the angular velocity tensor $[\Omega^{BE}]^B$, both in body coordinates. What is the velocity $[v_T^E]^B$ of the tip T of the control fin, which is displaced

$$[\overline{s_{TB}}]^B = [-2 \quad 0.5 \quad 0]\,\text{m}$$

from the missile c.m. B? The missile velocity is

$$[\overline{v_B^E}]^B = [350 \quad 0 \quad 0]\,\text{m/s}$$

and the angular rate $\omega = 100$ deg/s (*Solution:* $[\overline{v_T^E}]^B = [350\ 0\ 0.87]$ m/s).

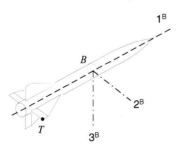

4.4   Tetragonal tensor is orthogonal.   Show that the tetragonal tensor $R_{90} = n\bar{n} + N$ is orthogonal, where n is the unit vector of rotation and N its skew-symmetric form.

4.5   Control surface forces.   A missile is steered by four control surfaces, arranged and numbered as shown in the figure, with positive deflections down as indicated. Wind-tunnel data on control effectiveness are not available, but the force f_1 on fin #1 was measured as a function of its deflection. If all four fins are deflected by the same amount in the positive direction, what is the total force f (missile incidence angles are zero)? Before you develop the component form, first obtain the invariant tensor.

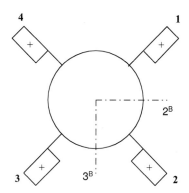

(a) You will need the tetragonal rotation tensor R_{90}. If b_1 is the base vector of the missile parallel to the missile centerline, show that the tetragonal rotation tensor is $R_{90} = P + B_1$, where $P = b_1\bar{b}_1$ and B_1 is the skew-symmetric form of b_1. Also, review the reflection tensor, given earlier $M = E - 2P$ (see Sec. 2.3.6).

(b) Express the total force vector f in terms of f_1, R_{90}, and M.

(c) The force on fin # 1, with a deflection of 10 deg, was measured as $[\bar{f}_1]^F = [-2.2 \quad 0 \quad -11]\ N$ in the fin coordinate system, defined by 1^F parallel to 1^B, 2^F parallel to the fin shaft and positive out, with the shaft rotated 45 deg from the 1^B, 3^B plane. Calculate the total force $[f]^B$ in missile body coordinates.

4.6   Missile launch.   A missile M separates from and aircraft A. The attitude of the missile centerline m_1 relative to the aircraft centerline a_1 can be expressed in terms of the rotation tensor R^{MA} with time-dependent elements: $m_1 = R(t)^{MA}a_1$. From telemetry data the yaw and pitch angles ψ, θ of the missile relative to the aircraft are known. Derive the rotation tensor $[R^{MA}]^A$ in aircraft axes and calculate the instantaneous axis of rotation $[n]^A$ and the rotation angle ε. Is the rotation axis defined at the start of separation?

4.7   Manipulator arm.   Initially, a satellite is stowed in the space shuttle bay with its symmetry axis s_1 parallel to the vehicle base vector b_1. The manipulator arm of the space shuttle deploys the satellite through an upward rotation θ about the point X, followed by a rotation about point Y, that brings the satellite's s_1 axis parallel and in the direction of the body base vector b_2. Determine the rotation tensor $[R^{SB}]^B$ of the satellite frame S wrt the body frame B, in body coordinates, and the axis of rotation $[n]^B$ and the angle of rotation ψ. As a numerical example use $\theta = 45$ deg (*Solution:* $\psi = 98.5$ deg, $[\bar{n}]^B = [0.357\ \ 0.357\ \ 0.863]$).

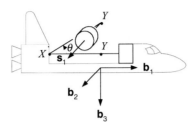

4.8   INS acceleration measurement.   Before the alignment of a strap-down INS, the accelerometer triad b_1, b_2, b_3, is misaligned wrt the local-level geographic triad l_1, l_2, l_3, north, east, and down. The alignment errors are azimuth $\varepsilon\psi = 1.0$ deg, elevation $\varepsilon\theta = 0.1$ deg, and roll $\varepsilon\phi = 0.05$ deg. Calculate the tilt vector $[\varepsilon r^{BE}]^L$ and the rotation tensor $[R^{BE}]^L$ of the body frame B wrt the Earth frame E, expressed in local-level axes $]^L$. In steady horizontal flight the aircraft is subjected to the normal load factor $[\bar{a}]^L = [0\ \ 0\ \ -1]$. What are the measurements of the three accelerometers?

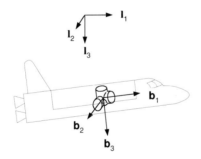

4.9   Captain's relative motion.   A boat B sails on rough water past the pier A. The linear velocity of its c.m. is v_B^A while its angular velocity is ω^{BA}. The captain C is moving aft in the boat at the velocity v_C^B. What is the velocity v_C^A of the captain relative to an observer A on the pier?

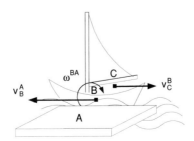

4.10   Air passenger's relative motion.   A passenger P moves forward in the cabin from point B at 1 m/s while the aircraft, cruising at constant speed $[\overline{v_B^E}]^B = [100 \ 0 \ 0]$ m/s, pitches up with $\theta(t) = \theta_0 + 0.001t$ rad.

(a) Show that the velocity of the passenger wrt the Earth frame E is $v_P^E = v_P^B + \Omega^{BE} s_{PB} + v_B^E$.

(b) Calculate the numerical values of $[v_P^E]^B$ and $[v_P^E]^L$ if, initially, the two coordinate systems $]^B$ and $]^L$ are aligned with each other. (*Solution:* $[v_P^E]^B = [101 \ \ 0 \ \ -0.001t]$ m/s.)

4.11   Intercept plane.   As a missile intercepts an aircraft, its fuse fires near the airframe. To simplify matters, we define an intercept plane that is normal to the differential velocity vector v_{MT}^E of the missile c.m. M wrt the target frame T and which contains the mass center T of the target. The miss distance of the missile is measured in this plane between T and M.

Derive the transformation matrix $[T]^{PT}$ of the intercept plane coordinates $]^P$ wrt the target aircraft coordinates $]^T$ using two methods.

(a) In the first method express the triad p_1, p_2, p_3 of the intercept plane in $]^T$ coordinates and thus construct $[T]^{PT}$. $[v_{MT}^E]^T$ and the target unit base vector $[t_3]^T$ are given in target coordinates. The intercept plane is oriented such that p_1 is parallel and in the direction of v_M^T and p_2 lies in the t_1, t_2 plane.

(b) The second method uses $[v_{MT}^E]^T$ to calculate the polar angles ψ_{PT}, θ_{PT} and derives $[T]^{PT}$ from the two consecutive transformations ψ_{PT} and θ_{PT}. Calculate the numerical values of $[T]^{PT}$ by both methods using $[v_{MT}^E]^T[-500 \ 100 \ 100]$ m/s.

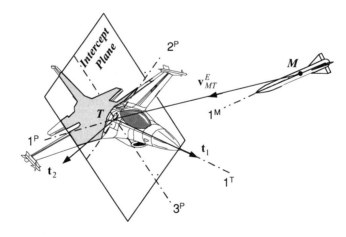

4.12   Line-of-sight rate kinematics.   A radar station R tracks a point target T. The line-of-sight (LOS) displacement vector s_{TR} is rotated from the radar's triad r_1, r_2, r_3 by the azimuth angle ψ and the elevation angle θ. We choose the inertial axes $]^I$ as the preferred coordinate system associated with the radar frame. For the LOS frame we define a coordinate system $]^O$ in the following way: The 1^O

direction is parallel and in the direction of s_{TR}; the 3° direction is normal to s_{TR}, lies in the plane subtended by r_3, s_{TR}, and for $0 < \theta < 90$ deg points toward r_3.

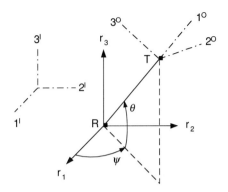

(a) Derive the transformation matrix $[T]^{OI}$ of the LOS axes with respect to the inertial axes. Sketch the orange peel diagram.

(b) Derive the equation for the LOS rates ω^{OR} that are based on the measured azimuth and elevation rates $\dot{\psi}$, $\dot{\theta}$ and express them in the inertial and LOS coordinates.

(c) Given the measured values $\psi = 40$ deg, $\theta = 30$ deg, and $\dot{\psi} = 10$ deg/s, $\dot{\theta} = 100$ deg/s, calculate the numerical values for $[\omega^{OR}]^I$ and $[\omega^{OR}]^O$ (Solution: $[\omega^{OI}]^I = [64.27\ -76.60\ 10]$, $[\overline{\omega^{OI}}]^O = [5.00\ -100\ 8.66]$ deg/s).

4.13   Missile acceleration in LOS coordinates. For the formulation of an advanced guidance law, the inertial acceleration of the missile $D^I v_M^I$ must be expressed in LOS coordinates $d/dt[v_M^I]^O$. The guidance law generates the acceleration command a_M^I, which through the autopilot results in the acceleration of the missile. For ideal autopilots $D^I v_M^I = a_M^I$. Provide the individual steps that lead to the desired relationship

$$\frac{d}{dt}\left[v_M^I\right]^O = -[\Omega^{OI}]\left[v_M^I\right]^O + [\bar{T}]^{MO}\left[a_M^I\right]^M$$

These are first-order differential equations in $[v_M^I]^O$ with the commanded acceleration $[a_M^I]^M$ in missile axes as inhomogeneous part. We have the following frames: $M = $ missile, $O = $ line of sight, and $I = $ inertial and the preferred coordinate systems $]^M$, $]^O$, $]^I$, respectively. Point M is the c.m. of the missile.

4.14   LOS rate. Flight simulators require the recording of angular LOS rates. The preferred components are 1) normal to the ground and 2) normal to the vertical plane containing the LOS. Given are the linear velocities of the missile and aircraft $[v_B^E]^L$, $[v_T^E]^L$, respectively, and the LOS displacement $[s_{BT}]^L$ in local-level coordinates (north, east, down). Derive the equations for the angular LOS rates ω^{OE} and their components ω_{l_3}, parallel to the vertical unit vector l_3 and ω_n parallel to the unit vector $\boldsymbol{n}$, which is normal to the vertical LOS plane.

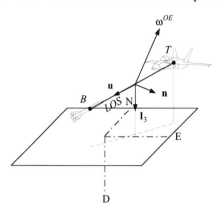

4.15   Rendezvous.   The frames are the following: inertial I, Earth E, space shuttle B, and satellite S. The points are the following: location of Earth tracking station E, satellite c.m. S, and shuttle c.m. B. The coordinate systems are the following: Earth $]^E$, geographic $]^G$, and shuttle body axes $]^B$.

A space shuttle is sent to orbit to service a satellite. Before it can use its terminal docking radar, it must maneuver into the vicinity of the satellite. The crew uses the $[s_{SB}]^B$ and $[v_S^B]^B$ displays in the cockpit to execute the approach. This position and velocity data are calculated by the onboard computer from $[s_{SE}]^G$, $[s_{BE}]^G$, $[v_S^E]^G$, and $[v_B^E]^G$ that are recorded by the Earth tracking station. The orientation $[R^{BI}]^B$ of the space shuttle wrt inertial space is determined by the onboard INS and the Earth's orientation $[R^{EI}]^E$ and transformation matrix $[T]^{EG}$ provided by almanac tables. Your task is to develop the software that computes the displays $[s_{SB}]^B$ and $[v_S^B]^B$.

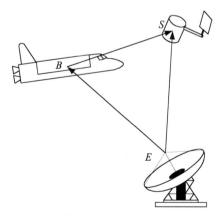

(a) Obtain the tensor relationship for s_{SB}. For v_S^B derive the result

$$v_S^B = v_S^E - v_B^E + (D^B R^{IB} \overline{R^{IB}} - D^E R^{IE} \overline{R^{IE}})(s_{SE} - s_{BE})$$

(b) Now derive the matrix equations for $[s_{SB}]^B$ and $[v_S^B]^B$, which are to be programmed for the onboard computer. As a check, I give you the velocity equation

$$[v_S^B]^B = [\overline{R^{BI}}]^B[R^{EI}]^E[T]^{EG}([v_S^E]^G - [v_B^E]^G) + ([\overline{\dot{R}^{BI}}]^B[R^{BI}]^B$$

$$- [\overline{R^{BI}}]^B[R^{EI}]^E[\overline{\dot{R}^{EI}}]^E[R^{BI}]^B)[[\overline{R^{BI}}]^B[R^{EI}]^E[T]^{EG}([s_{SE}]^G - [s_{BE}]^G)]$$

4.16 Rotation tensor from rotation quaternion. Derive the rotation tensors $[R_1]$ about the axis $[n_1] = [1\ 0\ 0]$, given the rotation quaternion

$$\{q\} = \left\{ \frac{\cos(\varepsilon/2)}{\sin(\varepsilon/2)[n]} \right\}$$

Do the same for $[R_2]$ and $[R_3]$ with $[n_2] = [0\ 1\ 0]$ and $[n_3] = [0\ 0\ 1]$, respectively.

4.17 Rotation vector from rotation quaternion. Given is the quaternion $\{\bar{q}\} = \{q_0\ q_1\ q_2\ q_3\}$. Provide the angle of rotation ε and the components of the unit vector of rotation $[\bar{n}] = [n_1\ n_2\ n_3]$ as a function of the quaternion elements $q_0, q_1, q_2,$ and q_3 only. Verify that the norm of $[n]$ is one.

4.18 Seeker gimbal kinematics—project. When a missile comes within the acquisition range of the target, its antenna axis must be pointed at the target based on the onboard INS navigation information. Derive the equations for the seeker gimbal's pitch and yaw angles, using the definitions displayed in the figure (the gimbal geometry is explained in more detail in Sec. 9.2.5).The INS provides the direction cosine matrix $[T]^{BL}$ and the LOS vector in local-level coordinates $[s_{TB}]^L$.

The definitions for the figure are as follows: body axes $1^B, 2^B, 3^B$, outer gimbal axes $1^P, 2^P, 3^P$, inner gimbal axes $1^S, 2^S, 3^S$, local-level axes $]^L$, yaw gimbal angle ψ_{SB}, and pitch gimbal angle θ_{SB}.

(a) Derive the equations for the seeker pointing angles θ_{SB} and ψ_{SB}, which are to be programmed for the onboard computer.

(b) Given the Euler angles $\psi_{BL} = 10$ deg, $\theta_{BL} = 3$ deg, $\phi_{BL} = 5$ deg, and the target LOS vector $\overline{[s_{TB}]}^L = [5\ 1\ -3]$ km, calculate by hand the seeker pointing angles θ_{SB} and ψ_{SB} in degrees.

(c) Program the equations for the seeker pointing angles in a subroutine INS POINT with the input arguments $[T]^{BL}$ and $[s_{TB}]^L$ and the output arguments θ_{SB} and ψ_{SB}.

(d) Check your subroutine by enclosing it into a program and using the input values of (b).

(e) Use your subroutine for a second example: $\psi_{BL} = -30$ deg, $\theta_{BL} = -10$ deg, $\phi_{BL} = -20$ deg and $\overline{[s_{TB}]}^L = [2\ -1\ -4]$ km and give the seeker pointing angles θ_{SB} and ψ_{SB} in degrees.

(f) Provide the source code listing.

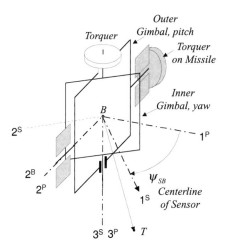

4.19   Target model—project.   Missile simulations require target models of aircraft, cruise missiles, and even other missiles. These models do not have to be full six-DoF representations, but should exhibit the center of mass accelerations and the approximate attitude of the vehicle airframe. Your task is to develop such a target model and check it out in a simple three-DoF simulation.

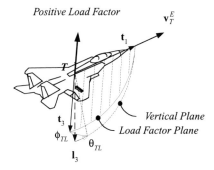

You model the target by its frame T and its center of mass, point T. Its *orientation* is described by the transformation matrix $[T]^{TL}$ of the target coordinate axes wrt the local-level axes. You can introduce a simplification by aligning the first base vector of the target t_1 with the velocity vector v_T^E of the target T wrt to the Earth frame E. The third base vector t_3 points in the opposite direction of the positive load factor. The two base vectors t_1 and t_3 span the *load factor plane*, while the local vertical l_3 together with t_1 define the *vertical* plane. Both planes are separated by the bank angle ϕ_{TL}. You calculate the transformation matrix $[T]^{TL}$ from the three Euler angle transformations: $\phi_{TL} \leftarrow \theta_{TL} \leftarrow \psi_{TL}$ (heading, flight-path angle, and bank angle). The *location* of the target is described by the displacement vector $[s_{TE}]^L$ of the c.m. of the target T from an arbitrary reference point on the Earth E expressed in local coordinates.

Three inputs generate the *target maneuvers*: the combined drag and thrust acceleration a_{X_C}, the load factor a_{N_C}, and the bank angle ϕ_{TL}. To simulate the time lag of the target, these three input commands are delayed by first-order transfer functions with their respective time constants T_{AX}, T_{AN}, T_ϕ.

(a) Derive the transformation matrix $[T]^{TL}$. Write down the nine first-order differential equations that govern the target dynamics (three input states, velocity vector $[v_T^E]^L$, and location vector $[s_{TE}]^L$). The target maneuvers in the maneuver plane, which differs from the load factor plane by the influence of the gravity vector. Derive the equations of the maneuver bank angle ϕ_{ML}, which is the angle between the projections of the total acceleration vector and I_3 on the plane normal to t_1 (between maneuver and vertical planes).

(b) Code the equations for the Module G1 of CADAC or in another simulation environment. Use worksheets.

(c) Build your own test runs for five cases: 1) straight and level flight, 2) planar climb, 3) horizontal 45-deg bank, 4) dive escape at 135-deg bank angle, and 5) launch of ballistic missile. You may also use staging (CADAC) to combine maneuvers.

(d) Document your results. Summarize the equations and add your worksheets and the code of Module G1 or your subroutine. Provide graphs of the five test cases.

5
Translational Dynamics

We studied first geometry, then kinematics, and now have amassed enough equipage to investigate dynamics, the effect of force on mass. In this chapter we concentrate on the motions of the center of mass (c.m.), assuming that all mass of a vehicle is localized at that point. The c.m., subjected to forces, describes trajectories in space, as recorded by an observer. Later in Chapter 6 we will model the vehicle as a body and watch it rotate under externally applied moments. Both combined, translation and attitude, convey the full six degrees of freedom of vehicle motions.

The physical law that governs translational motions is Newton's second law. After 300 years it still has maintained its preeminence in trajectory simulations. Most useful for engineering applications is the formulation: the time rate of change of linear momentum equals the applied external force. Therefore, before we treat Newton's law in detail we discuss the linear momentum of single and clustered bodies. Then, after deriving the translational equations of motions from Newton's law we discuss two transformations resulting from changes of reference frame and reference point and conclude the chapter with examples motivated by applications.

5.1 Linear Momentum

Have you experienced your linear momentum lately? Probably not if it remained constant. As an earthling of 80 kg mass, you were born with a linear momentum wrt the ecliptic of about 2.66×10^6 kg m/s and because the sun is part of the Milky Way galaxy, you can claim another 4.4×10^{10} kg m/s of linear momentum. That much should make anybody dizzy, but nobody takes notice. If, however, you travel in a car at 60 miles/h and hit a wall, your 2100 kg m/s change of linear momentum will kill you. It is the time rate of change of linear momentum that creates the force; Newton just reversed the order of cause and effect.

Before we apply Newton's law, we need to learn more about the linear momentum. First we define the linear momentum of a particle, then generalize it for a collection of particles (body) and single out its c.m. For clusters of bodies, we learn how to calculate their linear momentum, again with special emphasis on the common c.m.

Definition: The linear momentum p_i^R of a particle i relative to the reference frame R is defined by the rotational time derivative wrt to frame R of the displacement vector s_{iR} multiplied by its mass m_i (see Fig. 5.1).

$$p_i^R = m_i D^R s_{iR} = m_i v_i^R \tag{5.1}$$

Because the rotational derivative of the displacement vector yields the velocity

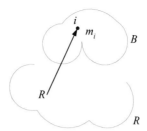

Fig. 5.1   Linear momentum of a particle.

$v_i^R = D^R s_{iR}$, we can also state

$$\text{linear momentum} = \text{mass} \times \text{linear velocity}$$

The linear velocity vector imparts the direction to the linear momentum and also its independence from a specific reference point. The nomenclature reflects that characteristic: subscript i is the particle point and superscript R an arbitrary reference frame.

Definition: The linear momentum of a collection of particles B relative to the reference frame R is the sum over all linear momenta of each particle

$$\sum_i p_i^R = \sum_i m_i D^R s_{iR} = \sum_i m_i v_i^R \tag{5.2}$$

This definition is awkward because it requires knowledge of every particle's velocity. We can simplify matters greatly, by defining the c.m. B of the collection of particles B with total mass m^B and specifying its velocity v_B^R. Then, the linear momentum p_B^R of c.m. B wrt the reference frame R is given by

$$p_B^R = m^B D^R s_{BR} = m^B v_B^R \tag{5.3}$$

Only the total mass and the c.m. of body B contribute to the linear momentum. The shape and attitude of the body are irrelevant. In effect, Eq. (5.3), in the light of Eq. (5.1), can be interpreted as the linear momentum of a particle with mass m^B and linear velocity of point B. One only should add that B has to be the c.m. The simplicity of the notation emphasizes this fact. Let us prove Eq. (5.3) by Eq. (5.2).

Proof: From Fig. 5.2 we deduce the displacement vector triangle

$$s_{iR} = s_{iB} + s_{BR}$$

Take the rotational derivative wrt the reference frame

$$D^R s_{iR} = D^R s_{iB} + D^R s_{BR}$$

Substitute it into the second term of Eq. (5.2), and, because the mass of each particle is constant, it can be brought inside the rotational derivative

$$\sum_i m_i D^R s_{iR} = \sum_i m_i D^R s_{iB} + \sum_i m_i D^R s_{BR} = \left[D^R \left(\sum_i m_i s_{iB} \right) \right] + m^B v_B^R$$

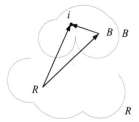

Fig. 5.2   Linear momentum of body B.

The next to the last term is zero because, if B is the common c.m., then $\sum_i m_i s_{iB} = \mathbf{0}$, and the rotational derivative of a null vector is zero. Therefore,

$$\sum_i m_i D^R s_{iR} = m^B v_B^R$$

The collection of particles, forming body B, does not have to be mutually fixed. As long as their c.m. and total mass are known, the linear momentum can be calculated.

We can extend the computation to a collection of bodies B_k, $k = 1, 2, 3 \ldots$ as shown in Fig. 5.3. Because bodies can be regarded as particles as long as their individual c.m. are used as representative points, we can use Eq. (5.2) to sum their individual contributions and calculate the total linear momentum:

$$p_{\Sigma B_k}^R = \sum_k p_{B_k}^R = \sum_k m^{B_k} D^R s_{B_k R} \tag{5.4}$$

Introducing the vector triangle for each body, with C the common c.m., and taking the rotational time derivative wrt the reference frame R

$$D^R s_{B_k R} = D^R s_{B_k C} + D^R s_{CR}$$

yields

$$p_{\Sigma B_k}^R = \sum_k m^{B_k} D^R s_{B_k C} + \sum_k m^{B_k} D^R s_{CR}$$

Because C is the common c.m., the first term vanishes as long as the mass of each

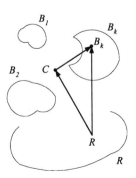

Fig. 5.3   Multiple bodies.

body remains unchanged:

$$\sum_k m^{B_k} D^R s_{B_k C} = D^R \left(\sum_k m^{B_k} s_{B_k C} \right) = D^R 0 = 0$$

Therefore, the linear momentum wrt a reference frame R of a collection of bodies ΣB_k, $k = 1, 2, 3, \ldots$, not necessarily rigid but with constant mass, is calculated from the total mass $m^{\Sigma B_k} = \sum_k m^{B_k}$ and the linear velocity v_C^R of the common center of mass wrt the reference frame

$$p_{\Sigma B_k}^R = \sum_k m^{B_k} D^R s_{CR} = m^{\Sigma B_k} v_C^R \tag{5.5}$$

We have two methods to calculate the linear momentum of clustered bodies. If we know the individual values, we just add them together [see Eq. (5.4)], or we use Eq. (5.5) if the cluster properties are known.

Example 5.1   Linear Momentum of Reentry Bodies

Problem.   A ballistic missile with 3000 kg of payload descends at 12,000 m/s and releases three nuclear warheads at 800 kg each and deploys six 100-kg decoys. Use two ways to calculate the linear momentum of the cluster of reentry bodies.

Solution.   Using individual bodies, the total linear momentum is

$$p_{\Sigma B_k}^I = \sum_k p_{B_k}^I = \sum_k m^{B_k} v_{B_k}^I = (3 \times 800 + 6 \times 100) \times 12,000 = 36 \times 10^6 \text{ kg m/s}$$

or using the cluster

$$p_{\Sigma B_k}^R = m^{\Sigma B_k} v_C^R = 3,000 \times 12,000 = 36 \times 10^6 \text{ kg m/s}$$

both methods lead to the same result.

The linear momentum is an essential ingredient of Newton's second law. As we add forces to the thought process, the kinematic world view becomes dynamic, and we can study the interactions of force, velocity, and mass in time.

5.2   Newtonian Dynamics

Sir Isaac Newton published in 1687 in the *Philosophiae Naturalis Principia Mathematica* three laws.[1] In plain English they postulate the following:

1) Every body continues in its state of rest or of uniform motion in a straight line unless it is compelled to change that state by forces impressed upon it.

2) The rate of change of linear momentum equals the impressed force and is in the direction in which the force acts.

3) To every action there is always opposed an equal reaction.

Newton used the word *motion* instead of *linear momentum* to define the second law, but the meaning is the same. Like any researcher he used the scientific method to arrive at this formulation. Observations pointed to hypothesis, and testing consolidated the theory. After more than 300 years of validation, we certainly are justified to call them natural laws.

However, even laws are constrained by assumptions. Newton's laws are valid in classical physics, where mass does not exceed that of natural substances and velocities are well below the speed of light. Any transgression requires relativistic expansion that Einstein has provided. Furthermore, as particles assume subatomic size, Heisenberg and Schroedinger[2] contributed the framework of quantum physics to explain emerging phenomena. All of these modern extensions however point back to Newton's classical dynamics, as extreme conditions are reduced to the level of our human experience. For any new dynamic theory to be acceptable, it must contain Newtonian dynamics as a limiting case.

The first law is validated by our experience. We do not notice our own linear momentum unless a wall stops us. The wall exerts the force that kills us (second law). Newton's third law is important in mechanics because it assures us that internal forces cancel in a collection of particles. However, it is the lesser of the three laws because it fails in classical electrodynamics. For modeling of aerospace vehicle dynamics, the second law is of greatest interest. We therefore refer to it mostly just as Newton's law.

Interestingly, Newton did not specify a frame of reference in formulating his second law. Others attempted to affix what was called the "luminiferous ether" to his law, until Michelson and Morley[2] in 1887 disproved the concept. Thus, wisely, Newton left it to the application to pick the proper reference frame, which we call today the *inertial* reference frame. From his first law we know that any nonaccelerating frame qualifies equally well; but do they exist? In Chapter 3, I suggested using the ecliptic of our sun as inertial reference frame, but we know that our solar system is located in the spiral arms of the Milky Way and therefore accelerating. Other theories suggest that all galaxies are fleeting with increasing speed. Where is the inertial frame? It probably does not exist in absolute terms.

Applications determine the inertial frame. Interplanetary travel requires the heliocentric frame; Earth satellite trajectories use the ecliptic oriented and Earth-center fixed frame, which we call plainly the *inertial frame*; and Earth-bound, low-speed flights can use the Earth frame. Whatever the accuracy requirement of your simulation is will determine the choice of inertial reference frame.

5.2.1 Newton's Second Law

Newton's second law is the most important tool for modeling aerospace vehicle dynamics. It governs the motions of the c.m. of the vehicle subjected to external forces. We apply it first to a single particle and then to a collection of particles and eventually arrive at a form that is most suitable for our modeling tasks.

For a particle i Newton's second law postulates

$$D^I p_i^I = f_i \qquad (5.6)$$

The equation states that the time rate of change wrt the inertial frame I of the linear momentum of particle i wrt the inertial frame equals the force acting on the particle. Introduce the definition of linear momentum from Eq. (5.1),

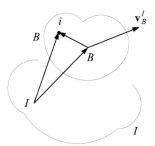

Fig. 5.4   Particle and c.m.

$p_i^I = m_i \, D^I s_{iI}$, into the left side of Eq. (5.6) so that

$$D^I p_i^I = D^I (m_i D^I s_{iI}) = m_i D^I D^I s_{iI} + D^I m_i D^I s_{iI}$$

where the last term is zero because the mass of a particle is time invariant. Now we sum over all of the particles of body B. The internal forces will cancel because of Newton's third law, but the external forces remain

$$\sum_i m_i D^I D^I s_{iI} = \sum_i f_i \tag{5.7}$$

Let us introduce the c.m. B of the collection of particles (see Fig. 5.4)

$$s_{iI} = s_{iB} + s_{BI}$$

and obtain

$$\sum_i m_i D^I D^I s_{iB} + \sum_i m_i D^I D^I s_{BI} = \sum_i f_i$$

With the mass of each particle constant, we can place it inside the rotational derivative and move the summation sign. Because B is the c.m. of all particles, the first term vanishes:

$$\sum_i m_i D^I D^I s_{iB} = \sum_i D^I D^I (m_i s_{iB}) = D^I D^I \underbrace{\sum_i m_i s_{iB}}_{=0} = 0$$

Abbreviating $\sum_i m_i = m^B$ and summing over all forces $\sum_i f_i = f$, we arrive at Newton's second law for body B relative to the inertial frame I:

$$m^B D^I D^I s_{BI} = f \tag{5.8}$$

Let us introduce the velocity $v_B^I = D^I s_{BI}$ of the c.m. B wrt the inertial frame I. Newton's law reads then as follows: The mass m^B of body B times the inertial acceleration $a_B^I = D^I v_B^I$ of its c.m. B equals the resultant external force f.

$$m^B D^I v_B^I = f \tag{5.9}$$

From this derivation we conclude that the c.m. B of a system of particles m_i moves like a single particle B whose mass is the total mass m^B, subject to the force f. Notice that we did not invoke a rigid body assumption, although most of our applications will do so.

We have formulated Newton's law in an invariant tensor form, which can be expressed in any allowable coordinate system. For instance, in coordinates $]^I$, associated with the inertial frame I, we get the ordinary time derivative

$$m^B[D^I v_B^I]^I = [f]^I \Rightarrow m^B \left[\frac{dv_B^I}{dt}\right]^I = [f]^I$$

For a noninertial coordinate system, say $]^B$, associated with the body frame B, we transform the rotational time derivative to the B frame via the Euler transformation to get the ordinary time derivative in the $]^B$ coordinates:

$$m^B[D^I v_B^I]^B = [f]^B$$

$$m^B[D^B v_B^I]^B + m^B[\Omega^{BI}]^B[v_B^I]^B = [f]^B$$

$$m^B[dv_B^I/dt]^B + m^B[\Omega^{BI}]^B[v_B^I]^B = [f]^B$$

The acceleration $[dv_B^I/dt]^B$ is the inertial speed, coordinated in body axes, with its components subjected to the ordinary time derivative. The additional term $[\Omega^{BI}]^B[v_B^I]^B$ is called the *tangent* acceleration.

As we developed our preferred formulation, Eq. (5.9), we assumed that the body holds onto all of its particles; in other words, the mass is time invariant. If particles are ejected, as for instance by a rocket motor, the linear momentum is changed, resulting in a thrust force. Traditionally, however, that force is moved to the right side of Newton's law and considered an external force (see Example 5.6). On the left side the mass has become a function of time.

Example 5.2   Trajectory Equations

The translational equations of an aerospace vehicle are directly derived from Newton's law, Eq. (5.9). With m^B the mass of the vehicle, v_B^I the linear velocity of the c.m. relative to the inertial frame I, and the external forces consisting of aerodynamic force f_a, propulsive force f_p, and gravitational force $m^B g$, the translational equations are

$$m^B D^I v_B^I = f_a + f_p + m^B g$$

This tensor equation is valid in any allowable coordinate system. The simplest implementation is in the inertial coordinate system $]^I$

$$m^B[D^I v_B^I]^I = [f_a]^I + [f_p]^I + m^B[g]^I$$

However, the aerodynamic and propulsive forces are most likely given in body coordinates $]^B$ and the gravitational acceleration in geographic coordinates $]^G$. With the two transformation matrices $[T]^{BI}$ and $[T]^{GI}$ we can formulate the differential equations for computer programming

$$m^B \left[\frac{dv_B^I}{dt}\right]^I = [\bar{T}]^{BI}[f_{a,p}]^B + m^B[\bar{T}]^{GI}[g]^G \qquad (5.10)$$

where the aerodynamic and propulsive forces have been lumped together. To calculate the trajectory, another integration is necessary for obtaining the displacement vector s_{BI} of the vehicle c.m. relative to an inertial reference point I

$$D^I s_{BI} = v_B^I$$

In inertial coordinates

$$\left[\frac{ds_{BI}}{dt}\right]^I = [v_B^I]^I \tag{5.11}$$

Equations (5.10) and (5.11) are the six differential equations that must be solved, starting from initial conditions, to generate the trajectory traces.

Example 5.3   Translating Inertial Frames

Problem.   Newton's first law states the experiential fact that, unless bodies are subjected to forces, they continue in a state of rest or uniform motion. From Newton's second law, $m^B D^I v_B^I = 0$, indeed we conclude that the inertial velocity does not change. This statement reflects also on the characteristic of the inertial frame. Newton's second law holds equally in an inertial frame at rest or at uniform motion in a straight line. Let us prove that fact.

Solution.   Figure 5.5 depicts the body B with c.m. B and two inertial frames I_1 and I_2 with two arbitrary reference points I_1 and I_2, respectively. The presupposition of both inertial frames not accelerating relative to each other implies that the mutual linear acceleration and angular velocity are zero:

$$D^{I_1} D^{I_1} s_{I_1 I_2} = 0, \quad \Omega^{I_2 I_1} = 0 \tag{5.12}$$

Introduce the vector triangle of Fig. 5.5

$$s_{BI_1} = s_{BI_2} + s_{I_2 I_1}$$

into the left side of Newton's second law, Eq. (5.8),

$$m^B D^{I_1} D^{I_1} s_{BI_1} = f \tag{5.13}$$

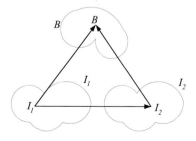

Fig. 5.5   Translating frames.

and obtain

$$m^B D^{I_1} D^{I_1} s_{BI_2} + m^B D^{I_1} D^{I_1} s_{I_2 I_1} = f$$

The second term on the left-hand side vanishes because of the first premise of Eq. (5.12). The first term is modified twice by Euler's transformation to the I_2 frame, and the second premise of Eq. (5.12) is used

$$m^B D^{I_1} D^{I_1} s_{BI_2} = m^B D^{I_1} \left(D^{I_2} s_{BI_2} + \Omega^{I_2 I_1} s_{BI_2} \right) = m^B D^{I_1} D^{I_2} s_{BI_2}$$

$$= m^B D^{I_2} \left(D^{I_2} s_{BI_2} \right) + \Omega^{I_2 I_1} \left(D^{I_2} s_{BI_2} \right) = m^B D^{I_2} D^{I_2} s_{BI_2}$$

Summarizing, from Eq. (5.13), with the assumptions of Eq. (5.12), we derived the alternate form

$$m^B D^{I_2} D^{I_2} s_{BI_2} = f$$

Both inertial reference frames lead to the same dynamic equation. Because either frame I_1 or I_2 is suitable, all inertial frames, not mutually accelerating, lead to the same formulation of Newton's second law.

Example 5.4   Two-Body Problem

Problem.   Given are two particles with masses m_1 and m_2, located at B_1 and B_2 and subject to the external forces f_1, f_2 (see Fig. 5.6).

1) Determine the displacement vector s_{BI} of the common c.m. B wrt an inertial reference point I. Express the location of the two masses $s_{B_1 I}$ and $s_{B_2 I}$ in terms of s_{BI} and $s_{B_1 B_2}$.

2) Apply Newton's second law to both masses individually and then show that $(m_1 + m_2) D^I D^I s_{BI} = f_1 + f_2$ and explain what the meaning of this reduction is.

3) If the point masses are not subject to external forces, then $f_1 + f_2 = 0$. What motion does the common center of mass B execute?

4) Let m_1 be the mass of a satellite and m_2 the Earth's mass. Assuming $m_2 \gg m_1$ and neglecting the acceleration of the common mass center, give the simplified equation of motion of a satellite orbiting the Earth. Summarize the assumption for this simplified single body problem.

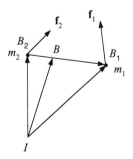

Fig. 5.6   Two particles.

Solution.   1) From the vector triangle in Fig. 5.6, we get the first relationship

$$s_{B_1 B_2} = s_{B_1 I} - s_{B_2 I}$$

Because B is the c.m. of both particles, the second condition is

$$(m_1 + m_2)s_{BI} = m_1 s_{B_1 I} + m_2 s_{B_2 I}$$

Solve both equations first for $s_{B_1 I}$ (multiply the first by m_2 and add both equations),

$$s_{B_1 I} = s_{BI} + \frac{m_2}{m_1 + m_2} s_{B_1 B_2} \tag{5.14}$$

and then similarly for $s_{B_2 I}$,

$$s_{B_2 I} = s_{BI} - \frac{m_1}{m_1 + m_2} s_{B_1 B_2} \tag{5.15}$$

2) Apply Newton's law to particles B_1 and B_2 and substitute Eqs. (5.14) and (5.15), respectively,

$$\begin{aligned}
m_1 D^I D^I s_{B_1 I} &= m_1 D^I D^I s_{BI} + \frac{m_1 m_2}{m_1 + m_2} D^I D^I s_{B_1 B_2} = f_1 \\
m_2 D^I D^I s_{B_2 I} &= m_2 D^I D^I s_{BI} - \frac{m_1 m_2}{m_1 + m_2} D^I D^I s_{B_1 B_2} = f_2
\end{aligned} \tag{5.16}$$

then add both equations

$$m D^I D^I s_{BI} = f, \quad m = m_1 + m_2, \quad f = f_1 + f_2 \tag{5.17}$$

If B is the common c.m., then Newton's law is applied as if all mass $m = m_1 + m_2$ were concentrated at the c.m. and acted upon by the resultant force $f = f_1 + f_2$.

3) Without external forces Eq. (5.17) becomes

$$m a_B^I = m D^I D^I s_{BI} = 0$$

i.e., the acceleration of the common c.m. is zero.

4) When $m_2 \gg m_1$, from Eq. (5.16) we obtain

$$m_1 D^I D^I s_{BI} + m_1 D^I D^I s_{B_1 B_2} = f_1$$

and if we assume that the common c.m. B is not accelerated,

$$m_1 D^I D^I s_{B_1 B_2} = f_1$$

The reduced single-body problem of a satellite B_1 orbiting Earth B_2 can be solved as if the common c.m. is centered in Earth's c.m. B_2. The assumptions are that 1) $m_2 \gg m_1$ and 2) the common c.m. is not under acceleration.

Example 5.5   Pulse Thruster

Problem.   The thruster of a satellite increases with a single pulse of the satellite speed from v_0 to v_f. The total particle count of the satellite is s, and the thruster ejects f number of particles at an exhaust velocity of v_e. Assuming that the pulse is instantaneous, what is the increase of the satellite speed $\Delta v = v_f - v_0$?

Solution. Let us start with Eq. (5.6) and sum over all particles and recognize that no external forces are applied:

$$\sum_{i=1}^{s+f} D^I p_i^I = 0$$

The total linear momentum is therefore conserved:

$$\sum_{i=1}^{s+f} p_i^I = \text{const}$$

Divide the particles up into satellite and ejected mass:

$$\sum_{i=1}^{s+f} p_i^I = \sum_{i=1}^{s} p_i^I + \sum_{i=s+1}^{s+f} p_i^I = \text{const} \tag{5.18}$$

We reduce the problem to one dimension in inertial coordinates, label the satellite mass

$$m^S = \sum_{i=1}^{s} m_i$$

and the mass of the ejected fuel

$$m^F = \sum_{i=s+1}^{s+f} m_i$$

Before pulse firing, the linear momentum is $(m^S + m^F)v_0$. The pulse is ejected in the opposite direction at $-v_e$ wrt the satellite and $(-v_e + v_0)$ wrt to the inertial frame. Afterward the satellite's linear momentum is $m^S v_f$ and that of the fuel $m^F(v_0 - v_e)$. Using Eq. (5.18) in one dimension delivers

$$(m^S + m^F)v_0 = m^S v_f + m^F(v_0 - v_e) = \text{const} \tag{5.19}$$

Solve for v_f

$$v_f = \frac{m^S v_0 + m^F v_e}{m^S}$$

and the velocity increase is

$$\Delta v = \frac{m^S v_0 + m^F v_e}{m^S} - v_0$$

The higher the exhaust speed v_e or the fuel mass m^F, then the greater the increase in satellite speed. What happens to the c.m. of the total particle count $s + f$?

Example 5.6   Rocket Propulsion

A rocket motor ejects fuel particles continuously. If we regard Eq. (5.19) per unit time dt and because satellite mass and exhaust velocity are constant, the second part of the equation becomes

$$m^S \frac{v_f}{dt} + \frac{m^F}{dt}(v_0 - v_e) = 0$$

Before we can write the thrust equation, we have to address a subtlety in sign exchange in the last term. In Eq. (5.19) the fact that the linear momentum of the exhaust is opposite to that of the satellite was expressed by the negative sign of v_e. Now, with the fuel loss derivative being negative, the exhaust velocity should be positive. Therefore, we redefine the exhaust velocity $c = -(v_0 - v_e)$, call the term thrust, and move it to the right side. (We do not have to distinguish any longer between satellite and fuel mass, and v_f becomes the satellite speed v.)

$$m \frac{dv}{dt} = -c \frac{dm}{dt} \tag{5.20}$$

This is Oberth's famous rocket equation, which can be solved by separation of variables:

$$\int dv = -c \int \frac{dm}{m}$$

Solving the integrals with v_0 and m_0 as the initial values, the increase in speed Δv is

$$\Delta v = v - v_0 = c \, \ln \frac{m_0}{m} \tag{5.21}$$

The rocket's burnout velocity v increases with increasing mass fraction m_0/m and exhaust velocity. In engineering applications fuel flow is usually taken positive and the rocket thrust calculated from Eq. (5.20):

$$F = \dot{m} c \tag{5.22}$$

Thus, I have demonstrated how the time rate of change of momentum of rocket propellant produces thrust. This force is moved to the right side of Newton's equation and portrayed as an external force.

Newton's second law suffices to model the trajectory of an aerospace vehicle. Deceptively simple to write down in inertial coordinates, it has many variants that become important for applications. We already encountered the formulation in body axis, which gives rise to the tangential acceleration term. Other variations consider noninertial reference frames and points that are displaced from the c.m. We consider such transformations next.

5.3 Transformations

Observing in the night sky a satellite still illuminated by the sun and an airplane flashing its strobe light, one may get the impression that both stay aloft by the same forces. However, we know better. Aerodynamic forces carry the airplane, but what holds up the satellite? You would answer, "the centrifugal force of course!" What is that centrifugal force? Is it a surface force, like aerodynamic lift, or a volume force, like gravity? Is it a force that should be included at the right-hand side of Newton's law? None of the above. It is all a matter of reference frame. Because you are not standing on an inertial reference frame, an apparent force, the centrifugal force, keeps the satellite from falling at your feet. However, if you were sitting on the ecliptic, you would marvel how the Earth's gravitational pull prevents the satellite from escaping the Earth's orbit.

Both observations are equally valid. So far we have taken the inertial perspective. Now I will derive the translational equations of motion for noninertial reference

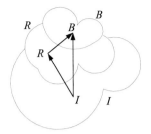

Fig. 5.7   Reference frame R.

frames. Besides the centrifugal force, we shall also encounter the Coriolis force, which gives this transformation its name.

Another situation arises when the c.m. is not the preferred reference point. Envision a large symmetrical space station with antennas on one side as appendages. It is sometimes advantageous to us the c.m. of the space station as reference point for the dynamic equations, rather than the common c.m. This point transformation is called the Grubin transformation.

5.3.1 Coriolis Transformation

Unequivocally, Newton's law must be referred to an inertial frame of reference I. Starting with such a frame however, we can use Euler's transformation of frames and shift the rotational derivatives over to another, noninertial reference frame R. The additional terms are moved to the right side as apparent forces to join the actual forces. Our goal is to write Newton's second law just like Eq. (5.9), but replace I by the noninertial reference frame R and include the additional terms on the right-hand side as corrections

$$m^B D^R v_B^R = f + \text{correction terms}$$

We begin with Newton's law in the form of Eq. (5.8) and introduce the vector triangle of displacement vectors shown in Fig. 5.7. B is the c.m. of body B, whereas the reference points R and I are any point of their respective frames

$$s_{BI} = s_{BR} + s_{RI}$$

Substituted into Eq. (5.8)

$$m^B D^I D^I s_{BR} + m^B D^I D^I s_{RI} = f \tag{5.23}$$

The second term represents the inertial acceleration of the reference frame and needs no further modification. However, both rotational derivatives in the first term must be shifted to the reference frame R. Let us work on this acceleration term alone:

$$D^I D^I s_{BR} = D^I \left(D^R s_{BR} + \Omega^{RI} s_{BR} \right)$$

$$= D^R \left(D^R s_{BR} + \Omega^{RI} s_{BR} \right) + \Omega^{RI} \left(D^R s_{BR} + \Omega^{RI} s_{BR} \right)$$

$$= D^R D^R s_{BR} + D^R \left(\Omega^{RI} s_{BR} \right) + \Omega^{RI} D^R s_{BR} + \Omega^{RI} \Omega^{RI} s_{BR}$$

$$= D^R D^R s_{BR} + 2\Omega^{RI} D^R s_{BR} + \Omega^{RI} \Omega^{RI} s_{BR} + \left(D^R \Omega^{RI} \right) s_{BR}$$

Substituting the definition of the relative velocity $v_B^R = D^R s_{BR}$ into Eq. (5.23) and moving all terms except the relative acceleration to the right yields the Coriolis form of Newton's second law:

$$m^B D^R v_B^R = f - m^B \begin{cases} 2\Omega^{RI} v_B^R & \text{Coriolis acceleration} \\ +\Omega^{RI}\Omega^{RI} s_{BR} & \text{centrifugal acceleration} \\ +(D^R \Omega^{RI}) s_{BR} & \text{angular acceleration} \\ +D^I D^I s_{RI} & \text{linear acceleration} \end{cases} \quad (5.24)$$

If the observer stands on a noninertial frame, he can apply Newton's law as long as he appends the correction terms. There are four additional terms. The first three involve the body, and the last one relates only to the reference frame. The Coriolis acceleration acts normal to the relative velocity v_B^R and the centrifugal acceleration outward. The angular and linear acceleration terms have no special name and appear only if the reference frame is accelerating.

Example 5.7 Earth as Reference Frame

Earth E is the most important noninertial reference frame for orbital trajectories. It has two characteristics that simplify the Coriolis transformation. Both the angular acceleration is zero, $D^I \Omega^{EI} = 0$, and the linear acceleration of Earth's center E vanishes, $D^I D^I s_{EI} = 0$.

Thus, the simplified Coriolis form of Newton's law emerges from Eq. (5.24):

$$m^B D^E v_B^E = f - m^B \left(2\Omega^{EI} v_B^E + \Omega^{EI}\Omega^{EI} s_{BE} \right) \quad (5.25)$$

with only the Coriolis and centrifugal forces to be included as apparent forces; s_{BE} is the displacement vector of the vehicle c.m. wrt the center of Earth, v_B^E is the vehicle velocity wrt Earth (geographic velocity), and Ω^{EI} Earth's angular velocity.

If you are the passenger in a balloon that hovers over a spot on Earth, you are only subject to the apparent centrifugal force. But when the balloon starts to move with v_B^E, the Coriolis force kicks in. The faster the geographic speed, the greater the force, except if you fly north or south from the equator, then the cross product $\Omega^{EI} v_B^E$ vanishes.

The Coriolis force is responsible for the counterclockwise movement of the air in a hurricane on the northern hemisphere. Newton's law governs the motions of the air molecules. As the atmospheric pressure drops, the depression draws in the air particles. Those south of the depression are moving north at velocity v_B^E and are deflected by the Coriolis force $m^B 2\Omega^{EI} v_B^E$ to the east. The northern air mass veers to the west as it is pulled south. These flow distortions set up the counterclockwise circulation of a hurricane.

5.3.2 Grubin Transformation

Now imagine that you are in the captain's chair of the fictional starship *Enterprise*. The ship's c.m. B is far behind your location B_r. Before you execute a maneuver, you want Mr. Spock to calculate the forces that you are exposed to because of your displacement s_{BB_r} from the c.m. of the spaceship.

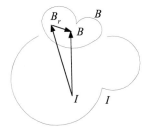

Fig. 5.8   Reference point B_r.

This problem was addressed by Grubin[3] and therefore carries his name. Similar to the Coriolis transformation, our goal is now to write Newton's law wrt an arbitrary reference point B_r of body B and move the additional terms to the right-hand side of the equation:

$$m^B D^I v^I_{B_r} = f + \text{correction terms}$$

From Fig. 5.8 derive the vector triangle

$$s_{BI} = s_{BB_r} + s_{B_r I}$$

and substitute it into Eq. (5.8):

$$m^B D^I D^I s_{BB_r} + m^B D^I D^I s_{B_r I} = f \qquad (5.26)$$

The second term is already in the desired form:

$$m^B D^I D^I s_{B_r I} = m^B D^I v^I_{B_r}$$

The first term, which will generate the apparent forces, is treated next. We will make use of the fact that $D^B s_{BB_r} = 0$ because both points belong to the body frame B and apply the chain rule

$$m^B D^I D^I s_{BB_r} = m^B D^I \left(D^B s_{BB_r} + \Omega^{BI} s_{BB_r} \right) = m^B D^I \left(\Omega^{BI} s_{BB_r} \right)$$

$$= m^B D^I \Omega^{BI} s_{BB_r} + m^B \Omega^{BI} D^I s_{BB_r}$$

$$= m^B D^I \Omega^{BI} s_{BB_r} + m^B \Omega^{BI} \left(D^B s_{BB_r} + \Omega^{BI} s_{BB_r} \right)$$

$$= m^B D^I \Omega^{BI} s_{BB_r} + m^B \Omega^{BI} \Omega^{BI} s_{BB_r}$$

Substitution into Eq. (5.26) leads to Grubin's form of Newton's second law:

$$m^B D^I v^I_{B_r} = f - m^B \begin{cases} \Omega^{BI} \Omega^{BI} s_{BB_r} & \text{centrifugal acceleration} \\ + (D^I \Omega^{BI}) s_{BB_r} & \text{angular acceleration} \end{cases} \qquad (5.27)$$

Sitting in your captain's chair you experience two additional forces caused by centrifugal and angular accelerations. If you move to the c.m. of the spaceship, both forces vanish as your displacement vector s_{BB_r} shrinks to zero.

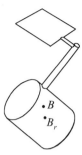

Fig. 5.9 Appendices.

Example 5.8 Satellite with Solar Array

Consider a space station with a long empennage of solar arrays (see Fig. 5.9). The geometric center of the space station B_r is a more important reference point than the rather obscure common c.m. B. Grubin's transformation shows us how to set up the equations of motion. If the space station is rotating with the angular velocity ω^{BI}, it is subject to the centrifugal acceleration $\Omega^{BI}\Omega^{BI}s_{BB_r}$ and the angular acceleration $D^I\Omega^{BI}s_{BB_r}$. To develop the trajectory equations for the geometric center, we assume that the gravitational force is given in inertial coordinates $[f]^I = m^B[g]^I$, as well as the angular velocity $[\Omega^{BI}]^I$, and the position vector in body axes $[s_{BB_r}]^B$. The transformation matrix $[T]^{BI}$ relates the body and inertial axes. Then, from Eq. (5.27) we obtain the equations of motion in matrix form:

$$m^B\frac{d}{dt}\left[v_{B_r}^I\right]^I = m^B[g]^I - m^B\left[[\Omega^{BI}]^I[\Omega^{BI}]^I[\bar{T}]^{BI}[s_{BB_r}]^B + \frac{d}{dt}\left([\Omega^{BI}]^I\right)[\bar{T}]^{BI}[s_{BB_r}]^B\right]$$

Another important application of Grubin's transformation is related to the specific force measurements of an INS. Seldom is the instrument cluster located at the c.m. of the missile. To determine the corrections that need be applied to the raw measurements, we use Grubin's transformation to express the vehicle's c.m. acceleration in terms of the center of the accelerometer cluster. The correction terms are the centrifugal and angular acceleration terms (see Problem 5.10).

We were able to derive the Coriolis and Grubin transformations of Newton's law in an invariant tensor form, valid in any allowable coordinate system. The last example gave an indication of the conversion process for computer implementation. In aerospace vehicle simulations you will encounter many different ways of modeling the translational equations of motions. In the next section I will summarize the most important ones, but reserve the details for Part 2.

5.4 Simulation Implementation

When implementing Newton's law on the computer, you have to answer many practical questions. What type of vehicle is being simulated: aircraft, missile, or satellite; is it flying near Earth or at great altitudes and hypersonic speeds; does the customer require high accuracy trajectory information or is he only interested in a quick, first-cut study? The answers determine the fidelity of your model.

The fidelity of a simulation is categorized according to the number of DoF it models. A rigid body, moving through air or space has six DoF, three translational and three rotational degrees. Newton's law models the three translational degrees of freedom of the vehicle's c.m., whereas Euler's law (see next chapter) governs the three rotational degrees of freedom. Both together provide the highest fidelity.

However, for preliminary trajectory studies it may be adequate to model the vehicle as a particle. Only the translational equations apply, and the simulation is called a three-DoF model. If attitude motions have to be included, but a complete database is lacking, an interim model, the so-called pseudo-five-DoF simulation is used to great advantage. Two attitude motions, either pitch–yaw or pitch–bank, augment the three translational degrees of freedom. However, the attitude motions are not derived from Euler's law, but from linearized autopilot responses. Ultimately, the full attitude motions, governed by Euler's law, joined by Newton's translational DoF form the full six-DoF simulations.

Besides fidelity requirements the form of the inertial frame categorizes a simulation. Interplanetary travel demands the heliocentric frame; Earth-orbiting or hypersonic vehicles use the J2000 inertial frame; and slow, Earth-bound vehicles can compromise with the Earth as an inertial frame.

I will summarize the more important versions of Newton's translational equations, as they are employed in aerospace simulations. I will give you a glimpse of each category: three, five, and six degrees of freedom later. Chapters 8, 9, and 10 will provide the details.

5.4.1 Three-Degree-of-Freedom Simulations

During preliminary design, system characteristics are very often not known in detail. The aerodynamics can only be given in trimmed form, and the autopilot structure can be greatly simplified. Fortunately, the trajectory of the c.m. of the vehicle is usually of greater interest than its attitude motions, and, therefore, the simple three-DoF simulations are very useful in the preliminary design of aerospace vehicles.

Newton's second law governs the three translational DoF. The aerodynamic, propulsive, and gravitational forces must be given. In contrast to six-DoF simulations, Euler's law is not used to calculate body rates and attitudes; therefore, there is no need to hunt for the aerodynamic and propulsive moments.

Suppose we build a three-DoF simulation for a hypersonic vehicle. We use the J2000 inertial frame of Chapter 3 for Newton's law. The inertial position and velocity components are directly integrated, but the aerodynamic forces of lift and drag are given in velocity coordinates. Therefore, we also need a TM of velocity wrt inertial coordinates to convert the forces to inertial coordinates.

The equations of motion are derived from Newton's law, Eq. (5.9):

$$m D^I v_B^I = f_{a,p} + m g \tag{5.28}$$

where m is the vehicle mass and v_B^I is the velocity of the missile c.m. B wrt the inertial reference frame I. Surface forces are aerodynamic and propulsive forces $f_{a,p}$, and the gravitational volume force is $m g$. Although v_B^I is the inertial velocity, we also need the geographic velocity v_B^E to compute lift and drag. Let us derive a relationship between the two velocities.

The position of the inertial reference frame I is oriented in the solar ecliptic, and one point I is collocated with the center of Earth. The Earth frame E is fixed with the geoid and rotates with the angular velocity ω^{EI}. By definition the inertial velocity is $v_B^I = D^I s_{BI}$, where s_{BI} is the location of the vehicle's c.m. wrt point I. To introduce the geographic velocity, we change the reference frame to E

$$D^I s_{BI} = D^E s_{BI} + \Omega^{EI} s_{BI} \tag{5.29}$$

and introduce a reference point E on Earth (any point), $s_{BI} = s_{BE} + s_{EI}$, into the first right-hand term

$$D^E s_{BI} = D^E s_{BE} + D^E s_{EI} = D^E s_{BE} \equiv v_B^E$$

where $D^E s_{EI}$ is zero because s_{EI} is constant in the Earth frame. Substituting into Eq. (5.29), we obtain a relationship between the inertial and geographic velocities

$$v_B^I = v_B^E + \Omega^{EI} s_{BI} \tag{5.30}$$

For computer implementation Eq. (5.28) is converted to matrices by introducing coordinate systems. The left side is integrated in inertial coordinates $]^I$, while the aerodynamic and propulsive forces are expressed in velocity coordinates and the gravitational acceleration in geographic coordinates $]^G$. The details of obtaining the TMs are given in Chapter 3. We just emphasize here that we have to distinguish the two velocity coordinate systems. The one associated with the inertial velocity v_B^I is called $]^U$, and the geographic velocity coordinate system is $]^V$. With these provisions we have the form of the translational equations of motion:

$$m \left[\frac{dv_B^I}{dt} \right]^I = [T]^{IG} \big([T]^{GU} [f_{a,p}]^U + m[g]^G \big) \tag{5.31}$$

These are the first three differential equations to be solved for the inertial velocity components $[v_B^I]^I$. The second set of differential equations calculates the inertial position

$$\left[\frac{ds_{BI}}{dt} \right]^I = [v_B^I]^I \tag{5.32}$$

Both equations are at the heart of a three-DoF simulation. You can find them implemented in the CADAC GHAME3 simulation of a hypersonic vehicle.

If you stay closer to Earth, like flying in the Falcon jet fighter, you can simplify your simulation by substituting Earth as an inertial frame. In Eqs. (5.31) and (5.32) you replace frame I and point I by frame E and point E. The distinction between inertial and geographic velocity disappears, and the geographic coordinate system is replaced by the local-level system $]^L$:

$$m \left[\frac{dv_B^E}{dt} \right]^L = [T]^{LV} [f_{a,p}]^V + m[g]^L \tag{5.33}$$

$$\left[\frac{ds_{BE}}{dt} \right]^L = [v_B^E]^L \tag{5.34}$$

These equations are quite useful for simple near-Earth trajectory work.

You will find more details in Chapter 8 with other useful information about the aerodynamic and propulsive forces. To experience an actual computer implementation, you should go to the CADAC Web site, download the GHAME3 simulation, and run the test case.

5.4.2 Five-Degree-of-Freedom Simulations

If the point-mass model of an aerospace vehicle, as implemented in three-DoF simulations, does not adequately represent the dynamics, one can expand the model by two more DoF. For a skid-to-turn missile pitch and yaw attitude dynamics are added, whereas for a bank-to-turn aircraft, pitch and bank angles are used. Euler's law could be used to formulate the additional differential equations. However, the increase in complexity approaches that of a full six-DoF simulation. To maintain the simple features of a three-DoF simulation and, at the same time, account for attitude dynamics, one adds the transfer functions of the closed-loop autopilot to the point-mass dynamics. This approach, with linearized attitude dynamics, is called a pseudo-five-DoF simulation.

The implementation uses the translational equations of motion, formulated from Newton's law and expressed in flight-path coordinates. The state variables and their derivatives are the speed of vehicle c.m. wrt Earth: $V = |v_B^E|$ and dV/dt; the heading angle and rate χ and $d\chi/dt$; and the flight-path angle and rate γ and $d\gamma/dt$. One key variable, the angular velocity of the vehicle wrt the Earth frame ω^{BE}, is not available directly because Euler's equations are not solved. Therefore, it must be pieced together from two other vectors

$$\omega^{BE} = \omega^{BV} + \omega^{VE}$$

where V is the frame associated with the geographic velocity vector v_B^E of the vehicle. The two angular velocities can be calculated because their angular rates and angles are available from the autopilot. The incidence rates are obtained from angle of attack α, sideslip angle β, and bank angle ϕ

$$\omega^{BV} = f(\alpha, \dot\alpha, \beta, \dot\beta) \quad \text{skid-to-turn}$$

$$\omega^{BV} = f(\alpha, \dot\alpha, \phi, \dot\phi) \quad \text{bank-to-turn}$$

and the flight-path angle rates

$$\omega^{VE} = f(\chi, \dot\chi, \gamma, \dot\gamma)$$

Thus, the solution of the attitude differential equations is replaced by kinematic calculations.

We formulate the translational equations for near-Earth trajectories, invoking the flat-Earth assumption and the local-level coordinate system. Application of Newton's law yields

$$m D^E v_B^E = f_{a,p} + mg$$

with aerodynamic, propulsive, and gravity forces as externally applied forces. The rotational time derivative is taken wrt the inertial Earth frame E. Using Euler's transformation, we change it to the velocity frame V

$$D^V v_B^E + \Omega^{VE} v_B^E = \frac{f_{a,p}}{m} + g$$

and use the velocity coordinate system to create the matrix equation

$$[D^V v_B^E]^V + [\Omega^{VE}]^V [v_B^E]^V = \frac{[f_{a,p}]^V}{m} + [g]^V \tag{5.35}$$

The rotational time derivative is simply $[\overline{D^V v_B^E}]^V = [\dot{V}\ 0\ 0]$. The aerodynamic and thrust forces are given in body coordinates, thus $[f_{a,p}]^V = [\bar{T}]^{BV}[f_{a,p}]^B$, whereas the gravity acceleration is best expressed in local-level coordinates $[g]^V = [T]^{VL}[g]^L$. With these terms and the angular velocity

$$[\omega^{VE}]^V = \begin{bmatrix} -\dot{\chi}\sin\gamma \\ \dot{\gamma} \\ \dot{\chi}\cos\gamma \end{bmatrix} \tag{5.36}$$

we can solve Eq. (5.35) for the three state variables V, χ, and γ

$$\begin{bmatrix} \dot{V} \\ \dot{\chi} V \cos\gamma \\ -\dot{\gamma} V \end{bmatrix} = [\bar{T}]^{BV} \frac{[f_{a,p}]^B}{m} + [T]^{VL}[g]^L \tag{5.37}$$

The vehicle's position is calculated from the differential equations

$$\begin{bmatrix} \dfrac{\mathrm{d}s_{BE}}{\mathrm{d}t} \end{bmatrix}^L = [\bar{T}]^{VL} \begin{bmatrix} V \\ 0 \\ 0 \end{bmatrix} \tag{5.38}$$

These are the translational equations of motion for pseudo-five-DoF simulations. The details, and particularly the derivation of Eq. (5.36), can be found in Chapter 9. Note that a singularity occurs at $\gamma = \pm 90$ deg.

Pseudo-five-DoF simulations have an important place in modeling and simulation of aerospace vehicles. They can easily be assembled from trimmed aerodynamic data and simple autopilot designs. Surprisingly, they give a realistic picture of the translational and rotational dynamics unless large angles and cross-coupling effects dominate the simulation. Trajectory studies, performance investigations, and guidance and navigation (outer-loop) evaluations can be executed successfully with pseudo-five-DoF simulations.

Chapter 9 is devoted to much more detail. There you find examples for aerodynamics, propulsion, autopilots, guidance, and navigation models, both for missile and aircraft. The CADAC Web site offers application simulations of air-to-air and cruise missiles AIM5, SRAAM5, and CRUISE5.

5.4.3 Six-Degree-of-Freedom Simulations

The ultimate virtual environment for aerospace vehicles is the six-DoF simulation. No compromises have to be made or shortcuts taken. The equations of motion model fully the three translational and three attitude degrees of freedom.

Any development program that enters flight testing requires this kind of detail for reliable test performance prediction and failure analysis. Fortunately, by that time the design is well enough defined so that detailed aerodynamics and autopilot data are available for modeling. Yet, the development and maintenance of such a six-DoF simulation consumes great resources. Industry dedicates their most talented engineers to this task and maintains elaborate computer facilities.

However, even in the conceptual phase of a program it can become necessary to develop a six-DoF simulation. This need is driven either by the importance and visibility of the program or by the highly dynamic environment that the vehicle may encounter. A good example is a short-range air-to-air missile intercepting a target at close range. Its velocity and attitude change rapidly, resulting in large incidence angles and control surface deflections.

Six-DoF simulations come in many forms. They can be categorized by the *inertial frame* (elliptical rotating Earth or stationary flat Earth), by the *type of vehicle* (missile, aircraft, spinning rocket, or spacecraft), or by the *architecture* (tightly integrated, modular, or object oriented). We derive here the general translational equations for elliptical and flat Earth and leave the detail to Chapter 10.

5.4.3.1 Round Earth. The translational equations for round Earth–be it spherical or elliptical–follow the same derivation used in three-DoF models. As we will discuss in the next chapter, even for six DoF, the trajectory can be calculated as if the vehicle were a particle. Therefore, we can be brief. Newton's law related to the J2000 inertial frame as applied to a vehicle with aerodynamic, propulsive, and gravitational forces is

$$m D^I v_B^I = f_{a,p} + mg$$

The integration is executed in inertial coordinates, but the aero/propulsion data are most likely given in body coordinates. We make those adjustments together with the expression of the gravitational acceleration in geographic coordinates:

$$m[D^I v_B^I]^I = [\bar{T}]^{BI}[f_{a,p}]^B + m[\bar{T}]^{GI}[g]^G \tag{5.39}$$

The main distinction with the three-DoF formulation, Eq. (5.31), lies in the handling of the aero/propulsive forces. Six-DoF simulations model the complex aerodynamic tables and propulsion decks in body coordinates, whereas their simple approximations in three-DoF simulations can be expressed in velocity coordinates.

Another set of differential equations provide the position traces

$$\left[\frac{ds_{BI}}{dt}\right]^I = [v_B^I]^I \tag{5.40}$$

which will have to be converted to more meaningful longitude, latitude, and altitude coordinates.

5.4.3.2 Flat Earth. Even in six-DoF simulations, with all of their emphasis on detail, the flat-Earth models are prevalent. All aircraft simulations that I know of are of that flavor, as well as cruise missiles and tactical air intercept and ground attack missiles. Earth E becomes the inertial frame, and the longitude/latitude grid is unwrapped into a plane. Newton's law takes on the form

$$mD^E v_B^E = f_{a,p} + mg$$

The majority of flat-Earth six-DoF models express the terms in body coordinates, save the gravitational acceleration. By this approach the geographic velocity $[v_B^E]^B$ in body coordinates can be used directly to calculate the incidence angles [see Eqs. (3.20–3.23)]. The conversion should be familiar to you by now. Transform the rotational derivative to the body frame

$$mD^B v_B^E + m\Omega^{BE} v_B^E = f_{a,p} + mg$$

and coordinate accordingly

$$\left[\frac{dv_B^E}{dt}\right]^B = -[\Omega^{BE}]^B[v_B^E]^B + \frac{1}{m}[f_{a,p}]^B + [T]^{BL}[g]^L \qquad (5.41)$$

This is the translational equation of motion for flat Earth, implemented in six-DoF simulations. One more integration completes the set:

$$\left[\frac{ds_{BE}}{dt}\right]^L = [\bar{T}]^{BL}[v_B^E]^B \qquad (5.42)$$

The body rates $[\Omega^{BE}]^B$ are provided by the rotational equations (see next chapter), and the direction cosine matrix $[T]^{BL}$ is calculated by one of the three options provided in Sec. 4.3.

On the CADAC Web site the GHAME6 simulation provides an example of the elliptical Earth implementation, and the flat-Earth model is used in SRAAM6. By running the sample trajectories, you can learn much about the world of six-DoF simulations.

Much more will be said in Chapters 8–10 about each of the three levels of simulation fidelity. At this point you can proceed directly to Chapters 8 and 9 to deepen your understanding of three- and five-DoF simulations. All of the necessary tools are in your possession. To tackle the six-DoF simulations, you first need to conquer the next chapter and its Euler law of rotational dynamics. Thereafter you are ready for the ultimate six-DoF experience of Chapter 10.

Let us pause and look at our newly acquired tools. The linear momentum, called motion variable by Newton, is related to mass and velocity by $p_B^I = m^B v_B^I$. It takes on the vector characteristics of the linear velocity, multiplied by the scalar mass of the vehicle. Our modeling elements, points and frames, are sufficient to define it completely. We could have carried over the superscript B of mass m^B to define p_B^{BI}, but I decided to drop it because of the particle nature of body B in Newton's law.

The other new vectors we encountered are the external forces. They consist of $f_{a,p}$, the aero/propulsion *surface* forces, and $m^B g$, the gravitational *volume* force. Both types must be applied at the c.m. of the vehicle. Only then can the vehicle be treated as a particle. In the next chapter, when we add the attitude motions to the translations of a body, we will derive the effect of shifting the forces to other reference points.

References

[1]Newton, I., *Mathematical Principles of Natural Philosophy* (reprint ed.), Univ. of California Press, Berkeley, CA, 1962.

[2]Franke, H., *Lexikon der Physik*, Frank'sche Verlagshandlung, Stuttgart, Germany, 1959.

[3]Grubin, C., "On Generalization of the Angular Momentum Equation," *Journal of Engineering Education*, Vol. 51, No. 3, Dec. 1960, pp. 237, 238, 255.

Problems

5.1   Linear momentum independent of reference point.   Show that the linear momentum p_B^I of a body B wrt the inertial frame I and referenced to the c.m. B depends only on frame I and not on a particular point I.

5.2   Transformation of body points.   The linear momentum $p_{B_1}^I$ of a rigid body B relative to the inertial frame I and referred to an arbitrary body point B_1 is shifted to another body point B_2. Prove the transformation equation

$$p_{B_2}^I = p_{B_1}^I + m^B \Omega^{BI} s_{B_1 B_2}$$

starting with the definition of the linear momentum of a collection of particles Eq. (5.2).

5.3   Satellite release.   The space shuttle B releases a satellite S with its manipulator arm at a velocity of

$$\left[\overline{v_S^B}\right]^B = [0 \quad 0 \quad -at]$$

with the constant acceleration a. Its circular orbit is in the 1^I, 3^I plane of the J2000 inertial coordinate system $]^I$ at an altitude of R and period of T, maintaining its 3^B axis pointed at Earth's center. Initially, the 1^I and 1^B axes are aligned, and the space shuttle flies toward the vernal equinox. Individual masses are m^B and m^S, respectively. Derive the equation for the inertial acceleration of the space shuttle first as a tensor $D^I v_B^I$, then coordinated $[dv_B^I/dt]^I$, and finally in components.

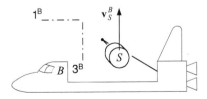

5.4   What's the difference?   The selection of the inertial frame for Newton's law is determined by the application. The statement was made that for near-Earth orbits the J2000 inertial frame I can be used. What error is incurred by using $m^B D^I v_B^I = f$ instead of the heliocentric reference frame H, $m^B D^H v_B^H = f$?

(a) Derive the error term in tensor form.

(b) Coordinate it in heliocentric coordinates.

(c) Give a maximum numerical value for the error. You can assume a circular orbit of Earth around the sun.

5.5   Planar trajectory equations of a missile.   Derive the planar point-mass equations of a missile in velocity coordinates $]^V$ with the dependent variables V as velocity magnitude and γ as flight-path angle. Lift L and drag D are given, as well as the thrust T in the opposite direction of D.

(a) Derive the translational equations in an invariant form consisting of the velocity and position differential equations.

(b) Coordinate the equations into matrix form.

(c) Multiply out the matrix equations, and write down the four component equations.

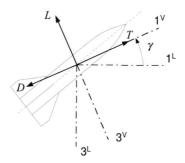

5.6   Centrifugal and coriolis forces, who cares?   An aircraft flies north with the velocity V at an altitude h above sea level. With the flat-Earth assumption we use Newton's law in the simple form $m^B D^E v_B^E = f$, but neglect on the right-hand side the Coriolis and centrifugal forces $m^B (2\Omega^{EI} v_B^E + \Omega^{EI}\Omega^{EI} s_{BE})$. What are the values of these accelerations and their directions at 60-deg latitude (use $h = 10,000$ m, $V = 250$ m/s)?

5.7   Hiking in the space colony.   Wernher von Braun dreamed of a large space colony S orbiting Earth in the shape of a wheel with spokes. For artificial gravity the wheel was to be revolving with ω_s about its 3^S axis, and its close link with Earth was maintained by keeping the spin axis pointing toward Earth. You are hiking from the hub through a spoke toward the rim along the 2^S axis. What are the Coriolis and centrifugal forces in $]^S$ coordinates that you have to counteract to prevent you from bumping into the walls?

5.8   Space station rescue.   A large, rigid, and force-free space station has a malfunctioning INS and begins to tumble in space. The chief engineer needs an alternate method to determine the angular velocity ω^{BI} of the station B wrt the inertial frame I in order to supply it to the stabilizing momentum wheels.

A radio navigation system R is located on a long boom and displaced from the space station c.m. B by s_{RB}. It measures its inertial velocity $[v_R^I]^B$ and time rate of change $[dv_R^I/dt]^B$ in space station coordinates $]^B$. Provide the chief engineer the matrix form of the differential equation $[\omega^{BI}]^B$ so that he can program it for the onboard computer. The mass of the boom and the navigation system may be neglected. (The orbital velocity $[v_B^I]^I$ remains unaffected and is known from the orbital elements of the space station.)

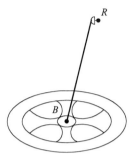

5.9   Kepler's law from Newton's law.   Derive Kepler's second law from Newton's second law by considering a particle (Earth E) acted upon by a central force $-\mu s_{ES}/|s_{ES}|^3$, where S is the center of the sun and μ the gravitational parameter in meters cubed per seconds squared ($\mu = Gm_{\text{sun}}$, G is the universal gravitational constant).

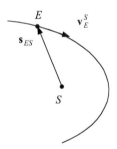

Hint: Kepler's second law states that the line joining Earth to the sun sweeps out equal areas in equal time. In other words, the area swept out in unit time is constant. With v_E^S the velocity of Earth wrt the sun, this statement translates into the vector product $S_{ES}v_E^S = \text{const}$. You should prove that Newton's second law reduces to this relationship.

5.10   Accelerometer compensation.   (a) An accelerometer triad A with its sensors mounted parallel to the missile body axes $]^B$ is displaced $[s_{AB}]^B$ from the missile c.m. B. What are the three specific force components $[a_B]^B$ acting on the missile, given the three accelerometer measurements $[a_A]^B$, the vehicle angular velocities

$$[\overline{\omega^{BI}}]^B = [p \quad q \quad r]$$

and their accelerations

$$d/dt[\overline{\omega^{BI}}]^B = [\dot{p} \quad \dot{q} \quad \dot{r}]$$

(b) The acceleration triad is displaced by $[\overline{s_{AB}}]^B = [1 \quad 0 \quad 0]$ m, and the missile executes steady coning type motions represented by

$$[\overline{\omega^{BI}}]^B = [0 \quad \sin t \quad \cos t]\,\text{rad/s}$$

What are the correction terms $[\Delta a_A]^B$ for the three accelerometer measurements?

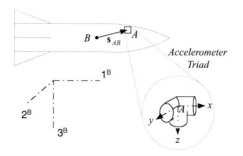

5.11   Burp of Graf Zeppelin.   In October 1924 the Graf Zeppelin crossed the Atlantic on its maiden voyage under the command of Dr. Hugo Eckener. Midway it encountered a gust that caused the ship to pitch up at a constant 10 deg/s and incremental load factor of 0.1 g. What was the linear acceleration that Dr. Eckener experienced in the gondola at point G in Zeppelin coordinates? The displacement from the ship's c.m. is $[\overline{s_{BG}}]^B = [-90 \quad 0 \quad -15]$ m.

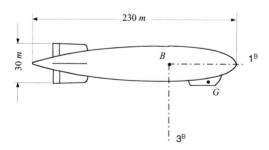

6
Attitude Dynamics

We have come a long way on the coattails of Sir Isaac Newton. His second law enables us to calculate the trajectory of any aerospace vehicle, provided we can model the external forces. For many applications we have to predict only the movement of the c.m. and can neglect the details of the attitude motions. Three- and even five-DoF simulations can be built on Newton's law only. If you were to stop here, what would you be missing?

It is like the difference between riding on a ferris wheel, which transports your c.m., vs the twists and turns you experience on the Kumba at Busch Gardens amusement park. Attitude dynamics brings excitement into the dullness of trajectory studies and three more dimensions to the modeling task. Do you see the affinity between geometry, kinematics, and dynamics? Chapter 3 dealt with geometry, describing position in terms of location and orientation; Chapter 4 dealt with kinematics; and now we characterize dynamics, consisting of translation (Chapter 5) and attitude motions (this chapter).

Attitude dynamics are the domain of Leonhard Euler, a Swiss physicist of the 18th century, whose name we have used before, but who now competes with Newton head on. We will study his law of attitude motion in detail. It has a strong resemblance to Newton's law. Newton's building blocks are mass, linear velocity, and force, whereas Euler's law uses moment of inertia, angular velocity, and moment. But it gets more complicated. Mass is a simple scalar, whereas the moment of inertia requires a second-order tensor as a descriptor.

To prepare the way, we start with the *moment-of-inertia* (MOI) tensor and derive some useful theorems that help us calculate its value for missiles and aircraft. The geometrical picture of a MOI ellipsoid will help us visualize the tensor characteristics. The concept of principal axes will be most useful in simplifying the attitude equations.

Combining the MOI tensor with the angular velocity vector will lead to the concept of *angular momentum*. We will learn how to calculate it for a collection of particles and a cluster of bodies. Again, we will see how important the c.m. is for simplified formulations.

From these elements we can formulate Euler's law. As Newton's law is often paraphrased as force = mass × linear acceleration, so can Euler's law be regarded as moment = MOI × angular acceleration. For freely moving bodies like missiles and aircraft, we use the c.m. as a reference point. Some gyrodynamic applications with a fixed point—for instance the contact point of a top—will lead to an alternate formulation of Euler's law.

Gyrodynamics is a fascinating study of rigid body motions, and we will devote some time to it, both in reverence to the giants of mechanics, like Euler, Poinsot, Klein, and Magnus, and because of its modern applications in INS and stabilization of spacecraft. I will introduce the kinetic energy of spinning bodies and the energy ellipsoid. Two integrals of motion are particularly fertile for studying the motions of force-free rigid bodies.

If you persevere with me through this chapter, you will have mastered a modern treatment of geometry, kinematics, and dynamics of Newtonian and Eulerian motions. The remaining chapters deal with a host of applications relevant for today's aerospace engineer with particular emphasis on computer modeling and simulation. So, with verve let us tackle a new tensor concept.

6.1   Inertia Tensor

We all have experienced the effect of mass, foremost as weight brought about by gravitational acceleration, or as inertia when we try to sprint. Yet, how do we sense MOI? If you are an ice dancer, you've had plenty of experience. Landing after a double or triple axel, you kick up plenty of ice to stop your turn. Actually, it is your angular momentum (MOI $\times$ angular velocity) that you have to catch, and the greater your MOI the greater the angular momentum.

As customary, we divide a body into individual particles and define the MOI of the body by summing over its particles. I shall introduce such familiar terms as *axial moments of inertia*, *products of inertia*, and *principal moments of inertia*. Huygen's theorem and the parallel axes theorem will show us how to change the reference point or the reference axis. Because we are dealing mostly with vehicles in three dimensions, the moment of inertia ellipsoid, its principal axes, and the radii of gyration will give us a geometrical picture of this elusive MOI tensor. We will conclude this section with some practical rules that take advantage of the symmetries inherent in missiles and aircraft.

6.1.1   Definition of Moment-of-Inertia Tensor

A material body is a three-dimensional differentiable manifold of particles possessing a scalar measure called *mass distribution*. Integrating the mass distribution over the volume of the body results in a scalar called *mass* (see Sec. 2.1.1). If the integration includes the distance of the particles relative to a reference point, then we obtain the first-order tensor that defines the location of the c.m. wrt the reference point. If the distance is squared, the integration yields a second-order tensor called the *inertia tensor*.

Definition: The inertia tensor of body B referred to an arbitrary point R is calculated from the infinite sum over all its mass particles m_i and their displacement vector s_{iR} according to the following definition:

$$I_R^B = \sum_i m_i (\bar{s}_{iR} s_{iR} E - s_{iR} \bar{s}_{iR}) = \sum_i m_i \bar{S}_{iR} S_{iR} \qquad (6.1)$$

where S_{iR} is the skew-symmetric form of the displacement vector s_{iR}.

The notation I_R^B reflects the reference point as subscript R, and the sum over all particles, the body frame as superscript B. The expression $\bar{s}_{iR} s_{iR} E - s_{iR} \bar{s}_{iR} = \bar{S}_{iR} S_{iR}$ is a tensor identity, which you can prove by substituting components and multiplying out the matrices.

For the body coordinates $]^B$ with $[\bar{s}_{iR}]^B = [s_{iR_1} \ s_{iR_2} \ s_{iR_3}]$, the MOI tensor has the component form

$$
[I_R^B]^B =
\begin{bmatrix}
\sum_i m_i \left(s_{iR_2}^2 + s_{iR_3}^2 \right) & -\sum_i m_i s_{iR_1} s_{iR_2} & -\sum_i m_i s_{iR_1} s_{iR_3} \\
-\sum_i m_i s_{iR_1} s_{iR_2} & \sum_i m_i \left(s_{iR_1}^2 + s_{iR_3}^2 \right) & -\sum_i m_i s_{iR_2} s_{iR_3} \\
-\sum_i m_i s_{iR_1} s_{iR_3} & -\sum_i m_i s_{iR_2} s_{iR_3} & \sum_i m_i \left(s_{iR_1}^2 + s_{iR_2}^2 \right)
\end{bmatrix}
\tag{6.2}
$$

The MOI tensor expressed in any allowable coordinate system is a real symmetric matrix and has therefore only six independent elements. Its diagonal elements are called *axial moments of inertia* and the off-diagonal elements *products of inertia*. They have the units meters squared times kilograms. Some examples should give you more insight.

Example 6.1   Axial Moment of Inertia

The *axial MOI* I_n of the MOI tensor I_R^B about a unit vector n through point R is the scalar

$$
I_n = \bar{n} I_R^B n
\tag{6.3}
$$

It has the same units of meters squared times kilograms as the elements of the MOI tensor. If we select the third-body base vector as axis and express it in body coordinates $[\bar{n}]^B = [0 \ 0 \ 1]$, then

$$
I_n = [0 \ 0 \ 1]
\begin{bmatrix}
I_{11} & I_{12} & I_{13} \\
I_{21} & I_{22} & I_{23} \\
I_{31} & I_{32} & I_{33}
\end{bmatrix}
\begin{bmatrix}
0 \\
0 \\
1
\end{bmatrix}
= I_{33}
$$

The 3,3 element was picked out by n, justifying the name axial moment of inertia.

Example 6.2   Lamina

A *lamina* is a thin body with constant thickness (see Fig. 6.1). If the lamina extends into the first and second direction, then the *polar moment of inertia* about

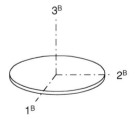

Fig. 6.1   Lamina.

the third axis is

$$I_{33} = I_{11} + I_{22} \tag{6.4}$$

For a proof we set $s_{iR_3} \approx 0$ in Eq. (6.2), then

$$I_{11} = \sum_i m_i s_{iR_2}^2; \quad I_{22} = \sum_i m_i s_{iR_1}^2; \quad I_{33} = \sum_i m_i \left(s_{iR_1}^2 + s_{iR_2}^2\right)$$

Substituting the first two relationships into the third completes the proof.

6.1.2   Displacement Theorems

The calculation of the MOI of a flight vehicle can be a tedious process. Only in recent times has it been automated by the use of CAD programs. However, you still may be challenged to provide rough estimates for prototype simulations. You can base these preliminary calculations on simplified geometrical representations and make use of two theorems that yield the MOI for shifted reference points and axes.

6.1.2.1   Point displacement theorem (Huygen's theorem).
The MOI of body B referred to an arbitrary point R is equal to the MOI referred to the c.m. B plus a term calculated as if all mass m^B were concentrated in the c.m.

$$I_R^B = I_B^B + m^B(\bar{s}_{BR}s_{BR}E - s_{BR}\bar{s}_{BR}) \tag{6.5}$$

or in the alternate form

$$I_R^B = I_B^B + m^B \bar{S}_{BR}S_{BR} \tag{6.6}$$

Compare the second terms on the right-hand sides with Eq. (6.1). They are the MOI of a particle with mass m^B and the displacement vector s_{BR} between the two reference points.

Proof: Introduce the vector triangle of Fig. 6.2

$$s_{iR} = s_{iB} + s_{BR}$$

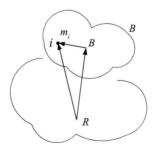

Fig. 6.2   Shifted reference point.

into Eq. (6.1):

$$I_R^B = \sum_i m_i \left[(\bar{s}_{iB} + \bar{s}_{BR})(s_{iB} + s_{BR})E - (s_{iB} + s_{BR})(\bar{s}_{iB} + \bar{s}_{BR}) \right]$$

$$= \sum_i m_i (\bar{s}_{iB} s_{iB} E - s_{iB}\bar{s}_{iB}) + (\bar{s}_{BR} s_{BR} E - s_{BR}\bar{s}_{BR}) \sum_i m_i$$

$$+ \sum_i m_i \bar{s}_{iB} s_{BR} + \bar{s}_{BR} \sum_i m_i s_{iB} - \sum_i m_i s_{iB} \bar{s}_{BR} - s_{BR} \sum_i m_i \bar{s}_{iB}$$

The last four terms are all individually zero because B is the c.m. The first term is according to Eq. (6.1)

$$\sum_i m_i (\bar{s}_{iB} s_{iB} E - s_{iB}\bar{s}_{iB}) = I_B^B$$

and the second term is already in the desired form of Eq. (6.5).

Huygen's theorem helps to build the total MOI of an aircraft from its individual parts. In this case R is the point of the overall c.m., whereas B is the c.m. of the individual part. We can modify Eq. (6.5) to encompass k number of individual bodies. Let B_k be the c.m. of body B_k. Then the total moment of inertia $I_R^{\Sigma B_k}$ of the cluster of k bodies B_k, $k = 1, 2, 3, \ldots$, referred to the common reference point R is

$$I_R^{\Sigma B_k} = \sum_k I_{B_k}^{B_k} + \sum_k m^{B_k} \left(\bar{s}_{B_k R} s_{B_k R} E - s_{B_k R} \bar{s}_{B_k R} \right) \tag{6.7}$$

and its alternate form

$$I_R^{\Sigma B_k} = \sum_k \left(I_{B_k}^{B_k} + m^{B_k} \bar{S}_{B_k R} S_{B_k R} \right) \tag{6.8}$$

According to Eq. (6.6), the right-hand side can also be expressed as the sum of individual MOIs:

$$I_R^{\Sigma B_k} = \sum_k I_R^{B_k} \tag{6.9}$$

An important conclusion follows: The total MOI of a cluster of bodies B_k, $k = 1, 2, 3, \ldots$, referred to the reference point R, can be calculated by adding the individual MOIs, also referred to R. In most practical cases, although not mandatory, R will be chosen as the overall c.m.

6.1.2.2 Parallel axes theorem.

The axial MOI I_{Rn} of a body B about any given axis n is the axial MOI about a parallel axis through the c.m. B plus an axial term calculated as if all of the mass of the body were located at the c.m. B (see Fig. 6.3):

$$I_{Rn} = \bar{n} I_B^B n + m^B \bar{n} \left(\bar{s}_{BR} s_{BR} E - s_{BR} \bar{s}_{BR} \right) n \tag{6.10}$$

Like any axial MOI, once the axis has been identified, it becomes a scalar with units in meters squared times kilograms.

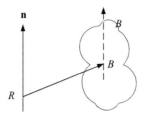

Fig. 6.3   Shifted reference axis.

Proof: Substitute Eq. (6.5) into Eq. (6.3) and obtain Eq. (6.10) directly.

Note that in Eqs. (6.5) and (6.10), the extra term can also be expressed by the tensor identity $\bar{s}_{BR} s_{BR} E - s_{BR}\bar{s}_{BR} = \bar{S}_{BR} S_{BR}$, which is usually simpler to calculate.

Example 6.3   Tilt Rotor MOI

Problem.   The axial MOI of the right tilt rotor is $I_{B_{33}}$ about the vertical axis $[\bar{n}]^B = [0 \ 0 \ 1]$ (see Fig. 6.4). What is its axial MOI $I_{R_{33}}$ wrt the aircraft c.m. R if the tilt rotor c.m. B is displaced by $[\overline{s_{BR}}]^B = [s_{BR_1} \ s_{BR_2} \ s_{BR_3}]$? What is the axial MOI of both rotors wrt R?

Solution.   We apply Eq. (6.10) directly and obtain for one rotor

$$I_{Rn} = [\bar{n}]^B [I_B^B]^B [n]^B + m^B \left([\bar{n}]^B [\overline{s_{BR}}]^B [s_{BR}]^B [n]^B [E] - [\bar{n}]^B [s_{BR}]^B [\overline{s_{BR}}]^B [n]^B \right)$$

$$I_{R_{33}} = I_{B_{33}} + m^B \left(s_{BR_1}^2 + s_{BR_2}^2 + s_{BR_3}^2 - s_{BR_3}^2 \right)$$

$$I_{R_{33}} = I_{B_{33}} + m^B \left(s_{BR_1}^2 + s_{BR_2}^2 \right)$$

The offset correction depends only on the square of the distance of the rotor axes from the aircraft c.m.; therefore, the axial MOI of both rotors is just twice the value of $I_{R_{33}}$.

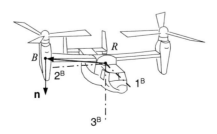

Fig. 6.4   Osprey moment of inertia.

6.1.3 Inertia Ellipsoid

The MOI tensor portrays a vivid geometrical interpretation, which is useful for the investigation of rigid-body dynamics. Being a real symmetric tensor, it has several important characteristics: it is positive definite, has three positive eigenvalues, has three orthogonal eigenvectors, and can be diagonalized by an orthogonal coordinate transformation with the eigenvalues as diagonal elements.

As we have seen, the axial MOI about axis n through reference point R is according to Eq. (6.3) in body coordinates

$$I_n = [\bar{n}]^B [I_R^B]^B [n]^B$$

$$= I_{11}n_1^2 + I_{22}n_2^2 + I_{33}n_3^2 + 2I_{12}n_1n_2 + 2I_{23}n_2n_3 + 2I_{31}n_3n_1 \quad (6.11)$$

Interestingly, this scalar equation in quadratic form has a geometric representation. Because the eigenvalues of $[I_R^B]^B$ are always real and positive, the geometrical surface, defined by Eq. (6.11), is an ellipsoid. If we introduce the normalized vector $[x]^B = [n]^B / \sqrt{I_n}$, we obtain the equation for the MOI ellipsoid:

$$1 = [\bar{x}]^B [I_R^B]^B [x]^B$$

$$= I_{11}x_1^2 + I_{22}x_2^2 + I_{33}x_3^2 + 2I_{12}x_1x_2 + 2I_{23}x_2x_3 + 2I_{31}x_3x_1 \quad (6.12)$$

Referring to Fig. 6.5, x is the displacement vector of a surface element relative to the center point R. A large value of $|x|$ means that the axial MOI I_n about this vector is small and vice versa.

If the body axes are principal axes, then $[I_R^B]^B$ is a diagonal matrix, and Eq. (6.12) simplifies to

$$1 = I_1 x_1^2 + I_2 x_2^2 + I_3 x_3^2 \quad (6.13)$$

where I_1, I_2, I_3 are the principal MOIs. They determine the lengths of the three semi-axes of the MOI ellipsoid

$$a = \frac{1}{\sqrt{I_1}}, \quad b = \frac{1}{\sqrt{I_2}}, \quad c = \frac{1}{\sqrt{I_3}} \quad (6.14)$$

The *radius of gyration* ρ_n is that distance from the axis at which all mass is concentrated such that the axial MOI can be calculated from $I_n = \rho_n^2 m^B$. We use

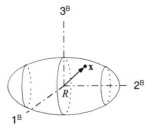

Fig. 6.5   Inertia ellipsoid.

it to get another expression for the surface vector:

$$[x]^B = \frac{[n]^B}{\left(\rho_n \sqrt{m^B}\right)}$$

Thus, the magnitude of the vector to a point on the inertia ellipsoid is inversely proportional to the radius of gyration about the direction of this vector. For example, in Fig. 6.5 the MOI about the third axis is greater than that about the second axis.

The directions of the eigenvectors e_1, e_2, e_3 are the principal axes. If they are known in an arbitrary coordinate system $[e_1]^A$, $[e_2]^A$, $[e_3]^A$, then the transformation matrix

$$[T]^{DA} = \begin{bmatrix} [\bar{e}_1]^A \\ [\bar{e}_2]^A \\ [\bar{e}_3]^A \end{bmatrix}$$

transforms the MOI tensor into its diagonal form

$$[I_{\text{diagonal}}]^D = [T]^{DA}[I]^A[\bar{T}]^{DA}$$

with the eigenvalues as principal MOIs.

Example 6.4   Shapes with Planar Symmetry

If a body with uniform mass distribution has a plane of symmetry, then one of its principal axes is normal to this plane. We validate this statement by the example of Fig. 6.6. The wing section has a plane of symmetry coinciding with the 1^B, 3^B axes. According to Eq. (6.2), the products of inertia containing the components s_{iR_2} are zero because their two components cancel. Thus

$$[I_R^B]^B = \begin{bmatrix} I_{11} & 0 & I_{13} \\ 0 & I_2 & 0 \\ I_{13} & 0 & I_{33} \end{bmatrix}$$

and I_2 is the principal MOI.

At no time did I assume the body to be rigid. Definitions and theorems of this section are valid for nonrigid as well as rigid bodies, and therefore, elastic structures are not excluded. However, a difficulty arises describing a frame for such an elastic body. Because, by definition, frame points are mutually fixed, we cannot use the particles of an elastic structure to make up the body frame. Instead, we have to idealize the structure and define the frame to coincide either with a no-load

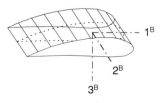

Fig. 6.6   Planar symmetry.

situation, the initial shape, or some average condition. Yet, do not be discouraged! The definition of the MOI does not rely on a body frame, but rather a collection of particles, mutually fixed or moving, which we designate as B. Only in the future, when we use the body as reference frame of the rotational derivative, do we need to specify a true frame. In those situations we will limit the discussion to rigid bodies.

The MOI joins the rotation tensors in our arsenal of second-order tensors. Both have distinctly different characteristics. Whereas the MOI tensor models a physical property of mass, the rotation tensor relates abstract reference frames. Their traits are contrasted by symmetrical vs orthogonal properties. However, both share the invariant property of tensors under any allowable coordinate transformation; and in both cases points and frames are sufficient to define them. Now we have reached the time to make the MOI come alive by joining it with angular velocity to form the angular momentum.

6.2 Angular Momentum

The angular momentum is the cousin of linear momentum. If you multiply the linear momentum by a displacement vector, you form the angular momentum. That at least is true for particles. By summing over all of the particles of a body, we define its total angular momentum. Again, introducing the c.m. will not only enable a compact formulation and simplify the change of reference points, but will also justify the separate treatment of attitude and translational motions. We close out this section with the formula for clusters of bodies, both for the common c.m. and an arbitrary reference point.

6.2.1 Definition of Angular Momentum

The definition of the angular momentum follows a pattern we have established for the linear momentum (Sec. 5.1). We start with a single particle and then embrace all of the particles of a particular body. Rigid-body assumptions and c.m. identification will lead to several useful formulations.

To define the angular momentum of a particle, we have to identify two points and one frame: the particle i, the reference point R, and its reference frame R (see Fig. 6.7).

Definition: The angular momentum l_{iR}^{R} of a particle i with mass m_i relative to the reference frame R and referred to reference point R is defined by the vector product of the displacement vector s_{iR} and its derivative $D^{R}s_{iR}$ multiplied by its

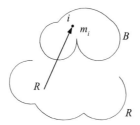

Fig. 6.7   Particle and references.

mass m_i:

$$l_{iR}^R = m_i S_{iR} D^R s_{iR} = m_i S_{iR} v_i^R \qquad (6.15)$$

Because the rotational derivative of s_{iR} is the linear velocity of the particle, $D^R s_{iR} = v_i^R$ and $m_i v_i^R = p_i^R$ is the linear momentum; we can express the angular momentum simply as the vector product of the displacement tensor and the linear momentum

$$l_{iR}^R = S_{iR} p_i^R \qquad (6.16)$$

The direction of the angular momentum is normal to the plane subtended by the displacement and the linear momentum vectors. (Any particle that is not at rest has a linear and angular momentum; it is just a matter of perspective. If the reference consists only of a frame, it exhibits linear momentum properties only. If a reference point is introduced, it displays also angular momentum characteristics.)

A body B, not necessarily rigid, can be considered a collection of particles i. The angular momentum of this body B relative to the reference frame R and referred to the reference point R is defined as the sum over the angular momenta of all particles

$$l_R^{BR} = \sum_i l_{iR}^R = \sum_i m_i S_{iR} D^R s_{iR} = \sum_i m_i S_{iR} v_i^R = \sum_i S_{iR} p_i^R \qquad (6.17)$$

Notice the shift of the subscript i in $\sum_i l_{iR}^R$ to a superscript B in l_R^{BR}, reflecting the gathering of all particles into body B.

6.2.2 Angular Momentum of Rigid Bodies

In most of our applications, the collection of particles can be assumed mutually fixed. This idealization, called a *rigid body*, is physically not realistic because molecules, even in solid matter, are oscillating. However, our macroscopic perspective permits this simplification. We need to be careful only when bending and vibrations (flutter) distort the airframe to such an extent that aerodynamic and mass properties are significantly changed. Here, we take advantage of the rigid-body concept.

Theorem: The angular momentum l_R^{BR} of a rigid body B wrt to any reference frame R and referred to reference point R can be calculated from two additive terms:

$$l_R^{BR} = I_B^B \omega^{BR} + m^B S_{BR} v_B^R \qquad (6.18)$$

The first term is the angular momentum l_B^{BR} of body B wrt to reference frame R and referred to its own c.m. B, $l_B^{BR} = I_B^B \omega^{BR}$, and the second term is a transfer factor accounting for the fact that R is not the c.m. Replacing the linear velocity by its definition $v_B^R = D^R s_{BR}$ results in another useful formulation:

$$l_R^{BR} = I_B^B \omega^{BR} + m^B S_{BR} D^R s_{BR} \qquad (6.19)$$

Proof: From Fig. 6.8 we derive the vector triangle and then take the rotational derivative wrt the reference frame R:

$$s_{iR} = s_{iB} + s_{BR} \Rightarrow D^R s_{iR} = D^R s_{iB} + D^R s_{BR}$$

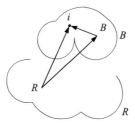

Fig. 6.8 Center of mass.

Substitute both into Eq. (6.17):

$$I_R^{BR} = \sum_i m_i \left[(S_{iB} + S_{BR})(D^R s_{iB} + D^R s_{BR}) \right]$$

Before we multiply out the terms, we use the Euler transformation to shift the rotational derivative of $D^R s_{iB}$ to the B frame $D^R s_{iB} = D^B s_{iB} + \Omega^{BR} s_{iB}$ and take advantage of the rigid-body assumption, i.e., $D^B s_{iB} = 0$ (all particles are fixed wrt the c.m.):

$$I_R^{BR} = \sum_i m_i \left[(S_{iB} + S_{BR})(\Omega^{BR} s_{iB} + D^R s_{BR}) \right]$$

$$= \sum_i m_i S_{iB} \Omega^{BR} s_{iB} + \left(\sum_i m_i \right) S_{BR} D^R s_{BR}$$

$$+ S_{BR} \Omega^{BR} \sum_i m_i s_{iB} + \left(\sum_i m_i S_{iB} \right) D^R s_{BR} \qquad (6.20)$$

The last two terms vanish because B is the c.m. The first term on the right-hand side is modified by first reversing the vector product and then transposing it to remove the negative sign:

$$\sum_i m_i S_{iB} \Omega^{BR} s_{iB} = -\sum_i m_i S_{iB} S_{iB} \omega^{BR} = \sum_i m_i \bar{S}_{iB} S_{iB} \omega^{BR} \qquad (6.21)$$

Eureka, we have unearthed the MOI tensor $\sum_i m_i \bar{S}_{iB} S_{iB} = I_B^B$ [see Eq. (6.1)]! The first term therefore becomes

$$\sum_i m_i S_{iB} \Omega^{BR} s_{iB} = I_B^B \omega^{BR} \qquad (6.22)$$

The second term of Eq. (6.20) is simply

$$\left(\sum_i m_i \right) S_{BR} D^R s_{BR} = m^B S_{BR} v_B^R$$

Substituting these terms into Eq. (6.20) yields

$$I_R^{BR} = I_B^B \omega^{BR} + m^B S_{BR} v_B^R$$

and proves the theorem.

The angular momentum of Eq. (6.18) consists of a rotary part $I_B^B \omega^{BR}$ with the angular velocity ω^{BR} of the body wrt the reference frame and a transfer term $m^B S_{BR} v_B^R$ with all mass concentrated at the c.m. If the reference point is the c.m. itself, $S_{BB} = 0$, and the transfer term vanishes:

$$l_B^{BR} = I_B^B \omega^{BR} \tag{6.23}$$

Because the displacement vector s_{BR} is not part of the calculations any longer, the angular momentum has become independent of the translational motion v_B^R of the body's mass center. What a welcome simplification! The c.m. as reference point separates the translational dynamics from the attitude motions.

Example 6.5   Change of Reference Frame

Problem.   Suppose the angular momentum l_B^{BI} of vehicle B wrt the J2000 inertial frame I and referred to the vehicle's c.m. B is known only wrt the Earth frame E, i.e., l_B^{BE}. What is the error if we neglect the difference?

Solution.   Expand the angular velocity $\omega^{BI} = \omega^{BE} + \omega^{EI}$ and substitute it into Eq. (6.23):

$$l_B^{BI} = I_B^B \omega^{BE} + I_B^B \omega^{EI}$$

The first term on the right-hand side is $I_B^B \omega^{BE} = l_B^{BE}$, and therefore the error is $I_B^B \omega^{EI}$.

Do you appreciate now the significance of the MOI? Because it is a second-order tensor, it acts like a transformation that converts the angular velocity vector into the angular momentum vector. However, the MOI being a symmetrical tensor alters not only the direction but also the magnitude of ω^{BR}.

6.2.3   Angular Momentum of Clusters of Bodies

Most aerospace vehicles consist of more than one body. Aircraft have, besides their basic airframe, rotating machinery like propellers, compressors, and turbines; and, as moving parts, control surfaces and landing gears. Missiles possess control surfaces and sometimes even spinning parts for stabilization. Certainly, you have heard of the Hubble telescope and its control momentum gyros, which point the aperture within a few microradians.

To calculate the total angular momentum, we could simply sum over all of the particles of all of the bodies in the cluster. This approach would bring no new insight. Instead, we derive a formula that takes the individual known angular momenta and combines them to form the total angular momentum (see Fig. 6.9).

Theorem: The angular momentum of a collection of rigid bodies $B_k, k = 1, 2, 3, \ldots$ (with their respective centers of mass B_k) relative to a reference frame R and referred to one of its points R is given by

$$l_R^{\Sigma B_k R} = \sum_k \left(I_{B_k}^{B_k} \omega^{B_k R} + m^{B_k} S_{B_k R} v_{B_k}^R \right) \tag{6.24}$$

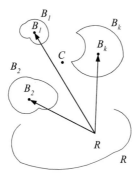

Fig. 6.9 Cluster of rigid bodies.

The individual points B_k can be moving relative to each other, but the bodies themselves must be rigid.

Proof: The proof follows from the additive properties of angular momenta [Eq. (6.17)], and the separation into rotary and particle terms [Eq. (6.18)]. To get the total angular momentum, we sum over all individual bodies

$$l_R^{\Sigma B_k R} = \sum_k l_R^{B_k R}$$

and adopt Eq. (6.18) for each body B_k

$$l_R^{B_k R} = I_{B_k}^{B_k} \omega^{B_k R} + m^{B_k} S_{B_k R} v_{B_k}^R$$

to prove the theorem

$$l_R^{\Sigma B_k R} = \sum_k \left(I_{B_k}^{B_k} \omega^{B_k R} + m^{B_k} S_{B_k R} v_{B_k}^R \right)$$

Equation (6.24) makes a general statement about clustered bodies. For many applications, like aircraft propellers, turbines, helicopter rotors, dual-spin satellites and flywheel stabilizers, this relationship can be simplified. In these cases the c.m. of the individual bodies are mutually fixed and so is the common c.m. Introducing this common c.m. C as reference point leads to a simpler formulation.

Theorem: If the common c.m. C of the cluster of bodies ΣB_k is introduced as reference point and if the individual c.m. B_k do not translate wrt C, then the angular momentum of the entire cluster wrt to the reference frame R and referred to the common c.m. C is

$$l_C^{\Sigma B_k R} = \sum_k I_{B_k}^{B_k} \omega^{B_k R} + \left(\sum_k m^{B_k} \bar{S}_{B_k C} S_{B_k C} \right) \omega^{CR} \qquad (6.25)$$

Compare Eqs. (6.25) and (6.24). The linear velocity $v_{B_k}^R$ does not appear any longer because we adopted the common c.m. (just as in the single-body case). Following earlier convention, we distinguish the two terms as rotary and transfer contributions. The second term concentrates the mass of body frame B_k in its c.m. B_k for the angular momentum calculation. According to Eq. (6.8), derived from the Huygen's theorem, the term in parentheses is the MOI of all of the individual body masses m^{B_k} referred to the common c.m. The vector ω^{CR} relates the angular velocity of frame C, consisting of the points B_k, to the reference frame R.

Proof: To prove this theorem, some stamina is required. The easier path is to accept the theorem and drop down to the example. For the proof we take three steps:

1) Introduce the vector triangle to include the total c.m. C

$$s_{B_k R} = s_{B_k C} + s_{CR}$$

into Eq. (6.24)

$$I_R^{\Sigma B_k R} = \sum_k I_{B_k}^{B_k} \omega^{B_k R} + \sum_k m^{B_k} (s_{B_k C} + s_{CR}) D^R (s_{B_k C} + s_{CR})$$

and execute the multiplications

$$I_R^{\Sigma B_k R} = \sum_k I_{B_k}^{B_k} \omega^{B_k R} + \sum_k m^{B_k} s_{B_k C} D^R s_{B_k C} + \sum_k m^{B_k} s_{CR} D^R s_{CR}$$

$$+ \left(\sum_k m^{B_k} s_{B_k C} \right) D^R s_{CR} + s_{CR} \left(\sum_k m^{B_k} D^R s_{B_k C} \right) \qquad (6.26)$$

The last two terms are zero because C is the common c.m. Let us demonstrate this fact for the last term. Because the body's mass is a constant scalar, it can be brought inside the rotational derivative, and the summation can be exchanged with the time derivative, resulting in

$$s_{CR} \left(\sum_k m^{B_k} D^R s_{B_k C} \right) = s_{CR} D^R \left(\sum_k m^{B_k} s_{B_k C} \right) = s_{CR} D^R 0 = 0$$

2) Now let the arbitrary reference point R be the common c.m., then $s_{CR} = s_{CC} = 0$, and the second term of Eq. (6.26) is zero. We are left with two terms:

$$I_C^{\Sigma B_k R} = \sum_k I_{B_k}^{B_k} \omega^{B_k R} + \sum_k m^{B_k} s_{B_k C} D^R s_{B_k C} \qquad (6.27)$$

The first term is the sum of all rotary angular momenta. The second term is expanded by transforming the rotational derivative to the frame C:

$$\sum_k m^{B_k} s_{B_k C} D^R s_{B_k C} = \sum_k m^{B_k} s_{B_k C} (D^C s_{B_k C} + \Omega^{CR} s_{B_k C})$$

$$= \sum_k m^{B_k} s_{B_k C} D^C s_{B_k C} + \sum_k m^{B_k} s_{B_k C} \Omega^{CR} s_{B_k C}$$

3) In addition, because the individual c.m. B_k and the common c.m. C are fixed in frame C, $D^C s_{B_k C} = 0$, and the first term is also zero, leaving Eq. (6.27) with

$$I_C^{\Sigma B_k R} = \sum_k I_{B_k}^{B_k} \omega^{B_k R} + \sum_k m^{B_k} s_{B_k C} \Omega^{CR} s_{B_k C}$$

The last term can be modified by a procedure we have used before [see Eq. (6.21)], and thus the proof is complete:

$$l_C^{\Sigma B_k R} = \sum_k I_{B_k}^{B_k} \omega^{B_k R} + \left(\sum_k m^{B_k} \bar{S}_{B_k C} S_{B_k C} \right) \omega^{CR}$$

Quite frequently, in aerospace applications one of the bodies is the main body, supporting all other spinning bodies. It takes on the function of frame C, but its own c.m. is not the common c.m. C. If that body is called B_1, then the theorem becomes

$$l_C^{\Sigma B_k R} = \sum_k I_{B_k}^{B_k} \omega^{B_k R} + \left(\sum_k m^{B_k} \bar{S}_{B_k C} S_{B_k C} \right) \omega^{B_1 R}$$

Example 6.6 Propeller Airplane

Problem. Determine the angular momentum of a single propeller-driven airplane wrt the inertial frame I. The propeller P with mass m^P and c.m. P is displaced from the reference point T at the tip of the airplane by s_{PT}. The c.m. B of the airframe B with mass m^B is displaced from T by s_{BT}. Their MOI are I_P^P and I_B^B and their angular velocities ω^{PB} and ω^{BI}, respectively.

The components of the tensors in airframe coordinates are for the propeller

$$[I_P^P]^B = \begin{bmatrix} I_{P1} & 0 & 0 \\ 0 & I_{P2} & 0 \\ 0 & 0 & I_{P3} \end{bmatrix}, \quad [\omega^{PB}]^B = \begin{bmatrix} \omega_P \\ 0 \\ 0 \end{bmatrix}, \quad [s_{PT}]^B = \begin{bmatrix} s_{PT1} \\ 0 \\ 0 \end{bmatrix}$$

and for the airframe

$$[I_B^B]^B = \begin{bmatrix} I_{B11} & 0 & I_{B13} \\ 0 & I_{B2} & 0 \\ I_{B13} & 0 & I_{B33} \end{bmatrix}, \quad [\omega^{BI}]^B = \begin{bmatrix} p \\ q \\ r \end{bmatrix}, \quad [s_{BT}]^B = \begin{bmatrix} s_{BT1} \\ 0 \\ 0 \end{bmatrix}$$

Solution. To determine the total angular momentum, we apply Eq. (6.25), referred to the inertial frame I:

$$l_C^{\Sigma B_k I} = \sum_k I_{B_k}^{B_k} \omega^{B_k I} + \left(\sum_k m^{B_k} \bar{S}_{B_k C} S_{B_k C} \right) \omega^{CI}$$

We are dealing with the propeller P and the airframe B serving as frame C:

$$l_C^{\Sigma B_k I} = I_P^P(\omega^{PB} + \omega^{BI}) + I_B^B \omega^{BI} + m^P \bar{S}_{PC} S_{PC} \omega^{BI} + m^B \bar{S}_{BC} S_{BC} \omega^{BI} \quad (6.28)$$

To determine the individual c.m. displacement vectors s_{PC} and s_{BC}, we first get the location of the common c.m. C from the reference point T

$$s_{CT} = \frac{m^B s_{BT} + m^P s_{PT}}{(m^B + m^P)}$$

and then the desired c.m. locations wrt the common c.m.

$$s_{PC} = s_{PT} - s_{CT}$$

$$s_{BC} = s_{BT} - s_{CT}$$

By eliminating s_{CT},

$$s_{PC} = \frac{m^B}{m^B + m^P}(s_{PT} - s_{BT})$$

$$s_{BC} = \frac{m^P}{m^B + m^P}(s_{BT} - s_{PT})$$

(6.29)

We have derived the solution in an invariant form, represented by Eqs. (6.28) and (6.29).

For developing the component form, we express Eq. (6.28) in airframe coordinates $]^B$

$$[l_C^{\Sigma B_k I}]^B = [I_P^P]^B([\omega^{PB}]^B + [\omega^{BI}]^B) + [I_B^B]^B[\omega^{BI}]^B + m^P[\bar{S}_{PC}]^B[S_{PC}]^B[\omega^{BI}]^B$$

$$+ m^B[\bar{S}_{BC}]^B[S_{BC}]^B[\omega^{BI}]^B$$

(6.30)

and then insert the components. Multiplying the matrices yields

$$[l_C^{\Sigma B_k I}]^B = \begin{bmatrix} I_{P1}(p + \omega_P) + I_{B11}p + I_{B13}r \\ (I_{P2} + I_{B2})q \\ (I_{P3} + I_{B33})r + I_{B13}p \end{bmatrix} + (m^P s_{PC1}^2 + m^B s_{BC1}^2)\begin{bmatrix} 0 \\ q \\ r \end{bmatrix}$$

(6.31)

where s_{PC1} and s_{BC1} are the first components of the vectors in Eq. (6.29). The second term affects only the pitch and yaw angular momenta.

Frequently, several simplifications are justified. With $\omega_p \gg p$ and $m^P s_{PC1}^2 \gg m^B s_{BC1}^2$ we can reduce Eq. (6.31) to

$$[l_C^{\Sigma B_k I}]^B = \begin{bmatrix} I_{P1}\omega_P + I_{B11}p + I_{B13}r \\ (I_{P2} + I_{B2})q \\ (I_{P3} + I_{B33})r + I_{B13}p \end{bmatrix} + m^P s_{PC1}^2 \begin{bmatrix} 0 \\ q \\ r \end{bmatrix}$$

More drastically, the second term and the product of inertia I_{B13} are sometimes dropped (only the principal MOIs I_{B1}, I_{B2}, and I_{B3} are left), and the MOI of the propeller is assumed much smaller than that of the airframe. Then we arrive at a popular representation that just adds the angular momentum of the propeller to that of the airframe

$$[l_C^{\Sigma B_k I}]^B = \begin{bmatrix} I_{P1}\omega_P + I_{B1}p \\ I_{B2}q \\ I_{B3}r \end{bmatrix}$$

You may have seen this form in the literature. It is quite adequate for most propeller airplanes, but you should be aware of the hidden assumptions.

Another entity, the angular momentum, has joined our collection of building blocks, but it is more sophisticated than the other items. It requires three defining super- and subscripts. The first frame represents the material body, followed by an arbitrary reference frame and a reference point. Frequently, the reference point is the c.m., and an inertial frame serves as reference. This situation arises in particular when we formulate the attitude equations of flight vehicles from Euler's law.

6.3 Euler's Law

Rapidly we reach the climax of Part 1. Its first pillar is Newton's second law, expressing the translational dynamics of aerospace vehicles using the linear momentum. With the angular momentum defined we are prepared to formulate Euler's law, the second pillar of flight dynamics.

We will begin with a historical argument that splits the dynamicists into two camps, the Newtonians and the Eulerians, though the consequences for modeling and simulation are zilch. The particle again will serve the elemental formulation, from which we derive two forms of Euler's law. Most important for us is the *free-flight* exposition, serving all aerospace vehicle applications. The other form, the *spinning top* with one point fixed, is more of historical and academic significance. Dealing with clustered bodies will be a venture for us. Fortunately, most air- and spacecraft contain spinning bodies with fixed mass centers. These arrangements can be treated in a straightforward manner. For moving bodies the formulation of Euler's equation gives us access to many challenging modeling tasks.

6.3.1 Two Approaches

Just as Newton's second law describes the translational degrees of freedom of a flight vehicle so does Euler's law govern the attitude degrees of freedom. Its origin is attributed to Euler and is considered either a consequence of Newton's law (Goldstein) or a fundamentally new principle of dynamics (Truesdell).

6.3.1.1 Euler's Law according to Truesdell.
Truesdell,[1] having conducted a thorough historical research, concluded that Euler's law in its embryonic form is based on a publication by Jakob Bernoulli (1686), predating the Newtonian laws by one year. Euler polished Bernoulli's ideas and formulated the angular momentum law as an independent principle of mechanics in 1744. In its elementary form we state it first for a particle (refer to Fig. 6.10).

The inertial time rate of change of angular momentum about a point is equal to and in the direction of the impressed moment about the same point. Consider a particle m_i, displaced from the reference point I by s_{iI} and moving with the linear velocity v_i^I wrt the inertial frame I. Its angular momentum is $l_{iI}^I = m_i S_{iI} v_i^I$, and the impressed moment relative to point I is $m_{iI} = S_{iI} f_i$, where f_i is the force acting on the particle.

Euler's law for such a particle states that the time rate of change wrt the inertial frame I of the angular momentum l_{iI}^I equals the external moment m_{iI}:

$$D^I l_{iI}^I = m_{iI} \tag{6.32}$$

and expanded

$$D^I \left(m_i S_{iI} v_i^I \right) = S_{iI} f_i \tag{6.33}$$

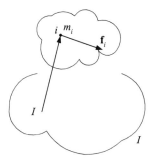

Fig. 6.10   Euler's law of a particle.

On each side of the equation is a vector product of the displacement vector s_{iI} with either the linear velocity v_i^I (related to the displacement vector by $v_i^I = D^I s_{iI}$) or the force f_i.

We introduced a new vector, the moment m_{iI} acting on particle i wrt a point I. It should not be confused with the scalar m_i, the mass of particle i. Now let us turn to the other interpretation.

6.3.1.2   Euler's law according to Goldstein.

The prevalent opinion of most books on classical mechanics or dynamics reflects the Newtonian viewpoint. I cite Goldstein[2] only as an example. Actually, it was Daniel Bernoulli who issued the first account coinciding with Euler's publication in 1744. Accordingly, the angular momentum equation can be derived from Newton's linear momentum law.

Starting with Eq. (5.6), premultiply Newton's law for a particle i by the skew-symmetric displacement vector S_{iI}:

$$S_{iI}D^I\left(m_i v_i^I\right) = S_{iI}f_i$$

If we can show that the left side equals that of Eq. (6.33), we have obtained Euler's law. Apply the chain rule to the left side of Eq. (6.33):

$$D^I\left(m_i S_{iI}v_i^I\right) = m_i D^I S_{iI}v_i^I + S_{iI}D^I\left(m_i v_i^I\right)$$
$$= m_i V_i^I v_i^I + S_{iI}D^I\left(m_i v_i^I\right) = S_{iI}D^I\left(m_i v_i^I\right)$$

Because the vector product of v_i^I with itself is zero, the equality is established. Therefore,

$$D^I\left(m_i S_{iI}v_i^I\right) = S_{iI}f_i$$

and with the angular momentum already introduced $l_{iI}^I = m_i S_{iI}v_i^I$ and moment $m_{iI} = S_{iI}f_i$ we get Euler's law:

$$D^I l_{iI}^I = m_{iI}$$

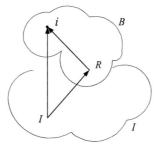

Fig. 6.11   Arbitrary reference point.

Again we are faced with the choice of the inertial frame. The options we considered for Newton's law are also pertinent here. Most often, for near-Earth simulations, we use the J2000 reference frame. If our vehicle is hugging the Earth, we can use the Earth itself.

I proceed now to derive two formulations that are most applicable to the modeling of aerospace vehicles. The first case represents Euler's law of a rigid body referred to its mass center. This is the basis for the attitude equations of flight vehicles. The other formulation, useful for gyro dynamics, is Euler's law of a rigid body referred to a point that is fixed both in the body and inertial frames and need not be the center of mass.

6.3.2   Free Flight

Let us begin by summing Euler's law Eq. (6.32) over all particles of a rigid body

$$\sum_i D^I l_{iI}^I = \sum_i m_{iI}$$

and do the same for its alternate form Eq. (6.33)

$$\sum_i D^I \left(m_i S_{iI} D^I s_{iI} \right) = \sum_i (S_{iI} f_i) \qquad (6.34)$$

where all internal moments cancel each other and only the external moments remain. The linear velocity was replaced by its time derivative of the displacement vector s_{iI}. Introduce for the time being an arbitrary reference point R (see Fig. 6.11) of the rigid body B into Eq. (6.34):

$$s_{iI} = s_{iR} + s_{RI}$$

We obtain six terms:

$$\sum_i D^I \left(m_i S_{iR} D^I s_{iR} \right) \text{ Term (1)} + \sum_i D^I \left(m_i S_{RI} D^I s_{RI} \right) \text{ Term (2)}$$

$$+ \sum_i D^I \left(m_i S_{iR} D^I s_{RI} \right) \text{ Term (3)} + \sum_i D^I \left(m_i S_{RI} D^I s_{iR} \right) \text{ Term (4)} \quad (6.35)$$

$$= \sum_i (S_{iR} f_i) \text{ Term (5)} + \sum_i (S_{RI} f_i) \text{ Term (6)}$$

At this point we split the treatment into the two cases. First, we confine the reference point R to the c.m. B and develop the free-flight attitude equations. Afterward, we let R be any point of body B and assign it also as a point of the inertial frame I, thus addressing the dynamics of the top.

Let us modify the six terms of Eq. (6.35). The inner rotational derivative of the first term is transformed to frame B, and because B is a point of frame B, $D^B s_{iB} = 0$.

Term (1):

$$\sum_i D^I (m_i S_{iB} D^I s_{iB}) = \sum_i D^I [m_i S_{iB} (D^B s_{iB} + \Omega^{BI} s_{iB})] = \sum_i D^I (m_i S_{iB} \Omega^{BI} s_{iB})$$

Referring back to Eq. (6.22), we conclude that the term in parentheses is the MOI I_B^B of the vehicle multiplied by its inertial angular velocity ω^{BI}, and therefore

$$\text{Term (1)} = D^I (I_B^B \omega^{BI})$$

Term (2):

$$D^I (m^B S_{BI} D^I s_{BI}) = S_{BI} D^I (m^B v_B^I) + m^B D^I S_{BI} v_B^I = S_{BI} D^I (m^B v_B^I)$$
$$+ m^B V_B^I v_B^I = S_{BI} D^I (m^B v_B^I)$$

because the cross product is zero.

Term (3):

$$D^I \left[\sum_i (m_i S_{iB}) D^I s_{BI} \right] = 0$$

because B is the c.m.

Term (4):

$$\sum_i D^I (m_i S_{BI} D^I s_{iB}) = D^I \left(S_{BI} \sum_i m_i D^I s_{iB} \right) = D^I \left[S_{BI} D^I \left(\sum_i m_i s_{iB} \right) \right] = 0$$

because m_i is constant and B is the c.m.

Term (5):

$$\sum_i (S_{iB} f_i) = m_B$$

total external moment.

Term (6):

$$\sum_i (S_{BI} f_i) = S_{BI} f$$

because all internal forces cancel. The modified Terms (2) and (6) express Newton's second law premultiplied by S_{BI} and are therefore satisfied identically (S_{BI} is generally not zero). From the remaining Terms (1) and (5) we receive our final result:

$$D^I\left(I^B_B\omega^{BI}\right) = m_B \tag{6.36}$$

where according to Eq. (6.23) $I^B_B\omega^{BI} = l^{BI}_B$ is the angular momentum of body B wrt the inertial frame and referred to the c.m. Euler's law for rigid bodies states therefore that the time rate of change relative to the inertial frame of the angular momentum l^{BI}_B of a rigid body referred to its c.m. is equal to the externally applied moment m_B with the c.m. as reference point

$$D^I l^{BI}_B = m_B \tag{6.37}$$

Equation (6.36) does not include any reference to the linear velocity or acceleration of the vehicle. What a fortuitous characteristic! Euler's law is applied as if the vehicle were not translating. This feature is referred to as the *separation theorem*. Just as linear and angular momenta can be calculated separately, then so can the translational equations of motion be formulated separately from the attitude equations. Newton's second law, Eq. (5.9), and Euler's law, Eq. (6.36), deliver the fundamental equations of aerospace vehicle dynamics

$$m^B D^I v^I_B = f, \quad D^I\left(I^B_B\omega^{BI}\right) = m_B$$

and with the compact nomenclature of linear and angular momenta

$$D^I p^I_B = f, \quad D^I l^{BI}_B = m_B \tag{6.38}$$

The key point is the c.m. B. It serves as the focal point for the linear momentum p^I_B, encompassing all mass of body B as if it were a particle. For the angular momentum l^{BI}_B it is that reference point which separates the attitude motions from the translational degrees of freedom. As I will show, without the c.m. as reference point the equations of motion of aerospace vehicles are more complex.

As a historical tidbit, I want to mention that the equations of motion (6.38), which we like to call today the six-DoF equations, have been known for quite some time. In 1924, while aviation was still in its infancy, R.v. Mises published the "Bewegungsgleichungen eines Flugzeuges," buried in his so-called *Motor Rechnung*.[3] He presented the translational and attitude equations in one compact formalism, already transformed to body coordinates, and identified the key external forces and moments. There we even find the inception of small perturbations and linearized equations of motion for an airplane.

Example 6.7 Aero Data Reference Point

Frequently, the aerodynamic and propulsive data are not given relative to the c.m. but to an arbitrary reference point of the aircraft or missile. If you have been involved in wind-tunnel testing, you have dealt with the moment center of the sting balance, which is usually nowhere close to the yet unknown c.m. of the flight vehicle. Or, as the space shuttle burns fuel during its ascent, large shifts of c.m. occur. In each case we need to modify the right side of Eq. (6.37).

Figure 6.12 shows the aerodynamic force f and moment m_{B_r} acting on the fixed reference center B_r. To calculate the moment m_B, referred to the c.m. B, we

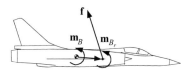

Fig. 6.12   Moment centers.

determine the torque $S_{B_r B} f$ caused by changing the origin of the force vector f, and add the free moment vector m_{B_r}:

$$m_B = m_{B_r} + S_{B_r B} f \tag{6.39}$$

Substituting Eq. (6.39) into Eq. (6.37) yields Euler's equation of motion referred to the c.m. B, but with the aerodynamics referenced to the arbitrary point B_r:

$$D^I l_B^{BI} = m_{B_r} + S_{B_r B} f \tag{6.40}$$

For an aircraft and missile the displacement vector $s_{B_r B}$ most likely will change in time, as the c.m. shifts during flight. Similar adjustments are made if the propulsion moment center changes.

Example 6.8   Attitude Equations for Six-DoF Simulations

Missile simulations use Euler's equation in a form that accommodates aerodynamic moment coefficients and the MOI tensor in body coordinates. We transfer the rotational time derivative of Eq. (6.36) to the body frame B

$$D^B \left(I_B^B \omega^{BI} \right) + \Omega^{BI} I_B^B \omega^{BI} = m_B$$

and pick body coordinates $]^B$

$$D^B \left([I_B^B]^B [\omega^{BI}]^B \right) + [\Omega^{BI}]^B [I_B^B]^B [\omega^{BI}]^B = [m_B]^B$$

Applying the chain rule to the first term and realizing that the MOI of a rigid body remains unchanged in time, $[dI_B^B/dt]^B = [0]$, we get the desired equations for programming:

$$[I_B^B]^B \left[\frac{d\omega^{BI}}{dt} \right]^B + [\Omega^{BI}]^B [I_B^B]^B [\omega^{BI}]^B = [m_B]^B \tag{6.41}$$

This is the attitude equation most frequently found in six-DOF simulations.

Euler's law, like Newton's second law, must be referred to an inertial frame and, for simplicity's sake, should be referred to the vehicle's c.m. Yet, just as in Sec. 5.3, we ask what are the correction terms if we change to a noninertial reference frame or an arbitrary body point.

6.3.2.1 Noninertial reference frame. Shifting the reference frame from inertial I to noninertial R, but maintaining the vehicles c.m. B as reference point, incurs two additional terms in Euler's equation. We start with Eq. (6.37) and transform the rotational derivative to the R frame:

$$D^R l_B^{BI} + \Omega^{RI} l_B^{BI} = m_B$$

The first term is modified first by replacing the angular momentum with Eq. (6.23) and introducing the angular velocity relationship between the three frames B, R, and I: $\omega^{BI} = \omega^{BR} + \omega^{RI}$,

$$D^R(I_B^B \omega^{BR}) + D^R(I_B^B \omega^{RI}) + \Omega^{RI} l_B^{BI} = m_B$$

where $I_B^B \omega^{BR} = l_B^{BR}$ is the angular momentum wrt the frame R. Now, the two correction terms are exposed on the right-hand side of Euler's equation:

$$D^R l_B^{BR} = m_B - \Omega^{RI} l_B^{BI} - D^R(I_B^B \omega^{RI}) \tag{6.42}$$

They consist of the precession term $\Omega^{RI} l_B^{BI}$ [see Eq. (6.57)] and the reference rate term $D^R(I_B^B \omega^{RI})$. For instance, if we used Earth E instead of the J2000 as inertial frame

$$D^E l_B^{BE} = m_B - \Omega^{EI} l_B^{BI} - D^E(I_B^B \omega^{EI})$$

the two terms $-\Omega^{EI} l_B^{BI}$ and $-D^E(I_B^B \omega^{EI})$ tell us whether the error is acceptable.

6.3.2.2 Arbitrary reference point. Euler's law takes on its simplest form if the vehicle's c.m. is used as reference point. Sometimes, however, it is desirable to use another point of the vehicle as reference. In Sec. 5.3.2 we used the example of a satellite with an asymmetric solar panel. It was more relevant to derive the translational equation relative to the geometrical center of the satellite B_r than the c.m. B. Now we force the attitude equation into the same mold by following Grubin.[4]

Beginning with Eq. (6.38), Newton's and Euler's equations are

$$m^B D^I D^I s_{BI} = f \tag{6.43}$$

$$D^I l_B^{BI} = m_B \tag{6.44}$$

We transform the angular momentum with the help of Eq. (6.19) to point B_r

$$l_B^{BI} = I_{B_r}^B \omega^{BI} - m^B S_{BB_r} D^I s_{BB_r}$$

and likewise shift the moment center to B_r using Eq. (6.39)

$$m_B = m_{B_r} - S_{BB_r} f$$

Both transformations are substituted into Eq. (6.44) and yield

$$D^I(I_{B_r}^B \omega^{BI}) - m^B D^I(S_{BB_r} D^I s_{BB_r}) = m_{B_r} - S_{BB_r} f$$

Applying the chain rule to the second term and using Eq. (6.43) for the last term provides

$$D^I(I_{B_r}^B \omega^{BI}) - m^B S_{BB_r} D^I D^I s_{BB_r} = m_{B_r} - m^B S_{BB_r} D^I D^I s_{BI}$$

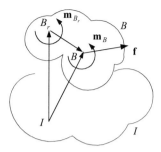

Fig. 6.13   Arbitrary reference point.

Introducing the vector triangle from Fig. 6.13 and taking the rotational derivative twice,

$$D^I D^I s_{BI} = D^I D^I s_{BB_r} + D^I D^I s_{B_r I}$$

and substituting into the last term provides, after canceling two terms,

$$D^I \left(I^B_{B_r} \omega^{BI} \right) = m_{B_r} - m^B S_{BB_r} D^I D^I s_{B_r I} \tag{6.45}$$

We have arrived at Euler's law referred to an arbitrary body point B_r:

$$D^I \left(I^B_{B_r} \omega^{BI} \right) = m_{B_r} - m^B S_{BB_r} D^I v^I_{B_r} \tag{6.46}$$

The last term adjusts for the fact that B_r is not the c.m. The linear velocity $v^I_{B_r}$ couples into the translational equation derived earlier [see Eq. 5.27)]:

$$m^B D^I v^I_{B_r} = f - m^B \begin{Bmatrix} \Omega^{BI} \Omega^{BI} s_{BB_r} & \text{centrifugal acceleration} \\ +(D^I \Omega^{BI}) s_{BB_r} & \text{angular acceleration} \end{Bmatrix}$$

and the angular velocity ω^{BI} connects back to the attitude equation. Both equations constitute the complete set of six-DoF equations of motion for an arbitrary reference point of body B. They are more complicated than the standard set Eq. (6.38). If B_r is the c.m. B, then $s_{BB_r} = 0$; the two equations uncouple and reduce to Eq. (6.38).

Example 6.9   Physical Pendulum with Moving Hinge Point

Problem.   You probably have solved this nontrivial problem before by the Lagrangian methodology. I will demonstrate here that Eq. (6.46) leads in a straight-forward manner to the solution.

The physical pendulum with mass m^B and MOI $I^B_{B_r}$ swings about the hinge point B_r, which in turn is excited by the forcing function $[s_{B_r I}]^I = [A \sin \omega t \quad 0 \quad 0]$ in inertial coordinates, as indicated in Fig. 6.14. What is the differential equation that governs the dynamics of the pendulum? The MOI is given in body coordinates

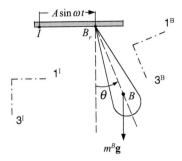

Fig. 6.14   Physical pendulum.

$[I_{B_r}^B]^B = [\text{diag}(I_1, I_2, I_3)]$, and the displacement of the c.m. of the pendulum B from the hinge point by $[s_{BB_r}]^B = [0 \ 0 \ l]$.

Solution.   To solve the problem, we express Eq. (6.45) in body coordinates with the exception of the inertial acceleration:

$$[I_{B_r}^B]^B \left[\frac{d\omega^{BI}}{dt}\right]^B + [\Omega^{BI}]^B [I_{B_r}^B]^B [\omega^{BI}]^B = [m_{B_r}]^B - m^B [S_{B,B}]^B [T]^{BI} \left[\frac{d^2 s_{B,I}}{dt^2}\right]^I$$

where $[\omega^{BI}]^B = [0 \ \dot\theta \ 0]$, $[m_{B_r}]^B = [0 \ -m^B gl \sin\theta \ 0]$, and $[d^2 s_{B,I}/dt^2]^I = [-A\omega^2 \sin\omega t \ 0 \ 0]$. Multiplying the matrices yields the equation of motion

$$I_2 \ddot\theta + m^B gl \sin\theta = m^B lA \omega^2 \sin\omega t \cos\theta$$

If you have tried to solve this problem before by the conventional method, you will agree that my method is easier.

After having dealt with the more important case, namely the free-flight attitude equations, we consider point R of Eq. (6.35) to be simultaneously a point of the body and the inertial reference frame, but not necessarily the c.m. A body with a contact point on the ground, the so-called top, can serve as an example.

6.3.3   Top

You may have played in your childhood with such a cone-shaped object and kept it spinning by lashing at it with the end of a whip. It made marvelous jumps, seemingly defying the law of gravity, as long as it spun fast enough. Now you realize that it is its angular momentum which stabilizes it.

Euler's law governs the dynamics of the top. We derive its specialized form by considering the reference point R a point of body B, which implies that for any particle i, $D^B s_{iR} = 0$. Furthermore, R is also the reference point I, thus $s_{RI} = s_{II} = 0$. Starting with the terms of Eq. (6.35), we modify them like before, except this time we cannot take advantage of the simplifications brought about by the c.m:

Term (1):

$$\sum_i D^I \left(m_i S_{iR} D^I s_{iR} \right) = \sum_i D^I \left[m_i S_{il} \left(D^B s_{il} + \Omega^{BI} s_{il} \right) \right]$$

$$= \sum_i D^I \left(m_i S_{il} \Omega^{BI} s_{il} \right) = D^I \left(\mathbf{I}_I^B \omega^{BI} \right)$$

Term (2):

$$\sum_i D^I \left(m_i S_{RI} D^I s_{RI} \right) = \mathbf{0}$$

because $s_{RI} = s_{II} = \mathbf{0}$.

Term (3):

$$\sum_i D^I \left(m_i S_{iR} D^I s_{RI} \right) = \sum_i D^I \left(m_i S_{iR} v_R^I \right) = \mathbf{0}$$

because $v_R^I = \mathbf{0}$.

Term (4):

$$\sum_i D^I \left(m_i S_{RI} D^I s_{iR} \right) = \mathbf{0}$$

because $S_{RI} = S_{II} = \mathbf{0}$.

Term (5):

$$\sum_i (S_{iR} f_i) = \sum_i (S_{il} f_i) = m_I$$

Term (6):

$$\sum_i (S_{RI} f_i) = S_{RI} f = \mathbf{0}$$

because $S_{RI} = S_{II} = \mathbf{0}$.

Only Terms (1) and (5) remain. Euler's law for a body spinning about a fixed point I is

$$D^I \left(\mathbf{I}_I^B \omega^{BI} \right) = m_I \qquad (6.47)$$

Compare both formulations, Eqs. (6.36) and (6.47). They are distinguishable only by the reference points. In both cases, whether it is the c.m. or a body/inertial reference point, Euler's law assumes the same simple form.

Example 6.10   Force-Free Top

A moment free symmetric body spins about its minor principal MOI axis and is supported at the bottom of its spin axis. Its MOI in body coordinates is

$$\left[I_I^B \right]^B = \begin{bmatrix} I_1 & 0 & 0 \\ 0 & I_2 & 0 \\ 0 & 0 & I_2 \end{bmatrix}$$

where the minor MOI is in the first direction and the two others are equal. The angular velocity of the top is a constant p_0. Its attitude equations are derived from Eq. (6.47) by transforming the rotational derivative to the body frame B and expressing the terms in body coordinates $]^B$

$$\left[I_I^B\right]^B \left[\frac{d\omega^{BI}}{dt}\right]^B + \left[\Omega^{BI}\right]^B \left[I_I^B\right]^B \left[\omega^{BI}\right]^B = [0]^B$$

With the angular velocity $\overline{[\omega^{BI}]}^B = [p_0 \ \ q \ \ r]$ the equations are in body coordinates

$$\begin{bmatrix} I_1 & 0 & 0 \\ 0 & I_2 & 0 \\ 0 & 0 & I_2 \end{bmatrix} \begin{bmatrix} 0 \\ \dot{q} \\ \dot{r} \end{bmatrix} + \begin{bmatrix} 0 & -r & q \\ r & 0 & -p_0 \\ -q & p_0 & 0 \end{bmatrix} \begin{bmatrix} I_1 & 0 & 0 \\ 0 & I_2 & 0 \\ 0 & 0 & I_2 \end{bmatrix} \begin{bmatrix} p_0 \\ q \\ r \end{bmatrix} = \begin{bmatrix} 0 \\ 0 \\ 0 \end{bmatrix}$$

and evaluated

$$I_2\dot{q} - (I_2 - I_1)p_0 r = 0$$
$$I_2\dot{r} + (I_2 - I_1)p_0 q = 0$$

These are two coupled linear differential equations with pitch rate q and yaw rate r as state variables. The terms $(I_2 - I_1)p_0 r$ and $(I_2 - I_1)p_0 q$ model the gyroscopic coupling between the pitch and yaw axes. You should be able to verify the oscillatory solution

$$q = A_0 \sin(\omega_0 t), \quad r = A_0 \cos(\omega_0 t) \quad \text{with} \quad \omega_0 = \frac{I_2 - I_1}{I_2}p_0$$

A_0 depends on the initial conditions.

6.3.4 Clustered Bodies

If you are looking for a challenge, go no further than the dynamics of clustered spinning bodies. You can go back to Eq. (6.32) and sum over all particles, just as we did for a single body. Executing all of these steps would blow the chapter. Fortunately, we do not have to start from scratch, but take advantage of the angular momentum of clustered bodies, Eqs. (6.24) and (6.25). These equations serve two distinctively different situations. Equation (6.24) represents the more general case of moving bodies, whereas Eq. (6.25) assumes that all bodies c.m. are mutually fixed.

The second case is more important and easier to deal with. It applies to air vehicles with spinning machinery, like turbines, rotors, propellers, or flywheels. I will deal with it first. If your stamina has not been exhausted by then, you may continue with the more general case that applies to rotating and translating objects within the vehicle. Imagine a jeep being pushed backward in a cargo aircraft for parachute drop, or the movement of the space shuttle's manipulator arm before release of a spinning satellite. I believe, however, both cases would be fun to explore.

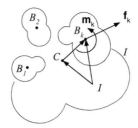

Fig. 6.15   Clustered bodies.

For both cases we begin with Euler's second law Eq. (6.32) and sum over all particles of rigid body B

$$\sum_i D^I l^I_{iI} = \sum_i m_{iI}$$

which can be abbreviated by

$$D^I l^{BI}_I = m_I$$

Now consider k rigid bodies B_k, $k = 1, 2, 3, \ldots$ with their external moments m_k and forces f_k (see Fig. 6.15). Summing over the entire cluster and shifting the reference point of the forces from their individual c.m. B_k to point I

$$m_I = \sum_k m_{B_k} + \sum_k s_{B_k I} f_k$$

yields

$$D^I \sum_k l^{B_k I}_I = \sum_k m_{B_k} + \sum_k s_{B_k I} f_k$$

where we abbreviate the left side by $D^I \sum_k l^{B_k I}_I = D^I l^{\Sigma B_k I}_I$

$$D^I l^{\Sigma B_k I}_I = \sum_k m_{B_k} + \sum_k s_{B_k I} f_k \tag{6.48}$$

We zero in on the angular momentum of clustered bodies $l^{\Sigma B_k I}_I$ using Eq. (6.24) with I as reference point

$$l^{\Sigma B_k I}_I = \sum_k \left(I^{B_k}_{B_k} \omega^{B_k I} + m^{B_k} s_{B_k I} D^I s_{B_k I} \right)$$

and introduce the common c.m. C of the cluster $s_{B_K I} = s_{B_K C} + s_{CI}$ into the last term

$$\sum_k m^{B_k} s_{B_k I} D^I s_{B_k I} = \sum_k m^{B_k} s_{B_k C} D^I s_{B_k C} + \sum_k m^{B_k} s_{CI} D^I s_{CI}$$

$$+ s_{CI} \sum_k m^{B_k} D^I s_{B_k C} + \sum_k m^{B_k} s_{B_k C} D^I s_{CI}$$

The last two terms vanish because C is the common c.m. Therefore

$$l_I^{\Sigma B_k I} = \sum_k I_{B_k}^{B_k} \omega^{B_k I} + \sum_k m^{B_k} S_{B_k C} D^I s_{B_k C} + \sum_k m^{B_k} S_{CI} D^I s_{CI} \quad (6.49)$$

At this juncture the two cases require separate treatment. For the fixed case $D^I s_{B_k C}$ can be simplified because $D^C s_{B_k C} = 0$. No such reduction is possible for the moving bodies.

6.3.4.1 Mass centers are mutually fixed.

Let us modify the second term on the right-hand side of Eq. (6.49) by transforming the rotational derivative to the C frame, which consists of the points B_k and the common c.m. C:

$$\sum_k m^{B_k} S_{B_k C} D^I s_{B_k C} = \sum_k m^{B_k} S_{B_k C} (D^C s_{B_k C} + \Omega^{CI} s_{B_k C}) = \sum_k m^{B_k} S_{B_k C} \Omega^{CI} s_{B_k C}$$

Reversing the vector product and transposing the skew-symmetric displacement vector yields

$$\sum_k m^{B_k} S_{B_k C} \Omega^{CI} s_{B_k C} = \sum_k m^{B_k} \bar{S}_{B_k C} S_{B_k C} \omega^{CI}$$

We arrive at an intermediate result if we substitute this expression into Eq. (6.49) and recognize that the first two terms on the right-hand side of Eq. (6.49) are in effect Eq. (6.25):

$$l_I^{\Sigma B_k I} = l_C^{\Sigma B_k I} + \sum_k m^{B_k} S_{CI} D^I s_{CI}$$

Substituting the angular momentum into Eq. (6.48) and introducing C as a reference point at the right-hand side, we obtain

$$D^I l_C^{\Sigma B_k I} + D^I \left(\sum_k m^{B_k} S_{CI} D^I s_{CI} \right) = \sum_k m_{B_k} + \sum_k S_{B_k C} f_k + S_{CI} \sum_k f_k$$

Applying the chain rule to the second term and combining it with the last term produces a familiar equation

$$\sum_k m^{B_k} S_{CI} D^I D^I s_{CI} = S_{CI} \sum_k f_k = S_{CI} f$$

which represents Newton's equation applied to the common c.m. It is satisfied identically, and therefore Euler's equation for bodies with mutually fixed mass centers consists of the remaining terms:

$$D^I l_C^{\Sigma B_k I} = \sum_k m_{B_k} + \sum_k S_{B_k C} f_k$$

For the final form, most useful for applications, we reintroduce Eq. (6.25):

$$\sum_k D^I (I_{B_k}^{B_k} \omega^{B_k I}) + \sum_k D^I (m^{B_k} \bar{S}_{B_k C} S_{B_k C} \omega^{CI}) = \sum_k m_{B_k} + \sum_k S_{B_k C} f_k \quad (6.50)$$

Given the MOIs of the individual bodies B_k, their displacements, angular rates, and their external moments m_k and forces f_k, we can model their attitude equation. Let us apply it to an important example.

Example 6.11   Dual-Spin Spacecraft

Problem.   A satellite, orbiting the Earth, is subject to perturbations that slowly change its attitude, unless thrusters correct the deviations. Such a control system is expensive to implement. Earlier in the space program, satellites would carry a rapidly spinning wheel that would maintain attitude just by the shear magnitude of its angular momentum.

The satellite consists of a cylindrical main body B_1 and a cylindrical rotor B_2, with their respective c.m. B_1 and B_2 and mass m^{B_1} and m^{B_2}. The rotor revolves about the third axis of the main body with the angular velocity $[\omega^{B_2 B_1}]^{B_1} = [0\ 0\ R]$, and the main body's inertial angular velocity is $[\omega^{B_1 I}]^{B_1} = [p\ q\ r]$. With the rotor placed at the common c.m., the points B_1, B_2, and C coincide. Both MOIs are referenced to B_1 and given in $]^{B_1}$ coordinates

$$\left[I_{B_1}^{B_2}\right]^{B_1} = \begin{bmatrix} I_T^{B_2} & 0 & 0 \\ 0 & I_T^{B_2} & 0 \\ 0 & 0 & I_Z^{B_2} \end{bmatrix} \quad \text{and} \quad \left[I_{B_1}^{B_1}\right]^{B_1} + \left[I_{B_1}^{B_2}\right]^{B_1} = \begin{bmatrix} I_T^{B_{1+2}} & 0 & 0 \\ 0 & I_T^{B_{1+2}} & 0 \\ 0 & 0 & I_Z^{B_{1+2}} \end{bmatrix}$$

Derive the scalar differential equations of the satellite, free of external forces and moments.

Solution.   Because the centers of mass are mutually fixed, Euler's law for clustered bodies [Eq. (6.50)] applies:

$$D^I\left(\sum_{k=1}^{2} I_{B_k}^{B_k}\omega^{B_k I}\right) = D^I\left(I_{B_1}^{B_1}\omega^{B_1 I} + I_{B_2}^{B_2}\omega^{B_2 I}\right) = 0$$

With $\omega^{B_2 I} = \omega^{B_2 B_1} + \omega^{B_1 I}$, point $B_2 = B_1$, and abbreviating $I_{B_1}^{B_1} + I_{B_1}^{B_2} = I_{B_1}^{B_1+B_2}$,

$$D^I\left(I_{B_1}^{B_1+B_2}\omega^{B_1 I} + I_{B_1}^{B_2}\omega^{B_2 B_1}\right) = 0$$

Transform the rotational derivative to the frame of the main body B_1:

$$D^{B_1}\left(I_{B_1}^{B_1+B_2}\omega^{B_1 I} + I_{B_1}^{B_2}\omega^{B_2 B_1}\right) + \Omega^{B_1 I}\left(I_{B_1}^{B_1+B_2}\omega^{B_1 I} + I_{B_1}^{B_2}\omega^{B_2 B_1}\right) = 0$$

The MOI of both the main body and the rotor (cylindrical symmetry) do not change wrt the main body; therefore, their rotational derivatives are zero, and we have arrived at the invariant formulation of the dual-spin spacecraft dynamics:

$$I_{B_1}^{B_1+B_2} D^{B_1}\omega^{B_1 I} + I_{B_1}^{B_2} D^{B_1}\omega^{B_2 B_1} + \Omega^{B_1 I} I_{B_1}^{B_1+B_2}\omega^{B_1 I} + \Omega^{B_1 I} I_{B_1}^{B_2}\omega^{B_2 B_1} = 0$$

$$(6.51)$$

Let us use the main body's coordinates $]^{B_1}$ to express the equation

$$[I_{B_1}^{B_1+B_2}]^{B_1}\left[\frac{d\omega^{B_1 I}}{dt}\right]^{B_1} + [I_{B_1}^{B_2}]^{B_1}\left[\frac{d\omega^{B_2 B_1}}{dt}\right]^{B_1} + [\Omega^{B_1 I}]^{B_1}[I_{B_1}^{B_1+B_2}]^{B_1}[\omega^{B_1 I}]^{B_1}$$

$$+[\Omega^{B_1 I}]^{B_1}[I_{B_1}^{B_2}]^{B_1}[\omega^{B_2 B_1}]^{B_1} = [0]^{B_1}$$

Substituting the components and multiplying the matrices yields the scalar differential equations of a dual-spin spacecraft:

$$I_T^{B_{1+2}}\dot{p} + \left(I_Z^{B_{1+2}} - I_T^{B_{1+2}}\right)qr + I_Z^{B_2}Rq = 0$$

$$I_T^{B_{1+2}}\dot{q} - \left(I_Z^{B_{1+2}} - I_T^{B_{1+2}}\right)pr - I_Z^{B_2}Rp = 0 \tag{6.52}$$

$$I_Z^{B_{1+2}}\dot{r} + I_Z^{B_2}\dot{R} = 0$$

The rotor's angular momentum $I_Z^{B_2}R$ dominates with its high spin rate R the term $(I_Z^{B_{1+2}} - I_T^{B_{1+2}})r$ and provides the stiffness for the satellite's stabilization.

As a historical note, the first U.S. satellite *Explorer I*, launched in February of 1958, was spin stabilized but started to tumble after a few orbits. NASA overlooked the known fact that an object with internal energy dissipation is only stable if it is spinning about its major moment of inertia axis.[5]

6.3.4.2 Mass centers are translating.
Clustered bodies whose c.m. are translating relative to each other are more difficult to treat because the common c.m. is also shifting. We start with Eq. (6.49). Substituting Eq. (6.49) directly into Eq. (6.48) and introducing $S_{B_k I} = S_{B_k C} + S_{CI}$ into the last term yields

$$\sum_k D^I\left(I_{B_k}^{B_k}\omega^{B_k I}\right) + \sum_k m^{B_k}S_{B_k C}D^I D^I s_{B_k C} + \sum_k m^{B_k}S_{CI}D^I D^I s_{CI}$$

$$= \sum_k m_{B_k} + \sum_k S_{B_k C}f_k + S_{CI}\sum_k f_k$$

where we expanded the second and third terms by the chain rule and took advantage of the vanishing vector product of like vectors. Embedded in this equation is Newton's second law premultiplied by S_{CI}:

$$\sum_k m^{B_k}S_{CI}D^I D^I s_{CI} = S_{CI}\sum_k f_k$$

These two terms are satisfied identically. Voilà, we have arrived at the Euler equation of mutually translating bodies referred to as the common c.m. C:

$$\sum_k D^I\left(I_{B_k}^{B_k}\omega^{B_k I}\right) + \sum_k m^{B_k}S_{B_k C}D^I D^I s_{B_k C} = \sum_k m_{B_k} + \sum_k S_{B_k C}f_k \tag{6.53}$$

where the right-hand side sums up the moments applied to the common c.m. C:

$$m_C = \sum_k m_{B_k} + \sum_k S_{B_k C}f_k$$

The equation of motion consists of the MOIs $I_{B_k}^{B_k}$ of the individual parts and their inertial angular rates $\omega^{B_k I}$ plus an extra transfer term $\sum_k m^{B_k} S_{B_k C} D^I D^I s_{B_k C}$ with a peculiar acceleration expression $D^I D^I s_{B_k C}$. This is the acceleration of the displacement vector $s_{B_k C}$ as observed from inertial space. It does not include the inertial acceleration of the common c.m. For clarification we introduce the vector triangle $s_{B_k C} = s_{B_k I} - s_{CI}$:

$$D^I D^I s_{B_k C} = D^I D^I s_{B_k I} - D^I D^I s_{CI} = a_{B_k}^I - a_C^I$$

As it turns out, it is the difference between the inertial accelerations of the individual c.m. B_k and the common c.m. C.

Example 6.12 Carrier Vehicle with Moving Appendage

A main vehicle B_1 carries an appendage B_2, whose c.m. is moving wrt the carrier. Typical examples are the deployment of a satellite from the space shuttle, the swiveling nozzle of a rocket, or the tilting nacelle of the Osprey-type aircraft. In each case the common c.m. C is not fixed in the vehicle. In these applications it would be more convenient if the equations of motions were referred to a fixed point, usually the c.m. of the main body of the aircraft or missile. We can make that switch by transferring Euler's equations to the c.m. B_1 of the carrier vehicle.

We derive the attitude equations from Eq. (6.53), specialized for two bodies:

$$D^I\left(I_{B_1}^{B_1}\omega^{B_1 I}\right) + D^I\left(I_{B_2}^{B_2}\omega^{B_2 I}\right) + m^{B_1} S_{B_1 C} D^I D^I s_{B_1 C} + m^{B_2} S_{B_2 C} D^I D^I s_{B_2 C}$$

$$= m_{B_1} + m_{B_2} + S_{B_1 C} f_1 + S_{B_2 C} f_2 \tag{6.54}$$

To replace C by B, we make use of two relationships, the moment arm balance and the displacement vector triangle,

$$m^{B_1} s_{B_1 C} + m^{B_2} s_{B_2 C} = 0$$

$$s_{B_1 C} - s_{B_2 C} = -s_{B_2 B_1}$$

Adding both equations, after the second one has been multiplied first by m^{B_2} and then by $-m^{B_1}$, yields the two relationships

$$s_{B_1 C} = -\frac{m^{B_2}}{m^{B_1} + m^{B_2}} s_{B_2 B_1}, \quad s_{B_2 C} = +\frac{m^{B_1}}{m^{B_1} + m^{B_2}} s_{B_2 B_1}$$

Substituting these two displacement vectors into the third and fourth terms of Eq. (6.54), and into the last two terms, removes the dependency on the common c.m. C. After two pairs of terms are combined, we have produced Euler's equation for a carrier vehicle B_1 with moving appendage B_2:

$$D^I\left(I_{B_1}^{B_1}\omega^{B_1 I}\right) + D^I\left(I_{B_2}^{B_2}\omega^{B_2 I}\right) + \frac{m^{B_1} m^{B_2}}{m^{B_1} + m^{B_2}} S_{B_2 B_1} D^I D^I s_{B_2 B_1}$$

$$= m_{B_1} + m_{B_2} + \frac{1}{m^{B_1} + m^{B_2}} S_{B_2 B_1}\left(-m^{B_2} f_1 + m^{B_1} f_2\right) \tag{6.55}$$

Do you recognize the two rotary terms, the transfer term that contains the inertial acceleration of the displacement vector, and the external moments and forces? It may be puzzling what the state variables are. We can take two perspectives. For the applications that I mentioned, the translational and angular motions of the appendage are known as a function of time. Therefore, only $\omega^{B_1 I}$ contributes three body rates as state variables. The differential equations are linear. If, however, the appendage has its own degrees of freedom, like the shifting cargo during aircraft maneuvers, $\omega^{B_2 I}$ and $s_{B_2 B_1}$ become also state variables. Additional equations must be adjoined to furnish a complete set of differential equations, which will couple the motions of the two bodies. The whole set of equations are nonlinear and, as you can imagine, difficult to solve. To become familiar with the solution process, you should attack Problems 6.14 and 6.15.

6.3.4.3 Summary.
With Euler's law firmly in your grasp, you are fully equipped to model all aspects of aerospace vehicle dynamics. Never mind whether it is derived from Newton's law or is a principle in its own right. What counts is that you are able to apply it correctly. From first principles I have built Euler's equation for rigid bodies, either referring it to the c.m. or another fixed point. The free-flight attitude equation uses the c.m. to detach itself from the trajectory parameters, enabling the separation of the translational and attitude degrees of freedom. You should have no problem to derive the full six-DoF equations of motion of an airplane, missile, or spacecraft. The difficulty lies in the modeling of aerodynamics, propulsion, and supporting subsystems. We will pick up this challenge in Part 2.

I also introduced you to the dynamics of clustered bodies. In most aerospace applications their mass centers are fixed among themselves. Under those circumstances the transfer term includes only one time differentiation. If the bodies are moving, second-order time derivatives of the displacement vectors appear. Particularly important are spinning rotors, which introduce desired (momentum wheels) or undesired (propeller, turbines) gyroscopic effects. Because of their significance, we devote a separate section to their treatment.

6.4 Gyrodynamics

Gyrodynamics is the study of spinning rigid bodies. It has many applications in modeling of aerospace vehicles. Just consider the gyroscopic devices in inertial navigation systems, gimbaled spin-stabilized sensors, dual-spin satellites, spin-stabilized projectiles and rockets, Magnus rotors, propellers, and turbojets.

The study of the Earth as a spinning object captured the interest of famous dynamicists like Poinsot, Klein, and others in the last centuries. During their time, it was the only practical application. Earth science and astronomy are benefiting to this day from their research.

Technical applications dominate today's interest. Millions of dollars are spent either improving the performance of gyroscopes or lowering their cost for mass production. They are an integral part of any INS, affecting the accuracy of its navigation solution. Wherever a body spins in machinery, technical problems surface because of imperfections. Tires wobble, motor bearings fail, and Hubble gyroscopes wear out and must be replaced.

For technical details, I refer you to the many excellent texts that are available. An early classic is the theoretical book by Klein and Sommerfeld.[6] One of the best treatments, both theoretical and practical, is given by Magnus.[7] Unfortunately, these books are written in German. The standard English reference is by Wrigley et al.[8] An older account is given by C. S. Draper et al.[9] Here, I will cover only some of the fundamental dynamic characteristics of gyroscopes. The mystery that surrounds the precession and nutation modes of fly wheels will be debunked. From the kinetic energy theorem we learn how a spinning body responds to external moments, and we will derive two integrals of motion for force-free bodies.

6.4.1   Precession and Nutation Modes

Euler's law governs the dynamics of rotating bodies. In general, its differential equations are of sixth order, with three angular rates and three attitude angles as state variables. For bodies with constant spin rate, we are only interested in the rates and attitudes normal to the spin axis. They are governed by four first-order differential equations. If linearized by small perturbations, their characteristic equation has two conjugate complex pairs of roots, giving rise to two dynamic modes called *precession* and *nutation*.

6.4.1.1   Precession.

Precession is the response of a gyroscope to a persistent external moment. Euler's law reveals the nature of that response and enables us to derive a relationship between precession rate and external moment.

Consider a gyro B with angular momentum l_B^{BI} and subjected to the external moment m_B. Euler's law, Eq. (6.37), states that $D^I l_B^{BI} = m_B$, i.e., the change of angular momentum is in the direction of the applied moment. Expressed in inertial coordinates and dropping the sub- and superscripts,

$$\left[\frac{dl}{dt}\right]^I = [m]^I \tag{6.56}$$

Integrated,

$$[l(t)]^I = [l(t_0)]^I + \int_{t_0}^t [m]^I \, dt$$

We evaluate the integral by dividing it into time increments Δt during which the moment can be considered constant:

$$[l(t)]^I = [l(t_0)]^I + \sum_k [m_k]^I \Delta t$$

With $[m_k]^I \Delta t = [\Delta l_k]^I$ the last term becomes

$$[l(t)]^I = [l(t_0)]^I + \sum_k [\Delta l_k]^I$$

Figure 6.16 shows the integration process. The incremental angular momentum $[\Delta l_k]^I$ is collinear with the instantaneous moment $[m_k]^I$. Overall, the angular momentum vector $[l(t)]^I$ lines up with the moment vector. For a fast gyro for which the spin axis, the angular velocity vector, and the angular momentum vector

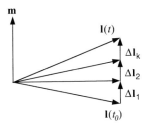

Fig. 6.16 Precession.

are close together, one can verbalize that the body axis tries to align itself with the momentum vector.

This motion is called precession. It is the slower one of the dynamic modes of a gyro and poorly understood. You probably have been at a science museum where you could not resist taking a seat on a turntable and grabbing a spinning flywheel by its handles. As you try to bank the flywheel, your seat starts to rotate on the turntable. You get off and explain to your son that this demonstrates the weird behavior of a gyroscope. It would be better to tell him that you experienced a precession in response to the torque you applied to the flywheel and encourage him to ask his physics teacher to fill in the details.

To get a quantitative relationship between moment and precession rate, we go back to Eq. (6.37) and introduce the precession frame P. This frame stays with the precessing angular momentum vector. Shifting the rotational derivative to P produces

$$D^P l_B^{BI} + \Omega^{PI} l_B^{BI} = m_B$$

If the magnitude of l_B^{BI} is constant and because the vector l_B^{BI} remains fixed in P, the rotational time derivative vanishes. The equation of the precession rate Ω^{PI} is therefore

$$\Omega^{PI} l_B^{BI} = m_B \tag{6.57}$$

This vector product establishes the right-handed rule of precession.

With Eq. (6.57) you can tell your son in advance how to apply the moment in order for the turntable to turn to the left. Turning to the left means the precession vector points up; and if the flywheel's angular velocity vector points right, the cross product tells you to generate a forward-pointing moment vector. Grab the wheel, push the right handle down and the left one up. You will be become an instant hero.

6.4.1.2 Nutation. Nutation is the response of a gyro to an impulse. Consider the free gyro in Fig. 6.17. We subject the gyro for a short time Δt to the moment $m_B = 2S_{AB}f$ and observe its reaction. According to Euler's law [Eq. (6.56)], the change of angular momentum as a result of the impulsive moment is

$$[\Delta l]^I \equiv [l(t)]^I - [l(t_0)]^I = [m_B]^I \Delta t \tag{6.58}$$

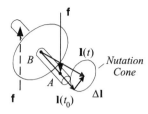

Fig. 6.17   Nutation.

During Δt, the angular momentum vector jumps from $[l(t_0)]^I$ to $[l(t)]^I$. The body axis, held back by the body's inertia, is still in its original position and starts to respond by revolving around the new location of the angular momentum, tracing out the half cone angle θ:

$$\tan \theta = \frac{|\Delta l|}{|l(t_0)|} = \frac{|m_B|}{|l(t_0)|} \Delta t \qquad (6.59)$$

The greater the impulse and the smaller the initial angular momentum are, the greater this nutation cone becomes. Initially, the body axis yields in the direction of f but returns to its original position through the nutation cycle. On the average the body axis evades the impulsive force perpendicularly. For many successive impulses a fast gyro with small nutational motions appears to move normal to the applied force. In the limit precession can be thought of as a sequence of infinitesimally small nutations caused by a sequence of impulses.

Let us play with the top, whose dynamic Eq. (6.47) we derived earlier. It is spinning on the ground about the vertical. We shake it out of its complacency by imparting an impulsive moment with our whip. The top starts to wobble, but refuses to fall down. The higher its spin rate (angular momentum) the smaller the nutation angle and the greater its resistance to our onslaught. We witness the inherent stability of spinning objects to perturbations. Several technical applications make use of this feature. I already introduced the dual-spin spacecraft in Example 6.11, and in Chapter 10 I will derive the equations of motion for spinning missiles and Magnus rotors.

6.4.2   Kinetic Energy

You may have heard of the flywheel-car project. Instead of using battery power as alternate energy, the car is driven by the kinetic energy stored in a massive flywheel. You drive up to a filling station, plug the drive motor into an outlet, and spin up the wheel. Supposedly, the stored energy could propel you 50 miles around town. How do we calculate the stored energy of a flywheel and, in more general terms, the kinetic energy of a freely spinning body or cluster of bodies? How does the time rate of change of kinetic energy relate to the applied external moment? We begin by defining kinetic energy.

The *kinetic energy* of the particle with mass m_i, translating with the velocity v_i^R relative to the arbitrary reference frame R, is defined by

$$T_i^R = \tfrac{1}{2} m_i \bar{v}_i^R v_i^R \qquad (6.60)$$

It is a scalar. Summing over the i particles of a body B, not necessarily rigid, establishes the kinetic energy of body B wrt reference frame R:

$$\sum_i T_i^R = \frac{1}{2} \sum_i m_i \bar{v}_i^R v_i^R \tag{6.61}$$

This formulation is not very useful because it requires knowledge of every particle's velocity. By introducing the c.m. B of the body, we can derive a much more practical relationship.

Theorem: With the c.m. B of a rigid body B known, the kinetic energy T^{BR} of body B wrt to reference frame R can be calculated from its rotational and translational parts:

$$T^{BR} = \frac{1}{2} \bar{\omega}^{BR} I_B^B \omega^{BR} + \frac{1}{2} m^B \bar{v}_B^R v_B^R \tag{6.62}$$

The *rotational* kinetic energy is a quadratic form of the angular velocity ω^{BR} of the body B wrt the reference frame R and the MOI tensor I_B^B of body B, referred to its c.m. B. The *translational* kinetic energy is patterned after Eq. (6.61), using the scalar product of the linear velocity of the c.m. multiplied by the total mass m^B of the body. Employing the c.m. of a body in calculating the kinetic energy is just a convenience yielding the most compact formula. Any other body point could be used. An additional term makes the adjustment and leads to the same numerical result (see Problem 6.11). Because the numerical value is independent of the reference point, the nomenclature T^{BR} refers only to the body and reference frames.

Proof: We start by expanding Eq. (6.61) and using the definition of the linear velocity $v_i^R = D^R s_{iR}$, with point R an element of frame R:

$$\frac{1}{2} \sum_i m_i \bar{v}_i^R v_i^R = \frac{1}{2} \sum_i m_i D^R \bar{s}_{iR} D^R s_{iR}$$

Now we introduce the displacement vector triangle $s_{iR} = s_{iB} + s_{BR}$ with B the c.m. of B:

$$2T^{BR} = \sum_i m_i \left(D^R \bar{s}_{iB} + D^R \bar{s}_{BR} \right) \left(D^R s_{iB} + D^R s_{BR} \right)$$

$$= \sum_i m_i D^R \bar{s}_{iB} D^R \bar{s}_{iB} + \left(\sum_i m_i D^R \bar{s}_{iB} \right) D^R s_{BR}$$

$$+ D^R \bar{s}_{BR} \left(\sum_i m_i D^R s_{iB} \right) + m^B D^R \bar{s}_{BR} D^R s_{BR}$$

Because B is the c.m., the second and the third terms vanish. Why is this so? First, m_i is constant, thus $\sum_i m_i D^R \bar{s}_{iB} = \sum_i D^R (m_i \bar{s}_{iB})$. Second, summation and differentiation may be exchanged; therefore, $\sum_i D^R (m_i \bar{s}_{iB}) = D^R (\sum_i m_i \bar{s}_{iB})$; but $\sum_i m_i \bar{s}_{iB} = \mathbf{0}$ is a null vector, and the rotational derivative of a null vector is zero. The last term, with the definition of the linear velocity $v_B^R = D^R s_{BR}$, provides the last term of Eq. (6.62). The first term needs some massaging to complete the proof. If B is rigid, $D^B s_{iB} = 0$, and

$$\sum_i m_i D^R \bar{s}_{iB} D^R s_{iB} = \sum_i m_i \left(D^B \bar{s}_{iB} + \overline{\Omega^{BR} s_{iB}} \right) \left(D^B s_{iB} + \Omega^{BR} s_{iB} \right)$$

$$= \sum_i m_i \overline{\Omega^{BR} s_{iB}} \Omega^{BR} s_{iB}$$

After some manipulations and with the definition of the MOI tensor Eq. (6.1), we produce

$$\sum_i m_i \overline{\Omega^{BR} s_{iB}} \Omega^{BR} s_{iB} = \bar{\omega}^{BR} \left(\sum_i m_i \bar{s}_{iB} s_{iB} \right) \omega^{BR} = \bar{\omega}^{BR} I_B^B \omega^{BR}$$

Moving the factor 2 to the right side confirms the first term of Eq. (6.62) and completes the proof.

For a cluster of rigid bodies B_k, $k = 1, 2, 3, \ldots$, we can superimpose the individual contributions of Eq. (6.62) and obtain the total rotational and translational kinetic energies

$$\sum_k T^{B_k R} = \frac{1}{2} \sum_k \bar{\omega}^{B_k R} I_{B_k}^{B_k} \omega^{B_k R} + \frac{1}{2} \sum_k m^{B_k} \bar{v}_{B_k}^R v_{B_k}^R \qquad (6.63)$$

The proof follows from the additive properties of kinetic energy.

Example 6.13 Flywheel Car

Problem.   A car with mass m^{B_1} stops to recharge its flywheel (mass m^{B_2} and MOI $I_{B_2}^{B_2}$) to the maximum permissible angular velocity $\omega^{B_2 R}$. If there were no losses, what would be the maximum speed the car could achieve?

Solution.   Initially, all energy is stored in the flywheel. To reach maximum velocity, the rotational energy must be fully converted into translational energy. From Eq. (6.63)

$$(m^{B_1} + m^{B_2}) \bar{v}_{B_1}^R v_{B_1}^R = \bar{\omega}^{B_2 R} I_{B_2}^{B_2} \omega^{B_2 R}$$

Let us introduce a coordinate system associated with reference frame R and the following components:

$$\left[v_{B_1}^R \right]^R = [V \ \ 0 \ \ 0], \quad \left[\omega^{B_2 R} \right]^R = [0 \ \ 0 \ \ r], \quad \text{and} \quad \left[I_{B_2}^{B_2} \right]^R = [\text{diag}\,(I_1, I_2, I_3)]$$

After substitution we obtain the maximum speed of the car:

$$V = r \sqrt{\frac{I_3}{m^{B_1} + m^{B_2}}}$$

A fast spinning, large wheel in a light car will provide maximum speed.

Applying a torque to a body increases its rotational kinetic energy. The energy theorem describes the phenomena.

Theorem: The time rate of change of rotational kinetic energy of a rigid body B wrt inertial frame I equals the scalar product of the body's angular velocity ω^{BI} and the applied moment m_B referred to the c.m.

$$\frac{\mathrm{d}T^{BI}}{\mathrm{d}t} = \bar{\omega}^{BI}m_B \tag{6.64}$$

To maximize the increase of kinetic energy, the external moment must be applied parallel to the angular velocity. Interestingly enough, the increase does not depend on the MOI of the body, but the current angular velocity.

Proof: Let us assume that point B is fixed in the inertial frame I so that we can concentrate on the rotational kinetic energy. From Eq. (6.62)

$$2T^{BI} = \bar{\omega}^{BI}I_B^B\omega^{BI}$$

Take the time derivative of the scalar T^{BI}, which is equivalent to the rotational derivative wrt any frame, and specifically the body frame. Then apply the chain rule

$$2\frac{\mathrm{d}T^{BI}}{\mathrm{d}t} = 2D^B T^{BI} = \overline{D^B\omega^{BI}}I_B^B\omega^{BI} + \bar{\omega}^{BI}D^B\left(I_B^B\omega^{BI}\right)$$

Recognize that the first term on the right equals the second term because 1) the term is a scalar and 2) the body B is rigid, which enables us to move I_B^B (symmetric tensor) under the rotational derivative

$$\overline{D^B\omega^{BI}}I_B^B\omega^{BI} = \bar{\omega}^{BI}\bar{I}_B^B D^B\omega^{BI} = \bar{\omega}^{BI}D^B\left(I_B^B\omega^{BI}\right)$$

Thus, introducing the angular momentum $l_B^{BI} = I_B^B\omega^{BI}$ we get

$$\frac{\mathrm{d}T^{BI}}{\mathrm{d}t} = \bar{\omega}^{BI}D^B\left(I_B^B\omega^{BI}\right) = \bar{\omega}^{BI}D^B l_B^{BI} \tag{6.65}$$

To replace the angular momentum term by the external moment, we substitute Euler's equations, transformed to the body frame $D^B l_B^{BI} + \Omega^{BI}l_B^{BI} = m_B$. Because the cross product with the same vectors vanishes, the proof is completed:

$$\frac{\mathrm{d}T^{BI}}{\mathrm{d}t} = \bar{\omega}^{BI}\left(m_B - \Omega^{BI}l_B^{BI}\right) = \bar{\omega}^{BI}m_B \qquad \text{QED}$$

An important question in gyrodynamics is the conditions for which the kinetic energy remains constant. Besides two trivial cases ($m_B = 0$ and $\omega^{BI} = 0$), the kinetic energy does not change if the external moment is applied normal to the spin axis. The energy theorem is useful for the study of gyroscopic responses and leads to one of the integrals of motions.

6.4.3 Integrals of Motion

Force-free motions of spinning rigid bodies have occupied the interest of researchers for centuries, with the Earth as their primary object of scrutiny. Yes, the Earth nutates and precesses, although at such miniscule amounts that our daily lives are unaffected. Astrophysicists, however, earn their living by analyzing and predicting these phenomena. The Frenchman Poinsot (1834) is particularly well known for his painstakingly geometrical description of the general motions of spinning bodies.

Modern technology has added other applications. Although gyros and spinning rockets are subject to moments and disturbances, much physical insight can be gained by studying their behavior in a force-free environment. Two integrals govern these motions. The *angular momentum* is constant in the absence of external moments, and the *kinetic rotational energy* is constant without work being applied to the body.

6.4.3.1   Angular momentum integral.

In the absence of external moments, Euler's law states that the inertial time rate of change of the angular momentum of a body B is zero:

$$D^I l_B^{BI} = 0$$

If integrated in inertial coordinates, we arrive at the first integral of motions

$$\left[l_B^{BI} \right]^I = [\text{const}]^I \tag{6.66}$$

Note that we had to choose a coordinate system to carry out the integration. Had we picked the body coordinates, the integral of motion would be more complicated:

$$\left[l_B^{BI} \right]^B = - \int [\Omega^{BI}]^B \left[l_B^{BI} \right]^B \, dt + [\text{const}]^B$$

Equation (6.66) is also called the theorem of *conservation of angular momentum*. In the absence of external stimuli, the angular momentum remains constant and fixed wrt the inertial frame.

6.4.3.2   Energy integral.

Without external moments no work is done on the spinning body, and therefore its kinetic energy remains constant, as confirmed by the energy theorem Eq. (6.64). Thus we conclude from Eq. (6.62), disregarding linear kinetic energy and substituting the angular momentum $l_B^{BI} = I_B^B \omega^{BI}$, that the energy integral is constant:

$$2T^{BI} = \bar{\omega}^{BI} I_B^B \omega^{BI} = \bar{\omega}^{BI} l_B^{BI} = \text{const} \tag{6.67}$$

Because we are dealing with a scalar product, Eq. (6.67) can be evaluated in any allowable coordinate system. Expressing it in body coordinates, which also serve as principal axes, we receive the energy ellipsoid

$$[\overline{\omega^{BI}}]^B [I_B^B]^B [\omega^{BI}]^B = [\omega_1 \quad \omega_2 \quad \omega_3] \begin{bmatrix} I_1 & 0 & 0 \\ 0 & I_2 & 0 \\ 0 & 0 & I_3 \end{bmatrix} \begin{bmatrix} \omega_1 \\ \omega_2 \\ \omega_3 \end{bmatrix}$$

$$= I_1 \omega_1^2 + I_2 \omega_2^2 + I_3 \omega_3^2 = 2T_B^{BI}$$

and normalized

$$\frac{\omega_1^2}{2T^{BI}/I_1} + \frac{\omega_2^2}{2T^{BI}/I_2} + \frac{\omega_3^2}{2T^{BI}/I_3} = 1 \tag{6.68}$$

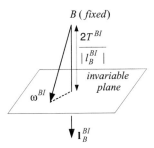

Fig. 6.18   Poinsot motions.

The energy ellipsoid is the locus of the endpoints of those vectors ω^{BI} that belong to an energy level T^{BI}. The three semi-axes are

$$a = \sqrt{2T^{BI}/I_1}, \quad b = \sqrt{2T^{BI}/I_2}, \quad c = \sqrt{2T^{BI}/I_3}$$

Comparison with the MOI ellipsoid Eq. (6.13) establishes the fact that both ellipsoids are similar, i.e., their principal axes are parallel and scaled by the constant factor $\sqrt{2T^{BI}}$.

6.4.3.3   Poinsot motions.

The two integrals of motion can be used to solve for the movements of spinning force-free bodies. Without the need for calculations, Poinsot has devised a geometrical method that visualizes the motions. For a detailed discussion you should consult Goldstein.[2] I will just provide the essentials here and use Fig. 6.18 to explain the geometry.

1) Equation (6.67), with $n = l_B^{BI}/|l_B^{BI}|$, defines a plane, whose normal form is $\bar{\omega}^{BI}n = 2T^{BI}/|l_B^{BI}| = \text{const}$ and which always contains the endpoint of ω^{BI}.

2) Equation (6.66) fixes l_B^{BI} in inertial space. The plane, defined by Eq. (6.67), is fixed in inertial space as well, and is called the *invariable plane*.

3) Equation (6.68), the energy ellipsoid, is the locus of all endpoints of ω^{BI}.

4) The general motion of a force-free gyro, rotating about a fixed point B, is described by the rolling of the energy ellipsoid on the invariable plane.

Example 6.14   Impulse Control

Problem.   A spin-stabilized missile with spin rate ω_0 and MOI I receives an impulsive torque $m_B \Delta t$ from its reaction control jet. What is the new direction of the missile and what is the roll rate $\dot{\phi}$ and nutation rate $\dot{\eta}$?

Solution.   Figure 6.19 shows the missile before the impulse is applied. Its spin rate ω_0 and angular momentum l_0 are still aligned with the body axis 1^B. Now the reaction jet fires, and the impulsive torque introduces a nutation of the 1^B axis of the missile (see Fig. 6.20). The new attitude of the missile is centered around the angular momentum vector l, displaced from its original position l_0 according to Eq. (6.58) by $\Delta l = m_B \Delta t$. Therefore,

$$l = l_0 + \Delta l = l_0 + m_B \Delta t \tag{6.69}$$

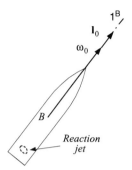

Fig. 6.19   Before impulse.

The angle between l_0 and l is the change in the mean attitude of the centerline of the missile. It is the nutation angle θ, calculated from Eq. (6.59) as

$$\theta = \arctan\left(\frac{|m_B|}{|l_0|}\Delta t\right)$$

(In an actual application the nutation is reduced to zero by aerodynamic damping). The motion of the missile can be visualized by the body and space cones of Fig. 6.20. The body cone is centered on the 1^B axis, and the space cone contains the angular momentum vector, which is fixed in space. As the body cone rolls on the space cone, the missile traces its path. The angular velocity vector ω consists of two components, the roll rate of the missile $\dot{\phi}$ and the nutation rate $\dot{\eta}$. They are calculated from the vector triangle, consisting of the absolute values p, $\dot{\phi}$, and $\dot{\eta}$ the half-cone angles λ and μ.

First, we determine λ, the angle between the angular momentum and angular velocity vectors. The angular momentum is given by Eq. (6.69) and the angular

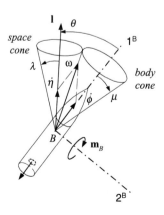

Fig. 6.20   After impulse.

velocity vector by $\omega = I^{-1}l$. The scalar product yields the desired relationship:

$$\lambda = \arccos \frac{\bar{\omega}l}{|\omega||l|}$$

The half-angle of the body cone is $\mu = \theta - \lambda$. From the law of sines, we can now calculate the roll rate

$$\dot{\phi} = \frac{\sin \lambda}{\sin(\pi - \theta)} \omega \qquad (6.70)$$

and the nutation rate

$$\dot{\eta} = \frac{\sin \mu}{\sin(\pi - \theta)} \omega \qquad (6.71)$$

The roll rate $\dot{\phi}$ increases with the widening of the space cone, and the nutation rate $\dot{\eta}$ gets larger with the increase of the body cone.

This short excursion into gyrodynamics should give you an appreciation for the "strange" behavior of spinning objects. Their main modes are nutation and precession. *Nutation* is a fast circular motion of the body axis, whereas *precession* is a slower movement of the angular velocity vector. I introduced a new term, the kinetic energy of a rigid body T^{BR}. It is a scalar that depends on the body B and an arbitrary reference frame R, consisting of rotational and translational kinetic energy. Flywheels are a good example for storing large amounts of rotational energy. Particularly simple to explain are the dynamics of force-free gyros. Two integrals of motion render Poinsot's graphical representation of the energy ellipsoid rolling on the invariable plane.

6.5 Summary

This chapter is dominated by Euler's law. We adopted the viewpoint that it is a basic principle which governs the attitude dynamics quite separately from Newton's law, although we recognize their kinship. For its formulation three new entities are needed. The moment of inertia is a second-order tensor, real and symmetric, can be diagonalized, and is represented geometrically by the inertia ellipsoid. The angular momentum vector derives from the linear momentum by the multiplication with a moment arm. For aerospace applications, however, it is more likely expressed as the product of the MOI tensor with the angular velocity vector. The externally applied moment or torque is the stimulus for the body dynamics. These are mostly aerodynamic and propulsive moments. Details will be discussed in Part 2.

The modeling techniques for flight dynamics are now completely assembled. You should be able to model the geometry of engagements, express vectors and tensors in a variety of coordinate systems, calculate linear and angular velocities, and derive the translational and rotational equations of motions of aerospace vehicles. I have consistently formulated first the invariant equations, and then expressed them in coordinate systems for computer implementation. We saw no need to deviate from the hypothesis that points and frames can model all entities which arise in flight mechanics. A consistent nomenclature sprang from this premise, enclosing all essential elements of a definition in sub- and superscripts.

You should be prepared now to face the cruel world of simulation, be it in three-, five- or six-DoF fidelity. In Part 2 I give you a running start with detailed examples and code descriptions that are available on the CADAC Web site. Before you proceed, however, I invite you to study one other topic, particularly important for engineering applications: the formulation of perturbation equations. Do not despise the "small" approximations, despite the raw computer power on your desk. Perturbation equations give insight into the dynamics of aerospace vehicles and are essential in the design of control systems.

References

[1]Truesdell, C., "Die Entwicklung des Drallsatzes," *Zeitschrift for Angewandte Mathematik und Mechanik*, Vol. 44, No. 4/5, 1964, pp. 149–158.

[2]Goldstein, H., *Classical Mechanics*, Addison Wesley, Longman, Reading, MA, 1965, p. 5,6.

[3]Mises, R. v., "Anwendungen der Motorrechnung," *Zeitschrift für Angewandte Mathematik und Mechanik*, Vol. 4, No. 3, 1924, pp. 209–211.

[4]Grubin, Carl, "On the Generalization of the Angular Momentum Equation," *Journal of Engineering Education*, Vol. 51, No. 3, 1960, p. 237.

[5]Bracewell, R. N., and Garriott, O. K., "Rotation of Artificial Earth Satellites," *Nature* Vol. 182, 20 Sept. 1958, pp. 760–762.

[6]Klein, F., and Sommerfeld, A., *Ueber die Theorie des Kreisels*, 4 Vols., Teubner, 1910–1922.

[7]Magnus, K., *Kreisel, Theorie und Anwendungen*, Springer-Verlag, Berlin, 1971.

[8]Wrigley, W., Hollister, W., and Denhard, W. G., *Gyroscopic Theory, Design and Instrumentation*, M.I.T. Press, Cambridge, MA, 1969.

[9]Draper, C. S., Wrigley, W., and Hovorka, J., *Inertial Guidance*, Pergamon, New York, 1960.

Problems

6.1 MOI of helicopter rotor. A three-bladed helicopter rotor revolves in the counterclockwise direction. Each blade has the same dimensions l = length, c = chord, and t = thickness. The mass center B_k of each blade k is displaced from

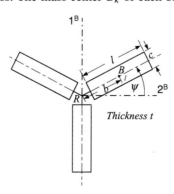

Thickness t

the hub's center R by b. Derive the MOI tensor of the three blades referred to R in helicopter coordinates $[I_R^{\Sigma B_k}]^B$. Assume constant density ρ of the blades.

6.2 Alternate formulation of axial MOI. The axial MOI I_n about unit vector n of MOI tensor I_R^B is

$$I_n = \bar{n} I_R^B n$$

or, with the definition of the MOI, summed over all particles i

$$I_n = \bar{n} \sum_i m_i (\bar{s}_{iR} s_{iR} E - s_{iR} \bar{s}_{iR}) n$$

Show that it can also be expressed as

$$I_n = \sum_i m_i \bar{s}_{iR} N s_{iR}$$

where $N = E - \bar{n}n$ is the planar projection tensor of the plane with normal n [see Eq. (2.25)].

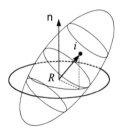

6.3 Mass properties of transport aircraft. For a six-DoF simulation of a transport aircraft, you need rough values for the mass m^B and the MOI tensor $[I_C^B]^B$ of the total body B referred to the common c.m. C and expressed in body axes $]^B$.

(a) Formulate the equations of the individual MOIs $[I_W^W]^B$, $[I_F^F]^B$, $[I_T^T]^B$ of wing, fuselage, and tail wrt their respective c.m. and express them in body axes. Afterward calculate their numerical values.

(b) Formulate the equation for the displacement vector $[s_{CR}]^B$ of system c.m. C wrt the reference point R at the nose of the aircraft. Afterward calculate its numerical value.

(c) Formulate the displacement vectors $[s_{WC}]^B$, $[s_{FC}]^B$, $[s_{TC}]^B$ for the subsystems c.m. wrt to C. Afterward calculate their numerical values.

(d) Now provide the equations of the vehicle's total mass properties m^B and $[I_C^B]^B$.

(e) Calculate the numerical values of the vehicle's total mass properties m^B and $[I_C^B]^B$.

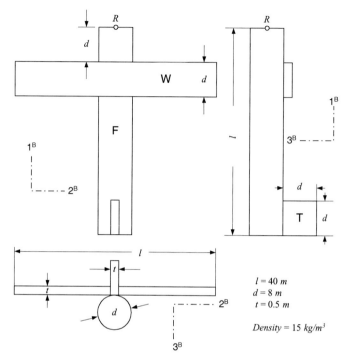

6.4   Change of reference point.   A helicopter rotor R has the angular momentum $[l_R^{RE}]^B$ wrt the Earth E, referred to the rotor's c.m. R, and expressed in the helicopter's body axes $]^B$. The helicopter flies over the spherical Earth at an altitude h with the velocity $[v_B^E]^B = [v_1 \ 0 \ 0]$. What is the rotor's angular momentum $[l_E^{RE}]^B$ wrt the center of Earth E?

6.5   Total angular momentum of a helicopter.   The airframe of a helicopter B with mass m^B has three rotary devices affixed: the main rotor with mass m^M and MOI I_M^M; the turbine m^T, I_T^T; and the tail rotor m^R, I_R^R. While the helicopter is hovering, you are to calculate the total angular momentum $[l_C^{\Sigma I}]^B$ of the helicopter wrt the inertial frame and referred to the common c.m. C in helicopter coordinates $]^B$. The individual MOI's and angular velocities are in air-frame coordinates $]^B$:

$$[I_M^M]^B = \begin{bmatrix} I_{M1} & 0 & 0 \\ 0 & I_{M2} & 0 \\ 0 & 0 & I_{M3} \end{bmatrix}, \quad [I_T^T]^B = \begin{bmatrix} I_{T1} & 0 & 0 \\ 0 & I_{T2} & 0 \\ 0 & 0 & I_{T3} \end{bmatrix},$$

$$[I_R^R]^B = \begin{bmatrix} I_{R1} & 0 & 0 \\ 0 & I_{R2} & 0 \\ 0 & 0 & I_{R3} \end{bmatrix}$$

$$[\overline{\omega^{MB}}]^B = [0 \ 0 \ \omega_M], \quad [\overline{\omega^{TB}}]^B = [\omega_T \ 0 \ 0], \quad [\overline{\omega^{RB}}]^B = [0 \ \omega_R \ 0]$$

6.6 Force-free body parameters. A force-free body B spins around its c.m. B with $[\overline{\omega^{BI}}]^I = [1 \ 0 \ 2]$ rad/s, having an MOI of $[I_B^B]^B = [\text{diag}(3, 2, 2)]$ kgm^2. What are its angular momentum $[l_B^{BI}]^I$ and kinetic energy T^{BI}?

6.7 Forces and moments. A B1 aircraft is subject to several forces and moments. The aerodynamic forces f_a and m_a are referred to reference point R, and the propulsive thrust of the right and left engines are f_p, respectively. To make a right turn, the pilot generates with the ailerons a couple m_c. What are the resultant force f and moment m_B wrt the aircraft mass center B? Introduce the necessary displacement vectors to make the equations self-defining.

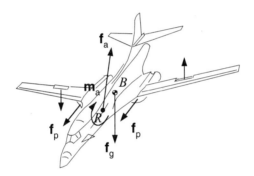

6.8 Symmetric gyro. A symmetric gyro executes motions characterized by the condition that three vectors always are coplanar: the angular momentum l_B^{BI}, the angular velocity ω^{BI}, and the unit vector b_1 of the symmetry axis. Use the following special components to verify this statement:

$$\left[I_B^B\right]^B = \begin{bmatrix} I_1 & 0 & 0 \\ 0 & I_1 & 0 \\ 0 & 0 & I_3 \end{bmatrix}, \quad [b_1]^B = \begin{bmatrix} 0 \\ 0 \\ 1 \end{bmatrix}$$

6.9 Free gyro. A two-axes gyro has two free gimbals. The inner gimbal supports the spin axis, and the outer gimbal rotates freely in the bearings attached to the vehicle B. The gimbal pitch angle θ_G and yaw angle ψ_G are indicated by their angular rate vectors $\dot{\theta}_G$ and $\dot{\psi}_G$.

(a) Derive the relationship between the gimbal angles and Euler angles of the vehicle. *Procedure:* The spin axis s is parallel to the 1^S axis and to the 1^I axis (assuming ideal gyro). Express its body coordinates both in inertial $]^I$ and inner gimbal coordinates $]^S$:

$$([s]^B =)[T]^{BI}[s]^I = [T]^{BS}[s]^S$$

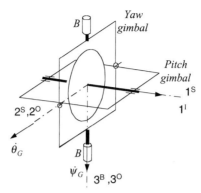

where $[T]^{BI}$ contains the Euler angles and $[T]^{BS}$ the gimbal angles. By comparing equal elements, the desired relationship is obtained.

(b) Derive the differential equations of the gimbal angles θ_G and ψ_G as functions of the body rates $[\omega^{BI}]^B = [p \quad q \quad r]$. *Procedure:* The angular velocity vector ω^{SB} of the inner gimbal frame S wrt the body frame B is expressed in terms of the inertial body rates ω^{BI} by the following steps. Take the rotational derivative of the spin axis s wrt the body frame B and transform it to the inner gimbal S and also to the inertial frame I:

$$(D^B s =)D^S s + \Omega^{SB} s = D^I s + \Omega^{IB} s$$

Because s is fixed in both the B and I frames, both derivatives are zero, and you get

$$\Omega^{SB} s = \Omega^{IB} s$$

For actual calculations it is most convenient to use the outer gimbal coordinates $]^O$

$$[\Omega^{SB}]^O [s]^O = [\Omega^{IB}]^O [s]^O$$

and express the spin axis $[s]^S$ in inner gimbal coordinates and the body rates $[\Omega^{BI}]^B$ in body coordinates

$$[\Omega^{SB}]^O [T]^{OS}[s]^S = [T]^{OB}[\Omega^{IB}]^O[T]^{BS}[s]^S$$

then reversing the angular velocities

$$[\Omega^{BS}]^O [T]^{OS}[s]^S = [T]^{OB}[\Omega^{BI}]^O[T]^{BS}[s]^S$$

The result is

$$\dot{\theta}_G = q \cos \psi_G + p \sin \psi_G$$

$$\dot{\psi}_G = r + (q \sin \psi_G - p \cos \psi_G) \tan \theta_G$$

Both methods can be used to calculate the gimbal angles. What are their respective advantages and disadvantages?

6.10   Gyro mass unbalance. The spin axis of a gimbaled gyro is subject to a mass unbalance Δm, located at a distance b from the c.m. The resulting moment is constant in the inner gimbal (precession frame P). Determine the precession angular velocity r_p if the spin velocity $\dot{\phi}$ and the MOI I of the rotor are given.

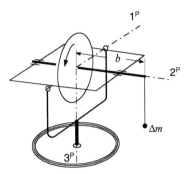

6.11   Change of reference point for kinetic energy. The kinetic energy T^{BR} of a rigid body finds its simplest expression if its c.m. B is used [see Eq. (6.62)]. If an arbitrary point B_1 of the body is introduced as reference point, show that the kinetic energy is calculated from the formula

$$T^{BR} = \tfrac{1}{2}\bar{\omega}^{BR} I^B_{B_1} \omega^{BR} + \tfrac{1}{2} m {}^B \bar{v}^R_{B_1} v^R_{B_1} + m {}^B \bar{v}^R_{B_1} \Omega^{BR} s_{BB_1}$$

where $m {}^B \bar{v}^R_{B_1} \Omega^{BR} s_{BB_1}$ is the supplementary term required because B_1 is not the c.m.

6.12   Bearing loads on turbine during pull-up. The Mirage jet fighter has a single turbine engine T located near the c.m. B of the aircraft B. As it pulls up, you are to calculate the forces and moments that the bearings have to support. The pull-up occurs in the vertical plane at a radius R and aircraft velocity v^I_B.

(a) Derive in invariant form the bearing forces counteracting centrifugal and gravity accelerations and the bearing torque opposing the gyroscopic moment.

(b) Express the bearing forces and moments in aircraft body coordinates while the Mirage is at the bottom of its pull-up. Besides R and $[\bar{v}^I_B]^B = [V\ 0\ 0]$, the following parameters are given for the turbine: mass m^T; MOI $[I^T_T]^B = [\text{diag}(I_1, I_2, I_3)]$; and angular velocity $[\omega^{TB}]^B = [\omega_1\ 0\ 0]$. You can assume that the two c.m. T and B coincide.

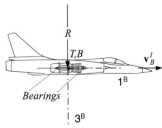

6.13   Control moment gyros of the Hubble telescope. The Hubble space telescope B_0 is stabilized by three control moment gyros (CMG) B_1, B_2, and B_3. The

CMG mass centers have the same distance x from the center B_0 and are equally spaced, starting with gyro #1 aligned with the 1^{B_0} axis of the telescope. The directions of the spin axes are shown in the accompanying figure. The following quantities are given: mass of telescope, m_0; mass of one CMG, m; spin MOI of CMG, I_s; transverse MOI of CMG, I; angular rate of GMC wrt B_0, ω; distance of CMG from B_0, x; MOI of telescope,

$$[I_{B_0}^{B_0}]^{B_0} = \begin{bmatrix} I_0 & 0 & 0 \\ 0 & I_0 & 0 \\ 0 & 0 & I_{03} \end{bmatrix}$$

velocity of telescope wrt inertial frame, $[\overline{v_{B_0}^I}]^I = [0 \ v_0 \ 0]$; angular velocity of telescope wrt inertial frame, $[\omega^{B_0 I}]^I = [0 \ 0 \ \omega_0]$.

(a) For the cluster $k = 0, 1, 2, 3$, determine in tensor format the linear momentum $p_{\Sigma B_k}^I$, the MOI $I_{B_0}^{\Sigma B_k}$, the kinetic energy $T^{\Sigma B_k I}$, and angular momentum $l_{B_0}^{\Sigma B_k I}$.

(b) Express the four quantities in the telescope's coordinates $]^{B_0}$ using the TM

$$[T]^{B_0 I} = \begin{bmatrix} \cos\psi & \sin\psi & 0 \\ -\sin\psi & \cos\psi & 0 \\ 0 & 0 & 1 \end{bmatrix}$$

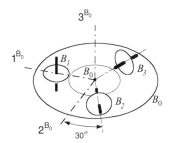

6.14   Shuttle pitch equations during release of satellite.   Derive the pitch attitude equations of the space shuttle B_0 as it launches a satellite B_1. Assume that the release is parallel and in the opposite direction of the space shuttle's 3 axis. The satellite's displacement vector from the shuttle's c.m B_0 is $[\overline{s_{B_1 B_0}}]^{B_0} = [-a \ 0 \ \eta]$, where a is a positive constant and $\eta(t)$ a known function of t. The mass of the manipulator's arm can be neglected, and the satellite treated as a particle with mass m^{B_1}. The mass properties of the shuttle are m^{B_0} and $[I_{B_0}^{B_0}]^{B_0} = [\text{diag}(I_1, I_2, I_3)]$. Determine the differential equation of motion of the shuttle's pitch angular velocity $[\omega^{B_0 I}]^{B_0} = [0 \ q \ 0]$. All external forces and moments can be neglected.

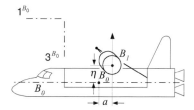

6.15   Missile pitch equations with swiveling motor.   The attitude of a missile B_0 is controlled by its swiveling rocket engine B_1 with thrust $[\bar{t}]^{B_1} = [T \quad 0 \quad 0]$ and known swivel angle $\delta(t)$. Neglecting all other forces and moments, determine the differential equation that governs the pitch angular velocity $[\omega^{B_0 I}]^{B_0} = [0 \quad q \quad 0]$ of the missile. The mass properties are given:

$$m^{B_0}, \left[I_{B_0}^{B_0}\right]^{B_0} = \mathrm{diag}(I_1, I_2, I_3)$$

$$m^{B_1}, \left[I_{B_1}^{B_1}\right]^{B_1} = \mathrm{diag}(J_1, J_2, J_3)$$

and assumed constant.

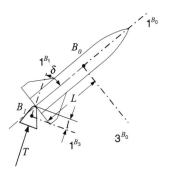

Selected Solutions

Solution 6.13

$$\left[I_{B_0}^{\Sigma B_k}\right]^{B_0} = \begin{bmatrix} I_0 + 2I + I_s + \dfrac{3}{2}mx^2 & 0 & 0 \\ 0 & I_0 + 2I + I_s + \dfrac{3}{2}mx^2 & 0 \\ 0 & 0 & I_{03} + 2I + I_s + 3mx^2 \end{bmatrix}$$

$$\left[p_{\Sigma B_k}^I\right]^{B_0} = (m_0 + 3m)v_0 \begin{bmatrix} \sin\psi \\ \cos\psi \\ 0 \end{bmatrix}$$

$$T^{\Sigma B_k I} = \tfrac{1}{2}I_{03}\omega_0^2 + \tfrac{1}{2}I_s(\omega + \omega_0)^2 + \left(I_s\omega^2 + I\omega_0^2\right) + \tfrac{1}{2}(m_0 + 3m)v_0^2$$

$$\left[l_{B_0}^{\Sigma B_k I}\right]^{B_0} = \begin{bmatrix} 0 \\ 0 \\ I_{03}\omega_0 + I_s(\omega + \omega_0) \end{bmatrix} + \frac{1}{2}\begin{bmatrix} -\left(1+\sqrt{3}\right)I_s\omega \\ -\left(1-\sqrt{3}\right)I_s\omega \\ 4I\omega_0 \end{bmatrix} + \begin{bmatrix} 0 \\ 0 \\ 3mx^2\omega_0 \end{bmatrix}$$

Solution 6.14

$$\left[\frac{I_2(m^{B_1} + m^{B_0})}{m^{B_1} m^{B_0}} + \eta^2(t) + a^2 \right] \dot{q} + 2\dot{\eta}(t)\eta(t)q = -a\ddot{\eta}(t)$$

Solution 6.15

$$\left[\frac{I_2 + J_2}{M} + l^2 + L^2 + 2lL\cos\delta(t) \right] \dot{q} - 2lL\sin\delta(t)\dot{\delta}(t)q$$

$$= -\left[\frac{J_2}{M} + l^2 + lL\cos\delta(t) \right] \ddot{\delta}(t) + lL\sin\delta(t)\dot{\delta}^2(t) - \frac{T}{m^{B_1}} L\sin\delta(t)$$

where

$$M = \frac{m^{B_1} m^{B_0}}{m^{B_1} + m^{B_0}}$$

7
Perturbation Equations

The last chapter completed the toolbox for modeling aerospace vehicle dynamics. You are now well acquainted with Newton's and Euler's laws as modeling tools for the equations of motion. In Chapters 8–10 we shall put them to work, simulating the dynamics of aircraft, hypersonic vehicles, missiles, and even Magnus rotors. Before pursuing that ambitious goal, I will address another important subject of modeling and simulation that deals with the linearization of the equations of motions.

Why should we, living in the computer age, still concern ourselves with the simplification of the dynamic equations? I can think of three reasons, and you may be able to add some more.

1) Stability investigations are an important part of any vehicle design. They require the linearization of the equations of motion in order to take advantage of linear stability criteria.

2) Control engineers will always need linearized representations of the plant, be they transfer functions or in state variable form.

3) For a basic understanding of the vehicle dynamics, the eigenvalues of the linear equations serve to indicate frequency and damping.

These simplifications are accomplished with perturbation techniques. There is the classical small perturbation method, developed to solve specific problems in atmospheric flight mechanics. It employs scalar perturbations and relates them, for each type of flight vehicle, to a special coordinate system. Instead of deriving the general perturbation equations first, restrictive assumptions are made, and, consequently, the perturbation equations are limited to steady flight regimes.

The objective of this chapter is to introduce the general perturbation equations of atmospheric flight mechanics that are valid even for unsteady flight regimes. To keep the derivation simple, the flight vehicles are assumed rigid bodies.

I will discuss three techniques, the scalar, total, and component perturbations and use the latter to derive the general perturbation equations of aerospace vehicles. They apply to any type of vehicle from aircraft to spinning missiles. Then I will address the expansion of the aerodynamic forces and moments into Taylor series. Taken together, they deliver the linear dynamic equations. Examples of pitch and roll linear state equations demonstrate practical applications. I will also venture into the realm of unsteady flight with nonlinear effects to challenge your imagination.

7.1 Perturbation Techniques

The classical perturbation technique, as outlined by Etkin,[1] proceeds as follows. First, an axis system is defined in relationship to physical quantities, such as the principal body axes or the relative wind velocity. The components of the state parallel to these axes are then identified. A particular steady flight regime is selected

217

with certain values for the reference components, e.g., x_{r1}, x_{r2}, x_{r3}, and a perturbed flight with x_{p1}, x_{p2}, x_{p3}. The scalar differences

$$\Delta x_i = x_{pi} - x_{ri}; \quad i = 1, 2, 3$$

are the perturbation variables. Because the perturbations are generated by a scalar subtraction, this technique is also called the *scalar perturbation method* (see Ref. 2). The disadvantage of this technique lies in the fact that all of the formulations are tied to one particular coordinate system. A change to other coordinate systems is very difficult to accomplish.

In theoretical work vectors are preferred over components, and perturbations are defined as the *vectorial* differences between the reference and perturbed vectors. No allusion is made to a particular coordinate system. Because this technique considers the total state variable rather than its components, it is called the *total perturbation method*. Denoting the state vectors during the reference and perturbed flights as x_r and x_p, respectively, the total perturbation is defined as

$$\delta x = x_p - x_r$$

The total perturbations have the advantage over the scalar perturbations that they hold for any coordinate system. In applications, however, numerical calculations require that vectors be expressed by their components, referred to a particular coordinate system. For instance, the MOI is given in body axes; vehicle acceleration and angular velocity are measured by accelerometers and rate gyros, mounted parallel to the body axes; wind-tunnel measurements are recorded in component form; and the whole framework of aerodynamics is based on force and moment components rather than total values.

To express the total perturbations in components, a transformation matrix must be introduced. In our notation the components of the δx perturbation, relative to any coordinate system, say $]^D$, become

$$[\delta x]^{Dp} = [x_p]^{Dp} - [T]^{DpDr}[x_r]^{Dr} \tag{7.1}$$

The subscripts r and p indicate reference and perturbed flights, respectively; $[x_r]^{Dr}$ and $[x_p]^{Dp}$ are the components as measured during reference and perturbed flights; and $[T]^{DpDr}$ is the transformation matrix of the coordinate system associated with the perturbed frame Dp relative to the coordinate system associated with the reference frame Dr.

Every numerical evaluation of equations based on the total perturbation method includes the transformation matrix $[T]^{DpDr}$. Consequently, the transformation angles and their trigonometric functions enter the calculations, increasing the complexity of the equations considerably.

Wouldn't you rather work with a perturbation methodology that combines the general invariance of the total perturbation method for theoretical investigations with the simple component presentation of the scalar perturbation method? We can formulate such a procedure by introducing the rotation tensor $[R^{DpDr}]$ of the Dp frame wrt the Dr frame in the following form:

$$\varepsilon x = x_p - R^{DpDr} x_r \tag{7.2}$$

The εx perturbation is obtained by first rotating the reference vector x_r through R^{DpDr} and then subtracting it from the perturbed vector x_p. It satisfies our first requirement of invariancy. To show that it reduces to a simple component form, we impose the $]^{Dp}$ coordinate system and transform the reference vector to the $]^{Dr}$ system:

$$[\varepsilon x]^{Dp} = [x_p]^{Dp} - [R^{DpDr}]^{Dp}[x_r]^{Dp}$$

$$= [x_p]^{Dp} - [R^{DpDr}]^{Dp}[T]^{DpDr}[x_r]^{Dr}$$

$$= [x_p]^{Dp} - [x_r]^{Dr}$$

The last equation follows from Eq. (4.6). Note that the transformation matrix of Eq. (7.1) is absent. Because this technique emphasizes the component form of a vector, Eq. (7.2) is referred to as the *component perturbation method* or alternately as the ε *perturbations*.

When you work with the component perturbation method, the choice of the R^{DpDr} tensor and thus the selection of the frame D is most important. As a general guideline, choose D so that the ε perturbation remains small throughout the flight. Especially in atmospheric flight, the selection of D is determined by the requirement of representing the aerodynamic forces as a function of small perturbations. Then a Taylor-series expansion is possible, and the difficult task of expressing the aerodynamic forces in simple analytical form can be achieved. I propose the designation *dynamic frame* for D because the dynamic equations of flight mechanics are solved in a coordinate system associated with frame D.

Let us discuss some examples. The dynamic frame of an aircraft is either the body frame B or the stability frame S. In both cases, for small disturbances, the rotation tensors are close to the unit tensor, expressing the fact that the frame Dp has been rotated by small angles from Dr. As will be outlined in more detail in Sec. 7.3, the dynamic frame plays also an important role in the aerodynamic force and moment expansions.

In missile dynamics the situation is similar except that the aeroballistic frame replaces the stability frame. However, for a spinning missile the body frame cannot serve as a dynamic frame because the perturbations of the aerodynamic roll angle can be large. To keep the perturbations small between the wind and dynamic frames, the nonrolling body frame is chosen as dynamic frame. The motions between the body frame and the dynamic frame thus are not explicitly included in the aerodynamic expansion, but rather the derivatives depend on them implicitly. To simplify the notation, I will use the abbreviated form R for R^{DpDr} whenever appropriate.

Perturbation techniques enable us to expand the aerodynamic forces in terms of small variables about the reference flight. Suppose $f(x)$ is the aerodynamic force vector with x representing a state vector. The force during the perturbed flight $f(x_p)$ is expressed in view of Eq. (7.2) by

$$f(x_p) = f(\varepsilon x + Rx_r)$$

Expanding about the reference flight ($\varepsilon x = 0$) yields

$$f(x_p) = f(Rx_r) + \frac{\partial f}{\partial x}\varepsilon x + \cdots \tag{7.3}$$

where $\partial f / \partial x$ is the Jacobian matrix. The Principle of Material Indifference, familiar to us from Sec. 2.1.3, states (see Ref. 3) that the physical process, generating fluid dynamic forces, is independent of spatial attitude. In other words, if x_r is rotated through R, the process of functional dependence remains the same. The only difference is that the force has also been rotated through R, i.e.,

$$Rf(x_r) = f(Rx_r)$$

Making use of this fact, Eq. (7.3) becomes

$$f(x_p) = Rf(x_r) + \frac{\partial f}{\partial x}\varepsilon x + \cdots \tag{7.4}$$

and f behaves like the ε perturbations, introduced by Eq. (7.2)

$$f_p = Rf_r + \varepsilon f \tag{7.5}$$

The component or ε perturbations satisfies both requirements of invariancy for theoretical derivations and simple component form for practical calculations. They are a generalization of the classical scalar perturbation method and are particularly well suited to formulate perturbations in a form invariant under time-dependent coordinate transformations.

7.2   Linear and Angular Momentum Equations

We use the component perturbation method to formulate the general perturbation equations of atmospheric flight. In this section I derive the perturbed linear and angular momentum equations and follow up with a detailed discussion of the aerodynamic force expansion in the next section.

The linear momentum of the body B with mass m relative to an inertial frame I is given by Eq. (5.3):

$$p_B^I = m v_B^I \tag{7.6}$$

where v_B^I is the linear velocity of the c.m. B relative to frame I. The angular momentum l_B^{BI} of body B relative to frame I and referred to the c.m. B is defined by the MOI tensor I_B^B of body B referred to the c.m. B and the angular velocity vector ω^{BI}:

$$l_B^{BI} = I_B^B \omega^{BI} \tag{7.7}$$

Using Eq. (7.2), the following ε perturbations of the state vectors are generated:

$$\varepsilon v_B^I = v_{Bp}^I - R v_{Br}^I \tag{7.8}$$

$$\varepsilon \omega^{BI} = \omega^{BpI} - R \omega^{BrI} \tag{7.9}$$

and for the linear and angular momenta

$$\varepsilon p_B^I = p_{Bp}^I - R p_{Br}^I \tag{7.10}$$

$$\varepsilon l_B^{BI} = l_{Bp}^{BpI} - R l_{Br}^{BrI} \tag{7.11}$$

Generalizing these equations for second-order tensors yields for the MOI tensor

$$\varepsilon I_B^B = I_{Bp}^{Bp} - R I_{Br}^{Br} \bar{R} \tag{7.12}$$

and the skew-symmetric form of the angular velocity vector

$$\varepsilon \Omega^{DI} = \Omega^{DpI} - R \Omega^{DrI} \bar{R} \tag{7.13}$$

Newton's and Euler's equation are replicated from Eq. (6.38):

$$D^I p_B^I = f = f_a + f_t + f_g \tag{7.14}$$

$$D^I l_B^{BI} = m = m_a + m_t \tag{7.15}$$

where f represents the forces and m the moments relative to the c.m. B. The subscripts a, t, and g refer to aerodynamics, propulsion, and gravity, respectively. Both equations are valid for the reference and perturbed flights.

To derive the linear momentum equations, let Eq. (7.14) describe the perturbed flight

$$D^I p_{Bp}^I = f_{ap} + f_{tp} + f_{gp}$$

and introduce the ε perturbations for each term

$$D^I \varepsilon p_B^I + D^I \left(R p_{Br}^I \right) = \varepsilon f_a + R f_{ar} + \varepsilon f_t + R f_{tr} + \varepsilon f_g + R f_{gr} \tag{7.16}$$

Let us modify the second term on the left side by applying the generalized Euler theorem, the chain rule, and the definition of the angular velocity vector Eq. (4.47). With Eq. (7.13) we obtain

$$D^I \varepsilon p_B^I + \varepsilon \Omega^{DI} R p_{Br}^I + \underline{R D^I p_{Br}^I}$$
$$= \underline{R f_{ar} + R f_{tr} + R f_{gr}} + \varepsilon f_a + \varepsilon f_t + \varepsilon f_g \tag{7.17}$$

The underlined terms are actually Eq. (7.14) applied to the reference flight and rotated through R. They are satisfied identically. The last term can be rewritten using the fact that the gravitational force is the same for the perturbed and reference flights $f_{gp} = f_{gr}$:

$$\varepsilon f_g = f_{gp} - R f_{gr} = (E - R) f_{gr}$$

The perturbation equation of the angular momentum is derived in the same way. Both equations are summarized as follows:

$$D^I \varepsilon p_B^I + \varepsilon \Omega^{DI} R^{DpDr} p_{Br}^I = \varepsilon f_a + \varepsilon f_t + (E - R^{DpDr}) f_{gr} \tag{7.18}$$

$$D^I \varepsilon l_B^{BI} + \varepsilon \Omega^{DI} R^{DpDr} l_{Br}^{BrI} = \varepsilon m_a + \varepsilon m_t \tag{7.19}$$

These are the *general perturbation equations of atmospheric flight mechanics*. No small perturbation assumptions have been made as yet. They are expressed in an invariant form, i.e., they hold for all coordinate systems. Two types of variables appear. The linear and angular momenta of the reference flight p_{Br}^I and l_{Br}^{BrI} are known as functions of time; and the component perturbations are marked by a preceding ε. The latter expressions εp_B^I and εl_B^{BI} represent the unknowns. The aerodynamic forces and moments will be discussed in Sec. 7.3. Evaluating the perturbational thrust and gravity forces is straightforward and will not be addressed.

The first terms on the left-hand sides of Eqs. (7.18) and (7.19) are the time rate of change of linear and angular momenta, whereas the second terms account for unsteady reference flights. Both equations are coupled nonlinear differential equations. To help you gain insight into the structure of the perturbation equations, I will derive two special cases: the all-important perturbations about a steady reference flight and the equations for turning reference flight.

7.2.1 Steady Reference Flight

I define *steady* as the nonaccelerated and nonrotating flight and choose the body frame B as the dynamic frame. With $\omega^{Brl} = 0$ (nonrotating reference flight) and therefore $l_{Br}^{Brl} = 0$, Eqs. (7.18) and (7.19) simplify to

$$D^I \varepsilon p_B^I + \varepsilon \Omega^{BI} R^{BpBr} p_{Br}^I = \varepsilon f_a + \varepsilon f_t + (E - R^{BpBr}) f_{gr} \qquad (7.20)$$

$$D^I \varepsilon l_B^{BI} = \varepsilon m_a + \varepsilon m_t \qquad (7.21)$$

To prepare for the use of the perturbed body coordinates $]^{Bp}$, we transform the rotational derivatives to the Bp frame. Let us start with Newton's equation Eq. (7.20) and use the fact that $\omega^{Brl} = 0$:

$$D^I p_B^I = D^{Bp} \varepsilon p_B^I + \Omega^{Bpl} \varepsilon p_B^I = D^{Bp} \varepsilon p_B^I + \Omega^{BpBr} \varepsilon p_B^I$$

Substitute the rotational derivative and use the definition of the linear momentum $p_B^I = m v_B^I$:

$$m\left(D^{Bp} \varepsilon v_B^I + \Omega^{BpBr} \varepsilon v_B^I + \varepsilon \Omega^{BI} R^{BpBr} v_{Br}^I\right) = \varepsilon f_a + \varepsilon f_t + (E - R^{BpBr}) f_{gr} \qquad (7.22)$$

Euler's equation is obtained by a similar transformation

$$D^I \varepsilon l_B^{BI} = D^{Bp} \varepsilon l_B^{BI} + \Omega^{Bpl} \varepsilon l_B^{BI} = D^{Bp} \varepsilon l_B^{BI} + \Omega^{BpBr} \varepsilon l_B^{BI}$$

and substituting it into Eq. (7.21):

$$D^{Bp} \varepsilon l_B^{BI} + \Omega^{BpBr} \varepsilon l_B^{BI} = \varepsilon m_a + \varepsilon m_t \qquad (7.23)$$

Modifying the perturbation εl_B^{BI} [Eq. (7.11)] by the definition of the angular momentum Eq. (7.7), and with $\omega^{Brl} = 0$, we obtain

$$\varepsilon l_B^{BI} = l_{Bp}^{Bpl} - R^{BpBr} l_{Br}^{Brl} = I_{Bp}^{Bp} \omega^{Bpl} - R^{BpBr} I_{Br}^{Br} \omega^{Brl} = I_{Bp}^{Bp} \omega^{BpBr}$$

and simplify Eq. (7.23) (with $D^{Bp} I_{Bp}^{Bp} = 0$):

$$I_{Bp}^{Bp} D^{Bp} \omega^{BpBr} + \Omega^{BpBr} I_{Bp}^{Bp} \omega^{BpBr} = \varepsilon m_a + \varepsilon m_t \qquad (7.24)$$

Equations (7.22) and (7.24) are the *perturbation equations of steady flight* in their invariant form. We select the perturbed body coordinates $]^{Bp}$ for the component formulation. First we deal with the gravitational term

$$([E]^{Bp} - [R^{BpBr}]^{Bp})[f_{gr}]^{Bp} = ([E]^{Bp} - [R^{BpBr}]^{Bp})[T]^{Bpl}[f_{gr}]^I$$

$$= ([T]^{BpBr} - [E])[T]^{Brl}[f_{gr}]^I \qquad (7.25)$$

then we express the linear momentum equations in $]^{Bp}$ coordinates

$$m\left(\left[\frac{d\varepsilon v_B^I}{dt}\right]^{Bp} + [\Omega^{BpBr}]^{Bp}[\varepsilon v_B^I]^{Bp} + \underline{[\Omega^{BpBr}]^{Bp}[v_{Br}^I]^{Br}}\right)$$

$$= [\varepsilon f_a]^{Bp} + [\varepsilon f_t]^{Bp} + \left([T]^{BpBr} - [E]\right)[T]^{Brl}[f_{gr}]^I \qquad (7.26)$$

and the angular momentum equation

$$[I_{Bp}^{Bp}]^{Bp}\left[\frac{d\omega^{BpBr}}{dt}\right]^{Bp} + [\Omega^{BpBr}]^{Bp}[I_{Bp}^{Bp}]^{Bp}[\omega^{BpBr}]^{Bp} = [\varepsilon m_a]^{Bp} + [\varepsilon m_t]^{Bp} \qquad (7.27)$$

These equations are nonlinear differential equations in the perturbation variables $[\varepsilon v_B^I]^{Bp}$ and $[\omega^{BpBr}]^{Bp}$. Eq. (7.26) is coupled with Eq. (7.27) through $[\omega^{BpBr}]^{Bp}$. In addition, the underlined term of Eq. (7.26) also couples the angular velocity perturbations via the reference velocity. With small perturbation assumptions and therefore neglecting terms of second order, we can linearize the left-hand sides:

$$m\left(\left[\frac{d\varepsilon v_B^I}{dt}\right]^{Bp} + [\Omega^{BpBr}]^{Bp}[v_{Br}^I]^{Br}\right)$$

$$= [\varepsilon f_a]^{Bp} + [\varepsilon f_t]^{Bp} + \left([T]^{BpBr} - [E]\right)[T]^{Brl}[f_{gr}]^I \qquad (7.28)$$

$$[I_{Bp}^{Bp}]^{Bp}\left[\frac{d\omega^{BpBr}}{dt}\right]^{Bp} = [\varepsilon m_a]^{Bp} + [\varepsilon m_t]^{Bp} \qquad (7.29)$$

As you see, the translational equation (7.28) is still coupled with the rotational equation (7.29) through the angular velocity perturbations $[\omega^{BpBr}]^{Bp}$. Equation (7.29) would be uncoupled from Eq. (7.28) were it not for the aerodynamic moment $[\varepsilon m_a]^{Bp}$, which is a function of the linear velocity. Both equations are still nonlinear differential equations through their aerodynamic functions.

The perturbation equations for steady flight are the workhorse for linear stability analysis. They apply equally to aircraft and missiles and have been used as far back as Lanchester, that great British aerodynamicist who introduced the stability derivative. A more intriguing challenge is the modeling of perturbations for unsteady flight. Much of our hard-earned tools will have to be put to use. With them we can study such exotic problems as the stability of cruise missiles in pitch-over dive and the dynamics of agile missile intercepts.

7.2.2 Unsteady Reference Flight

Return to Eqs. (7.18) and (7.19), the general perturbation equations, and keep the unsteady term $\varepsilon\Omega^{DI}R^{DpDr}l_{Br}^{Brl}$. They model the perturbations of aerospace vehicles in maneuvering flight. *Unsteady* means that the reference flight is rotating, like the pull-up maneuver of an aircraft, the circular intercept path of an air-to-air missile, or the pushdown trajectory of a cruise missile during terminal attack. If the parameters in the differential equations are functions of time, like the Mach dependence of aerodynamic coefficients, I call these terms *nonautonomous*.

Because we concentrate on nonspinning vehicles, the body frame is chosen as the dynamic frame, and we modify Eq. (7.18)

$$D^I \varepsilon p_B^I + \varepsilon \Omega^{BI} R^{BpBr} p_{Br}^I = \varepsilon f_a + \varepsilon f_t + (E - R^{BpBr}) f_{gr} \qquad (7.30)$$

and Eq. (7.19)

$$D^I \varepsilon l_B^{BI} + \varepsilon \Omega^{BI} R^{BpBr} l_{Br}^{BrI} = \varepsilon m_a + \varepsilon m_t \qquad (7.31)$$

To simplify these perturbation equations, second-order terms in ε are neglected. Such terms will now be identified. First, the rotational time derivatives are transposed to frame Bp via the Euler transformation:

$$D^I \varepsilon p_B^I = D^{Bp} \varepsilon p_B^I + \Omega^{BpI} \varepsilon p_B^I$$

$$D^I \varepsilon l_B^{BI} = D^{Bp} \varepsilon l_B^{BI} + \Omega^{BpI} \varepsilon l_B^{BI}$$

then, Eq. (7.13) is used to replace Ω^{BpI}. Finally, substituting back into Eqs. (7.30) and (7.31) yields the second-order terms $\varepsilon \Omega^{BI} \varepsilon p_B^I$ and $\varepsilon \Omega^{BI} \varepsilon l_B^{BI}$. Neglecting these terms reduces Eqs. (7.30) and (7.31) to

$$D^{Bp} \varepsilon p_B^I + R^{BpBr} \Omega^{BrI} \overline{R^{BpBr}} \varepsilon p_B^I + \varepsilon \Omega^{BI} R^{BpBr} p_{Br}^I$$
$$= \varepsilon f_a + \varepsilon f_t + (E - R^{BpBr}) f_{gr} \qquad (7.32)$$

$$D^{Bp} \varepsilon l_B^{BI} + R^{BpBr} \Omega^{BrI} \overline{R^{BpBr}} \varepsilon l_B^{BI} + \varepsilon \Omega^{BI} R^{BpBr} l_{Br}^{BrI} = \varepsilon m_a + \varepsilon m_t \qquad (7.33)$$

The second terms on the left-hand sides are the vestiges from the Euler transformations. They couple the reference rotation Ω^{BrI} with the perturbations εp_B^I and εl_B^{BI}.

We continue with the introduction of the linear velocity perturbation εv_B^I and the angular velocity perturbation $\varepsilon \omega^{BI}$, using the definition of Eqs. (7.6) (mass does not change from the reference to the perturbed flight)

$$\varepsilon p_B^I = \varepsilon (m v_B^I) = \varepsilon m v_B^I + m \varepsilon v_B^I = m \varepsilon v_B^I \qquad (7.34)$$

and Eq. (7.7)

$$\varepsilon l_B^{BI} = \varepsilon (I_B^B \omega^{BI}) = \varepsilon I_B^B \omega^{BI} + I_B^B \varepsilon \omega^{BI} = I_B^B \varepsilon \omega^{BI}$$

We use the fact that the perturbation of the MOI tensor is also zero. This follows from the definition of the MOI perturbation Eq. (7.12), where the rotated MOI tensor I_{Br}^{Br}, now coinciding with I_{Bp}^{Bp}, is subtracted from I_{Bp}^{Bp}:

$$\varepsilon I_B^B = I_{Bp}^{Bp} - R^{BpBr} I_{Br}^{Br} \overline{R^{BpBr}} = 0$$

With the definition of Eq. (7.11) and replacing I_{Br}^{Br} by I_{Bp}^{Bp}, the angular momentum perturbations evolve with the definition of $\varepsilon \omega^{BI}$ [Eq. (7.9)] as follows:

$$\varepsilon l_B^{BI} = I_{Bp}^{Bp} \omega^{BpI} - R^{BpBr} I_{Br}^{Br} \omega^{BrI}$$
$$= I_{Bp}^{Bp} (\omega^{BpI} - R^{BpBr} \omega^{BrI}) \qquad (7.35)$$
$$= I_{Bp}^{Bp} \varepsilon \omega^{BI}$$

Substituting Eqs. (7.34) and (7.35) into Eqs. (7.32) and (7.33) produces the *perturbation equations of unsteady flight* in tensor form suitable for applications

$$m D^{Bp} \varepsilon v_B^I + m R^{BpBr} \Omega^{BrI} \overline{R^{BpBr}} \varepsilon v_B^I + m \varepsilon \Omega^{BI} R^{BpBr} v_{Br}^I$$

$$= \varepsilon f_a + \varepsilon f_t + (E - R^{BpBr}) f_{gr} \qquad (7.36)$$

$$I_{Bp}^{Bp} D^{Bp} \varepsilon \omega^{BI} + R^{BpBr} \Omega^{BrI} \overline{R^{BpBr}} I_{Bp}^{Bp} \varepsilon \omega^{BI} + \varepsilon \Omega^{BI} R^{BpBr} I_{Br}^{Br} \omega^{BrI}$$

$$= \varepsilon m_a + \varepsilon m_t \qquad (7.37)$$

The perturbation variables are the linear velocity εv_B^I and angular velocity $\varepsilon \omega^{BI}$. The perturbation attitude angles ψ, θ, ϕ are contained in the small rotation tensor R^{BpBr}. Look at the terms on the left-hand sides of both equations, going from left to right: first, the time derivative wrt the perturbed body frame in anticipation of using perturbed body coordinates; second, the unsteady term caused by the rotating reference flight. The last terms of the left sides have a different purpose in each equation. In the first equation it is the coupling term with the angular momentum equation through $\varepsilon \omega^{BI}$. In the second equation this term makes an unsteady contribution of ω^{BrI}, similar to the preceding term.

To use the equations in numerical calculations, we express them in body coordinates associated with the perturbed frame Bp. The rotation tensor $[R^{BpBr}]^{Bp}$ disappears, as we transform the reference variables $[v_{Br}^I]^{Bp}$, $[\omega^{BrI}]^{Bp}$, and $[I_{Br}^{Br}]^{Bp}$ to the reference body axes $]^{Br}$. With the gravitational term expressed in inertial axes according to Eq. (7.25), Eqs. (7.36) and (7.37) become

$$m \left[\frac{d\varepsilon v_B^I}{dt}\right]^{Bp} + m[\Omega^{BrI}]^{Br}[\varepsilon v_B^I]^{Bp} + m[\varepsilon \Omega^{BI}]^{Bp}[v_{Br}^I]^{Br}$$

$$= [\varepsilon f_a]^{Bp} + [\varepsilon f_t]^{Bp} + ([T]^{BpBr} - [E])[T]^{BrI}[f_{gr}]^I \qquad (7.38)$$

$$[I_{Bp}^{Bp}]^{Bp} \left[\frac{d\varepsilon \omega^{BI}}{dt}\right]^{Bp} + [\Omega^{BrI}]^{Br}[I_{Bp}^{Bp}]^{Bp}[\varepsilon \omega^{BI}]^{Bp} + [\varepsilon \Omega^{BI}]^{Bp}[I_{Br}^{Br}]^{Br}[\omega^{BrI}]^{Br}$$

$$= [\varepsilon m_a]^{Bp} + [\varepsilon m_t]^{Bp} \qquad (7.39)$$

We succeeded in expressing all perturbation and references variables in perturbed and reference coordinates, respectively. The transformation matrix $[T]^{BpBr}$ consists of the attitude perturbations ψ, θ, ϕ, whereas $[T]^{BrI}$ establishes the coordinates of the gravitational force in reference body axes. Frequently, you will choose the Earth as inertial frame and the associated local-level coordinate system (see Sec. 3.2.2.7). Then, the gravitational force will take a particular simple form $[\overline{f_{gr}}]^L = m[0\ 0\ g]$.

How can we apply these equations? Imagine an air-to-air engagement. The target aircraft pulls a high-g maneuver, and the missile goes for the kill in a circular trajectory. Both execute unsteady circular trajectories. Record the reference values of $[v_{Br}^I]^{Br}$, $[\omega^{BrI}]^{Br}$, and $[T]^{BrI}$ for the aircraft and the missile. To analyze the dynamics of either vehicle, insert these reference values into Eqs. (7.38) and (7.39) and provide the appropriate mass and aerodynamic and propulsive parameters.

Equations (7.38) and (7.39) are the starting point for the two examples of Sec. 7.5. But before these equations can be derived, we have to deal with the subject of aerodynamic modeling and linearization.

7.3 Aerodynamic Forces and Moments

The most difficult problem in atmospheric flight mechanics is the mathematical modeling of the aerodynamic forces in a form that can be analyzed and evaluated quantitatively. Because the functional form is not known, the aerodynamic force functions are expanded in Taylor series in terms of the state variables relative to a reference flight. Even for digital computer simulations, restrictions for storage and computer time require that the number of independent variables in the aerodynamic tables be kept to a minimum. The dependency on the other variables then is expressed analytically by Taylor-series expansions.

For analytical studies a complete expansion is carried out for all state variables. There are two requirements that must be met. First, the partial derivatives of the expansions must be continuous—a condition that is usually satisfied; and second, the expansion variables must be small. In generating the aerodynamic forces three frames are involved: the *atmosphere-fixed* frame A, the *body* frame B, and the *relative wind* frame W. If the air is in uniform rectilinear motion relative to an inertial frame, A itself is an inertial frame. The wind frame has the c.m. of the vehicle as one of its points.

Usually it is postulated that the aerodynamic forces depend on external shape and size (represented by length l), atmospheric density ρ, and pressure p, the linear velocity of the airframe, c.m. relative to the atmosphere v_B^A, the angular velocity of the body relative to atmosphere ω^{BA}, the acceleration of the c.m. wrt the atmosphere $D^A v_B^A$ and, finally, the control surface deflections η. In summary, the functional form is

$$f_a = f_f\left(l, p, \rho, v_B^A, \omega^{BA}, D^A v_B^A, \eta\right) \qquad (7.40)$$

The same functional relationship holds for the aerodynamic moment.

$$m_a = f_m\left(l, p, \rho, v_B^A, \omega^{BA}, D^A v_B^A, \eta\right) \qquad (7.41)$$

The expansions, called *force expansion* according to Hopkin,[2] are carried out in the form of Eqs. (7.40) and (7.41). Variables that remain small throughout the perturbed flight must be identified. If the body frame does not yield these variables, the dynamic frame of the preceding section is introduced. As an example, a spinning missile requires a nonrotating body frame as dynamic frame.

7.3.1 Aerodynamic Symmetry of Aircraft and Missiles

The number of aerodynamic derivatives in the Taylor series increases vastly with higher-order terms. Even the linear derivatives add up to $12 \times 6 = 72$, more than the aerodynamicist would like to deal with. Fortunately, the configurational symmetries of aircraft and missiles reduce the number of nonzero derivatives drastically.

Maple and Synge[4] investigated the vanishing of aerodynamic derivatives in the presence of rotational and reflectional symmetries. They considered the dependence of the aerodynamic forces on linear and angular velocities only and employed complex variables to derive the results. The Maple–Synge theory contributed to the solution of many nonlinear ballistic problems in the past. However, with the advent of guided missiles the dependency of the aerodynamic forces on unsteady flow effects and control effectiveness has gained in importance.

In my dissertation and later in a paper[5] I derived, starting with the Principle of Material Indifference, rules of vanishing derivatives for aircraft and guided missiles. The aerodynamic forces are assumed functions of linear and angular velocities, linear accelerations, and control surface deflections. I will summarize the results with enough detail so that you can apply the rules successfully, but spare you the derivations. For the curious among you, my paper provides the details.

The functional form of Eqs. (7.40) and (7.41) will be used, but subscript notations will be substituted for the dependent and independent variables. The kth-order derivative of the Taylor-series expansion will be formulated in these subscripts. After reviewing the planar and tetragonal symmetry tensors, thought experiments are conducted that engage the Principle of Material Difference in discarding zero derivatives. Rules will be given for vanishing derivatives by adding up sub- and superscripts. For ease of application, two charts are presented that sift out the vanishing derivatives up to second order for missiles and up to third order for aircraft.

7.3.1.1 Taylor-series expansion.
We begin with the aerodynamic functionals of Eqs. (7.40) and (7.41), select the dynamic coordinate system $]^D$, and introduce components for the forces, moments, and dependent variables:

$$\begin{bmatrix} X \\ Y \\ Z \end{bmatrix}^D = \left[f_f \left(l, p, \rho; \begin{bmatrix} u \\ v \\ w \end{bmatrix}^D, \begin{bmatrix} p \\ q \\ r \end{bmatrix}^D, \begin{bmatrix} \dot{u} \\ \dot{v} \\ \dot{w} \end{bmatrix}^D, \begin{bmatrix} \delta p \\ \delta q \\ \delta r \end{bmatrix}^D \right) \right]^D$$

$$\begin{bmatrix} L \\ M \\ N \end{bmatrix}^D = \left[f_m \left(l, p, \rho; \begin{bmatrix} u \\ v \\ w \end{bmatrix}^D, \begin{bmatrix} p \\ q \\ r \end{bmatrix}^D, \begin{bmatrix} \dot{u} \\ \dot{v} \\ \dot{w} \end{bmatrix}^D, \begin{bmatrix} \delta p \\ \delta q \\ \delta r \end{bmatrix}^D \right) \right]^D$$

$$(7.42)$$

The acceleration components require additional comments. The $[D^A v_B^A]^D$ derivative must be transferred to the D frame before it can be expressed as $[\dot{v}_B^A]^D = [\dot{u}, \dot{v}, \dot{w}]$ components

$$[D^A v_B^A]^D = [D^D v_B^A]^D + [\Omega^{DA}]^D [v_B^A]^D = [\dot{v}_B^A]^D + [\Omega^{DA}]^D [v_B^A]^D$$

The additional term $[\Omega^{DA}]^D [v_B^A]^D$ is absorbed in the $[v_B^A]$ and $[\omega^{BA}]$ dependencies. Now we introduce the subscripted independent variables

$$z_j = \{u, v, w, p, q, r, \dot{u}, \dot{v}, \dot{w}, \delta p, \delta q, \delta r\}, \quad j = 1, 2, \ldots, 12 \qquad (7.43)$$

The two velocity components v and w, if expressed in body coordinates, can also be viewed as angle of attack $\alpha = \arctan(w/u)$ and side-slip angle $\beta = \arcsin(v/\sqrt{u^2 + v^2 + w^2})$. The variables $\delta p, \delta q, \delta r$ represent the missile controls—roll, pitch, and yaw—or the aircraft effectors—aileron, elevator and rudder. The dependent variables are abbreviated by

$$y_i = \{X, Y, Z, L, M, N\}; \quad i = 1, 2, \ldots, 6 \tag{7.44}$$

With these abbreviations Eq. (7.42) can be summarized as

$$y_i = d_i(z_j); \quad i = 1, 2, \ldots, 6; \quad j = 1, 2, \ldots, 12 \tag{7.45}$$

The aerodynamic functional is expanded into a Taylor series in terms of the 12 state variable components z_j, relative to the reference state $\bar{z}_j$. The Taylor expansion is mathematically justified if the partial derivatives in the expansion are continuous and the expansion variables $\Delta z_j = z_j - \bar{z}_j$ are small. For aircraft and missiles the aerodynamic forces are continuous functions of their states for most flight maneuvers. However, unsteady effects, such as vortex shedding, can introduce discontinuities that cannot be presented accurately by this method. In subscript notation the Taylor series assumes the form

$$y_i = d_i\{\bar{z}_j\} + \left(\frac{\partial d_i}{\partial z_{j_1}}\right)\Delta z_{j_1} + \frac{1}{2}\left(\frac{\partial^2 d_i}{\partial z_{j_1}\partial z_{j_2}}\right)\Delta z_{j_1}\Delta z_{j_2} + \cdots$$

$$+\frac{1}{k!}\left(\frac{\partial^k d_i}{\partial z_{j_1}\cdots\partial z_{j_k}}\right)\Delta z_{j_1}\cdots\Delta z_{j_k} + \cdots$$

$$i = 1, 2, \ldots, 6; \quad j_1, j_2, \ldots, j_k = 1, 2, \ldots, 12$$

The partial derivatives, evaluated at the reference flight conditions, are the *aerodynamic derivatives*. The kth derivative is a $k + 1$ order tensor and is abbreviated by

$$D_i^{j_1 j_2 \cdots j_k} = \frac{1}{k!}\left(\frac{\partial^k d_i}{\partial z_{j_1}\cdots\partial z_{j_k}}\right) \tag{7.46}$$

It is a function of the implicit variables M and Re. As an example, the third-order rolling moment derivative with $i = 4$, $j_1 = 1$, $j_2 = 5$, $j_3 = 11$ becomes, by correlating the subscripts with Eqs. (7.43) and (7.44),

$$\frac{\partial^3 d_4}{\partial z_1 \partial z_5 \partial z_{11}} = \frac{\partial^3 L}{\partial u \partial q \partial(\delta q)} = L_{uq\delta q} \tag{7.47}$$

This is the rolling moment derivative caused by the forward velocity component u, the pitch rate q, and the pitch control deflection δq.

7.3.1.2 Configurational symmetries.

Most aircraft and guided missiles have a planar or cruciform external shape. The planar configuration dominates among aircraft and cruise missiles, while missiles that execute rapid terminal maneuvers have cruciform airframes. Two types of symmetry are, therefore, considered: reflectional and tetragonal (90 deg rotational) symmetries.

To derive the conditions of vanishing derivatives, precise definitions of these symmetries are required. In the case of reflectional symmetry, the existence of a plane, satisfying certain conditions, is required, whereas tetragonal symmetry calls for an axis with specific characteristics.

In Chapter 2, I introduced the reflection tensors M and in Chapter 4 the tetragonal symmetry tensor R_{90}. In body coordinates they have the form

$$[M]^B = \begin{bmatrix} 1 & 0 & 0 \\ 0 & -1 & 0 \\ 0 & 0 & 1 \end{bmatrix} \quad \text{and} \quad [R_{90}]^B = \begin{bmatrix} 1 & 0 & 0 \\ 0 & 0 & -1 \\ 0 & 1 & 0 \end{bmatrix}$$

$[R_{90}]^B$, with a determinant of $+1$, is a proper rotation, whereas $[M]^B$ is improper because its determinant value is -1. For an aircraft the displacement vectors s_{SP}, originating from the symmetry plane and extending to the surface, occur in pairs, related by

$$s'_{SP} = M s_{SP}$$

and similarly, for a missile, the displacement vectors s_{SA}, reaching from the symmetry axis to the surface, also occur in pairs related by

$$s'_{SA} = R_{90} s_{SA}$$

These relationships together with the PMI, already encountered in earlier chapters, lead us to the desired conditions for vanishing derivatives. Noll[3] has provided a precise mathematical formulation. Applied to the aerodynamic problem at hand, the PMI asserts that the physical process of generating aerodynamic forces d_i from the variables z_j is independent of spatial attitude. For any rotation tensor R_{in}, in tensor subscript notation and summation over repeated indices, it states

$$R_{in} d_n \{z_j\} = d_i \{R_{jp} z_p\} \tag{7.48}$$

Read Eq. (7.48) with me from left to right: the vector valued function d_n of the state vector z_j, rotated through the rigid rotation R_{in}, equals the same vector valued function of the state variables rotated through the same tensor R_{jp}. A functional with the properties expressed by Eq. (7.48) is called an *isotropic* function. The rotation is allowed to be proper or improper; i.e., its determinant can be plus or minus one.

Let us apply the PMI first to planar vehicles. Suppose Eq. (7.45) describes the aerodynamics of a particular wind-tunnel test result:

$$y_i = d_i \{z_j\}$$

Consider a second test under the same conditions, but with flow variables z_j mirrored by the reflection tensor M_{jm}

$$y'_i = d_i \{M_{jp} z_p\}$$

The resulting aerodynamics y' should also be mirrored.

$$y'_i = M_{in} y_n$$

Therefore, equating the last two relationships, and with Eq. (7.45), we obtain

$$M_{in} d_n \{z_j\} = d_i \{M_{jp} z_p\} \tag{7.49}$$

just like the PMI, Eq. (7.48) states. But if the external configuration of the test object possesses planar symmetry, the aerodynamics is indistinguishable in the two tests

$$d_i\{z_j\} = d_i\{M_{jp}z_p\}$$

and therefore substituting into Eq. (7.49) we obtain the condition for vanishing derivatives

$$M_{in}d_n\{z_j\} = d_i\{z_j\} \tag{7.50}$$

We expanded both sides in Taylor series. In body coordinates the elements of M_{in} consist of $+1$ and -1 terms only. Those derivatives that exhibit different signs because of M_{in} must be zero! If you read my paper, you will see that the derivation is somewhat more complicated. Yet Eq. (7.50), with the abbreviation of Eq. (7.46), leads eventually to the relationship between the derivatives:

$$D_i^{j_1 j_2 \cdots j_k} = (-1)^{\Sigma j_k + k + i + 1} D_i^{j_1 j_2 \cdots j_k} \tag{7.51}$$

Rule 1:   The aerodynamic derivatives $D_i^{j_1 j_2 \cdots j_k}$ of a vehicle with reflectional symmetry vanish if the sum $\Sigma j_k + k + i + 1$ is an odd number.

When the exponent of (-1) is odd, a negative sign will appear at the right-hand side of Eq. (7.51). The same derivatives with different signs can only be equal if their values are zero. The subscript i indicates the force or moment components and the superscripts $j_1, j_2, \ldots, j_k$ designate the components of the state vector of the of the kth partial derivative. To convert from the derivatives with physical variables to their subscript notation $D_i^{j_1 j_2 \cdots j_k}$, use Table 7.1.

Let us apply Rule 1 to the example, Eq. (7.47): $\Sigma j_k + k + i + 1 = (1 + 5 + 11) + 3 + 4 + 1 = 25$. The derivative does not exist; a result you would have predicted if you are an aerodynamicist.

To derive the condition for vanishing aerodynamic derivatives of vehicles with tetragonal symmetry, we make use of the fact that a cruciform vehicle has two

Table 7.1   Association of dependent and independent variables with subscripts and superscripts

i, j	j	i
1	u	X
2	v	Y
3	w	Z
4	p	L
5	q	M
6	r	N
7	$\dot{u}$	——
8	$\dot{v}$	——
9	$\dot{w}$	——
10	δp	——
11	δq	——
12	δr	——

planes of reflectional symmetry. The two planes are rotated into each other by the tetragonal symmetry tensor $\boldsymbol{R}_{90}$, and they intersect at the axis of symmetry. The PMI is applied twice to the two symmetry planes. The first one we carried out already for the reflectional symmetry plane. Therefore, Rule 1 applies also to cruciform vehicles. We derive the second condition by rotating the original experiment through 90 deg and applying the PMI the second time. I will spare you the details. The result is the relationship

$$C_p^{q_1 q_2 \cdots q_k} = (-1)^{\Sigma q_k + k + p + 1} C_p^{q_1 q_2 \cdots q_k} \tag{7.52}$$

where C is related to the D derivative by simply exchanging every second or third subscript. Thus the rule for vanishing derivatives for cruciform vehicles is stated as follows.

Rule 2: The aerodynamic derivative $D_i^{j_1 j_2 \cdots j_k}$ of a vehicle with tetragonal symmetry vanishes if the sum $\Sigma j_k + k + i + 1$ is an odd number (Rule 1) *or* if $\Sigma q_k + k + p + 1$ is an odd number as well.

The relationship of the subscripts between $D_i^{j_1 j_2 \cdots j_k}$ and $C_p^{q_1 q_2 \cdots q_k}$ is given by Table 7.2.

As a test case, do you expect $N_{wp\delta q}$ to exist for an aircraft or a missile? It is the control-coupling derivative of pitch control δq, contributing to the yawing moment N, in the presence of a vertical velocity component w and roll rate p. For an aircraft we have $N_{wp\delta q} = D_6^{3\,4\,11}$. Applying Rule 1, $\Sigma j_k + k + i + 1 = 3 + 4 + 11 + 3 + 6 + 1 = 28$, we get an even number, and therefore the derivative is nonzero. For a missile, with Rule 2, $C_5^{2\,4\,12} = D_6^{3\,4\,11}$, and $\Sigma q_k + k + p + 1 = 2 + 4 + 12 + 3 + 5 + 1 = 27$ is an odd number, and the derivative vanishes. Did you guess correctly?

Let us try another example: $Y_{w\delta r} = D_2^{3\,12}$ is the yawing force derivative Y caused by rudder control δr in the presence of downwash w. It survives the test for planar

Table 7.2 Subscript and
superscript relationship
between the D and
C derivatives

i, j_k	p, q_k
1	1
2	3
3	2
4	4
5	6
6	5
7	7
8	9
9	8
10	10
11	12
12	11

vehicles (from Rule 1: $\Sigma j_k + k + i + 1 = 3 + 12 + 2 + 2 + 1 = 20$), indicating that, for aircraft, the derivative is linearly dependent on the downwash. For missiles, however, with $C_3^{2\,11} = D_3^{3\,12}$ (Rule 2: $\Sigma q_k + k + p + 1 = 2 + 11 + 2 + 3 + 1 = 19$), the derivative does not exist. Physically speaking, the downwash is symmetrical for cruciform configurations. It affects the side force not linearly, which would result in a sign change, but quadratically, as shown by the existence of the derivative $Y_{w^2 \delta r} = D_2^{3\,3\,12} = C_3^{2\,2\,11} : \Sigma j_k + k + i + 1 = 3 + 3 + 12 + 3 + 2 + 1 = 24$, and $\Sigma q_k + k + p + 1 = 2 + 2 + 11 + 3 + 3 + 1 = 22$.

These rules are quite helpful not only for modeling but also for investigating nonlinear effects. I put them to good use in my dissertation, describing the nonlinear aerodynamic phenomena of Magnus rotors with higher-order derivatives. The real challenge of course is the extraction of these derivatives from wind-tunnel or free-flight tests, which we leave to the expert.

I do not have space here to discuss the physical interpretation of aerodynamic derivatives in any more detail. You will find the linear derivatives explained by Pamadi[6] or Etkin.[1] For nonlinear phenomena you have to search the specialist literature that applies to your particular modeling problem.

7.3.1.3 Derivative maps.

As you build your aerodynamic model, you have to apply the vanishing-derivative rules numerous times. Just for the linear derivatives it would be 72 times. To save you time, I supply maps that let you determine the existence of derivatives by inspection of their grid pattern. They apply for up to third-order derivatives for aircraft and up to second-order derivatives for missiles.

Figure 7.1 graphically patterns Eq. (7.51) and the associated Rule 1 for planar vehicle derivatives up to third order. In the following discussion, however, rather than referring to the vanishing derivatives, I will emphasize those that survive the sifting process.

Depending on the force components i, the order of the derivative, and the even or odd integer of the third superscript, the existence of the derivative is indicated by two symbols—cruciform or box—in the top table of Fig. 7.1. For instance, for the first-order derivative X_u the table assigns a cruciform symbol to the force component X. To determine existence, refer to the single row array. Because X_u is associated with a cruciform symbol, it exists. However, X_v, having a box symbol in the array, vanishes because it does not show the required cruciform symbol of the table. Moving into the next column of the table, the first order derivative $L_{\delta p}$ must have a box symbol. The single array confirms its existence. You can use this array to determine quickly, which derivatives you must include in you linear aerodynamic model.

For second-order derivatives $D_i^{j_1 j_2}$ the symbols are reversed in the table of Fig. 7.1, and the 12×12 array is used to determine their existence. The array is symmetric because the order of taking partial derivatives is irrelevant (assuming continuous functions). Therefore, you can start with either rows or columns. For example, $Z_{w\delta q}$, requiring the box symbol, exists according to the array, but $Y_{w\delta q}$, associated with the cruciform pattern, vanishes. About 198 second-order derivatives exist. It is up to the aerodynamicist to determine their significance and magnitude. Hopefully, if called to model nonlinear effects, you can neglect most of them, but only after you have reasoned through all exclusions.

Third-order derivatives must be separated into two groups, depending on an even or odd third-order superscript (even or odd refers to the position number of the variable in the state vector). If the last superscript is even, e.g., v, the cruciform

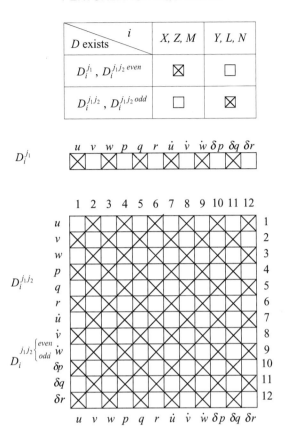

Fig. 7.1   Aerodynamic derivatives of planar vehicles.

symbol is associated with a derivative such as Z_{wrv}. Entering the square array with w and r indicates existence of that derivative. Let us check out our first example $L_{uq\delta q} = D_4^{1511}$ of Eq. (7.47) for planar vehicles. Its third superscript is odd, and because L is in the second column, the cruciform symbol applies. The square array entry with u and q requires the box symbol; therefore, the derivative vanishes.

For vehicles with tetragonal symmetry, a compact graphic display is possible only for first- and second-order derivatives. Figure 7.2 summarizes both Eqs. (7.51) and (7.52) or Rules 1 and 2. The table in Fig. 7.2 assigns different symbols for the existence of four groups of derivatives. For instance, X_u exists, and X_v vanishes; Z_{uw} survives, but $Z_{u\delta p}$ does not. I am sure by now you have caught on to my scheme.

The graphical aids of both figures can be used to determine uniquely the existence or nonexistence of aerodynamic derivatives. A significant number of derivatives can be eliminated by symmetry alone. Reflectional symmetry eliminates about half of the linear candidates, and because the square array is symmetrical, only approximately a quarter of the second- and third-order derivatives need be considered. For vehicles with tetragonal symmetry, these numbers are further reduced by a factor of one-half.

I already mentioned earlier that some of the state variables could also be replaced by other relevant quantities. Particularly, the substitutions of α for w and β for v are

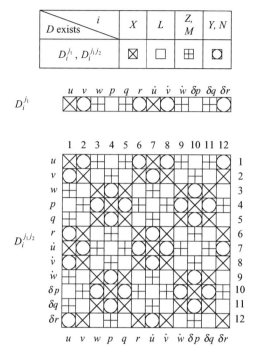

Fig. 7.2   Aerodynamic derivatives of cruciform vehicles.

quite common. Also u is often replaced by the Mach-number dependence. Similar alternatives are used for $\dot{w} \to \dot{\alpha}$ and $\dot{v} \to \dot{\beta}$. The controls δp, δq, δr refer either to the missile's roll, pitch, and yaw or the aircraft's aileron, elevator, and rudder.

The coordinate system of the expansion variables is the dynamic system. In most cases the body coordinates serve as the dynamic system. For an aircraft in steady flight, the reference body axes are the inertial axes, and during its perturbed flight the body axes become the coordinate system for the aerodynamic expansion. Frequently, the stability axes (special body axes) are used. However, other possibilities must also be considered. For a spinning missile the dynamic coordinates are associated with the nonspinning body frame. Because this glove-like frame also has rotational symmetry, the derivatives are expressed in these coordinates, and Rule 2 applies. A similar situation exists for Magnus rotors (see Sec. 10.1.1.4) or spinning golf balls. Their spin axes, however, are essentially normal to the velocity vector. Thus the nonspinning frame exhibits planar symmetry, and Rule 1 should be used.

The modeling of the aerodynamics for computer simulations frequently includes tabular look-up for variables with large variations, and the Taylor expansion is only carried out for those variables that remain small. So far, we have dealt with complete expansions of all 12 components of the state vector. With minor modifications the results are applicable also to these incomplete expansions.

For instance, if the aerodynamics is expressed as tabular functions of the velocity component u, the Taylor series is carried out in terms of the state variable components 2 through 12 only. All derivatives remain implicit functions of u, and the order of the derivatives is reduced by one. For example, an aircraft's $X_{w\delta q}(u)$ derivative is modeled by a one-dimensional table. Instead of a table, it also could be completely expanded in powers of u, provided the polynomial fits the data:

$$X_{w\delta q}(u) = X_{uw\delta q}u + X_{u^2w\delta q}u^2 + X_{u^3w\delta q}u^3 + \cdots$$

Please confirm the existence of the derivatives on the left- and the right-hand sides.

This procedure applies to any derivative and any state variable component. Also more than one variable can be replaced by implicit functions. We will use this approach in several instances. In Sec. 10.2.1 you will see it applied to aircraft and missile six-DoF models. For the CADAC FALCON6 simulation I will introduce reduced derivatives that are implicit functions of Mach, angle of attack, and, in some cases, also of sideslip angle. The CADAC SRAAM6 air-to-air missile model, using aeroballistic instead of body axes, can also be pressed into this scheme, and you will see that most derivatives are implicit functions of Mach and total angle of attack. Finally, the CADAC GHAME6 hypersonic vehicle is a straight expansion of derivatives with Mach and angle of attack as implicit variables.

The most frequently encountered task, however, is the linear expansion of the aerodynamic derivatives. I will demonstrate the procedure for the linear perturbation equations of steady flight and specifically derive some simple state equations that are needed for our autopilot designs in Sec. 10.2.2. A further sophistication is the extension to unsteady flight like missiles in pushover and terminal dive or in lateral turns.

7.4 Perturbation Equations of Steady Flight

After this excursion into aerodynamic modeling, let us pick up the discussion from Sec. 7.2. The equations of motion, Eqs. (7.28) and (7.29), must be completed by the aerodynamic expansions of the right-hand sides. The linear terms of the Taylor expansion can be grouped according to the state variables

$$[\varepsilon f_a]^{Bp} = \left[\frac{\partial m_a}{\partial v_B^A}(M, Re)\right]^{Bp}[\varepsilon v_B^A]^{Bp} + \left[\frac{\partial f_a}{\partial \omega^{BA}}(M, Re)\right]^{Bp}[\varepsilon \omega^{BA}]^{Bp}$$

$$+ \left[\frac{\partial f_a}{\partial \dot{v}_B^A}(M, Re)\right]^{Bp}[\varepsilon \dot{v}_B^A]^{Bp} + \left[\frac{\partial f_a}{\partial \eta}(M, Re)\right]^{Bp}[\varepsilon \eta]^{Bp} \qquad (7.53)$$

and

$$[\varepsilon m_a]^{Bp} = \left[\frac{\partial m_a}{\partial v_B^A}(M, Re)\right]^{Bp}[\varepsilon v_B^A]^{Bp} + \left[\frac{\partial m_a}{\partial \omega^{BA}}(M, Re)\right]^{Bp}[\varepsilon \omega^{BA}]^{Bp}$$

$$+ \left[\frac{\partial m_a}{\partial \dot{v}_B^A}(M, Re)\right]^{Bp}[\varepsilon \dot{v}_B^A]^{Bp} + \left[\frac{\partial m_a}{\partial \eta}(M, Re)\right]^{Bp}[\varepsilon \eta]^{Bp} \qquad (7.54)$$

where we have lumped the scaling and flow variables l, p, ρ into the implicit Mach and Reynolds numbers of the reference flight (we will neglect the Reynolds-number dependency). Combined with Eqs. (7.28) and (7.29) and assuming no wind [i.e., frame A = frame I results in $[\varepsilon v_B^A]^{Bp} = [\varepsilon v_B^I]^{Bp}$ according to Eq. (7.8), and $[\varepsilon \omega^{BA}]^{Bp} = [\omega^{BpBr}]^{Bp}$ according to Eq. (7.9)], Newton's perturbation equation for steady flight becomes then

$$
m \left[\frac{d\varepsilon v_B^I}{dt} \right]^{Bp} + m [\Omega^{BpBr}]^{Bp} [v_{Br}^I]^{Br}
$$

$$
= \left[\frac{\partial f_a}{\partial v_B^I}(M) \right]^{Bp} [\varepsilon v_B^I]^{Bp} + \left[\frac{\partial f_a}{\partial \omega^{BpBr}}(M) \right]^{Bp} [\omega^{BpBr}]^{Bp}
$$

$$
+ \left[\frac{\partial f_a}{\partial \dot{v}_B^I}(M) \right]^{Bp} [\varepsilon \dot{v}_B^I]^{Bp} + \left[\frac{\partial f_a}{\partial \eta}(M) \right]^{Bp} [\varepsilon \eta]^{Bp}
$$

$$
+ [\varepsilon f_t]^{Bp} + ([T]^{BpBr} - [E])[T]^{BrI} [f_{gr}]^I \tag{7.55}
$$

and Euler's equation

$$
[I_{Bp}^{Bp}]^{Bp} \left[\frac{d\omega^{BpBr}}{dt} \right]^{Bp} = \left[\frac{\partial m_a}{\partial v_B^I}(M) \right]^{Bp} [\varepsilon v_B^I]^{Bp} + \left[\frac{\partial m_a}{\partial \omega^{BpBr}}(M) \right]^{Bp} [\omega^{BpBr}]^{Bp}
$$

$$
+ \left[\frac{\partial f_a}{\partial \dot{v}_B^I}(M) \right]^{Bp} [\varepsilon \dot{v}_B^I]^{Bp} + \left[\frac{\partial m_a}{\partial \eta}(M) \right]^{Bp} [\varepsilon \eta]^{Bp} + [\varepsilon m_t]^{Bp} \tag{7.56}
$$

These *fundamental stability equations of steady flight* are fully coupled by the state variables $[\varepsilon v_B^I]^{Bp}$ and $[\omega^{BpBr}]^{Bp}$, appearing in the aerodynamic terms and in addition by $[\omega^{BpBr}]^{Bp}$ being present on the left-hand side of Eq. (7.55). They are now linear differential equations, possibly with time-dependent coefficients through the variable Mach number. The controls $[\varepsilon \eta]^{Bp}$ are the inhomogeneous input to the equations.

The acceleration variable $[\varepsilon \dot{v}_B^I]^{Bp}$ appears on both sides of Eq. (7.55). For a state-space representation they would have to be combined. Yet frequently in the aerodynamic expansions, $[\varepsilon \dot{v}_B^I]^{Bp}$ is expressed as $\dot{\alpha}, \dot{\beta}$ and, particularly in missile dynamics, often combined with the damping effect caused by $[\omega^{BpBr}]^{Bp}$, thus eliminating the acceleration term on the right-hand sides.

For even greater simplicity the coupling term $m[\Omega^{BpBr}]^{Bp}[v_{Br}^I]^{Br}$ is usually neglected in the development of plant dynamics for autopilot designs, and so are the thrust perturbations and the gravitational term. We will follow this practice as we derive transfer and state equations for roll, acceleration, rate, and flight-path angle autopilots.

7.4.1 Roll Transfer Function

The simplest application is probably the roll channel, uncoupled from the other degrees of freedom. From Eq. (7.56) we select the first component equation and scan Fig. 7.1 for the existing roll derivatives

$$
I_{11} \dot{p} = L_v v + L_p p + L_r r + L_{\dot{v}} \dot{v} + L_{\delta p} \delta p + L_{\delta r} \delta r
$$

Because we neglect cross-coupling, there remains just

$$I_{11}\dot{p} = L_p p + L_{\delta p}\delta p$$

and the transfer function is

$$\frac{p(s)}{\delta p(s)} = \frac{L_{\delta p}/I_{11}}{s - L_p/I_{11}}$$

The mass properties are commonly lumped into the derivative, and the nomenclature LL is used for L to avoid confusion with the designation for lift. Furthermore, because we emphasize here the aircraft, we replace δp by the aileron symbol δa:

$$\frac{p(s)}{\delta a(s)} = \frac{LL_{\delta a}}{s - LL_p} \qquad (7.57)$$

I shall use this transfer function for the roll autopilot design in Sec. 10.2.2.2. $LL_{\delta a}$ and LL_p are the *dimensioned* roll control and damping derivatives, evaluated at the reference flight. They are related to the *nondimensional* derivatives $C_{l_{\delta a}}$ and C_{l_p} by $LL_{\delta a} = (\bar{q}Sb/I_{11})C_{l_{\delta a}}$ and $L\,L_p = (\bar{q}Sb/I_{11})(b/2V)C_{l_p}$, where $\bar{q}$ is the dynamic pressure, S the reference area, b the wing span, and V the vehicle speed.

7.4.2 Pitch Dynamic Equations

The uncoupled pitch dynamics consist of the pitching moment equation, second of Eq. (7.56), and the normal force equation, third of Eq. (7.55). This example is for tetragonal missiles. We therefore use Fig. 7.2 to write down the nonvanishing linear derivatives

$$I_2\dot{q} = M_w w + M_q q + M_{\dot{w}}\dot{w} + M_{\delta q}\delta q$$

Instead of w we prefer the expansion in terms of the angle of attack α and merge $M_{\dot{a}}$ into M_q (quite common for missiles). Furthermore, to conform to conventions the derivatives are divided by the pitch moment of inertia I_2 but retain their letter symbol M:

$$\dot{q} = M_\alpha \alpha + M_q q + M_{\delta q}\delta q \qquad (7.58)$$

We model the normal force dynamics by the third component of Eq. (7.55). Neglecting the gravitational term, we have

$$m\dot{w} = Z_w w + Z_q q + Z_{\dot{w}}\dot{w} + Z_{\delta q}\delta q \qquad (7.59)$$

Again, we replace w by α, neglect all damping derivatives, and furthermore follow missile conventions, replacing X by $-N$ (normal force) and the vertical acceleration $\dot{w}$ by the normal acceleration $-a$. Redefining the normal force derivatives N by including the mass in the denominator, we formulate

$$a = N_\alpha \alpha + N_{\delta q}\delta q \qquad (7.60)$$

Equations (7.58) and (7.60) will be used for an acceleration autopilot design with inner rate-loop damping. Commonly, only pitch gyros and accelerometers are available as sensors, but not angle of attack. The α dependency must therefore be eliminated. You accomplish this feat by first taking the derivative of Eq. (7.60):

$$\dot{a} = N_\alpha \dot{\alpha} + N_{\delta q}\delta \dot{q} \qquad (7.61)$$

Then recalling that the normal acceleration a is proportional to the flight-path-angle rate $\dot{\gamma}$ (for small α).

$$a = V\dot{\gamma} \tag{7.62}$$

and with the kinematic relationship $\dot{\gamma} = q - \dot{\alpha}$

$$a = V\dot{\gamma} = V(q - \dot{\alpha}) \tag{7.63}$$

Solving for $\dot{\alpha}$ and substituting into Eq. (7.61)

$$\dot{a} = N_\alpha q - \frac{N_\alpha}{V}a + N_{\delta q}\delta\dot{q} \tag{7.64}$$

Now, premultiply Eq. (7.58) by N_α and (7.60) by M_α, subtract them from each other, and then solve for $\dot{q}$. You derived the important result without α dependency:

$$\dot{q} = M_q q + \frac{M_\alpha}{N_\alpha}a + \left(M_{\delta q} - \frac{M_\alpha N_{\delta q}}{N_\alpha}\right)\delta q \tag{7.65}$$

Equations (7.64) and (7.65) are the state equations for a and q. The derivative of the pitch control $\delta\dot{q}$ of Eq. (7.64) is acquired from a first-order actuator, as shown in Fig. 7.3. It provides the third state equation

$$\delta\dot{q} = \lambda u - \lambda\delta q \tag{7.66}$$

with u the commanded input and $1/\lambda$ the actuator time constant. Substituting this $\delta\dot{q}$ into Eq. (7.64) yields

$$\dot{a} = N_\alpha q - \frac{N_\alpha}{V}a - \lambda N_{\delta q}\delta q + N_{\delta q}\lambda u \tag{7.67}$$

Collecting the three equations (7.65–7.67) yields the desired result:

$$\begin{bmatrix} \dot{q} \\ \dot{a} \\ \delta\dot{q} \end{bmatrix} = \begin{bmatrix} M_q & \dfrac{M_\alpha}{N_\alpha} & M_{\delta q} - \dfrac{N_{\delta q}M_\alpha}{N_\alpha} \\ N_\alpha & -\dfrac{N_\alpha}{V} & -\lambda N_{\delta q} \\ 0 & 0 & -\lambda \end{bmatrix} \begin{bmatrix} q \\ a \\ \delta q \end{bmatrix} + \begin{bmatrix} 0 \\ \lambda N_{\delta q} \\ \lambda \end{bmatrix} u \tag{7.68}$$

These state equations in pitch rate q, normal acceleration a, and pitch control δq are quite useful for autopilot design. Particularly, we succeeded in replacing α by a, therefore replacing the difficult to implement angle-of-attack sensor by the readily available accelerometer from the INS.

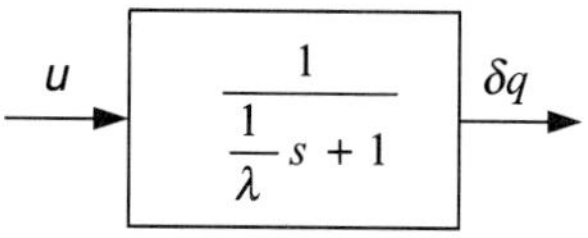

Fig. 7.3   Actuator dynamics.

The dimensional derivatives N_α, $N_{\delta q}$, M_α, M_q, $M_{\delta q}$ are related to the nondimensional derivatives C_{N_α}, $C_{N_{\delta q}}$, C_{m_α}, C_{m_q}, $C_{m_{\delta q}}$ by

$$N_\alpha = \frac{\bar{q}S}{m}C_{N_\alpha}; \quad N_{\delta q} = \frac{\bar{q}S}{m}C_{N_{\delta q}}; \quad M_\alpha = \frac{\bar{q}Sd}{I_2}C_{m_\alpha};$$

$$M_q = \frac{\bar{q}Sd^2}{2I_2V}C_{m_q}; \quad M_{\delta q} = \frac{\bar{q}Sd}{I_2}C_{m_{\delta q}}$$

where d is the missile diameter and S the maximum cross section.

Sometimes autopilots are designed without consideration of actuator dynamics. For these simplified circumstances we set $\delta\dot{q} = 0$; thus, from Eq. (7.66) $\lambda u = \lambda\delta q$. Neglecting $N_{\delta q}M_\alpha/N_\alpha$ against the significantly larger $M_{\delta q}$, we gained the *reduced-order* state equations

$$\begin{bmatrix} \dot{q} \\ \dot{a} \end{bmatrix} = \begin{bmatrix} M_q & \dfrac{M_\alpha}{N_\alpha} \\ N_\alpha & -\dfrac{N_\alpha}{V} \end{bmatrix}\begin{bmatrix} q \\ a \end{bmatrix} + \begin{bmatrix} M_{\delta q} \\ 0 \end{bmatrix}\delta q \qquad (7.69)$$

These pitch-plane equations, which depend on pitch rate q and normal acceleration a as state variables only and have the pitch control δq as input, play an important role in the design of air-to-air missile autopilots. Because of their simplicity as plant descriptor, a self-adaptive autopilot can be constructed around them. I present the details in Sec. 10.2.2.4.

The discussion would be incomplete, however, without also reintroducing the angle of attack as one of the state variables. Substituting Eq. (7.63) into Eq. (7.61) eliminates the acceleration a completely, and we are left with

$$\dot{\alpha} = q - \frac{N_\alpha}{V}\alpha - \frac{N_{\delta q}}{V}\delta q$$

which we use to replace the $\dot{a}$ equation in Eq. (7.69):

$$\begin{bmatrix} \dot{q} \\ \dot{\alpha} \end{bmatrix} = \begin{bmatrix} M_q & M_\alpha \\ 1 & -\dfrac{N_\alpha}{V} \end{bmatrix}\begin{bmatrix} q \\ \alpha \end{bmatrix} + \begin{bmatrix} M_{\delta q} \\ -\dfrac{N_{\delta q}}{V} \end{bmatrix}\delta q \qquad (7.70)$$

We will make use of this format when we design the rate autopilot in Sec. 10.2.2.1. There, we will derive the $q(s)/\delta q(s)$ transfer function by eliminating α, thus bypassing the need for an α sensor.

Similarly, the lateral acceleration equations with the state variables' yaw rate r and sideslip angle β, and the control input δr are (see Problem 7.1)

$$\begin{bmatrix} \dot{r} \\ \dot{\beta} \end{bmatrix} = \begin{bmatrix} LN_r & LN_\beta \\ -1 & \dfrac{Y_\beta}{V} \end{bmatrix}\begin{bmatrix} r \\ \beta \end{bmatrix} + \begin{bmatrix} LN_{\delta r} \\ \dfrac{Y_{\delta r}}{V} \end{bmatrix}\delta r \qquad (7.71)$$

where LN designates the yawing moment derivative (to avoid confusion with the normal force derivative N).

7.4.3 Flight-Path-Angle State Equations

For the design of a flight-path-angle tracker, we need the pitch dynamics expressed in flight path angle γ, pitch rate q, and its integral, pitch angle θ. Because this autopilot function is primarily for an aircraft, we take a slightly different approach than in the preceding section. The lift force replaces the normal force. Therefore, the left-hand side of the third component of Eq. (7.55) $m\dot{w}$ is replaced by $mV\dot{\gamma}$, using the relationship [Eq. (7.62)]

$$m\dot{w} = ma = mV\dot{\gamma}$$

and the right-hand aerodynamics is formulated in terms of the lift force L. The symmetry conditions of the expansion Rule 1 apply as well for the lift as the normal force, but, alas, we have to put up with confusing notation. Aircraft aerodynamicists prefer lift to normal force and use L instead of Z or N. They designate the rolling moment as LL, the convention we used earlier. From Fig. 7.1 we derive the linear derivatives and neglect the coupling and gravitational terms

$$mV\dot{\gamma} = L_u u + L_\alpha \alpha + L_q q + L_{\dot{u}}\dot{u} + L_{\dot{\alpha}}\dot{\alpha} + L_{\delta e}\delta e \qquad (7.72)$$

Comparison with Eq. (7.59) shows that a tetragonal missile lacks the u derivatives. We neglect these u-dependent derivatives for the aircraft, and the damping effects $L_{\dot{u}}$ and $L_{\dot{\alpha}}$ as well. By replacing α with $\alpha = \theta - \gamma$, we succeed in deriving one of the state equations, after having absorbed m in the denominator of the derivatives:

$$\dot{\gamma} = \frac{L_\alpha}{V}\theta - \frac{L_\alpha}{V}\gamma + \frac{L_{\delta e}}{V}\delta e$$

The other state equations follow directly form Eq. (7.58):

$$\dot{q} = M_q q + M_\alpha \theta - M_\alpha \gamma + M_{\delta e}\delta e$$

Both equations combined yield the desired flight-path-angle formulation

$$\begin{bmatrix} \dot{q} \\ \dot{\theta} \\ \dot{\gamma} \end{bmatrix} = \begin{bmatrix} M_q & M_\alpha & -M_\alpha \\ 1 & 0 & 0 \\ 0 & L_\alpha/V & -L_\alpha/V \end{bmatrix} \begin{bmatrix} q \\ \theta \\ \gamma \end{bmatrix} + \begin{bmatrix} M_{\delta e} \\ 0 \\ L_{\delta e}/V \end{bmatrix} \delta e \qquad (7.73)$$

where the dimensional derivatives are calculated from the nondimensional derivatives according to

$$L_\alpha = \frac{\bar{q}S}{m}C_{L_\alpha}; \quad L_{\delta e} = \frac{\bar{q}S}{m}C_{L_{\delta e}}; \quad M_\alpha = \frac{\bar{q}Sc}{I_2}C_{m_\alpha};$$

$$M_q = \frac{\bar{q}Sc^2}{2I_2 V}C_{m_q}; \quad M_{\delta e} = \frac{\bar{q}Sc}{I_2}C_{m_{\delta e}}$$

These equations, in state variable format, are used in Sec. 10.2.2.6 to develop a self-adaptive flight-path-angle tracker. I am always amazed how plant equations, as simple as Eq. (7.73), produce useful models for autopilot designs, which can be implemented in six-DoF simulations.

These examples should be enough to help you develop other linear models for aerospace vehicles. We covered the roll transfer function, pitch dynamics expressed in normal acceleration, and flight-path-angle dynamics. You should be able to derive the yawing equations, the roll/yaw coupled dynamics, and the full linear stability equations of steady flight. We turn now to applications of unsteady flight.

7.5 Perturbation Equations of Unsteady Flight

With a good understanding of perturbed steady flight, we can branch out and derive the perturbation equations for some important unsteady maneuvers. We focus on those flight conditions that maintain significant angular velocities, like the pull-up maneuver of an aircraft and the circular engagement of an air-to-air missile. In both cases we start with Eqs. (7.38) and (7.39), the general perturbation equations of unsteady flight and use the expansions of aerodynamic derivatives, Eq. (7.53) and (7.54). The resulting equations, expressed in state-variable format, are quite useful for specialized dynamic investigations of unsteady flight.

7.5.1 Aircraft Executing Vertical Maneuvers

Pull-ups of aircraft or push-down attacks of cruise missiles are maneuvers with sustained pitch angular velocities. They occur in the vertical plane and are symmetrical maneuvers in the sense that the yawing and rolling rates are near zero. We proceed with expressing the perturbation equations in components, starting with Eq. (7.38).

The state variables are the linear and angular velocity perturbations

$$\left[\varepsilon v_B^I\right]^{Bp} = [u \quad v \quad w], \ \left[\varepsilon \omega^{BI}\right]^{Bp} = [p \quad q \quad r], \ \left[\varepsilon \Omega^{BI}\right]^{Bp} = \begin{bmatrix} 0 & -r & q \\ r & 0 & -p \\ -q & p & 0 \end{bmatrix}$$

and for vehicles with planar symmetry, like aircraft or cruise missiles, the MOI tensor (same in reference and perturbed body coordinates) is

$$\left[I_{Bp}^{Bp}\right]^{Bp} = \left[I_{Br}^{Br}\right]^{Br} = \begin{bmatrix} I_{11} & 0 & I_{13} \\ 0 & I_2 & 0 \\ I_{13} & 0 & I_{33} \end{bmatrix}$$

The transformation matrix simplifies, under the assumption of small angles ψ, θ, and ϕ,

$$[T]^{BpBr} = \begin{bmatrix} 1 & \psi & -\theta \\ -\psi & 1 & \phi \\ \theta & -\phi & 1 \end{bmatrix}$$

Now we specify the components of the reference maneuver. Because it is executed in the vertical plane, only the linear velocity components u_r and w_r and pitch rate q_r are nonzero:

$$\left[v_{Br}^I\right]^{Br} = [u_r \quad 0 \quad w_r], \quad \left[\omega^{BrI}\right]^{Br} = [0 \quad q_r \quad 0], \quad \left[\Omega^{BrI}\right]^{Br} = \begin{bmatrix} 0 & 0 & q_r \\ 0 & 0 & 0 \\ -q_r & 0 & 0 \end{bmatrix}$$

and the transformation matrix consists only of the reference pitch angle θ_r

$$[T]^{BrI} = \begin{bmatrix} \cos\theta_r & 0 & -\sin\theta_r \\ 0 & 1 & 0 \\ \sin\theta_r & 0 & \cos\theta_r \end{bmatrix} \tag{7.74}$$

Substituting these components into Eq. (7.38), multiplying out the matrix products, and rearranging terms, yields the translational equations

$$
\begin{bmatrix} \dot{u} \\ \dot{v} \\ \dot{w} \end{bmatrix} = \begin{bmatrix} -q_r w - w_r q \\ -u_r r + w_r p \\ q_r u + u_r q \end{bmatrix} + g \begin{bmatrix} -\cos\theta_r \theta \\ \sin\theta_r \psi + \cos\theta_r \phi \\ -\sin\theta_r \theta \end{bmatrix} + \frac{1}{m} \begin{bmatrix} (\varepsilon f_a)_1^{Bp} \\ (\varepsilon f_a)_2^{Bp} \\ (\varepsilon f_a)_3^{Bp} \end{bmatrix} \quad (7.75)
$$

The same substitutions into Eq. (7.39) lead to the rotational equations

$$
\begin{bmatrix} I_{11}\dot{p} + I_{13}\dot{r} \\ I_2\dot{q} \\ I_{13}\dot{p} + I_{33}\dot{r} \end{bmatrix} = q_r \begin{bmatrix} -I_{13}p + (I_2 - I_{33})r \\ 0 \\ (I_{11} - I_2)p + I_{13}r \end{bmatrix} + \begin{bmatrix} (\varepsilon m_a)_1^{Bp} \\ (\varepsilon m_a)_2^{Bp} \\ (\varepsilon m_a)_3^{Bp} \end{bmatrix}
$$

which, however, require further modifications. The left-hand side consists of more than one state vector derivative in the first and third components. The excess must be removed to arrive at a true state-variable representation. The terms $I_{13}\dot{r}$ and $I_{13}\dot{p}$ can be eliminated by the following manipulations: 1) multiply the third row by I_{13} and subtract it from the first row multiplied by I_{33}, and 2) multiply the first row by I_{13} and subtract it from the third row multiplied by I_{11}. Neglecting terms that are multiplied by the small factors I_{13}/I_{33} and I_{13}/I_{11} furnishes finally the desired format:

$$
\begin{bmatrix} \dot{p} \\ \dot{q} \\ \dot{r} \end{bmatrix} = q_r \begin{bmatrix} -\dfrac{I_{33}(I_{33} - I_2) + I_{13}^2}{I_{11}I_{33} - I_{13}^2} r \\ 0 \\ \dfrac{I_{11}(I_{11} - I_2) + I_{13}^2}{I_{11}I_{33} - I_{13}^2} p \end{bmatrix} + \begin{bmatrix} \dfrac{I_{33}}{I_{11}I_{33} - I_{13}^2}(\varepsilon m_a)_1^{Bp} \\ \dfrac{1}{I_2}(\varepsilon m_a)_2^{Bp} \\ \dfrac{I_{11}}{I_{11}I_{33} - I_{13}^2}(\varepsilon m_a)_3^{Bp} \end{bmatrix} \quad (7.76)
$$

Equations (7.75) and (7.76) are the perturbation equations for vertical maneuvers. On the left-hand sides remain only the time derivatives of the state variables. The effect caused by the unsteady reference flight is arranged in the first terms on the right-hand sides of the equations.

The gravity term of Eq. (7.75) is a function of the reference pitch angle, which can vary between 0 and 90 deg. Its perturbation variables are the Euler angles. The last terms of the dynamic equations contain the aerodynamic force and moment derivatives. We turn now to their assessment.

7.5.1.1 Aerodynamic expansions.
The aerodynamic force is chosen as axial, side, and normal components in agreement with standard wind-tunnel conventions and the appropriate sign changes. There is no sign change in the moments. To prevent confusion with the normal force, the yawing moment is labeled LN.

$$
\begin{bmatrix} (\varepsilon f_a)_1^{Bp} \\ (\varepsilon f_a)_2^{Bp} \\ (\varepsilon f_a)_3^{Bp} \end{bmatrix} = \begin{bmatrix} -A \\ Y \\ -N \end{bmatrix} ; \qquad \begin{bmatrix} (\varepsilon m_a)_1^{Bp} \\ (\varepsilon m_a)_2^{Bp} \\ (\varepsilon m_a)_3^{Bp} \end{bmatrix} = \begin{bmatrix} L \\ M \\ LN \end{bmatrix}
$$

As practiced before, we expand the aerodynamic perturbations into Taylor series. From Fig. 7.1 we determine the linear derivatives for planar vehicles as

$$A = \frac{\partial A}{\partial u}u + \frac{\partial A}{\partial w}w + \frac{\partial A}{\partial q}q + \underline{\frac{\partial A}{\partial \dot{u}}\dot{u}} + \underline{\frac{\partial A}{\partial \dot{w}}\dot{w}} + \frac{\partial A}{\partial \delta e}\delta e$$

$$Y = \frac{\partial Y}{\partial v}v + \frac{\partial Y}{\partial p}p + \frac{\partial Y}{\partial r}r + \underline{\frac{\partial Y}{\partial \dot{v}}\dot{v}} + \underline{\underline{\frac{\partial Y}{\partial \delta a}\delta a}} + \frac{\partial Y}{\partial \delta r}\delta r$$

$$N = \frac{\partial N}{\partial u}u + \frac{\partial N}{\partial w}w + \frac{\partial N}{\partial q}q + \underline{\frac{\partial N}{\partial \dot{u}}\dot{u}} + \underline{\frac{\partial N}{\partial \dot{w}}\dot{w}} + \frac{\partial N}{\partial \delta e}\delta e$$

$$L = \frac{\partial L}{\partial v}v + \frac{\partial L}{\partial p}p + \frac{\partial L}{\partial r}r + \underline{\frac{\partial L}{\partial \dot{v}}\dot{v}} + \underline{\underline{\frac{\partial L}{\partial \delta a}\delta a}} + \frac{\partial L}{\partial \delta r}\delta r$$

$$M = \frac{\partial M}{\partial u}u + \frac{\partial M}{\partial w}w + \frac{\partial M}{\partial q}q + \underline{\frac{\partial M}{\partial \dot{u}}\dot{u}} + \underline{\underline{\frac{\partial M}{\partial \dot{w}}\dot{w}}} + \frac{\partial M}{\partial \delta e}\delta e$$

$$LN = \frac{\partial LN}{\partial v}v + \frac{\partial LN}{\partial p}p + \frac{\partial LN}{\partial r}r + \underline{\frac{\partial LN}{\partial \dot{v}}\dot{v}} + \underline{\underline{\frac{\partial LN}{\partial \delta a}\delta a}} + \frac{\partial LN}{\partial \delta r}\delta r$$

Based on experimental evidence, the underlined terms are neglected, and the assumption is made that the acceleration terms, underlined twice, can be absorbed in the corresponding damping derivatives.

Wind-tunnel test data are usually presented as functions of Mach number M, angle of attack α, and sideslip angle β, rather than the velocity components u, v, w. As before, we can expand the derivatives in terms of α and β and then replace $\alpha = w/V_r$ and $\beta = v/V_r$ in order to reintroduce the state variables w and v (with small-angle assumption and V_r the reference speed).

Replacing the u dependence by the Mach number is more complicated and takes a few steps. Let us start with the definition of the nondimensional axial derivative

$$A = \tfrac{1}{2}\rho V^2 SC_A$$

The partial derivative relative to u has two terms, evaluated at reference conditions

$$\left.\frac{\partial A}{\partial u}\right|_r = \rho_r V_r SC_{A_r}\left.\frac{\partial V}{\partial u}\right|_r + \frac{1}{2}\rho_r V_r^2 S\frac{\partial C_A}{\partial M}\left.\frac{\partial M}{\partial u}\right|_r$$

The first partial $\partial V/\partial u = 1$ because V and u are changing by the same amount (small-angle assumption) and similarly $\partial M/\partial u = \partial V/(a\partial u) = 1/a$ (a is sonic speed). Therefore, with the dynamic pressure designated $\bar{q}_r$

$$\left.\frac{\partial A}{\partial u}\right|_r = \rho_r V_r SC_{A_r} + \bar{q}_r S\frac{\partial C_A}{a_r \partial M}$$

Now we are in a position to express the aerodynamic perturbations in a form most suitable to receive experimental data

$$A = \left(\rho_r V_r S C_{A_r} + \frac{\bar{q}_r S}{a_r} C_{A_M} \right) u + \frac{\bar{q}_r S}{V_r} C_{A_\alpha} w + \bar{q}_r S C_{A_{\delta e}} \delta e$$

$$Y = \frac{\bar{q}_r S}{V_r} C_{Y_\beta} v + \bar{q}_r S C_{Y_{\delta r}} \delta r \tag{7.77}$$

$$N = \left(\rho_r V_r S C_{N_r} + \frac{\bar{q}_r S}{a_r} C_{N_M} \right) u + \frac{\bar{q}_r S}{V_r} C_{N_\alpha} w + \bar{q}_r S C_{N_{\delta e}} \delta e$$

$$L = \frac{\bar{q}_r S b}{V_r} C_{l_\beta} v + \frac{\bar{q}_r S b^2}{2V_r} C_{l_p} p + \frac{\bar{q}_r S b^2}{2V_r} (C_{l_r} + C_{l_\beta}) r + \bar{q}_r S b C_{l_{\delta a}} \delta a + \bar{q}_r S b C_{l_{\delta r}} \delta r$$

$$M = \left(\rho_r V_r S c C_{m_r} + \frac{\bar{q}_r S c}{a_r} C_{m_M} \right) u + \frac{\bar{q}_r S c}{V_r} C_{m_\alpha} w$$

$$+ \frac{\bar{q}_r S c^2}{2V_r} (C_{m_q} + C_{m_\alpha}) q + \bar{q}_r S c C_{m_{\delta e}} \delta e \tag{7.78}$$

$$LN = \frac{\bar{q}_r S b}{V_r} C_{n_\beta} v + \frac{\bar{q}_r S b^2}{2V_r} C_{n_p} p + \frac{\bar{q}_r S b^2}{2V_r} (C_{n_r} + C_{n_\beta}) r$$

$$+ \bar{q}_r S b C_{n_{\delta a}} \delta a + \bar{q}_r S b C_{n_{\delta r}} \delta r$$

All derivatives were nondimensionalized with the help of the reference area S, the chord c, and span b for longitudinal and lateral derivatives, respectively. The subscript r indicates the variables associated with the reference trajectory. They are in general a function of time. As an exception, however, the subscript r in the derivatives C_{l_r} and C_{n_r} refers to the perturbed yaw rate of the vehicle.

The coefficients of the reference trajectory C_{Ar}, C_{Nr}, and C_{mr} show up in the longitudinal aerodynamics of A, N, and M in conjunction with the forward velocity perturbation u. They are known functions of time. The terms in which they occur account for the fact that the dynamic pressure is increased (to the first order) by $\rho_r V_r u$, as we conclude from the following:

$$\bar{q}_r = \frac{\rho_r}{2}(V_r + u)^2 \approx \frac{\rho_r}{2}(V_r^2 + 2V_r u) = \frac{\rho_r}{2} V_r^2 + \rho_r V_r u$$

For steady reference flight the reference pitching moment C_{mr} is zero. However, for our pull-up and pushdown maneuvers, a reference pitching moment exists, and therefore the C_{mr} term should be included in the equations.

Herewith, we completed the modeling of the aerodynamics for the unsteady perturbation equations. During the derivation, I made quite a few simplifying shortcuts. Only linear terms describe the model, and even some of these derivatives were assumed negligible based on some rather sketchy test results. Furthermore, the bending of the lifting surfaces, which could be quite substantial during high load factors, was not discussed, but could have been included in the derivatives C_{N_α}, C_{m_α}, and C_{l_β}. Yet we did accomplish our objective, i.e., to derive the unsteady perturbation equations in state-variable form for the vertical pull-up and push-down maneuvers for aircraft or cruise missiles. What remains is the summary of the results.

7.5.1.2 State equations of vertical maneuvers.

The perturbation equations of unsteady vertical maneuvers can finally be written in the desired matrix form by combining Eqs. (7.75) with (7.77) and (7.76) with (7.78). The six differential equations are augmented by the integrals ϕ, θ, and ψ of the angular rates p, q, and r. As we inspect the nine differential equations, we discover that they can be decoupled into two sets of longitudinal and lateral equations. The longitudinal state vector is composed of u, w, q, and θ, and the lateral state vector consists of the remaining v, p, r, ϕ, and ψ variables.

Longitudinal equations.

$$
\begin{bmatrix} \dot{u} \\ \dot{w} \\ \dot{q} \\ \dot{\theta} \end{bmatrix} =
\begin{bmatrix}
-\frac{\rho_r V_r S}{m}C_{Ar} - \frac{\bar{q}_r S}{ma_r}C_{AM} & -\frac{\bar{q}_r S}{mV_r}C_{A\alpha} - q_r & -w_r & -g\cos\theta_r \\
-\frac{\rho_r V_r S}{m}C_{Nr} - \frac{\bar{q}_r S}{ma_r}C_{NM} + q_r & -\frac{\bar{q}_r S}{mV_r}C_{N\alpha} & u_r & -g\sin\theta_r \\
\frac{\rho_r V_r Sc}{I_2}C_{mr} + \frac{\bar{q}_r Sc}{I_2 a_r}C_{mM} & \frac{\bar{q}_r Sc}{I_2 V_r}C_{m\alpha} & \frac{\bar{q}_r Sc^2}{2I_2 V_r}(C_{mq}+C_{m\dot{\alpha}}) & 0 \\
0 & 0 & 1 & 0
\end{bmatrix}
$$

$$
\times \begin{bmatrix} u \\ w \\ q \\ \theta \end{bmatrix} +
\begin{bmatrix} \frac{\bar{q}_r S}{m}C_{A\delta e} \\ \frac{\bar{q}_r S}{m}C_{N\delta e} \\ \frac{\bar{q}_r Sc}{I_2}C_{m\delta e} \\ 0 \end{bmatrix} \delta e \tag{7.79}
$$

The four differential equations are linear in the velocity perturbations u and w and the pitch rate q and, of course, also in $\dot{\theta} = q$. Their coefficients would be constant if we froze the maneuver at the time instance when u_r, w_r, q_r, and θ_r are constant. Otherwise, if we study the maneuver as it unfolds, but still maintain constant q_r, the matrix elements that contain u_r, w_r, and θ_r are variable. You should deduce for yourself the vanishing terms for an aircraft that points straight up or down.

Lateral equations.

$$
\begin{bmatrix} \dot{v} \\ \dot{p} \\ \dot{r} \\ \dot{\phi} \\ \dot{\psi} \end{bmatrix} =
\begin{bmatrix}
\frac{\bar{q}_r S}{mV_r}C_{Y\beta} & w_r & -u_r & g\cos\theta_r & g\sin\theta_r \\
\frac{\bar{q}_r Sb}{I'_{11}V_r}C_{l\beta} & \frac{\bar{q}_r Sb^2}{2I'_{11}V_r}C_{lp} & -\frac{(I_{33}-I_2)+I'^2_{13}}{I'_{11}}q_r + \frac{\bar{q}_r Sb^2}{2I'_{11}V_r}(C_{lr}+C_{l\beta}) & 0 & 0 \\
\frac{\bar{q}_r Sb}{I'_{33}V_r}C_{n\beta} & \frac{(I'_{11}-I'_2)+I'^2_{13}}{I'_{33}}q_r + \frac{\bar{q}_r Sb^2}{2I'_{33}V_r}C_{np} & \frac{\bar{q}_r Sb^2}{2I'_{33}V_r}(C_{nr}+C_{n\beta}) & 0 & 0 \\
0 & 1 & 0 & 0 & 0 \\
0 & 0 & 1 & 0 & 0
\end{bmatrix}
$$

$$
\times \begin{bmatrix} v \\ p \\ r \\ \phi \\ \psi \end{bmatrix} +
\begin{bmatrix} \frac{\bar{q}_r S}{m}C_{Y\delta r}\delta r \\ \frac{\bar{q}_r Sb}{I'_{11}}C_{l\delta a}\delta a \\ \frac{\bar{q}_r Sb}{I'_{11}}C_{l\delta r}\delta r \\ \frac{\bar{q}_r Sb}{I'_{33}}C_{n\delta r}\delta r \\ \frac{\bar{q}_r Sb}{I'_{33}}C_{n\delta a}\delta a \\ 0 \\ 0 \end{bmatrix} \tag{7.80}
$$

with $I'_{11} = (I_{11}I_{33} - I^2_{13})/I_{33}$, $I'_{33} = (I_{11}I_{33} - I^2_{13})/I_{11}$. Now we have five linear differential equations in the lateral velocity component v, the roll and yaw rates p and r, and the derivative roll and yaw angles $\dot{\phi} = p$, $\dot{\psi} = r$. Notice how the reference pitch rate q_r multiplied by a rather complicated ratio of the MOI components affects the rolling and yawing equations. Furthermore, the aileron and rudder controls couple into the yawing and rolling equations.

The matrix of Eq. (7.80) has the undesirable characteristic that its determinant is singular. You can verify that fact by multiplying the last two columns by $g \sin\theta_r$ and $g \cos\theta_r$, respectively, and subtracting column 5 from column 4. The last column of the determinant is zero. A rank deficient matrix like this will cause you problems when you calculate its eigenvalues or try to control all of the state variables.

Fortunately, we can avoid that hazard by replacing the roll and yaw angles with the horizontal perturbation roll angle σ. I will show you how to proceed. Because the angular perturbations are small angles, we can write them as an angular vector with the components ϕ, θ, and ψ in the reference body coordinates. Transforming this angular vector into inertial coordinates (local level coordinates), using Eq. (7.74), we obtain the horizontal component σ:

$$\begin{bmatrix} \sigma \\ \theta \\ \tau \end{bmatrix} = [\bar{T}]^{Brl} \begin{bmatrix} \phi \\ \theta \\ \psi \end{bmatrix} = \begin{bmatrix} \cos\theta_r & 0 & \sin\theta_r \\ 0 & 1 & 0 \\ -\sin\theta_r & 0 & \cos\theta_r \end{bmatrix} \begin{bmatrix} \phi \\ \theta \\ \psi \end{bmatrix}$$

$$= \begin{bmatrix} \phi\cos\theta_r + \psi\sin\theta_r \\ \theta \\ -\phi\sin\theta_r + \psi\cos\theta_r \end{bmatrix}; \quad \sigma = \phi\cos\theta_r + \psi\sin\theta_r$$

After taking the time derivative

$$\dot{\sigma} = \dot{\phi}\cos\theta_r + \dot{\psi}\sin\theta_r = p\cos\theta_r + r\sin\theta_r$$

we replace the last two rows in Eq. (7.80), delete the last column, and replace the first element in the fourth column by g. The nonsingular lateral equations are now

$$\begin{bmatrix} \dot{v} \\ \dot{p} \\ \dot{r} \\ \dot{\sigma} \end{bmatrix} = \begin{bmatrix} \frac{\bar{q}_r S}{mV_r}C_{Y_\beta} & w_r & -u_r & g \\ \frac{\bar{q}_r Sb}{I'_{11}V_r}C_{l_\beta} & \frac{\bar{q}_r Sb^2}{2I'_{11}V_r}C_{l_p} & -\frac{(I_{33}-I_2)+I^2_{13}}{I'_{11}}q_r + \frac{\bar{q}_r Sb^2}{2I'_{11}V_r}(C_{l_r} + C_{l_\beta}) & 0 \\ \frac{\bar{q}_r Sb}{I'_{33}V_r}C_{n_\beta} & \frac{(I_{11}-I_2)+I^2_{13}}{I'_{33}}q_r + \frac{\bar{q}_r Sb^2}{2I'_{33}V_r}C_{n_p} & \frac{\bar{q}_r Sb^2}{2I'_{33}V_r}(C_{n_r} + C_{n_\beta}) & 0 \\ 0 & \cos\theta_r & \sin\theta_r & 0 \end{bmatrix}$$

$$\times \begin{bmatrix} v \\ p \\ r \\ \sigma \end{bmatrix} + \begin{bmatrix} \frac{\bar{q}_r S}{m}C_{Y_{\delta r}}\delta r \\ \frac{\bar{q}_r Sb}{I'_{11}}C_{l_{\delta a}}\delta a \\ \frac{\bar{q}_r Sb}{I'_{11}}C_{l_{\delta r}}\delta r \\ \frac{\bar{q}_r Sb}{I'_{33}}C_{n_{\delta r}}\delta r \\ \frac{\bar{q}_r Sb}{I'_{33}}C_{n_{\delta a}}\delta a \\ 0 \end{bmatrix} \tag{7.81}$$

Equations (7.79) and (7.81) are the preferred set of perturbation equations for vertical maneuvers, although we have to come to grips with the horizontal roll angle σ.

The lateral and longitudinal equations are formulated in body-fixed coordinates. Their state vectors are the linear and angular velocity components parallel to the perturbed body axes. This form is most convenient for feedback control analysis because the state vector components and their time derivatives can be directly measured by onboard sensors. Note, however, that the state vector represents the perturbations from the reference trajectory. Therefore, only for the lateral equations can the state vector be measured directly. The measurements of the longitudinal variables contain both the reference and perturbation values. For flight-test data matching the reference values must be subtracted from the measurements to obtain the longitudinal perturbations.

Equations (7.79–7.81) can also serve as perturbation equations of steady flight. All you have to do is set $q_r = 0$. Moreover, if the reference flight is horizontal, $w_r = 0, \sin \theta_r = 0, \cos \theta_r = 1$, the equations are particularly simple. Finally, I hope that the derivation of the equations of motions for aircraft and cruise missiles during pull-up or push-down maneuvers gives you a good perception of the modeling of flight vehicle perturbations.

7.5.2 Perturbation Equations of Agile Missile

Agile missiles execute severe maneuvers for target intercepts. Seldom do they experience steady flight conditions. To study their flight dynamics, the perturbation equations of unsteady flight are derived, and the aerodynamic forces and moments expanded into derivatives. The second-order derivatives disclose the strong coupling between the yaw, pitch, and roll channels.

A six-DoF air-to-air missile simulation, stripped of the flight control system, is used to investigate the aerodynamic and inertial couplings. We will see that the aerodynamically induced rolling moment couples into the pitch channel during a yaw maneuver, and the yaw channel is excited by rolling motions in the presence of a pitch maneuver.

7.5.2.1 Equations of motion.
Of particular interest is the intercept trajectory of a short-range air-to-air missile. Its speed is constantly changing during thrusting, and it is turning in response to maneuvering targets. This reference trajectory is characterized by the linear velocity components u_r, v_r, w_r and angular velocity components p_r, q_r, r_r in body coordinates. The perturbation variables consist of the linear velocities u, v, w and angular velocities p, q, r.

We start with the general perturbation equations of unsteady flight, derived earlier as Eqs. (7.38) and (7.39) and repeated here for convenience. The first equation governs the translational and the second the rotational perturbations:

$$m\left[\frac{d\varepsilon v_B^l}{dt}\right]^{Bp} + m[\Omega^{Brl}]^{Br}\left[\varepsilon v_B^l\right]^{Bp} + m[\varepsilon\Omega^{Bl}]^{Bp}\left[v_{Br}^l\right]^{Br}$$

$$= [\varepsilon f_a]^{Bp} + [\varepsilon f_t]^{Bp} + ([T]^{BpBr} - [E])[T]^{Brl}[f_{gr}]^l$$

$$[I_{Bp}^{Bp}]^{Bp}\left[\frac{d\varepsilon\omega^{BI}}{dt}\right]^{Bp} + [\Omega^{BrI}]^{Br}[I_{Bp}^{Bp}]^{Bp}[\varepsilon\omega^{BI}]^{Bp} + [\varepsilon\Omega^{BI}]^{Bp}[I_{Br}^{Br}]^{Br}[\omega^{BrI}]^{Br}$$

$$= [\varepsilon m_a]^{Bp} + [\varepsilon m_t]^{Bp}$$

These are two sets of first-order differential equations, each having three state variables: u, v, w modeling the translational and p, q, r the rotational degrees of freedom. The body rates p, q, r couple into the translational equation by the term $m[\varepsilon\Omega^{BI}]^{Bp}[v_{Br}^I]^{Br}$ of the attitude equation, and the aerodynamic forces and moments impart additional cross coupling.

Unsteady reference flights are characterized by curved trajectories, i.e., their reference angular velocity $[\omega^{BrI}]^{Br}$ is nonzero. Three terms of the perturbation equations contain unsteady terms. The translational equation is affected by $m[\Omega^{BrI}]^{Br}$ $[\varepsilon v_B^I]^{Br}$, a centrifugal force, and the attitude equations by the two terms

$$[\Omega^{BrI}]^{Br}[I_{Bp}^{Bp}]^{Bp}[\varepsilon\omega^{BI}]^{Bp}$$

and

$$[\varepsilon\Omega^{BI}]^{Bp}[I_{Br}^{Br}]^{Br}[\omega^{BrI}]^{Br}$$

that produce centrifugal moments.

The last term of the translational equations represents the gravity perturbations.

$$m[g]^{Bp} = ([T]^{BpBr} - [E])[T]^{BrI}[f_{gr}]^I$$

It consists of the direction cosine matrix associated with the small perturbation angles: yaw ψ, pitch θ, and roll ϕ

$$[T]^{BpBr} = \begin{bmatrix} 1 & \psi & -\theta \\ -\psi & 1 & \phi \\ \theta & -\phi & 1 \end{bmatrix}$$

and the transformation matrix of the reference flight ($\phi_r = 0$)

$$[T]^{BrI} = \begin{bmatrix} \cos\psi_r\cos\theta_r & \sin\psi_r\cos\theta_r & -\sin\theta_r \\ -\sin\psi_r & \cos\psi_r & 0 \\ \cos\psi_r\sin\theta_r & \sin\psi_r\sin\theta_r & \cos\theta_r \end{bmatrix}$$

Multiplying the matrices yields the gravity perturbation

$$[g]^{Bp} = g\begin{bmatrix} -\theta\cos\theta_r \\ \psi\sin\theta_r + \phi\cos\theta_r \\ -\theta\sin\theta_r \end{bmatrix}$$

Notice that $[g]^{Bp}$ is, as it should be, independent of the reference heading angle ψ_r.

Now let us write the translational equations in component form

$$
\begin{bmatrix} \dot{u} \\ \dot{v} \\ \dot{w} \end{bmatrix} = \begin{bmatrix} r_r v - q_r w - w_r q + v_r r \\ -r_r u + p_r w + w_r p - u_r r \\ q_r u - p_r v - v_r p + u_r q \end{bmatrix} + \frac{1}{m} \begin{bmatrix} X \\ Y \\ Z \end{bmatrix} + \frac{1}{m} \begin{bmatrix} \varepsilon f_{t1} \\ \varepsilon f_{t2} \\ \varepsilon f_{t3} \end{bmatrix}
$$

$$
+ g \begin{bmatrix} -\theta \cos \theta_r \\ \psi \sin \theta_r + \phi \cos \theta_r \\ -\theta \sin \theta_r \end{bmatrix} \tag{7.82}
$$

and the attitude equations

$$
\begin{bmatrix} \dot{p} \\ \dot{q} \\ \dot{r} \end{bmatrix} = \begin{bmatrix} \dfrac{I_2 - I_3}{I_1}(r_r q + q_r r) \\ \dfrac{I_3 - I_1}{I_2}(r_r p + p_r r) \\ \dfrac{I_1 - I_2}{I_3}(q_r p + p_r q) \end{bmatrix} + \begin{bmatrix} \dfrac{L}{I_1} \\ \dfrac{M}{I_2} \\ \dfrac{N}{I_3} \end{bmatrix} + \begin{bmatrix} \dfrac{\varepsilon m_{t1}}{I_1} \\ \dfrac{\varepsilon m_{t2}}{I_2} \\ \dfrac{\varepsilon m_{t3}}{I_3} \end{bmatrix} \tag{7.83}
$$

The terms with p_r, q_r, and r_r are the contributions of the unsteady reference flight.

For our application we consider missiles with tetragonal symmetry only, thus $I_3 = I_2$; furthermore, we neglect any propulsive perturbation terms $[\varepsilon f_t]^{Bp}$ and $[\varepsilon m_t]^{Bp}$. Because the reference roll rate ϕ_r is also zero, the *equations of motion of the perturbed unsteady flight for missiles* become

$$
\dot{u} = r_r v - q_r w - w_r q + v_r r + \frac{X}{m} - g\theta \cos \theta_r
$$

$$
\dot{v} = -r_r u + w_r p - u_r r + \frac{Y}{m} + g(\psi \sin \theta_r + \phi \cos \theta_r) \tag{7.84}
$$

$$
\dot{w} = q_r u - v_r p + u_r q + \frac{Z}{m} - g\theta \sin \theta_r
$$

$$
\dot{p} = \frac{L}{I_1}
$$

$$
\dot{q} = \frac{r_r p(I_2 - I_1)}{I_2} + \frac{M}{I_2} \tag{7.85}
$$

$$
\dot{r} = \frac{q_r p(I_1 - I_2)}{I_2} + \frac{N}{I_2}
$$

We are surprised by the roll rate p appearing in four equations. It couples through w_r and v_r into the translational equations and through r_r and q_r into the attitude equations.

It is a well-known phenomenon in missile dynamics that once the missile begins to roll, the yawing and pitching channels are adversely affected. Yet, what causes the missile to roll? We have to look for aerodynamic phenomena that induce a rolling moment on the missile.

Table 7.3   First- and second-order derivatives

Component	Linear derivatives	Roll-rate coupling	Roll-control coupling	
Y	$Y_v, Y_r,$ $Y_{\dot v}, Y_{\delta r}$	$Y_{wp}, Y_{\dot w p},$ $Y_{qp}, Y_{\delta qp}$	$Y_{w\delta p}, Y_{\dot w\delta p},$ $Y_{q\delta p}, Y_{\delta q\delta p}$	
Z	$Z_w, Z_q,$ $Z_{\dot w}, Z_{\delta q}$	$Z_{vp}, Z_{\dot v p},$ $Z_{rp}, Z_{\delta rp}$	$Z_{v\delta p}, Z_{\dot v\delta p},$ $Z_{r\delta p}, Z_{\delta r\delta p}$	
M	$M_w, M_q,$ $M_{\dot w}, M_{\delta q}$	$M_{vp}, M_{\dot v p},$ $M_{rp}, M_{\delta rp}$	$M_{v\delta p}, M_{\dot v\delta p},$ $M_{r\delta p}, M_{\delta r\delta p}$	
N	$N_v, N_r,$ $N_{\dot v}, N_{\delta r}$	$N_{wp}, N_{\dot w p},$ $N_{qp}, N_{\delta qp}$	$N_{w\delta p}, N_{\dot w\delta p},$ $N_{q\delta p}, N_{\delta q\delta p}$	
Component	Linear derivatives	Incidence coupling	Rate coupling	Control coupling
L	$L_p, L_{\delta p},$ $L_p, L_{\delta p}$	$L_{wv}, L_{\dot w\dot v},$ $L_{\dot wv}, L_{w\dot v}$	$L_{vq}, L_{\dot vq}, L_{wr},$ $L_{\dot wr}, L_{rq}$	$L_{v\delta q}, L_{\dot v\delta q}, L_{r\delta q},$ $L_{w\delta r}, L_{\dot w\delta r},$ $L_{q\delta r}, L_{\delta q\delta r}$

7.5.2.2   Aerodynamic cross coupling.   Just as for aircraft, it is difficult to model the aerodynamic forces of missiles in a form that can be analyzed and evaluated quantitatively. Because the functional form is not known, the aerodynamic functions are expanded in a Taylor series in terms of the state variables relative to a reference flight. To capture the rolling and other cross-coupling effects, we have to include at least second-order derivatives.

Let us go back to Sec. 7.3.1 to sort out the existence of the derivatives. The tetragonal symmetry of our missile implies that certain derivatives are vanishing. The expansion of the aerodynamic force and moment perturbations Y, Z, L, M, N is carried out up to second order. Those that survive the symmetry test are listed in Table 7.3. The derivatives in the x direction are disregarded because we maintain the Mach-number dependency in tabular form.

The second column of Table 7.3 displays all of the familiar linear derivatives, whereas the remaining columns show the second-order derivatives. Notice that all of the second-order derivatives of the yaw and pitch channels are dependent either on roll rate p or roll control deflection δp. The rolling moment itself is a function of incidence angles (represented by v and w), yaw, pitch rate (r, q) coupling, and the effects of yaw, pitch control (δr, δq).

As a practical example, consider an air-to-air missile, executing a lateral maneuver toward an intercept (see Fig. 7.4). To generate the lateral acceleration, the airframe develops v or, equivalently, sideslip angle. Although the missile does not roll to execute the maneuver, the roll channel will be excited by the aerodynamic coupling, assuming the vertical channel is also active. There will be vertical channel transients because w or, equivalently, angle of attack is necessary to maintain altitude. Once the roll DoF is stirred, it couples back into the longitudinal and lateral channels.

In the example all of the first- and second-order derivatives play a part, with the control effectiveness derivatives excited by the control fin deflections. It is easy to

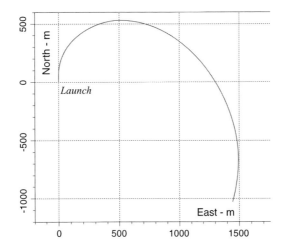

Fig. 7.4   50-g lateral maneuver.

cogitate about these yaw/pitch/roll cross-coupling effect, but the difficult part is to flesh out the skeleton of Table 7.3 with numerical values. Particularly challenging are wind-tunnel tests that measure unsteady derivatives associated with the body rates p, q, r and the incidence rates $\dot{v}, \dot{w}$.

For our limited discussion we focus on the important L_{wv} derivative that causes roll torques in the presence of angle of attack and sideslip angle. Once the roll rate is stirred, we investigate the inertial coupling into the yaw channel in the presence of pitch rate.

To test the coupling effects, I use the CADAC SRAAM6 simulation, a generic air-to-air missile. I keep its rather complex aerodynamic model and the propulsion subroutine, but bypass the guidance and control loops completely. Fortunately, the missile airframe is aerodynamically stable—though somewhat oscillatory—thus enabling open-loop computer runs.

7.5.2.3 Aerodynamically induced rolling moment.

The open-loop horizontal turn of Fig. 7.4 will serve as a test case. It is generated by 10-deg yaw control and is typical of a 50-g (peak) intercept trajectory.

First, we study the aerodynamic rolling moment and its effects on the roll excursions of the missile. Then we trace the inertial coupling of roll into the pitch channel, and finally create a hypothetical case without the induced rolling moment to highlight the lack of transient dynamics.

The rolling-moment equation is the first of the attitude equations, Eq. (7.85):

$$\dot{p} = \frac{1}{I_1}L = \frac{1}{I_1}(L_p p + L_{\delta p}\delta p + L_{wv} wv + \cdots)$$

with the key nonlinear derivative L_{wv} coupling vertical and lateral perturbations into the roll channel. The nondimensional equivalent of this derivative is $C_{l_{\alpha\beta}}$. For our prototype missile the C_{l_α} derivative is plotted against β for various Mach numbers in Fig. 7.5. As β increases, so does the roll coupling, particularly in the transonic regime.

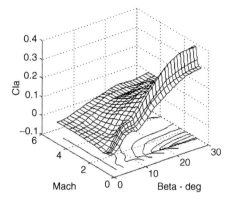

Fig. 7.5 C_{l_α} **derivative vs** β **and Mach.**

For the test trajectory of Fig. 7.4, the incidence angles and pitch and roll rates are plotted in Fig. 7.6. The severe lateral turn is executed with sideslip angle β as high as −40 deg. In the presence of even small α, a large roll rate builds up, leveling out at −600 deg/s.

Inertial coupling enters through the second of the attitude equations, Eq. (7.85):

$$\dot{q} = \frac{r_r p(I_2 - I_1)}{I_2} + \frac{M}{I_2}$$

The sustained yaw rate r_r of the lateral turn multiplies with the roll-rate perturbations p and generates pitch-rate perturbations q that grow to 250 deg/s.

Figure 7.6 traces both the aerodynamic and inertial coupling from the roll to the pitch channel. As a hypothetical exercise, we can ask the question, what happens

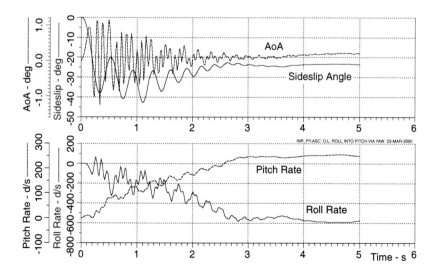

Fig. 7.6   Induced aerodynamic roll-rate coupling into pitch channel.

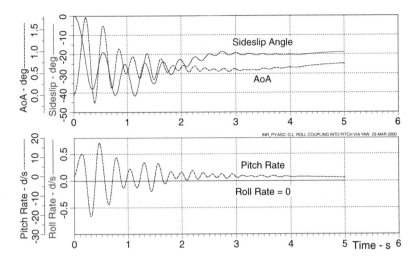

Fig. 7.7   Same as Fig. 7.6 but without aerodynamic roll coupling.

if the aerodynamic rolling moment is absent? I deleted the roll coupling terms in the simulation, repeated the test case, and plotted the result in Fig. 7.7.

Without the induced rolling moment the roll rate is zero, and the pitch rate reflects only the small angle-of-attack transients without any inertial cross coupling.

7.5.2.4   *Roll/yaw inertial coupling.*   Finally, we use the third of the attitude equations, Eq. (7.85),

$$\dot{r} = \frac{q_r p(I_1 - I_2)}{I_2} + \frac{N}{I_2}$$

and study the inertial coupling of roll rate p into yaw rate r in the presence of a sustained pitch maneuver q_r. A new reference trajectory is needed. With a pitch control of 1 deg, we generate a planar trajectory whose pitch rate q_r peaks at -150 deg/s and its angle of attack α at -30 deg. Now we excite the roll channel with a sinusoidal roll control input of 0.1-deg amplitude and 1-s period. The coupling of roll via pitch into yaw is shown in Fig. 7.8. In the presence of pitch-rate transients, the yaw rate is oscillatory with the period of the roll perturbations and peaks at 15 deg/s. If the pitch control is zeroed, the missile flies ballistic, and the coupling of roll into yaw disappears.

In summary, the perturbation equations of missiles during unsteady flight were derived from the general six-DoF equations of motion, and the aerodynamic forces and moments expanded in the Taylor series up to second-order terms. With these equations we interpreted the results of a six-DoF air-to-air missile simulation.

The aerodynamically induced rolling moment couples into the pitch channel during a yaw maneuver. The yaw channel is excited by rolling motions in the presence of a pitch maneuver. All three attitude channels are mutually linked by aerodynamic and inertial coupling.

These perturbation equations give insight into the dynamics of agile missiles. They shed light on the stability characteristic and should guide the control engineer in suppressing unwanted coupling effects.

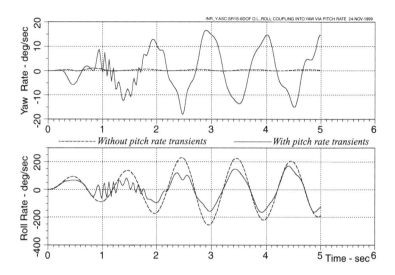

Fig. 7.8   Yaw-rate response caused by roll oscillations in the presence and absence of reference pitch rate.

With these rather sophisticated dynamic escapades, I conclude Part 1. I hope you enjoyed the ride and did not fall asleep earlier from sheer exhaustion. We solved simple geometrical problems, wrestled with the kinematics of translation and rotation, and derived the equations of motion thanks to Newton and Euler. We followed our motto "from tensor modeling to matrix coding" faithfully and saw no reason to deviate from the hypothesis that all dynamic problems can be formulated by points and frames. If you solved the majority of the problems at the end of each chapter, you should have reached the pinnacle of dynamic bliss.

Now has come the time to put all theoretical knowledge to practical use. As the structure follows the blueprint, so do simulations proceed from modeling. In Part 2 you will be challenged to put your new-found skills to work and develop three-, five-, and six-DoF simulations.

References

[1]Etkin, B., *Dynamics of Atmospheric Flight*, Wiley, New York, 1972.

[2]Hopkin, H. R., "A Scheme of Notation and Nomenclature for Aircraft Dynamics and Associated Aerodynamics," Royal Aircraft Establishment, TR66200, Farnborough, U.K., June 1966.

[3]Noll, W., "On the Continuity of the Solid and Fluid States," *Journal of Rational Mechanical Analysis*, Vol. 4, No. 1, 1955, p. 17.

[4]Maple, C. G., and Synge, J. L., "Aerodynamic Symmetry of Projectiles," *Quarterly of Applied Mathematics*," Vol. 6, Jan. 1949, pp. 315–366.

[5]Zipfel, P. H., "Aerodynamic Symmetry of Aircraft and Guided Missiles," *Journal of Aircraft*, Vol. 13, No. 7, 1976, pp. 470–475.

[6]Pamadi, B. N., *Performance, Stability, Dynamics, and Control of Airplanes*, AIAA Education Series, AIAA, Reston, VA, 1998, Chap. 4.

Problems

7.1   Yaw stability equations.   To support the design of a simple yaw damper for an aircraft, you are requested to derive the state equations of yaw rate r and sideslip angle β. Start with the linear perturbation equations (7.55) and (7.56) and use linear aerodynamic derivatives. The desired form is

$$\begin{bmatrix} \dot{r} \\ \dot{\beta} \end{bmatrix} = \begin{bmatrix} LN_r & LN_\beta \\ -1 & \dfrac{Y_\beta}{V} \end{bmatrix} \begin{bmatrix} r \\ \beta \end{bmatrix} + \begin{bmatrix} LN_{\delta r} \\ \dfrac{Y_{\delta r}}{V} \end{bmatrix} \delta r$$

State clearly the assumptions that lead to this formulation.

7.2   Fill in the details.   Apply your tensor manipulative skills and fill in the intermediate steps between Eqs. (7.16) and (7.17).

7.3   Missile linear perturbation equations over the flat Earth.   Start with Eqs. (7.55) and (7.56) and derive the full linear dynamic equations of a missile with tetragonal symmetry. Which equations are coupled? Can you group them into two uncoupled sets? Compare your equations with the stability equations for aircraft in Pamadi's book.[6]

7.4   Aircraft pitching moment derivatives.   Expand the M derivative of an aircraft up to second order.

7.5   Six-DoF aerodynamic model.   You are asked to develop the aerodynamic model for a six-DoF aircraft simulation. Mach number and angle of attack vary extensively so that you are directed to build your aerodynamic coefficients as tables in M and α. The other variables $\beta, \dot{\alpha}, \dot{\beta}, p, q, r$ and the controls $\delta a, \delta e, \delta r$ experience only small excursions and are therefore to be modeled by linear derivatives. The aircraft reference area is S, its mean chord c, its span b, and its speed V. Write down the aerodynamic force and moments in terms of the nondimensional stability and control derivatives.

Part 2
Simulation of Aerospace Vehicles

8
Three-Degree-of-Freedom Simulation

What a journey it has been so far! Provided you have not skipped the first seven chapters, you have reached Part 2 with a tool chest full of gadgets that aspire to be used for challenging simulation tasks. You are trained in coordinate systems, translational and rotational kinematics, and are able to apply Newton's and Euler's laws to the dynamics of aerospace vehicles.

If you skimmed over the first part, because of your maturity in such matters, you are also welcome to join us. Make sure, however, that you understand my notation and the invariant formulation of dynamic equations. Then it should be easy for you to follow us. To make the following three chapters self-contained, I will derive the equations of motion from first principles.

Let us ease into the world of simulation with simple three-DoF, point-mass models. They are suitable for trajectory studies of rockets, missiles, and aircraft. All you need is an understanding of Newton's second law and basic aerodynamic and propulsion data. In no time will you be productive, churning out time histories of key flight parameters. The more sophisticated five- and six-DoF simulations are left for the following chapters.

In preliminary design, vehicle characteristics are often sketchy and aerodynamics and propulsion data only known approximately. There may be just enough information to build simple three-DoF simulations. Fortunately, the trajectory of the c.m. of the vehicle is of greater interest than its attitude motions. Therefore, these three-DoF simulations are very useful for initial performance estimates and trade studies.

Newton's second law governs the three translational degrees of freedom of three-DoF simulations. Aerodynamic, propulsive, and gravitational forces must be known. In contrast to six-DoF simulations, Euler's law is not used, and body rates and attitudes are not calculated. Therefore, there is no requirement for aerodynamic and propulsive moments.

In deriving the equations of motion, three different perspectives can be taken according to the state variables selected for integration. Vinh[1] takes the direct approach and derives the equations for the following state variables: geographic speed, flight-path angle, heading angle, radial distance, and longitude and latitude. The isolation of the state variables on the left side of the differential equations requires complicated manipulations that are not documented in his book. (Vinh's equations are implemented by the TEST case, supplied with CADAC-Studio.)

The second method, the so-called *Cartesian* approach, formulates the equations in Cartesian coordinates. The state variables are the vehicle's inertial velocity and position components $[v_B^I]^I$ and $[s_{BI}]^I$, expressed in inertial coordinates.

The third method takes the perspective of the missile's velocity vector wrt Earth $[v_B^E]^V$ in velocity coordinates. Its polar components $|v_B^E|$, χ, γ are the velocity state variables and $[s_{BI}]^E$ the position states, expressed in Earth coordinates. We refer to it as the *polar* approach.

The derivations of the Cartesian and polar equations of motion will be provided first as invariant tensor form and then expressed as matrices for programming. Before you can code up the simulation, you need to have some elementary understanding of the atmosphere, gravitational acceleration, aerodynamics, and propulsion. To help you take the first steps, I include two example simulations of a three-stage rocket and a hypersonic vehicle. The complete code is provided on the CADAC Web site as ROCKET3 and GHAME3.

8.1   Equations of Motion

Before Newton's law can be applied, the choice of the inertial frame must made. In Chapter 5 we discussed the options and Chapter 3 defined the frames. Most Earth-bound simulations use the J2000 inertial frame (see Sec. 3.1.2.2) or sometimes the Earth itself for vehicles that hug the ground. We shall focus on the J2000 frame to support the equations of motion for rockets and hypersonic vehicles. En passant, I will point out the simplifications that lead to the formulation over a flat Earth.

The derivation of the Cartesian approach is straightforward. The inertial position and velocity coordinates are directly integrated, and the forces, given in flight-path coordinates, are transformed to inertial coordinates.

The polar approach is used to track the effects of the Coriolis and centrifugal accelerations. Newton's law is transferred to the rotating Earth frame and expressed in flight-path coordinates. No transformation is necessary for the forces.

8.1.1   Cartesian Equations

We begin with Newton's second law, as expressed by Eq. (5.9), and expand the right-hand side to include the aerodynamic and propulsive forces $f_{a,p}$ and the weight mg:

$$mD^I v_B^I = f_{a,p} + mg \qquad (8.1)$$

On the left side is the rotational derivative relative to the inertial frame I, operating on the velocity v_B^I of vehicle c.m. B wrt the inertial frame. This inertial velocity $[v_B^I]^I$, expressed in inertial coordinates, is suitable as state variable, but not for formulating the aerodynamic forces. It is the movement of the vehicle relative to the atmosphere that determines the air loads. Because the atmosphere is attached to the Earth (no wind assumption), the aerodynamics depends on the geographic velocity v_B^E of the vehicle wrt Earth E. The relationship between inertial and geographic velocities will be derived first.

The position of the inertial reference frame I is oriented in the solar ecliptic, and one of its points I is collocated with the center of the Earth. The Earth frame E, fixed with the geoid, rotates with the angular velocity ω^{EI}. By definition, the inertial velocity is $v_B^I = D^I s_{BI}$, where s_{BI} is the location of the vehicle's c.m. wrt point I. To introduce the geographic velocity, we change the reference frame to E:

$$D^I s_{BI} = D^E s_{BI} + \Omega^{EI} s_{BI} \qquad (8.2)$$

and introduce a reference point E on Earth (any point), $s_{BI} = s_{BE} + s_{EI}$, into the first right-hand term

$$D^E s_{BI} = D^E s_{BE} + D^E s_{EI} = D^E s_{BE} \equiv v_B^E$$

where $D^E s_{EI}$ is zero because s_{EI} is constant in the Earth frame. Substituting into Eq. (8.2), we obtain the relationship between the inertial and geographic velocities:

$$v_B^I = v_B^E + \Omega^{EI} s_{BI} \tag{8.3}$$

The difference between the absolute values is approximately 465 m/s at the equator.

Now we are prepared to coordinate Eq. (8.1) for computer programming. The aerodynamic and propulsive forces are expressed in flight-path coordinates $]^V$, whereas the gravitational acceleration is given in geographic axes $]^G$. Because the state variables are to be calculated in inertial coordinates, we introduce the transformation matrices $[T]^{GV}$ and $[T]^{IG}$ and write the component form

$$m \left[\frac{dv_B^I}{dt} \right]^I = [\bar{T}]^{GI} \left([\bar{T}]^{VG} [f_{a,p}]^V + m[g]^G \right) \tag{8.4}$$

These are the first three differential equations to be solved for the inertial velocity components $[v_B^I]^I$. The second set consists of the inertial position coordinates $[s_{BI}]^I$:

$$\left[\frac{ds_{BI}}{dt} \right]^I = [v_B^I]^I \tag{8.5}$$

Don't forget the initial conditions. You need to specify the velocity and position vectors at launch.

Two transformation matrices must be programmed. The geographic wrt the inertial coordinates TM is composed of the TMs $[T]^{GI} = [T]^{GE}[T]^{EI}$, provided by Eqs. (3.13) and (3.12), respectively, whereas $[T]^{VG}$ is given by Eq. (3.25).

For aircraft and tactical missiles you can simplify your simulation by substituting Earth as inertial frame. In Eqs. (8.4) and (8.5) you replace frame I and point I by frame E and point E. The distinction between inertial and geographic velocity disappears and the geographic coordinate system is replaced by the local-level system $]^L$:

$$m \left[\frac{dv_B^E}{dt} \right]^L = [\bar{T}]^{VL} [f_{a,p}]^V + m[g]^L \tag{8.6}$$

$$\left[\frac{ds_{BE}}{dt} \right]^L = [v_B^E]^L \tag{8.7}$$

Only one TM $[T]^{VL}$ is required and is given by Eq. (3.29). These equations of motion are quite useful for simple near-Earth trajectory work.

The Cartesian formulation is the easiest to implement among the three options. It only suffers from a lack of intuitiveness. Who can compose the inertial velocity $[v_B^I]^I$ and position $[s_{BI}]^I$ components into a mental picture? Because of this deficiency, the polar equations are sometimes preferred. They formulate the equations of motion in terms of the geographic velocity v_B^E, which is much easier to visualize.

8.1.2 Polar Equations

The derivation of the polar equations is more complicated. We have to transfer the rotational derivative of Newton's equation first to the Earth frame and then to the reference frame associated with the geographic velocity. Only then have we created the intuitive format we desire.

The point of departure is again Eq. (8.1). First, we deal with Newton's acceleration $a_B^I = D^I v_B^I$ and transform the rotational derivative to the frame E:

$$D^I v_B^I = D^E v_B^I + \Omega^{EI} v_B^I$$

Then we substitute Eq. (8.3) into both terms on the right-hand side:

$$D^I v_B^I = D^E v_B^E + D^E \left(\Omega^{EI} s_{BI}\right) + \Omega^{EI} v_B^E + \Omega^{EI} \Omega^{EI} s_{BI} \qquad (8.8)$$

Apply the chain rule, realizing that Earth's angular velocity is constant

$$D^E \left(\Omega^{EI} s_{BI}\right) = D^E \Omega^{EI} s_{BI} + \Omega^{EI} D^E s_{BI} = \Omega^{EI} D^E s_{BI}$$

and expand the term further by the vector triangle $s_{BI} = s_{BE} + s_{EI}$, making use of the facts that $s_{EI} = 0$ and that the geographic velocity is defined by $v_B^E = D^E s_{BE}$:

$$\Omega^{EI} D^E s_{BI} = \Omega^{EI} \left(D^E s_{BE} + D^E s_{EI}\right) = \Omega^{EI} D^E s_{BE} = \Omega^{EI} v_B^E$$

Substituting into Eq. (8.8) and collecting terms yields

$$D^I v_B^I = D^E v_B^E + 2\Omega^{EI} v_B^E + \Omega^{EI} \Omega^{EI} s_{BI}$$

Now we introduce the velocity frame V, which is defined by its three base vectors v_1, v_2, v_3. The direction of the geographic velocity vector v_B^E defines v_1, while v_2 is normal to it and horizontal and v_3 completes the triad. Transforming the rotational derivative $D^E v_B^E$ to the V frame yields the final form of the inertial acceleration:

$$D^I v_B^I = D^V v_B^E + \Omega^{VE} v_B^E + 2\Omega^{EI} v_B^E + \Omega^{EI} \Omega^{EI} s_{BI}$$

We are ready now to replace the rotational derivative in Newton's law and thus obtain the translational equation of motion

$$D^V v_B^E + \Omega^{VE} v_B^E = \frac{1}{m} f_{a,p} + g - 2\Omega^{EI} v_B^E - \Omega^{EI} \Omega^{EI} s_{BI} \qquad (8.9)$$

It contains the famous Coriolis $2\Omega^{EI} v_B^E$ and centrifugal $\Omega^{EI} \Omega^{EI} s_{BI}$ accelerations, which we encountered already in Sec. 5.3.1. Another integration yields the position s_{BI} of the missile c.m. B wrt the center of Earth I (recall that points I and E coincide):

$$D^E s_{BI} = v_B^E \qquad (8.10)$$

These are six coupled differential equations. As intended, the missile's acceleration is referred to the velocity frame. This shift in reference frame introduced the angular velocity ω^{VE} of the velocity frame wrt Earth.

For a nonrotating Earth the equations are uncoupled and reduce to the flat-Earth three-DoF model

$$D^V v_B^E + \Omega^{VE} v_B^E = \frac{1}{m} f_{a,p} + g$$

$$D^E s_{BE} = v_B^E$$

where Earth's reference point E can be any fixed point on Earth.

Equations (8.9) and (8.10) are valid in any coordinate system. We pick the flight-path coordinates for ease of expressing the geographic velocity vector simply as

$$[\overline{v_B^E}]^V = [V \quad 0 \quad 0]; \qquad V = |v_B^E|$$

However, Earth's angular velocity and the vehicle's position are best expressed in Earth coordinates and the gravity vector in geographic coordinates:

$$\left[\frac{dv_B^E}{dt}\right]^V + [\Omega^{VE}]^V [v_B^E]^V = \frac{1}{m}[f_{a,p}]^V + [T]^{VG}[g]^G$$

$$- 2[T]^{VE}[\Omega^{EI}]^E [\bar{T}]^{VE}[v_B^E]^V - [T]^{VE}[\Omega^{EI}]^E [\Omega^{EI}]^E [s_{BI}]^E \qquad (8.11)$$

$$\left[\frac{ds_{BI}}{dt}\right]^E = [\bar{T}]^{VE}[v_B^E]^V \qquad (8.12)$$

Some of these terms have simple components:

$$\left[\overline{\frac{dv_B^E}{dt}}\right]^V = [\dot{V} \quad 0 \quad 0], \quad [\bar{g}]^G = [0 \quad 0 \quad \text{grav}(|s_{BI}|)], \quad [\overline{\omega^{EI}}]^E = [0 \quad 0 \quad \omega^{EI}]$$

The transformation matrices that need to be programmed are $[T]^{VG}$ from Eq. (3.25) and $[T]^{VE} = [T]^{VG}[T]^{GE}$ with $[T]^{GE}$ from Eqs. (3.13).

The angular velocity ω^{VE} of the velocity wrt the Earth frame needs special attention. We derive its component form $[\omega^{VE}]^V$ from Fig. 8.1. It consists of the

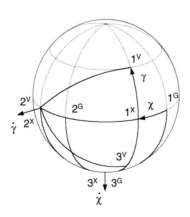

Fig. 8.1   Heading and flight-path rates.

vector addition of the angular rates $\dot{\chi}$ and $\dot{\gamma}$ times their respective unit vectors x_3 and v_2:

$$\omega^{VE} = \dot{\chi} x_3 + \dot{\gamma} v_2$$

Coordinated

$$[\omega^{VE}]^V = \dot{\chi}[T]^{VX}[x_3]^X + \dot{\gamma}[v_2]^V$$

and expressed in component form

$$[\omega^{VE}]^V = \dot{\chi} \begin{bmatrix} \cos\gamma & 0 & -\sin\gamma \\ 0 & 1 & 0 \\ \sin\gamma & 0 & \cos\gamma \end{bmatrix} \begin{bmatrix} 0 \\ 0 \\ 1 \end{bmatrix} + \dot{\gamma} \begin{bmatrix} 0 \\ 1 \\ 0 \end{bmatrix} = \begin{bmatrix} -\dot{\chi}\sin\gamma \\ \dot{\gamma} \\ \dot{\chi}\cos\gamma \end{bmatrix}$$

The left side of Eq. (8.11) becomes then

$$\left[\frac{dv_B^E}{dt} \right]^V + [\Omega^{VE}]^V [v_B^E]^V = \begin{bmatrix} \dot{V} \\ 0 \\ 0 \end{bmatrix} + \begin{bmatrix} 0 & -\dot{\chi}\cos\gamma & \dot{\gamma} \\ \dot{\chi}\cos\gamma & 0 & \dot{\chi}\sin\gamma \\ -\dot{\gamma} & -\dot{\chi}\sin\gamma & 0 \end{bmatrix} \begin{bmatrix} V \\ 0 \\ 0 \end{bmatrix}$$

$$= \begin{bmatrix} \dot{V} \\ \dot{\chi} V \cos\gamma \\ -V\dot{\gamma} \end{bmatrix}$$

with the three state variables: speed $\dot{V}$, heading rate $\dot{\chi}$, and flight-path rate $\dot{\gamma}$.

The flat-Earth implementation again is easier to program because the last two terms of Eq. (8.11) vanish and the local-level axes $]^L$ replace the geographic and Earth coordinates

$$\begin{bmatrix} \dot{V} \\ \dot{\chi} V \cos\gamma \\ -V\dot{\gamma} \end{bmatrix} = \frac{1}{m}[f_{a,p}]^V + [T]^{VL}[g]^L \tag{8.13}$$

$$\left[\frac{ds_{BE}}{dt} \right]^L = [\bar{T}]^{VL}[v_B^E]^V \tag{8.14}$$

Comparison of Eq. (8.11) with Eq. (8.4) clearly shows the greater complexity of the polar formulation. However, the three state variables V, χ, and γ are easier to visualize than the inertial velocity components $[v_B^I]^I$. Under the flat-Earth assumption, both the polar equations (8.13) and (8.14) and the Cartesian equations (8.6) and (8.7) are programmed with the same ease. Equation (8.13) actually has the advantage that the angles for the TM $[T]^{VL}$ are computed directly.

You may have been deafened by my silence over the right-hand side of Newton's law. The impressed forces consist of surface forces (aerodynamic and propulsive) and volume forces (gravity). Dividing the surface forces by the vehicle's mass generates what is called the *specific force*, but actually it is an acceleration. Accelerometers of INS measure this specific force. They cannot sense the gravitational volume force. In the following sections I delve deeper into the modeling of aerodynamic propulsive and gravitational forces.

8.2   Subsystem Models

8.2.1   Atmosphere

Without atmosphere, life on Earth would be miserable. I particularly appreciate the 21% of oxygen mixed in with 78% of nitrogen and 1% of argon and carbon dioxide. For engineers the atmosphere has other important attributes like density, temperature, and pressure that affect the trajectory of an aerospace vehicle. Air density determines the aerodynamic forces and moments, temperature is linked to the speed of sound, and air pressure modulates the thrust of a rocket engine.

Air temperature changes with altitude, but it does not consistently decrease with greater heights. Discrete changes in its gradient are used to divide the atmosphere into several layers. The *troposphere*, characterized by a decreasing temperature gradient, reaches from sea level up to 11 km, followed by the *tropopause* region with constant temperature until 20 km. From then on, in the *stratosphere* the temperature increases first as a result of the absorption of infrared radiation from Earth and solar ultraviolet radiation and then decreases. Its upper boundary at 80 km coincides with the somewhat arbitrary upper limit of the *endo-atmosphere*, above which the density is so low that it cannot deliver any significant aerodynamic lift. Beyond the stratosphere lies the *exo-atmosphere*, characterized by the dissociation of oxygen and ionization of nitrogen. In that region, also called the *thermosphere*, the temperature increases more strongly until 400 km, where the free-path lengths between molecules and atoms are so large that the definition of temperature becomes meaningless.

Much effort has been invested exploring the atmosphere. In 1955, President Eisenhower proclaimed the International Geophysical Year, which focused on upper atmospheric research with sounding rockets. Many measurements were taken and distilled into so-called standard atmospheres. The 1959 ARDC (Air Research Development Command)[2] atmosphere was used exclusively for many years until it was supplemented by the U.S. Standard Atmosphere in 1976.[3]

For simulations that do not require high fidelity, or are limited to the lower regions of the atmosphere, simple functions are used. For the three-DoF simulations of this chapter and the five-DoF simulations of Chapter 5, we use the 1962 International Standard Atmosphere or ISO 2533, summarized here.

Troposphere (altitude $H < 11$ km):

$$\text{Temperature (}^\circ\text{K)} \quad T = 288.15 - 0.0065H$$

$$\text{Pressure (Pa)} \quad p = 101325 \left(\frac{T}{288.15} \right)^{5.2559}$$

where H is in meters.

Tropopause-stratosphere (altitude 11 km $< H <$ 80 km):

$$\text{Temperature (}^\circ\text{K)} \quad T = 216$$

$$\text{Pressure (Pa)} \quad p = 22630 e^{-0.00015769(H - 11000)}$$

where H is in meters.

Endo-atmosphere (altitude 0 km < H < 80 km):

$$\text{Density (kg/m}^3)\ \rho = \frac{p}{RT}$$

$$\text{Sonic speed (m/s)}\ a = \sqrt{\gamma RT}$$

Dynamic pressure:

$$\bar{q} = \frac{\rho}{2}V^2$$

Mach number:

$$M = \frac{V}{a}$$

Temperature and pressure are the primary variables, approximated by functions, whereas density is derived from the perfect gas law (R gas constant) and sonic speed from the adiabatic flow formula ($\gamma = 1.4$ ratio of specific heat for air).

The three atmospheres are almost indistinguishable in the troposphere. Only in the stratosphere do we see differences in temperature (see Fig. 8.2), whereas density and pressure are already so low that any distinction is washed out. The ARDC atmosphere is used in space ascent and hypersonic vehicle simulations that require high fidelity, e.g., the CADAC GHAME6 six-DoF model. The U.S. 76 atmosphere is recommended for in-atmosphere and stratosphere simulations that do not exceed 86 km altitude. You find it implemented in the CADAC FALCON6 simulation.

Standard atmospheres are useful for comparative trajectory studies, but, of course, no airplane ever encounters it. There is so much seasonal and diurnal variability that any exhaustive analysis should include an atmospheric sensitivity study. You can consult the literature, like the *Handbook of Geophysics*,[4] and

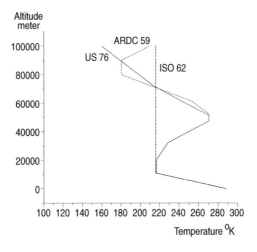

Fig. 8.2   Comparison of standard atmospheres.

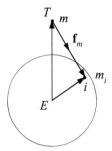

Fig. 8.3   Gravitational attraction.

find the so-called hot and cold atmospheres at various latitudes—the CADAC weather deck allows you to enter them in tabular form. Or, because the U.S. 76 atmosphere is normalized to standard sea-level conditions, you simply change the sea-level values for temperature, pressure, and density and build your own generic variations.

8.2.2   Gravitational Attraction

What is gravity? Nobody knows for sure; we only can measure its effects. Long ago the ancient Greeks wrote about it and speculated about its origin. For us engineers, Isaac Newton has provided all we have to know. His inverse square law says it succinctly: The gravitational pull between two bodies is proportional to the product of their masses and inversely proportional to the square of their distance. Suppose two particles with mass m and m_i are separated by r, then the force of attraction F is, with G as the proportionality constant,

$$F = G\frac{mm_i}{r^2} \tag{8.15}$$

If the body consists of a collection of particles $m_i, i = 1, 2, 3, \ldots$, then the test mass m is affected by the attraction of each particle. As we place the test mass at different locations, we can calculate the gravitational field by measuring the force and dividing it by the test mass m. Introduce the displacement vectors s_{TE} of the test mass T wrt a reference point E of the massive body E, and the displacement vector s_{iE} of a particle i wrt the same reference point E (see Fig. 8.3). The gravitational field is described by

$$g\{s_{TE}\} = \frac{f_m}{m} = G\sum_i \frac{m_i(s_{iE} - s_{TE})}{|s_{iE} - s_{TE}|^3} \tag{8.16}$$

where the force vector of the test mass f_m has the same direction as the displacement vector $s_{iT} = s_{iE} - s_{TE}$. The gravitational field is the *gradient of the gravitational potential* $\nabla U\{s_{TE}\} = g\{s_{TE}\}$. The gravitational potential itself defines the equipotential surfaces

$$U\{s_{TE}\} = G\sum_i \frac{m_i}{|s_{iE} - s_{TE}|}$$

A unit test mass has the same potential energy anywhere on an equipotential surface against the collection of particles m_i, $i = 1, 2, 3, \ldots$.

The equipotential surface of the Earth is called the *geoid*, an irregularly shaped envelope. An important approximation of the geoid is the *geocentric equipotential ellipsoid of revolution*. Its most recently updated dimensions are given in the Defense Mapping Agency's "U.S. World Geodetic System 1984 (WGS84)".[5] I have more to say about it in Chapter 10, when we derive the equations of motion for an elliptical Earth.

In this chapter we confine ourselves to the approximation of a spherical Earth, defined by a gravitational field with spherical equipotential surfaces. Its gravitational field simplifies after having summed Eq. (8.16) over the homogenous sphere

$$g\{s_{TE}\} = \frac{f_m}{m} = G \sum_i m_i \frac{-s_{TE}}{|s_{TE}|^3} = -GM \frac{s_{TE}}{|s_{TE}|^3}$$

Suppose the test mass is a missile T with mass m. The gravitational acceleration on the missiles is in the opposite direction of the displacement vector s_{TE} of the missile wrt the center of Earth E, and the magnitude of the gravitational force is with $|s_{TE}| = r$

$$F = G \frac{Mm}{r^2} \tag{8.17}$$

The product GM is an important constant for planet Earth with the value $GM = 3.986005 \times 10^{14}$ m^3/s^2.

Living on Earth subjects us to a gravitational acceleration of $g = 9.82023$ m/s^2. Yet we do not feel the full brunt of Earth's pull because it is opposed by the centrifugal force of our "merry-go-round." Both accelerations vectorially added results in the so-called *gravity acceleration*

$$g_g\{s_{TE}\} = g - \Omega^{EI}\Omega^{EI} s_{TE} = -GM \frac{s_{TE}}{|s_{TE}|^3} - \Omega^{EI}\Omega^{EI} s_{TE}$$

with Earth's angular rate of $\omega^{EI} = 7.292115 \times 10^{-5}$ rad/s. The value of the gravity acceleration depends on your latitude and altitude. At sea level (mean radius of Earth = $6,371,005$ m) standing on equator, $g_g = 9.7864$ m/s^2, and at $45°$ latitude $g_g = 9.8033$ m/s^2. The accepted standard average value is $g = 9.8066$ m/s^2. Come, join me at the equator, life is lighter there! Before you leave, however, make sure you understand the difference between gravitational and gravity acceleration.

8.2.3 Parabolic Drag Polar

Modeling aerodynamic forces and moments of aerospace vehicles can be a formidable task. Multimillion-dollar wind-tunnel facilities have been built and supercomputers put to work to measure, calculate, and predict the flow phenomena. A mathematical framework must be found to express these data in a form that can be programmed for the computer. In Chapters 9 and 10 you will encounter models for missiles and aircraft at increasing levels of sophistication. For three-DoF simulations, we can confine ourselves to expressions that relate drag and lift by simple polynomials.

We go back to the latter part of the 19th century and find Otto Lilienthal experimenting with hang gliders. He took his hobby very seriously and is credited with relating lift and drag by what he called "*die Flugpolare*" (the drag polar). After his accidental death in 1896, the Wright brothers[6] credited him in 1901 with laying the foundation of flight by experimentation.

The lift force L is normal to the velocity vector of the aircraft wrt the air and is contained in the plane of symmetry of the aircraft. The drag force D is parallel and in the opposite direction of the velocity vector. Nondimensional aerodynamic coefficients are formed from the dynamic pressure $\bar{q}$ and the reference area S (airplanes use wing area and missiles employ body cross section).

Lift coefficient:

$$C_L = \frac{L}{\bar{q}S}$$

Drag coefficient:

$$C_D = \frac{D}{\bar{q}S}$$

with $\bar{q} = (\rho/2)V^2$, ρ the air density, and V the speed of the aircraft relative to air. Both coefficients are assumed to be functions of the following parameters:

$$C_L, C_D = f\{\text{Mach, angle of attack, power on/off, shape}\}$$

Mach number, the ratio of vehicle velocity over sonic speed, can have a significant effect on the coefficients during the transonic and supersonic flight regimes. The main effector however is the angle of attack, which, with only small variations, changes the lift coefficient decisively. Depending on the installation of the propulsion unit, the airflow around the wing or tail modifies the drag characteristics. Particularly for missiles with boost motors, the drag increases significantly during the coast phase. Naturally, the shape of the vehicle determines the overall aerodynamic performance, but the size of the vehicle has only a minor influence on the coefficients. This insensitivity to scale justifies much of the considerable wind-tunnel investments.

When drag data are plotted against lift, a near parabolic curve emerges for any given Mach number. What a break for the aerodynamicist! He can model the functional relationship by a second-order polynomial, called the *parabolic drag polar*:

$$C_D = C_{D_0} + k\left(C_L - C_{L_0}\right)^2 \tag{8.18}$$

The parabola is shown in Fig. 8.4. Not surprisingly, drag is never zero even at zero lift. It has a minimum value of C_{D_0}, which may occur at a nonzero lift value C_{L_0}. A parabolic drag polar thus shifted upward is called an *offset polar*. The factor k determines the drag increase caused by deviation from minimum drag and is referred to as the *induced drag* coefficient.

If the minimum drag occurs at zero lift, the function simplifies to a *centered polar* (see Fig. 8.5):

$$C_D = C_{D_0} + kC_L^2 \tag{8.19}$$

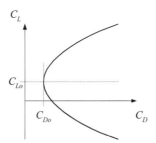

Fig. 8.4   Offset drag polar.

The drag polar is centered on the drag axis if the vehicle has two planes of symmetry—like a conventional missile—or if a wing with symmetrical airfoil is the dominant lifting surface. The parameter of the parabola is the angle of attack α. The higher the angle of attack is, the greater the lift and drag forces, up to the point when the flow starts to separate from the main lifting surface. Thereafter, lift breaks down, but drag keeps increasing, and the parabolic model has lost its usefulness.

The noninduced drag coefficient C_{D_0} models such phenomena as surface friction, profile drag, and supersonic wave drag. The second term represents the induced drag caused by lift. For vehicles that traverse through more than one Mach region—subsonic, transonic, supersonic, or hypersonic—the coefficients C_{D_0}, C_{L_0}, C_L, k must be modeled as functions of Mach number.

The drag polar presupposes that lift is given and drag is derived. In simulations, however, one prefers to specify the angle of attack as input rather than lift. A relationship must therefore be established between the lift coefficient and angle of attack. Fortunately, experimental evidence points to a linear relationship (see Fig. 8.6):

$$C_L = C_{L\alpha0} + C_{L_a}\alpha \qquad (8.20)$$

That linearity extends to the onset of flow separation, when lift brakes down rapidly. It is present over all Mach regimes. For vehicles with a centered drag polar, the lift slope goes through the origin, i.e., $C_{L\alpha0} = 0$.

The parabolic drag polar is an aerodynamic model suitable for simple point–mass three-DoF simulations from subsonic to hypersonic flight regimes. Be careful,

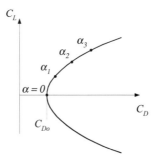

Fig. 8.5   Centered drag polar.

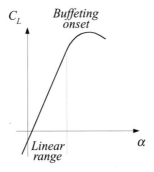

Fig. 8.6   Linear lift slope.

however, and do not expect too much accuracy from the results. Imposing a second-order polynomial curve washes out minimum drag cups near the cruise conditions and, as already noted, does not account for the onset of buffeting. A parabola also assumes lift symmetry for positive and negative angles of attack—hardly the case for airplane wings with high-lift airfoils. Furthermore, we also neglected Reynolds-number dependency and skin-friction changes with altitude.

All of these shortcuts were taken to get you started with simple simulations. As you gather more data, you can abandon the parabolic fit in favor of higher-order polynomials or use tables to accurately model the functional relationship between the lift and drag coefficients. However, there are inherent restrictions that come with the point–mass approach.

One supposition is the neglect of the control surface effects on lift and drag. Their contribution could be included as so-called *trimmed* values, if we had a full force and moment model available for data reduction. However, in that fortunate case we may as well build a full six-DoF simulation.

Another assumption restricts the lateral maneuver to coordinated turns only, i.e., the aircraft banks without sideslipping. The same limitation applies to missiles, unless they possess rotational symmetry, in which case the lift and drag forces always lie in the load factor plane, irrespective of the body bank attitude. In effect, for both missiles and aircraft the drag polar applies to the aerodynamic forces in the load factor plane.

Figure 8.7 helps us to define the load factor plane. It coincides with the 1^B, 3^B symmetry plane of the aircraft and contains the velocity vector v_B^E of the aircraft wrt to Earth. The bank angle ϕ establishes the orientation of the load factor plane relative to the vertical plane, which contains the 1^V, 3^V axes (1^V coincides with v_B^E). The angle of attack α positions the aircraft centerline above the velocity vector in the load factor plane. It is useful to introduce the load factor coordinate system. Its 1^M axis is parallel and in the direction of the velocity vector v_B^E and 2^M coincides with 2^B. In load factor coordinates the resultant aerodynamic force possesses lift and drag as its two components:

$$[\bar{f}_a]^M = \bar{q}S[-C_D \quad 0 \quad -C_L] \tag{8.21}$$

The transformation matrix of the load factor wrt velocity coordinates is determined

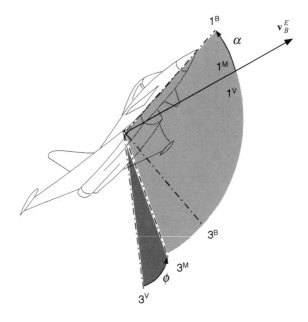

Fig. 8.7   Load factor plane of aircraft.

by the bank angle ϕ

$$[T]^{MV} = \begin{bmatrix} 1 & 0 & 0 \\ 0 & \cos\phi & \sin\phi \\ 0 & -\sin\phi & \cos\phi \end{bmatrix} \tag{8.22}$$

and the aerodynamic force in velocity coordinates is therefore

$$[f_a]^V = [\bar{T}]^{MV}[f_a]^M = \bar{q}S \begin{bmatrix} -C_D \\ C_L \sin\phi \\ -C_L \cos\phi \end{bmatrix}$$

As expected, the drag force opposes the aircraft velocity directly, and the lift force, modulated by bank angle ϕ, generates the horizontal maneuver force $C_L \sin\phi$ and the vertical force $-C_L \cos\phi$.

Modeling the aerodynamics of airplanes and missiles with a parabolic drag polar is a quick way to get preliminary performance estimates of new concepts when the database is still scant. Furthermore, this simple approach is also quite useful for mission-level simulations with their frugal trajectory models. There, the aircraft and missiles are well defined; but because of the large number of participating vehicles, their fly-out simulations must be kept artless.

So far, we have dealt with the gravitational and aerodynamic forces. To complete the right-hand side of Newton's law, we must address the force that overcomes gravity and drag, namely thrust. I shall discuss rocket and airbreathing propulsion in a form not just suitable for three-DoF models, but also quite applicable to five- and six-DoF simulations.

8.2.4 Propulsion

Unless you are a glider enthusiast, you value propulsion as the means of keeping missiles and aircraft in the air. The thrust vector overcomes drag and gravity and maintains the speed necessary for lift generation. It is usually directed parallel to the vehicle's centerline, although helicopters and the V-22 Osprey display their individuality by thrusting in other directions as well. For our simulations we deal only with body-fixed propulsion systems whose thrust vector is essentially in the positive direction of the body 1^B axis, possibly slanted by a fixed angle.

You will be surprised how far just basic physics will take us in modeling missile and aircraft propulsion. However, you should not bypass the solid foundations laid in the classic book by Zucrow[7] and the excellent textbook by Cornelisse et al.[8] Some recent and up-to-date compendiums were published by AIAA for missile propulsion,[9] hypersonic airbreathers,[10] and aircraft propulsion.[11] Even the control book by Stevens and Lewis[12] has some useful information on turbojet engine modeling.

Most missiles are rocket propelled with the oxidizer carried onboard. Some supersonic missiles use the oxygen of the air for combustion in their ramjet or scramjet propulsion units. The air is captured by the inlet, retarded and compressed, fuel is injected and ignited, and the mixture exhausted through the nozzle. No rotary machinery is employed. Aircraft and cruise missiles, on the other hand, employ rotating compressors and turbine machinery for propulsion. Based on simple physics, I will derive the thrust equations for rockets, turbojets, and combined-cycle engines.

Newton's second law will serve us well, both for missile and aircraft propulsion. In each case the time-rate-of-change of momentum generates the propulsive thrust. We first derive the thrust equation for rockets.

8.2.4.1 Rocket propulsion.

The principle of rocket thrust goes back to the ancient Chinese and their brilliant firework displays. Then and now it is based on Newton's second law, applied to the exhaust stream with the velocity c and the mass flow $\dot{m}$ (see Example 5.6):

$$F = \dot{m}c$$

Instead of providing the exhaust velocity, usually the specific impulse I_{sp} is given. It is defined as the ratio of the impulse delivered, divided by the propellant weight consumed

$$I_{sp} = \frac{F\Delta t}{\dot{m}g_o \Delta t} = \frac{F}{\dot{m}g_o} \tag{8.23}$$

where g_o is Earth's gravity acceleration referenced to the fixed, standard value $g_o = 9.80665 \text{ m/s}^2$. Solving for F yields the alternate thrust equation

$$F = I_{sp}\dot{m}g_o \tag{8.24}$$

The exhaust velocity is therefore related to the specific impulse by $c = I_{sp}g_o$.

Specific impulse provides an important characterization of the rocket engine and its propellant. Typical values are given in Table 8.1 for double-based solid propellants like nitrocellulose (NC) and nitroglycerin (NG) and liquid bipropellants like hydrazine ($N_2 H_4$) and oxygen (O_2).

Table 8.1   Typical values for solid and liquid propellants

Propellant	Density, kg/m^3	I_{sp}, s	Burn rate functional dependence
Solid	1700	250	$f(p, T_{fuel},$ surface area)
Liquid	1200	350	f (pump feeding)

The thrust of a missile is usually given at sea level. A correction has to be made for the thrust at altitude. Suppose F_{SL} is the sea-level thrust; a term is added that corrects for the fact that the pressure at altitude p_{Alt} is less than the pressure at sea level p_{SL}. With the exhaust nozzle area A_e the thrust at altitude is

$$F_{Alt} = F_{SL} + (p_{SL} - p_{Alt})A_e \qquad (8.25)$$

For solid propellant the sea-level thrust is most likely given as a function of burn time and possibly of propellant temperature. A simple table loop-up routine will suffice. If only specific impulse and propellant burn rate are known, Eq. (8.24) will provide the thrust. This equation also serves the liquid propellant rocket motor. You just need to include a multiplying factor that represents the throttle ratio (values between zero and one). To complete the propulsion model, the expended propellant is monitored for updating the vehicle's mass.

8.2.4.2   Turbojet propulsion.   The physical principle of airbreathing propulsion again derives from Newton's second law. However, the time rate of change of momentum is now based on the velocity increase of the airflow $\dot{m}_a$ through the turbine. With V the flight velocity and V_e the exhaust velocity the thrust is (neglecting fuel mass and assuming ideal expansion)

$$F = \dot{m}_a(V_e - V)$$

The faster the exhaust velocity (turbine output) or the greater the airflow (high bypass), the greater the thrust F. In general, the thrust depends on several parameters:

$$F = f(\text{Mach, altitude, power setting, angle of attack})$$

For some of our applications, we neglect the angle-of-attack dependency.

The specific fuel consumption (SFC) b_F is an important indicator for the efficiency of the turbojet. It is defined by the ratio of fuel flow to thrust

$$b_F = \frac{\dot{m}_f}{F}$$

The units of b_F are usually given as kilograms/(deka-Newton hour), where dN can be written 10N, and $\dot{m}_f$ is the fuel flow in kilograms/hour. The strange use of dN is justified by the approximate numerical equivalency of metric and English units

$$1[\text{kg/(dN h)}] = 0.980665 \ [\text{lbm/(lbf h)}]$$

Typical values of SFC are between 1.0 to 0.3 kg/(dN h), with turbojets being less efficient than high-bypass turbofans.

8.2.4.3 Combine-cycle propulsion. In this section we focus on the high-speed regime of airbreathing engines. It is an area of vigorous research and development, spurred on by the National Aerospace Plane (NASP), the single-stage-to-orbit (SSTO) requirement, and various X vehicles. We are also motivated to look into high-speed propulsion because the GHAME3 and GHAME6 simulations use turbojet, ramjet, and scramjet engines to ascend through all Mach regimes into the stratosphere.

Turbojets and *turbofans* are particularly suited for the low-speed portions of the mission and have adequate performance up to Mach 3. The upper limit is imposed by the thermal constraints of their materials. Designs tend to have low overall pressure ratios and low rotor speeds at takeoff. With cryogenic fuels like liquid hydrogen, precooling can increase the maximum Mach-number regime to beyond 4, provided the engine operates above stoichiometric conditions.

Ramjets have no rotating machinery and start to operate above Mach 2. The internal flow remains subsonic, although they may perform up to Mach 6, limited by dissociation and material temperatures. Using hydrogen as a fuel and thus eliminating the need for flameholders can alleviate some of the material constraints.

Turboramjets combine turbojets and ramjets in wraparound or tandem designs. A high efficiency intake is combined with an ejector nozzle. It matches the full intake capture area demanded during transonic flight. The excess capture flow is passed down a duct, concentric with the engine, which also serves to bypass the turbomachinery in the ramjet mode. Turboramjets operate from static conditions at sea level up to Mach 6 at high altitudes.

Turborockets use hydrogen/oxygen combustors to produce the working fluid for the turbine. The combustion is fuel rich so that the turbine entry temperature is kept within the capability of uncooled materials. The excess hydrogen is burned in the fan stream air, with secondary hydrogen injected to produce an overall stoichiometric mixture. Fan materials limit the upper Mach number to about 4.

Scramjets are similar to ramjets, except that their combustion occurs at supersonic speeds. Although they can operate at lower speeds, they become more efficient than ramjets only above Mach 6.

The *turboramjet/scramjet* is a three-cycle variable inlet geometry design, capable of providing thrust from static sea-level conditions to hypersonic atmospheric exit. NASA uses it for their GHAME concept. The breakpoints for the cycles are 1) turbojet from Mach 0 to Mach 2, 2) ramjet until Mach 6, and 3) scramjet beyond.

The authors of the GHAME propulsion package[13] apologize for the simplicity of their approach, but I find their model quite lucid. They start with the basic thrust equation (8.24) $F = I_{sp}\dot{m}g_o$; i.e., thrust equals specific impulse times weight flow rate through the engine. Because we deal with airbreathers, the weight flow rate is essentially the amount of air sucked through the intake area A_c. Therefore, given the speed of the vehicle $V = Ma$ and the air density ρ, the weight flow rate is

$$\dot{m}g_o = g_o \rho Ma A_c$$

which assumes that the air enters the cowl uniformly. However, the intake flow of a turboramjet/scramjet engine is very intricate, influenced by engine cycle, Mach number, and angle of attack. This complexity is distilled into a capture-area coefficient C_a, which is dependent on Mach number and angle of attack. Subsonically, it starts with values near 1, drops to 0.2 in the transonic region, then rises slightly until the ramjet takes over at Mach 2. Thereafter, it increases beyond 1 and tops out under the scramjet cycle at about 5. As the angle of attack increases, the effective capture area grows, and the C_a value almost doubles at 21 deg. With this correction factor the effective weight flow rate is

$$\dot{m}g_o = g_o \rho Ma\, C_a(M, \alpha) A_c$$

The pilot controls the fuel flow and the variable intake by the throttle setting thr. Indirectly, the pilot adjusts the fuel/air ratio of the engine to the stoichiometric ratio, which equals $0.029 \times$ thr, by adjusting thr between the values zero and two. The specific impulse I_{sp} is a function of the engine cycle, the throttle setting, and Mach number. It increases with throttle setting and decreases with Mach number.

Now we have assembled all of the elements for the thrust equation (8.24):

$$F = 0.029\, \text{thr}\, I_{sp}(M, \text{thr}) g_o \rho Ma\, C_a(M, \alpha) A_c \qquad (8.26)$$

As the vehicle takes off with full throttle, low Mach, and high angle of attack, its thrust is at a near maximum. During the climb-out, in the transonic region with decreasing α the effective capture area is significantly reduced so that the pilot will maintain max throttle until the ramjet regime is reached. Then the pilot begins to throttle back to conserve fuel.

You will be surprised and may be disappointed to learn that this is all I have to say about propulsion. If you have to build a propulsion simulation, you should assemble a team of experts that will model such effects as inlet flow, thermodynamics, combustion efficiency, exhaust, installed drag, and stall. You can then provide the aerodynamics, mass properties, and the simulation environment. Here, I emphasize a general treatment that enables you to build three-, five-, and six-DoF simulations quickly without resorting to specialists.

You will find examples throughout the family of CADAC simulations. ROCKET3 models liquid-fueled, three-stage rockets; GHAME3 and GHAME6 mimic the NASA combined-cycle engine; CRUISE5 and FALCON6 use simple subsonic turbojet models; and AIM5, SRAAM5, and SRAAM6 are propelled by solid rocket motors. We will turn now to the description of the two, three-DoF simulations GHAME3 and ROCKET3.

8.3 Simulations

Building simulations is best learned by example. Study the venues, which others have trodden before you. For this reason I provide you with several simulations on the CADAC Web site. You should supplement them with examples from your own work environment. Become a simulation glutton! In this section I document the two, three-DoF simulations GHAME3 and ROCKET3 that you find on the CADAC Web site.

You may perceive my documentation lacking in detail. I challenge you to combine the source code, which is well interspersed with comments, with the following figures and derive the complete flow diagram of the two simulations on your own.

If this is your first exposure to CADAC, you have to lay some groundwork. First read the two volumes of CADAC Studio: *Quick Start* and *Programmer's Manual* and print them out. You will need them as constant companions. Then run the test

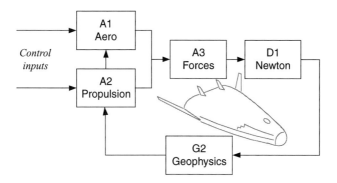

Fig. 8.8   GHAME3 simulation modules.

case TEST to make sure your computer is set up properly. Appendix B will launch you with the CADAC Primer.

8.3.1   GHAME3: Hypersonic Vehicle Simulation

Thanks to NASA's forethought, we have a complete data package of a hypersonic vehicle, called Generic Hypersonic Aerodynamic Model Example (GHAME). You will encounter it again in Chapter 10 as an example of a complex six-DoF simulation.

The simple three-DoF structure of the CADAC GHAME3 simulation is shown in Fig. 8.8. The three external force modules, gravity (G2), aerodynamics (A1), and propulsion (A2) are combined in the A3 Force Module to serve Newton's law (D1). In addition, the Geophysics Module G2 provides also atmospheric density pressure and sonic speed.

8.3.1.1   Aerodynamics.
For the purpose of this chapter, I have reduced the aerodynamics of the six-DoF GHAME model to an offset parabolic drag polar by curve fitting the full data set (see Fig. 8.9). Because the lift coefficient C_L is the independent variable, automatic plotting programs place it on the abscissa and use the drag coefficient as ordinate. Notice the change of the drag polar with Mach number. The minimum zero-lift drag C_{D_0} occurs at subsonic and hypersonic speeds and peaks near Mach 1. The lift-over-drag ratio, indicated by the flatness of the parabola, decreases with increasing Mach number. Do you see the slight bias of the parabola centerlines toward positive lift values? This shift is more evident in the lift slopes of Fig. 8.10. The zero-lift points occur between 1–2 deg angle of attack at all Mach numbers. Both sets of curves are implemented as tables in the A1 Module.

8.3.1.2   Propulsion.
The combined-cycle engine of GHAME is programmed in Module A2. From Eq. (8.26)

$$F = 0.029 \, \text{thr} \, I_{\text{sp}}(M, \, \text{thr}) g_o \rho \, Ma C_a(M, \alpha) A_c \qquad (8.27)$$

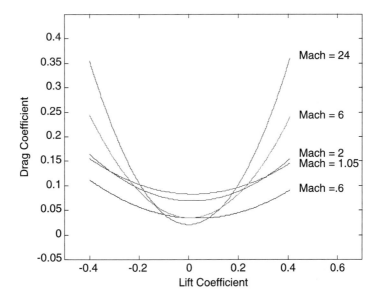

Fig. 8.9   GHAME parabolic drag polar.

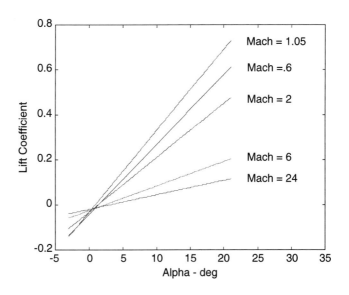

Fig. 8.10   GHAME lift slopes.

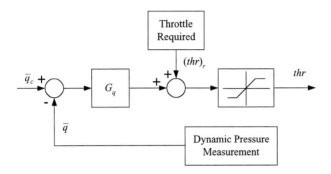

Fig. 8.11   Autothrottle for constant dynamic pressure control.

we calculate thrust F from two tables, specific impulse $I_{sp}(M, \text{thr})$ and capture-area coefficient $C_a(M, \alpha)$. To keep track of the remaining fuel, the fuel rate is monitored based on Eq. (8.24)

$$\dot{m} = \frac{F \times \text{thr}}{I_{sp} g_0} \tag{8.28}$$

and integrated to provide the expended fuel mass.

One of the crucial factors of a hypersonic vehicle is the optimum throttle setting for best climb at minimum heating. An ascent with constant dynamic pressure approximates these requirements. Thus, we incorporate an automatic throttle feedback loop into the propulsion module that maintains constant dynamic pressure and call it *autothrottle*. Figure 8.11 shows the control loop. Measured dynamic pressure $\bar{q}$ is compared with the desired input $\bar{q}_c$, processed through a gain G_q, and summed with the required throttle setting $(\text{thr})_r$ overcoming the drag force. After limiting, the throttle setting for the thrust of the engine is obtained.

To synthesize the autothrottle gain G_q, we complete the control loop as shown in Fig. 8.12. The thrust F from the engine with throttle setting thr accelerates the vehicle and, after integration, provides the velocity V that determines the dynamic pressure $\bar{q}$.

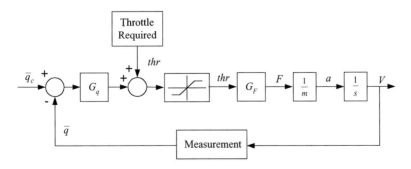

Fig. 8.12   Autothrottle feedback loop.

Several steps determine the gain G_q. First, calculate the required thrust F_r by setting it equal to the drag, projected into the body 1^B axis by the angle of attack α

$$F_r = \frac{C_D S}{\cos \alpha} q_c$$

With Eq. (8.27) we calculate the required throttle setting

$$(\text{thr})_r = \frac{F_r}{0.029 I_{sp} g_0 \rho V C_a A_c}$$

The thrust gain is therefore

$$G_F = \frac{F_r}{(\text{thr})_r} = 0.029 I_{sp} g_0 \rho V C_a A_c \tag{8.29}$$

The autothrottle control loop is essentially a first-order lag transfer function

$$T(s) = \frac{1}{T_q s + 1}$$

with the autothrottle time constant

$$T_q = \frac{2m}{\rho V G_q G_F} \tag{8.30}$$

To determine the autothrottle gain, pick a reasonable time lag between the commanded and achieved dynamic pressure and calculate the gain from

$$G_q = \frac{2m}{\rho V G_F} \frac{1}{T_q} \tag{8.31}$$

where m is the mass of the vehicle. Clearly, the control loop is always stable, just make sure to select a realistic time constant, possibly making the time constant dependent on air density.

8.3.1.3 Forces.   The aerodynamic and propulsive forces are combined in Module A3, coordinated first in load factor and then in velocity axes, divided by vehicle mass, and sent to the $D1$ Module as specific force $[f_{sp}]^V$.

Refer back to Fig. 8.7 to visualize the geometry. To get a better understanding of the angles and coordinate axes, Fig. 8.13 displays all of the relevant information. It shows the heading and flight-path angles χ and γ, the bank angle ϕ, and the angle of attack α. These angles reflect the transformation sequence

$$]^B \xleftarrow{\alpha}]^M \xleftarrow{\phi}]^V \xleftarrow{\gamma, \chi}]^G$$

We already derived the TM $[T]^{MV}$ and the aerodynamic force $[f_a]^V$ in Sec. 8.2.3.

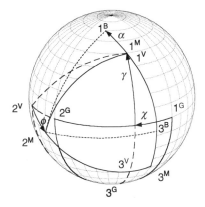

Fig. 8.13   Geographic, velocity, load factor, and body axes.

Now we combine it with the thrust force F from Eq. (8.26):

$$[f_{a,p}]^M = \begin{bmatrix} F\cos\alpha - \bar{q}SC_D \\ 0 \\ -F\sin\alpha - \bar{q}SC_L \end{bmatrix}$$

and transform it to velocity coordinates

$$[f_{a,p}]^V = [\bar{T}]^{MV}[f_{a,p}]^M = \begin{bmatrix} F\cos\alpha - \bar{q}SC_D \\ \sin\phi(F\sin\alpha + \bar{q}SC_L) \\ -\cos\phi(F\sin\alpha + \bar{q}SC_L) \end{bmatrix}$$

Given α and ϕ as input, the so-called contact forces (nongravitational forces) can be evaluated and provided as specific force f_{sp} to Newton's equation in velocity coordinates

$$[f_{sp}]^V = \frac{1}{m}[f_{a,p}]^V$$

where m is the vehicle mass.

8.3.1.4   Specific force.   We have arrived at a convenient situation to summarize the four Modules G2, A1, A2, and A3 (see Fig. 8.14). Given the aerodynamic and propulsive characteristics, the inputs α, ϕ, and thr produce the specific force that is sent to the D1 Module.

8.3.1.5   Newton's law.   For the equations of motion, let us go back to Eq. (8.4) and divide both sides by the vehicle mass m

$$\left[\frac{dv_B^I}{dt}\right]^I = [\bar{T}]^{GI}\left([\bar{T}]^{VG}[f_{sp}]^V + [g]^G\right)$$

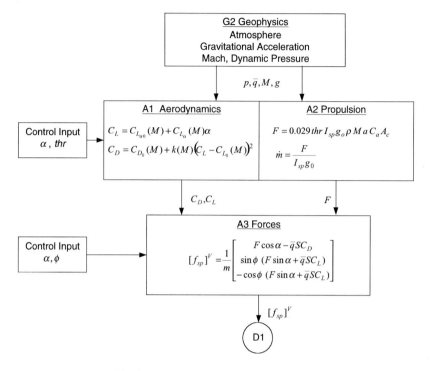

Fig. 8.14   Generating the specific force.

and adjoin the position equations (8.5):

$$\left[\frac{ds_{BI}}{dt}\right]^{I} = \left[v_{B}^{I}\right]^{I}$$

As convenient these equations are for integration, their interpretation is as difficult. Who wants to input trajectory parameters in inertial coordinates? We would much rather use geographic variables as input and output. Therefore, we need to develop code that transforms the input variables' geographic speed $|v_{B}^{E}|$, heading angle χ, flight-path angle γ, longitude l, latitude λ, and altitude h into inertial position $[s_{BI}]^{I}$ and velocity $[v_{B}^{I}]^{I}$. Figure 8.15 shows the equations of motion and the conversion process.

Two types of utility subroutines are employed: the matrix utilities MATxxx and CADAC utilities CADxxx. You can find the MATxxx routines described in the CADAC user documentation. These subroutines abide by FORTRAN call conventions. Writing them as $[v_{B}^{E}]^{G} = MATCAR(|v_{B}^{E}|, \chi, \gamma)$ emphasizes the input/output relationship, but they are coded actually as CALL MATCAR ($[v_{B}^{E}]^{G}, |v_{B}^{E}|, \chi, \gamma$). There are three CADxxx subroutines appended to the D1 Module. CADTEI produces $[T]^{EI}$, CADTGE calculates the transformation matrix Eq. (3.13), and

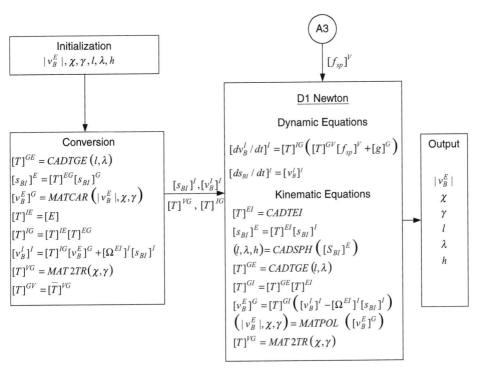

Fig. 8.15 Calculating the trajectory parameters.

CADSPH is the inverse transaction. If you are wondering how to get $[s_{BI}]^G$, remember that its third coordinate is just the distance to the center of the Earth $[\overline{s_{BI}}]^G = [0 \ \ 0 \ \ -(R_\oplus + h)]$.

Now it is your turn to make this hypersonic vehicle soar into the stratosphere. Review the MODULE.FOR code, the INPUT.ASC and HEAD.ASC files; compile, link, and run the test case. It should produce an output that looks like the traces in Figs. 8.16 and 8.17. I produced both figures with the CADAC Studio plot programs KPLOT/2DIM and KPLOT/GLOBE. As you see from the weaving trajectory, cruising at constant angle of attack does not deliver a constant altitude trajectory.

Now change the input parameters and observe a variety of trajectories. Get a feel for the sensitivity of the vehicle to various modifications. Afterward do the SSTO project or turn to the ROCKET3 simulation.

8.3.2 ROCKET3: Three-Stage Rocket Simulation

ROCKET3 is a derivative of GHAME3. Figure 8.18 shows the module structure. Only the shaded Modules A1 and A2 are different. I will be very brief in my description. After having thoroughly explored the GHAME3 model, you should have no problem deciphering the FORTRAN code of ROCKET3.

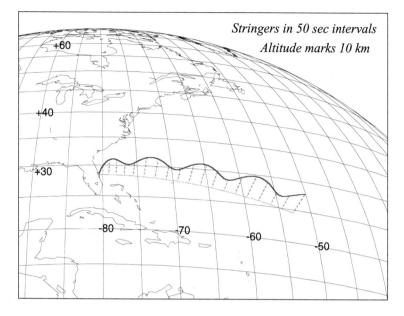

Fig. 8.16   GHAME3 hypersonic vehicle trajectory.

The aerodynamics is modeled by simple polynomials in Mach number with linear dependency on angle of attack. The thrust for each stage is calculated from Eq. (8.24), then substituted into Eq. (8.25). Because we are dealing with liquid rockets, a throttle factor thr is inserted:

$$F_{\text{Alt}} = I_{\text{sp}} \dot{m}_0 \, \text{thr} + (p_{\text{SL}} - p_{\text{Alt}}) A_e \qquad (8.32)$$

You may be puzzled by the fact that the A3 Module remains unchanged. For the earlier planar symmetry case, the airplane executes a perfect bank maneuver, maintaining its plane of symmetry in the load factor plane. Now look at Fig. 8.19. Conventional rockets and missiles, having tetragonal symmetry, do not bank to turn, but can generate a maneuver by pitch and lateral force. As a result, however, the load factor plane forms a roll angle wrt the vertical plane. This roll angle is

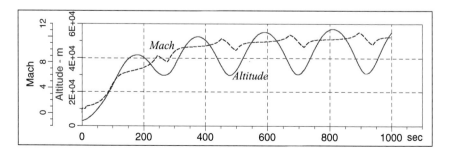

Fig. 8.17   GHAME3 trajectory parameters.

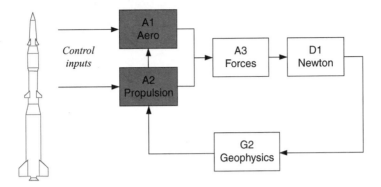

Fig. 8.18   ROCKET3 simulation modules.

the same as the bank angle of the aircraft. Therefore, ϕ serves as control input for both the aircraft and the missile, and we have no reason to change the A3 Module.

You have become the master of simple point–mass, three-DoF simulations. I only carried through the Cartesian form of the equations of motion. It is left for you to implement the polar equations (see Problem 8.3). You should have an understanding of the standard atmosphere, gravitational attraction, and gravity acceleration. Simple drag polars model the aerodynamic forces, and the rate of change of linear momentum produces thrust. These elements of point–mass simulations are the basis for further development of more sophisticated five- and six-DoF models.

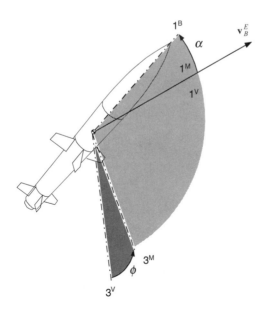

Fig. 8.19   Load factor plane of rocket.

References

[1]Vinh, Nguyen, *Optimal Trajectories in Atmospheric Flight*, Elsevier, Amsterdam, 1981, pp. 50–59.

[2]Minzer, R. A., "The ARDC Model Atmosphere," Air Force Cambridge Research Center, Air Research Development Command, Bedford, MA, 1959.

[3]U.S. Standard Atmosphere 1976, NOAA–S/T 76-1562, U.S. Government Printing Office, Washington, DC.

[4]*Handbook of Geophysics and the Space Environment*, U.S. Air Force Geophysics Lab., National Technical Information Center, ADA 16700, Cameron Station, Alexandria, VA, 1985.

[5]"Department of Defense World Geodetic System 1984, Its Definition and Relationships with Local Geodetic Systems," 3rd ed., NIMA WGS 84 Update Committee, NIMA TR 8350.2, 4 July 1997.

[6]Wright, W., "Some Aeronautical Experiments," *Journal of the Western Society of Engineers*, Dec. 1901.

[7]Zucrow, M. J., *Aircraft and Missile Propulsion*, Vols. I and II, Wiley, New York, 1958.

[8]Cornelisse, J. W., Schöyer, H. F. R., and Wakker, K. F., *Rocket Propulsion and Spaceflight Dynamics*, Pitman, 1997.

[9]Jensen, G. E., *Tactical Missile Propulsion*, Progress in Astronautics and Aeronautics, AIAA Washington, DC, 1993.

[10]Heiser, W. H., *Hypersonic Airbreathing Propulsion*, AIAA Education Series, AIAA, Washington, DC, 1994.

[11]Oates, G. C. (ed.), *Aircraft Propulsion Systems Technologies and Design*, AIAA Education Series, AIAA, Washington, DC, 1989.

[12]Stevens, Brian L., and Lewis, Frank L., *Aircraft Control and Simulation*, Wiley, New York, 1992.

[13]White, D., and Sofge, D., *Handbook of Intelligent Control*, Van Nostrand Reinhold, New York, 1992, Chap. 11.

Problems

8.1   Three-stage rocket ascent to 300-km orbit. *Task 1*: Download the ROCKET3 simulation form the CADAC Web site and run the test case INLAUNCH.ASC. Using CADAC-KPLOT, plot altitude, geographic and inertial speed, Mach number, dynamic pressure, heading, and flight-path angles vs time. Has the rocket reached orbital conditions?

Task 2:   Now is your turn to lift the rocket to a 300-km near-circular orbit by scheduling angle of attack. Build the input file IN300.ASC. Can you achieve orbital conditions? Again plot altitude, geographic and inertial speed, Mach number, dynamic pressure, heading and flight-path angles, and angle of attack vs time.

Task 3:   Summarize your findings in a brief ROCKET3 Trajectory Report. Include all plots and the input file IN300.ASC.

8.2   SSTO vehicle simulation.   If you followed the CADAC Primer from the CADAC Web site, you have already flown the GHAME3 simulation with the

INPUT.ASC file, but could not reach orbital conditions. With the rocket-propelled SSTO, launched from a Super Boeing 747, you can achieve a low-Earth orbit.

Task 1:   Modify the A1 and A2 modules of the GHAME3 simulation, using the data SSTO3 from the CADAC Web site. The A1 and A2 modules are much simpler for the SSTO.

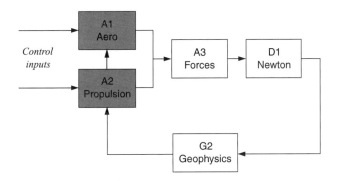

Task 2:   Now, launch the SSTO from the Super B747 12 km above Cape Canaveral, Florida, horizontally in an easterly direction with $|v_B^E| = 253$ m/s. Building the input file INCAPE.ASC with the following control commands for the ascent:

t, s	α, deg	thr
< 200	22.93	0.9
200–400	5.73	0.9
>400	5.73	0.5

At burn-out what are the values of $|v_B^E|, \chi, \gamma, l, \lambda, h$? What is the inertial speed $[v_B^I]^I, |v_B^I|$? (*Solution*: $t = 658$ s, $|v_B^E| = 7442$ m/s, $\chi = 102°$, $\gamma = 6.4°$, $l = -1.074$ rad, $\lambda = 0.461$ rad, and $h = 106$ km.)

Task 3:   Next, repeat Task 2 for Vandenberg, California, but launch in a westerly direction. Build the input file INVAN.ASC and provide the same output.

Task 4:   Repeat Tasks 2 and 3 for a nonrotating Earth.

Task 5:   Write a summary report SSTO3 Ascent Trajectories. Provide all burn-out conditions in one summary table. Include the input files. For Task 2 plot altitude, geographical and inertial speeds, flight-path angles, and fuel mass vs time.

8.3   SSTO3 simulation with polar equations of motion.   In Sec. 8.1.2 I derived the equations of motion with the Earth as reference, while maintaining J2000 as the inertial frame. These equations should lead to the same results as the Cartesian formulation of Problem 8.2.

Task 1:   Review Sec. 8.1.2 and code a new Module D1 with the polar equations of motion, Eqs. (8.11) and (8.12). Keep all other modules of the SSTO3 unchanged. Verify that all changes are made correctly by using MKHEAD3.EXE.

Task 2:   Use the input file INCAP.ASC from Problem 8.2 and run your polar SSTO3 simulations. The endpoint parameters should agree with less than 1% error. Plot the Coriolis and centrifugal accelerations and compare them to the gravitational term. Plot these three variables vs time. What conclusions do you draw?

Task 3:   Summarize your work in the SSTO Polar Simulation Report. Document your D1 Module, show your plots, and discuss your findings.

9
Five-Degree-of-Freedom Simulation

Frequently, three-DoF models, as described in the preceding chapter, do not model in sufficient detail the vehicle dynamics. Hence we may add two attitude degrees of freedom to the three translational equations and call the composite a five-DoF simulation. For a vehicle that executes skid-to-turn maneuvers (an intercept missile), pitch and yaw attitude dynamics are incorporated. For a bank-to-turn aircraft, the yaw angle of the missile is replaced by the bank angle. Euler's law formulates the differential equations for the two attitude angles. However, the increase in complexity is significant and approaches that of a full six-DoF simulation. To maintain the simple features of a three-DoF simulation and at the same time account for the attitude dynamics, the transfer functions of the closed-loop autopilot replace Euler's equations. This implementation is called a pseudo-five-DoF simulation. The word *pseudo* conveys the meaning of approximating the attitude dynamics with the linear differential equations of the transfer functions.

Pseudo-five-DoF simulations are popular models for concepts that are only loosely defined. During preliminary design, the vehicle's aerodynamics may be sketchy, the autopilot design rudimentary, and the guidance and navigation implementations uncertain. These are good reasons to match these notional systems with the simple pseudo-five-DoF models. If you want to find out whether a simulation has this pseudo characteristic, look for these telltales: trimmed aerodynamics, angle-of-attack as the output from a transfer function, body rates not obtained by solving the Euler's equations, and the absence of controls and actuator models.

Using the CADAC environment (see Appendix B), I have built such simulations for medium range air-to-air missiles, air-to-ground guided bombs, cruise missiles, airplanes, antisatellite interceptors, and reentry vehicles. These simulations were in support of either concept evaluations or man-in-the-loop simulators. It is amazing how useful these bare-bones models are. They make trade studies feasible, yield quick results for those hurried marketers, and are easily modified for other applications. One feature is particularly important: the integration step can be one or even two orders of magnitude greater than that of a six-DoF simulation. When execution time is critical as in air combat simulators, these pseudo-five-DoF models may be the only feasible approach. What enables the greater time steps is the disregard of high-frequency phenomena, like attitude motions, fast autopilots, actuators, and sensor dynamics.

Some modelers are more ambitious and would like to create a six-DoF showpiece. They add the rolling transfer function of missiles or the yawing transfer function of aircraft to the dynamics and thus create a pseudo-six-DoF simulation. This expansion is easily accomplished and may be beneficial when the attitude dynamics are emphasized. However, the pseudo limitations still apply, and it is doubtful that much fidelity is gained without the modeling of controls and higher-order dynamics.

Finally, a pseudo-five- or six-DoF simulation can become the trailblazer for the full six-DoF masterpiece. The aerodynamics is replaced by untrimmed data including aerodynamic moments and control effectiveness. Euler's equations are introduced to solve the three attitude degrees of freedom, and autopilot details and actuator dynamics increase model fidelity. If your pseudo-five-DoF had a complete guidance loop, you may be able to transfer it directly. I took that shortcut for several air-to-air missile simulations. The sensor and guidance algorithms developed earlier during the conceptual phase worked perfectly well in the six-DoF simulation.

In this chapter we will concentrate on the pseudo-five-DoF simulations for rotating round Earth (strategic missiles, hypersonic aircraft, and orbital vehicles) and for flat Earth (tactical missile and aircraft applications). The equations of motion are based on Newton's second law and supplemented by kinematic equations that calculate the attitude angles. If you need the sixth pseudo-DoF, you should be able to add it yourself. On the other hand, if you want to develop a full five-DoF simulation you should turn to Chapter 10, and reduce your model by one degree of freedom.

My plan is to derive the equations of motion in tensor form, provide the relevant coordinate transformations, and express them in matrix form for programming. The right-hand sides of these equations consist of the externally applied forces. We will develop these forces from the inside out, beginning with the trimmed aerodynamics for missiles or aircraft, the propulsive forces of rockets or turbojets, and the gravitational acceleration. Then we enlarge the circle and discuss how autopilots control these aerodynamic forces through acceleration and altitude commands for both skid-to-turn and bank-to-turn vehicles. Finally, the guidance law places demands on the autopilot to achieve certain trajectory objectives. We will discuss proportional navigation for target intercept and line guidance for trajectory shaping (waypoint guidance and automatic landing approaches). We conclude by addressing electro-optical or microwave sensors that provide the target line of sight to the guidance processor.

You can visit the CADAC Web site and download several examples of pseudo-five-DoF simulations. Besides the simple and more complex air-to-air missile simulations AIM5 and SRAAM5, you can find a generic cruise missile CRUISE5. With the material covered in this chapter, you should be able to decipher their source code, make some test runs, and adapt them to your own needs.

9.1   Pseudo-Five-DoF Equations of Motion

According to our game plan, the derivation of the equations of motion will proceed from general tensor formulation to specific matrix equations. First, we formulate Newton's second law wrt the flight-path reference frame. Second, we pick either the inertial coordinates $]^I$ for the round rotating Earth model or the local level coordinates $]^L$ for the flat-Earth simplification. Finally, we develop the kinematic equations that mimic the attitude dynamics.

Attitude information is important even in pseudo-five-DoF simulations. We must calculate the angular velocity ω^{BI} of body B wrt the inertial frame I (in six-DoF models ω^{BI} is the output of Euler's equations) and the direction cosine matrix $[T]^{BI}$ of body frame B wrt inertial frame I. Both are vitally important for the modeling of homing seekers, inertial measuring unit (IMU) sensors, and coordinate transformations. To construct the body rates, we will use the flight-path-angle rates and the incidence angle rates. Their integrals build the direction cosine matrix.

The key to this venue is the *inertial velocity frame U*, which is the frame that is associated with the velocity vector v_B^I of the vehicle's c.m. B wrt the inertial frame. When Newton's equations are expressed in this frame, the three state variables become *inertial heading* angle, *inertial flight-path* angle, and *inertial speed*, ψ_{UI}, θ_{UI}, $|v_B^I|$, with their derivatives $d\psi_{UI}/dt$, $d\theta_{UI}/dt$, $d|v_B^I|/dt$. From the first two derivatives we build the angular velocity ω^{UI} of the velocity frame wrt the inertial frame. However, to extract the complete body rate ω^{BI}, we need to calculate the angular velocity ω^{BU} of the vehicle wrt the velocity frame. Then we have

$$\omega^{BI} = \omega^{BU} + \omega^{UI} \tag{9.1}$$

Let us pause here and preempt a possible quandary. In Sec. 5.4.2 we derived the pseudo-five-DoF equations for flat Earth and used the velocity frame V of the geographic velocity v_B^E. Now we derive the pseudo-five-DoF equations for a round rotating Earth, still using a velocity frame, but associate it with the inertial velocity v_B^I. Both velocities are mutually related by Eq. (5.30):

$$v_B^I = v_B^E + \Omega^{EI} s_{BI}$$

Therefore, the inertial velocity frame U and geographic velocity frame V are separated by Earth's angular velocity. Only when we accept Earth as the inertial frame do U and V become the same.

The missing link ω^{BU} of Eq. (9.1) is provided by the incidence angular rates that are computed by the autopilot transfer function. Skid-to-turn missiles use the angle of attack and sideslip angle rates $d\alpha/dt$, $d\beta/dt$, and bank-to-turn aircraft employ next to the angle of attack also the bank angle rate $d\alpha/dt$, $d\phi_{UI}/dt$.

Before we can express the body rates in matrix form, we must deal with the direction cosine matrix $[T]^{BI}$ of vehicle coordinates $]^B$ wrt inertial coordinates $]^I$. By factoring, we will reach the objective

$$[T]^{BI} = [T]^{BU}[T]^{UI} \tag{9.2}$$

recognizing that $[T]^{BU}$ is a function of α, β or α, ϕ_{UI} and $[T]^{UI}$ a function of ψ_{UI}, θ_{UI}.

9.1.1 Derivation of the Pseudo-Five-DoF Equations

Now we are ready to proceed with the derivation. First, let us develop the pseudo-five-DoF equations for the round rotating Earth and then simplify them for the flat Earth. Newton's second law, Eq. (5.9), applied to a vehicle of mass m^B, with external aerodynamic and propulsive forces $f_{a,p}$, and gravitational force f_g yields

$$m^B D^I v_B^I = f_{a,p} + f_g \tag{9.3}$$

We shift to the velocity frame U using Euler's transformation

$$D^U v_B^I + \Omega^{UI} v_B^I = \frac{1}{m^B}(f_{a,p} + f_g)$$

and express the equation in inertial velocity coordinates

$$[D^U v_B^I]^U + [\Omega^{UI}]^U [v_B^I]^U = \frac{1}{m^B} ([f_{a,p}]^U + [f_g]^U) \tag{9.4}$$

The rotational time derivative is simply

$$[\overline{D^U v_B^I}]^U = \left[\frac{\mathrm{d}}{\mathrm{d}t} |v_B^I| \quad 0 \quad 0 \right] \equiv [\dot{U} \quad 0 \quad 0]$$

Because the aerodynamic and propulsive forces are usually modeled in body co-ordinates, they must be converted to velocity axes $[f_{a,p}]^U = [\bar{T}]^{BU} [f_{a,p}]^B$, as well as the gravity force, which is given in geographic coordinates $[f_g]^U = [T]^{UG} [f_g]^G$. Before we can program the equations, we have to determine the coordinate trans-formation matrices $[T]^{UI}$, $[T]^{UG}$, and $[T]^{BU}$.

9.1.2 Coordinate Transformation Matrices and Angular Rates

At this point I advise you to review Chapter 3. It will lubricate your understanding of the abbreviated derivations that follow. Besides the transformations, I will also deal with the angular velocity vectors ω^{BU} and ω^{UI} because they can be derived directly from our orange peel diagrams.

9.1.2.1 Transformation matrix of velocity wrt inertial coordinates.
The inertial coordinates are defined in Chapter 3. Figure 9.1 turns the world upside down so that the heading and flight-path angles take their conventional orientation, and we can readily switch later to flat-Earth approximation. However, ψ_{UI} and θ_{UI} are at this point not the usual heading and flight-path angles. They are referenced to the Earth-centered inertial (J2000) coordinate system for the sole purpose of for-mulating Newton's equations wrt the inertial velocity frame. We call them *inertial* heading and flight-path angles to distinguish them from the standard heading and flight-path angles, which we will derive later.

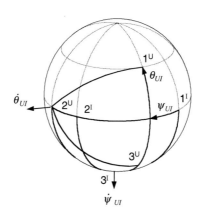

Fig. 9.1 $[T]^{UI}$ transformation.

The inertial coordinate system $]^I$ is associated with the inertial frame I. Its axes are defined as follows: 1^I is the direction of vernal equinox, and 3^I is the Earth rotation axis. The inertial velocity axes $]^U$ are associated with the inertial velocity frame and given by the following: 1^U is the direction of velocity vector, 2^U is in 1^I, 2^I plane; ψ_{UI} is the inertial heading angle; and θ_{UI} is the flight-path angle. The standard sequence of transformation is

$$]^U \xleftarrow{\;\theta_{UI}\;}] \xleftarrow{\;\psi_{UI}\;}]^I$$

It is similar to the transformation sequence in Sec. 3.2.2.6 of the flight-path coordinates wrt geographic coordinates. Only here we start with the inertial coordinates $]^I$ and end up with the velocity coordinates $]^U$. The transformation matrix is

$$[T]^{UI} = \begin{bmatrix} \cos\theta_{UI}\cos\psi_{UI} & \cos\theta_{UI}\sin\psi_{UI} & -\sin\theta_{UI} \\ -\sin\psi_{UI} & \cos\psi_{UI} & 0 \\ \sin\theta_{UI}\cos\psi_{UI} & \sin\theta_{UI}\sin\psi_{UI} & \cos\theta_{UI} \end{bmatrix} \quad (9.5)$$

Let us take the opportunity and derive the angular velocity of the velocity frame wrt the inertial frame. In Fig. 9.1 the angular rates of the inertial heading and flight-path angles are indicated. Combining them with their respective unit vectors and adding them vectorially yields

$$\boldsymbol{\omega}^{UI} = \dot{\psi}_{UI}\boldsymbol{i}_3 + \dot{\theta}_{UI}\boldsymbol{u}_2 \quad (9.6)$$

Later we will need their component form in the velocity coordinate system. So let us express the inertial unit vector in its preferred coordinate system $]^I$ and convert it to the $]^U$ coordinates

$$[\omega^{UI}]^U = \dot{\psi}_{UI}[T]^{UI}[i_3]^I + \dot{\theta}_{UI}[u_2]^U$$

and multiplied out with the help of Eq. (9.5)

$$[\omega^{UI}]^U = \begin{bmatrix} -\dot{\psi}_{UI}\sin\theta_{UI} \\ \dot{\theta}_{UI} \\ \dot{\psi}_{UI}\cos\theta_{UI} \end{bmatrix} \quad (9.7)$$

The angular velocity of the inertial velocity frame wrt the inertial frame is a function of angular rates and the flight-path angle but not the heading angle. Both the angular rates and the flight-path angle are obtained by solving the equations of motion.

We now turn to the incidence angle transformation matrices and their angular rates. As already discussed, we must distinguish between the skid-to-turn and the bank-to-turn cases for missiles and aircraft, respectively.

9.1.2.2 Skid-to-turn incidence angles and rates.

In the skid-to-turn case the angle of attack and sideslip angles determine the deviation of the velocity vector from the centerline of the vehicle. However, we must use the velocity vector of the vehicle relative to the air mass instead of the inertial frame because incidence angles are used in conjunction with the aerodynamics of the vehicle. We name

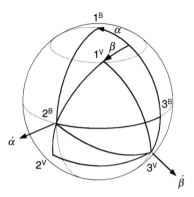

Fig. 9.2   Skid-to-turn $[T]^{BV}$ transformation.

this velocity frame V and the associated geographic velocity vector v_B^E, i.e., the velocity of the c.m. of the vehicle B wrt the Earth E.

The body axes $]^B$ are associated with the body frame B, and their positive direction is defined as follows: 1^B is the body centerline, 2^B the right wing, and 3^B points down.

The geographic velocity axes $]^V$ are associated with the geographic velocity frame V and given by 1^V as the velocity vector, 2^V in 1^G, 2^G horizontal plane.

The incidence angles are α as the angle of attack and β as the sideslip angle. Refer to Fig. 9.2 and compare it to Fig. 3.17 in Chapter 3 to confirm that the incidence angles are the same. The sequence of transformation is $]^B \xleftarrow{\alpha}]\xleftarrow{-\beta}]^V$. Notice the negative sense of the transformation of the sideslip angle. The transformation matrix between the body and geographic velocity coordinates is

$$[T]^{BV} = \begin{bmatrix} \cos\alpha\cos\beta & -\cos\alpha\sin\beta & -\sin\alpha \\ \sin\beta & \cos\beta & 0 \\ \sin\alpha\cos\beta & -\sin\alpha\sin\beta & \cos\alpha \end{bmatrix} \qquad (9.8)$$

It is the same for our pseudo-five-DoF treatment as for the full-up six-DoF simulations (see Chapter 10).

The angular velocity of the body frame wrt the geographic velocity frame is derived from Fig. 9.2. Combining the incidence rates with their respective unit vectors and adding them vectorially yields

$$\omega^{BV} = \dot{\beta}\boldsymbol{u}_3 + \dot{\alpha}\boldsymbol{b}_2 \qquad (9.9)$$

Expressed in body coordinates

$$[\omega^{BV}]^B = \dot{\beta}[T]^{BV}[u_3]^V + \dot{\alpha}[b_2]^B$$

and evaluated with the help of Eq. (9.8)

$$[\omega^{BV}]^B = \begin{bmatrix} \dot{\beta}\sin\alpha \\ \dot{\alpha} \\ -\dot{\beta}\cos\alpha \end{bmatrix} \qquad (9.10)$$

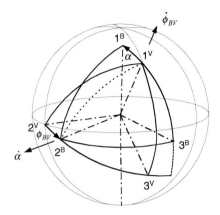

Fig. 9.3   Bank-to-turn $[T]^{BV}$ transformation.

The angular velocity of the body frame wrt the geographic velocity frame is a function of the incidence angular rates and the angle of attack, but not of the sideslip angle. Both the angular rates and the angle of attack are given by the transfer functions of the autopilot.

9.1.2.3   Bank-to-turn incidence angles and rates.

Now we use two different incidence angles. The included angle between the geographic velocity vector v_B^E and the first unit vector of the body b_1 is the total angle of attack α (we maintain the same symbol as in the skid-to-turn case). It is contained in the 1^B, 3^B vertical body plane, which is also called the *normal load factor plane*. The banking of this plane from the vertical plane 1^V, 3^V is designated by the bank angle ϕ_{BV} (see Fig. 9.3). Distinguish carefully between the Euler roll angle ϕ_{BG} (body axes, see Chapter 3), the aerodynamic roll angle ϕ' (aeroballistic axes, see Chapter 3), and our bank angle ϕ_{BV} (velocity axes).

The body axes $]^B$ and the geographic velocity axes $]^V$ are defined as before. However, the total angle of attack α lies now in the normal load factor plane, and the bank angle ϕ_{BV} is obtained by rotating about the velocity vector v_B^E, which is parallel to the base vector v_1. The sequence of rotation is

$$]^B \xleftarrow{\alpha}] \xleftarrow{\phi_{BV}}]^V$$

and the transformation matrix of the body coordinates wrt the geographic velocity coordinates is

$$[T]^{BV} = \begin{bmatrix} \cos \alpha & \sin \alpha \sin \phi_{BV} & -\sin \alpha \cos \phi_{BV} \\ 0 & \cos \phi_{BV} & \sin \phi_{BV} \\ \sin \alpha & -\cos \alpha \sin \phi_{BV} & \cos \alpha \cos \phi_{BV} \end{bmatrix} \qquad (9.11)$$

Here you can see the difference between the aeroballistic and bank angle treatments. Compare this transformation with the transformation matrix of the body wrt the aeroballistic wind coordinates $[T]^{BV}$, Eq. (3.19) of Chapter 3, and you will recognize the difference.

The angular velocity of the body frame wrt the geographic velocity frame is derived from Fig. 9.3. Combining the incidence rates with their respective unit vectors and adding them vectorially yields

$$\omega^{BV} = \dot{\phi}_{BV} \boldsymbol{v}_1 + \dot{\alpha} \boldsymbol{b}_2 \tag{9.12}$$

Expressed in body coordinates

$$[\omega^{BV}]^B = \dot{\phi}_{BV}[T]^{BV}[v_1]^V + \dot{\alpha}[b_2]^B$$

and evaluated with the help of Eq. (9.11),

$$[\omega^{BV}]^B = \begin{bmatrix} \dot{\phi}_{BV}\cos\alpha \\ \dot{\alpha} \\ \dot{\phi}_{BV}\sin\alpha \end{bmatrix} \tag{9.13}$$

The angular velocity of the body frame wrt the geographic velocity frame is a function of the incidence angular rates and the angle of attack (redefined) but not of the bank angle. Both the angular rates and the angle of attack are given by the transfer functions of the autopilot.

9.1.3   More Kinematics

Let us take stock of the progress we have made since laying out our requirements toward expressing Eq. (9.4) in component form. Equation (9.5) provides the transformation matrix $[T]^{UI}$, and Eqs. (9.8) and (9.11) deliver $[T]^{BV}$ for the skid-to-turn missile and bank-to-turn aircraft, respectively. We build $[T]^{UG}$ from

$$[T]^{UG} = [T]^{UI}[\bar{T}]^{GI} \tag{9.14}$$

with $[T]^{GI}$, the transformation matrix of the geographic wrt inertial coordinates, given by the longitude and latitude angles (see Chapter 3).

To come to grips with the $[T]^{BU}$ transformation matrix, we string it out

$$[T]^{BU} = [T]^{BV}[T]^{VG}[\bar{T}]^{UG} \tag{9.15}$$

The challenge is to calculate $[T]^{VG}$, the TM of the geographic velocity wrt geographic frame. This will take several steps. We first calculate the geographic velocity $\boldsymbol{v}_B^E$ from the definition of the inertial velocity $\boldsymbol{v}_B^I$ using the inertial position of the vehicle s_{BI} and the Euler transformation

$$\boldsymbol{v}_B^I \equiv D^I s_{BI} = D^E s_{BI} + \Omega^{EI} s_{BI} = \boldsymbol{v}_B^E + \Omega^{EI} s_{BI}$$

Solve for $\boldsymbol{v}_B^E$ and express it in geographic coordinates

$$[v_B^E]^G = [T]^{GI}\left([v_B^I]^I - [\Omega^{EI}]^I[s_{BI}]^I\right)$$

From $[v_B^E]^G$ we calculate the geographic heading and flight-path angles ψ_{VG} and θ_{VG}, recognizing the fact that they are the polar angles of the velocity vector in geographic axes (CADAC utility MATPOL). Finally, from these angles we obtain $[T]^{VG}$ (CADAC utility MAT2TR). We are now able to calculate Eq. (9.15).

In a moment we also will need

$$[T]^{BI} = [T]^{BU}[T]^{UI} \tag{9.16}$$

which we construct from Eqs. (9.15) and (9.5).

Back to Eq. (9.4), the angular velocity $[\Omega^{UI}]^U$ or its vector counterpart $[\omega^{UI}]^U$ is given by Eq. (9.7). Furthermore, we also need the body rates $[\omega^{BI}]^B$ for various modeling tasks. We build them up from

$$[\omega^{BI}]^B = [\omega^{BV}]^B + [\omega^{VU}]^B + [\omega^{UI}]^B \tag{9.17}$$

$[\omega^{BU}]^B$ is given by Eq. (9.10) for skid-to-turn missiles and by Eq. (9.13) for bank-to-turn aircraft. The second term $[\omega^{VU}]^B$ is the angular velocity vector of the geographic velocity frame wrt the inertial velocity frame. These two frames differ by the Earth's angular velocity ω^{EI}. Expressed in inertial coordinates

$$[\omega^{VU}]^I = [\omega^{EI}]^I = \begin{bmatrix} 0 \\ 0 \\ \omega^{EI} \end{bmatrix}$$

Using Eqs. (9.16), (9.6), and (9.15), we can calculate the body rates

$$[\omega^{BI}]^B = [\omega^{BV}]^B + [T]^{BI}[\omega^{EI}]^I + [T]^{BU}[\omega^{UI}]^U \tag{9.18}$$

Rejoice! Our kinematic construction set is complete, and we can turn to the more profitable task of formulating the equations of motions.

9.1.4 Equations of Motion over Round Rotating Earth

Returning to Eq. (9.4), we multiply out the matrices on the left-hand side and express on the right side the aerodynamic/propulsive forces in body coordinates and the gravitational force in geographic coordinates:

$$\begin{bmatrix} \dot{U} \\ \dot{\psi}_{UI} U \cos\theta_{UI} \\ -\dot{\theta}_{UI} U \end{bmatrix} = [\bar{T}]^{BU} \frac{1}{m^B}[f_{a,p}]^B + [T]^{UG} \frac{1}{m^B}[f_g]^G \tag{9.19}$$

$\dot{U}$ is an abbreviation for $d/dt|v_B^I|$, $[T]^{BU}$ is given by Eq. (9.15), and $[T]^{UG}$ by Eq. (9.14). These are three first-order nonlinear differential equations with the states U, ψ_{UI}, θ_{UI}. Solving for the state derivatives, we discover that in the second equation $U \cos\theta_{UI}$ appears in the denominator. Therefore, these equations cannot be solved if $U = 0$ or $\theta_{UI} = \pm 90$ deg, which we avoid by programming around it. The designation *specific force* $[f_{\mathrm{sp}}]^B$ is assigned to the term $[f_{a,p}]^B/m^B$, although it has the units of acceleration. (Remember: accelerometers measure specific force.) The gravitational term is simply

$$\frac{1}{m^B}[f_g]^G = [g]^G = \begin{bmatrix} 0 \\ 0 \\ g \end{bmatrix}$$

We have succeeded in expressing the equations of motion in matrix form. They are now ready for programming.

9.1.5 Equations of Motion over Flat Earth

As you may have suspected all along, Earth is flat, at least for many engineers who develop simulations for aircraft and short-range missiles. They make Earth the inertial frame and unwrap the curved longitude and latitude grid into a local plane tangential to Earth near liftoff. What a helpful assumption! It eliminates several coordinate transformations, simplifies the calculation of the body rates, and eliminates the distinction between inertial and geographic velocity.

If in Chapter 5, Eq. (5.25), the terms of Coriolis and transport acceleration are neglected, we obtain a form of Newton's second law that assumes the Earth frame E is the inertial frame. We replace in Eq. (9.4) the inertial reference frame I by E and reinterpret the inertial velocity frame U as the geographic velocity frame V. The flat-Earth equations of motions are then

$$\left[D^V v_B^E\right]^V + [\Omega^{VE}]^V [v_B^E]^V = \frac{1}{m^B}\left([f_{a,p}]^V + [f_g]^V\right) \qquad (9.20)$$

with $[\overline{v_B^E}]^V = [V \ 0 \ 0]$, the geographic velocity of the c.m. wrt Earth expressed in geographic velocity coordinates, and $[\Omega^{VE}]^V$ the skew-symmetric equivalent of the angular velocity $[\omega^{VE}]^V$ of the geographic velocity frame wrt Earth. The rotational time derivative is simply

$$\left[\overline{D^V v_B^E}\right]^V = \left[\frac{d}{dt}|v_B^E| \quad 0 \quad 0\right] \equiv [\dot{V} \quad 0 \quad 0]$$

On the right-hand side of (9.20), we must convert $[f_{a,p}]^V = [\bar{T}]^{BV}[f_{a,p}]^B$ because the aeropropulsive forces are usually given in body coordinates. The gravity force is expressed best in local geographic coordinates, which for the round-Earth case were designated by $]^G$. With the flat-Earth assumption they are renamed *local-level* coordinates with the label $]^L$. Finally, we calculate the gravity force from $[f_g]^V = [T]^{VL}[f_g]^L$ with $[\bar{f}_g]^L = m^B[0 \ 0 \ g]$ and g the gravity acceleration.

Before Eq. (9.20) can be programmed, we have to convert the kinematic relationships to the flat-Earth case. We need the transformations $[T]^{BV}$, $[T]^{VL}$, and $[T]^{BL}$ the angular velocity vectors $[\omega^{VE}]$, $[\omega^{BV}]$, and $[\omega^{BE}]$.

The transformation matrix $[T]^{VL}$ of the velocity wrt the local-level coordinates derives directly from Eq. (9.5). Just replace the angles ψ_{UI}, θ_{UI} by the geographic heading angle ψ_{VL} and flight-path angle θ_{VL} both referenced to the local-level coordinate axes. Alas, we are back on familiar ground. The obscure ψ_{UI}, θ_{UI} angles have become the tried and true heading and flight-path angles on the Earth, which we called in Chapter 3 χ and γ, but prefer to designate here ψ_{VL} and θ_{VL}. Figure 9.1 still applies with these changes, and Eq. (9.5) becomes

$$[T]^{VL} = \begin{bmatrix} \cos\theta_{VL}\cos\psi_{VL} & \cos\theta_{VL}\sin\psi_{VL} & -\sin\theta_{VL} \\ -\sin\psi_{VL} & \cos\psi_{VL} & 0 \\ \sin\theta_{VL}\cos\psi_{VL} & \sin\theta_{VL}\sin\psi_{VL} & \cos\theta_{VL} \end{bmatrix} \qquad (9.21)$$

The incidence angle transformation matrix $[T]^{BV}$ is retained for both the skid-to-turn and bank-to-turn vehicles. They are given by Eqs. (9.8) and (9.11), respectively. The direction cosine matrix $[T]^{BL}$ of body wrt local-level coordinates is the composition of

$$[T]^{BL} = [T]^{BV}[T]^{VL} \qquad (9.22)$$

Now we convert the angular velocity vectors to the flat-Earth case. The angular velocity $[\omega^{VE}]^V$ of the velocity frame wrt the Earth frame and expressed in the velocity coordinates is obtained from Eq. (9.7) by redefining the inertial frame I to become the Earth frame E and the angles ψ_{UI}, θ_{UI} to change to ψ_{VL}, θ_{VL}:

$$[\omega^{VE}]^V = \begin{bmatrix} -\dot{\psi}_{VL} \sin\theta_{VL} \\ \dot{\theta}_{VL} \\ \dot{\psi}_{VL} \cos\theta_{VL} \end{bmatrix} \qquad (9.23)$$

The incidence angular rates are transcribed from Eq. (9.10) for skid-to-turn missiles

$$[\omega^{BV}]^B = \begin{bmatrix} \dot{\beta} \sin\alpha \\ \dot{\alpha} \\ -\dot{\beta} \cos\alpha \end{bmatrix} \qquad (9.24)$$

and for bank-to-turn aircraft from Eq. (9.13)

$$[\omega^{BV}]^B = \begin{bmatrix} \dot{\phi}_{BV} \cos\alpha \\ \dot{\alpha} \\ \dot{\phi}_{BV} \sin\alpha \end{bmatrix} \qquad (9.25)$$

$\dot{\phi}_{BV}$ is the bank angle of the normal load factor plane rotated about the $[v_B^E]$ vector from the vertical plane. Finally, the body rates $[\omega^{BE}]$ are the vectorial addition

$$\omega^{BE} = \omega^{BV} + \omega^{VE}$$

in body axes

$$[\omega^{BE}]^B = [\omega^{BV}]^B + [T]^{BV}[\omega^{VE}]^V \qquad (9.26)$$

Notice how much simpler the calculation of the body rates is with the flat-Earth assumption than for the round rotating Earth, Eq. (9.17). The simplification occurred because geographic and inertial velocities are undistinguishable. The kinematic conversions are now complete!

For programming, the left-hand side of the dynamic equations, Eq. (9.20), must be expressed in component form. We go back to the round-Earth component equations, Eq. (9.19), and modify them for the flat-Earth case:

$$\begin{bmatrix} \dot{V} \\ \dot{\psi}_{VL} V \cos\theta_{VL} \\ -\dot{\theta}_{VL} V \end{bmatrix} = [\bar{T}]^{BV} \frac{1}{m^B} [f_{a,p}]^B + [T]^{VL} \frac{1}{m^B} [f_g]^L \qquad (9.27)$$

$\dot{V}$ is the abbreviation for $d/dt|v_B^E|$, $[T]^{BV}$ is given by Eqs. (9.8) or (9.11), and $[T]^{VL}$ by Eq. (9.21). The state variables of the linear differential equations are $V, \psi_{VL}, \theta_{VL}$. For the flat-Earth case the second and third equations become singular if $V \cos\theta_{VL}$ or V are zero. Therefore, we cannot simulate a missile that takes off vertically, an aircraft that dives straight to the ground, or, for that matter, a hovering helicopter. Fudging the initial conditions can help us get started, and once underway we program around the singular values until the equations are again well behaved. The errors incurred may be tolerable, as experience has shown.

Let us expand the right-hand side of Eq. (9.27). The aerodynamic and propulsive term can be expressed in velocity coordinates directly. Lift and drag are referred to these coordinates, and thrust, usually parallel to the 1^B axis, is projected by the angle of attack into the 1^V axis. With these conventions we can adopt the three-DoF aerodynamic model of Sec. 8.2.3 to five-DoF simulations. From Fig. 8.14, A3 Forces, we borrow the formulation of the specific force:

$$[f_{sp}]^V = \frac{1}{m^B} \begin{bmatrix} F\cos\alpha - \bar{q}SC_D \\ \sin\phi_{BV}(F\sin\alpha + \bar{q}SC_L) \\ -\cos\phi_{BV}(F\sin\alpha + \bar{q}SC_L) \end{bmatrix}$$

where F is the thrust, $\bar{q}$ the dynamic pressure, and S the aerodynamic reference area. In five-DoF simulations the lift and drag coefficients may now be more complicated functions of Mach and α than the simple parabolic flight polars of Chapter 8. More will be said about this subject in Sec. 9.2.1.

The gravity term of Eq. (9.27) is multiplied by the transformation matrix of Eq. (9.21). Combining the right-hand components with the left-hand side of Eq. (9.27) yields

$$\begin{bmatrix} \dot{V} \\ \dot{\psi}_{VL} V \cos\theta_{VL} \\ -\dot{\theta}_{VL} V \end{bmatrix} = \frac{1}{m^B} \begin{bmatrix} F\cos\alpha - \bar{q}SC_D \\ \sin\phi_{BV}(F\sin\alpha + \bar{q}SC_L) \\ -\cos\phi_{BV}(F\sin\alpha + \bar{q}SC_L) \end{bmatrix} + g \begin{bmatrix} -\sin\theta_{VL} \\ 0 \\ \cos\theta_{VL} \end{bmatrix}$$

$$(9.28)$$

Before you program these equations, clear the left-hand side of anything but the state variable derivatives $\dot{V}$, $\dot{\psi}_{VL}$, and $\dot{\theta}_{VL}$. You have three first-order differential equations with angle of attack α and bank angle ϕ_{BV} as input. How this input is generated is the subject of the following sections.

9.2 Subsystem Models

For a missile or aircraft to fly effectively, many components must work together harmoniously. Just to name the most important ones: airframe, propulsion, controls, autopilot, sensors, guidance, navigation, and, in the case of a manned aircraft, the pilot. As we build a simulation, these subsystems must be modeled mathematically or included as hardware. In a flight simulator the pilot has the privilege to represent himself.

The shape of the airframe determines the aerodynamic forces and moments, and its structure determines the mass properties and deflections under loads. For our simplified pseudo-five-DoF approach we assume that the airframe is rigid. As mentioned earlier, the moments are balanced, and the drag as a result of steady-state control deflections is included in the aerodynamic forces. We model the aerodynamics using the so-called *trimmed* force approach.

Propulsion can be delivered by a simple rocket, a turbojet, a ramjet, or some other exotic device. It provides the force that overcomes drag and gravity and accelerates the vehicle. We simplify its features by employing tables for thrust and specific fuel consumption and introduce first-order lags for delays in the system.

In pseudo-five-DoF simulations, because of the treatment of the aerodynamics, controls are not modeled explicitly. What a simplification! Actuators need not be included, and you can forget about hinge moments. However, we give up the opportunity to study the dynamics of the controls and the effect of saturation of the control rates and deflections.

The stability and controllability of the vehicle are governed by the autopilot. The outer-loop feedback variable categorizes the type of autopilot. We distinguish between rate, acceleration, altitude, bank, flight path, heading, and incidence angle hold autopilots. Make sure, however, that for any type of autopilot, the incidence angle and its rate are computed and provided for the kinematic calculations (see Sec. 9.1). The autopilots are simplified models of the control-loop dynamics, and their responses simulate the vehicle's attitude dynamics.

Sensors measure the states of the vehicle wrt other frames, like inertial, Earth, body, or target frames. They may be part of the air data system, inertial navigation system, autopilot, propulsion, landing, or targeting systems. Those with high bandwidth, like gyros and accelerometers, are modeled by gains without dynamics, and their output is corrupted by noise. Gimbaled homing seekers, on the other hand, exhibit transients near the autopilot bandwidth and should, therefore, be modeled dynamically, although for simple applications we also use kinematic seekers.

The smarts of a missile reside in its guidance laws. Given the state of the missile relative to the target, it sends the steering commands to the autopilot for intercept. We distinguish between pursuit, proportional, line, parabolic, and arc guidance. Guidance occupies the outermost control loop of the vehicle. In pseudo-five-DoF simulations, with the autopilot providing the vehicle response, this loop may be modeled with sufficient detail to be representative of six-DoF performance. Experience has shown that you could design the guidance loop initially in five-DoF and later include it in your full six-DoF simulation without modifications.

Finally, the navigation subsystem furnishes the vehicle with its position and velocity relative to the inertial or geographic frames. The core is the IMU with its accelerometers and gyros. Once their measurements are converted into navigation information, it becomes the INS. Errors in the measurements and computations corrupt the navigation solution. Therefore, navigation aids are employed to update the INS. Loran, Tacan, GPS or just overflight of a landmark can provide the external stimuli. Uncertainties are modeled by the INS error equations and by the noisy updates and filter dynamics.

9.2.1 Trimmed Aerodynamics

Aerodynamics simulates the forces and moments that shape the flight trajectory. To model these effects, the designer can resort to many references, computer prediction codes, and wind-tunnel data. A two-volume set of missile aerodynamics,[1] updated in 1992, is a compendium of experimental and theoretical results, quite suitable for aerodynamic analysis. Semi-empirical computer codes, like Missile DATCOM,[2] can make your life much easier and generate aerodynamic tables quickly, but at the expense of insight into the physical underpinning of the data. If you venture into the hypersonic flight regime, the industry standard is the Supersonic Hypersonic Arbitrary Body Program (S/HABP)[3] for missiles and reentry vehicles. For aircraft, the old faithful DATCOM is still available[4] and made more palatable by Roskam.[5] Of recent vintage are two books by Stevens and Lewis[6] and Pamadi[7] that treat aerodynamics as part of the control problem. Finally, let us not forget the venerable book by Etkin[8] that served two generations of engineers.

In missile and aircraft simulations the emphasis is more on performance rather than on stability and control. The autopilot, controlling the vehicle, is already designed before building the simulation and hopefully performs well throughout the flight regime. Therefore, the focus is on tabular modeling of the forces and moments and not on stability derivatives. Angle of attack and Mach number are the primary independent variables, sometimes supplemented by sideslip angle and altitude dependency (skin-friction effects).

Pseudo-five-DoF simulations are content with simple aerodynamic representations. Because we assume that the moments are always balanced and that the trim drag of the control surfaces is included in the overall drag table, we need only two tables: normal and axial forces, or alternatively, lift and drag forces. If power on/off influences the drag, we have to double up the drag table, and, if the c.m. shifts significantly during the flight, we have to interpolate between changing trim conditions.

We shall proceed from general aerodynamic principles. Aerodynamic forces and moments are, in general, dependent on the following parameters:

aero forces and moments

$$
= f \left\{ \underbrace{M, \text{Re}}_{\substack{\text{flow} \\ \text{characteristics}}}, \quad \underbrace{\alpha, \beta, \dot{\alpha}, \dot{\beta}}_{\substack{\text{incidence angles} \\ \text{and rates}}}, \quad \underbrace{p, q, r}_{\text{body rates}}, \quad \underbrace{\delta p, \delta q, \delta r}_{\substack{\text{control surface} \\ \text{deflections}}}, \quad \text{shape, scale, power} \right\}
$$

where the Mach number is velocity/sonic speed and the Reynolds number is inertia forces/frictional forces.

The forces and moments are nondimensionalized by the parameters $\bar{q}$ (dynamic pressure), S (reference area), and l (reference length). The resulting coefficients are independent of the scale of the vehicle. If a missile flies a steady course, exhibiting only small perturbations, the dependence on the unsteady parameters $\dot{\alpha}$, $\dot{\beta}$, p, q, r may be neglected. For the trimmed approximation the moments are balanced, and their net effect is zero. Thus, only the lift and drag coefficients remain nonzero. With the effects of the trimmed control surface deflections implicitly included, the lift and drag coefficients are the following.

Lift coefficient:

$$
C_L = \frac{L}{\bar{q}S}
$$

Drag coefficient:

$$
C_D = \frac{D}{\bar{q}S}
$$

where L and D are the lift and drag forces, respectively. Their dependencies are reduced to

$$
C_L \text{ and } C_D = f\{M, \text{Re}, \alpha, \beta, \text{ shape, power on/off}\}
$$

The Reynolds number primarily expresses the dependency of the size of the vehicle and skin friction as a function of altitude. With size and shape of a particular vehicle fixed and altitude dependency neglected, the coefficients simplify further:

$$C_L \text{ or } C_D = f\{M, \alpha, \beta, \text{ power on/off}\}$$

Now let us treat skid-to-turn missiles and bank-to turn aircraft separately. By the way, I am using the term *missile* and *aircraft* somewhat loosely. A short-range air-to-air missile most likely will have tetragonal symmetry (configuration replicates every 90-deg rotation) and execute skid-to-turn maneuvers; but a cruise missile or a hypersonic vehicle, with planar symmetry, behaves like a bank-to-turn aircraft.

9.2.1.1 Tetragonal missiles.
A tetragonal missile's aerodynamics is only weakly dependent on the roll orientation of the body. For simple pseudo-five-DoF simulations, we neglected that effect altogether and are left only with the total incidence angle α':

$$C_L \text{ or } C_D = f\{M, \alpha', \text{ power on/off}\}$$

If the missile executes skid-to-turn maneuvers, the autopilot provides α and β information, which is converted to aeroballistic incidence angles by Eq. (3.24):

$$\alpha' = \arccos\{\cos \alpha \cos \beta\}, \quad \phi' = \arctan\left\{\frac{\tan \beta}{\sin \alpha}\right\} \qquad (9.29)$$

The lift vector is normal to the velocity vector in the load factor plane and is a function of α'. Many missile simulations require the forces to be expressed in body coordinates. We make the conversion in two steps. First, we transform lift and drag to normal force and axial force coefficients in aeroballistic wind coordinates through the angle α'

$$
\begin{aligned}
C_{A'} &= -C_L \sin \alpha' + C_D \cos \alpha' \\
C_{N'} &= C_L \cos \alpha' + C_D \sin \alpha'
\end{aligned}
\qquad (9.30)
$$

followed by the rotation through the angle ϕ' to body fixed axes

$$
\begin{aligned}
C_A &= C_{A'} \\
C_Y &= -C_{N'} \sin \phi' \\
C_N &= C_{N'} \cos \phi'
\end{aligned}
\qquad (9.31)
$$

Let us pause and point out the difference between the aeroballistic ϕ' of Fig. 9.4 and the bank angle ϕ_{BV} of Fig. 9.3. Both are transformation angles about a 1 axis, but in the case of ϕ' it is the body axis and for ϕ_{BV} it is the velocity vector.

The aerodynamic force vector for Newton's equations is now in body axes

$$[f_a]^B = \bar{q} S \begin{bmatrix} -C_A \\ C_Y \\ -C_N \end{bmatrix} \qquad (9.32)$$

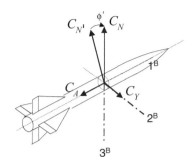

Fig. 9.4   Transformation of aeroballistic wrt body axes.

Notice the negative directions of C_A and C_N relative to the positive directions of the body axes. This convention is universally used for missiles and has its origin in the definition of positive lift and drag.

The aerodynamic tables of a typical air-to-air missile are a function of Mach, total angle of attack α', and power on/off. For the sample simulation CADAC SRAAM5, I also included skin-friction drag corrections caused by altitude changes. To improve the realism even more and recognizing the large change in mass properties during fly-out, I included the tables for three c.m. positions: fore, middle, and aft. They represent the changing trim drag caused by control deflections and are interpolated during thrusting.

In some applications you may be given the aerodynamic tables directly in normal and axial force coefficients. If that is the case, you just bypass Eq. (9.30) and continue by converting the coefficients to body axes with the transformation Eq. (9.31).

We have defined the load factor plane as the plane containing the lift vector of a tetragonal skid-to-turn missile. Turning to planar aircraft, including cruise missiles, this arrangement is particularly advantageous.

9.2.1.2 Planar aircraft.   In five-DoF simulations an aircraft, executing bank-to-turn maneuvers, is assumed to do so at zero sideslip angle. The lift and drag

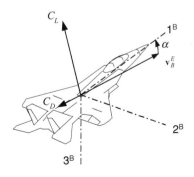

Fig. 9.5   Lift and drag coefficients.

vectors lie in the load factor plane (see Fig. 9.5), and their aerodynamic tables are, therefore, only a function of the incidence angle α.

$$C_L \text{ or } C_D = f\{M, \alpha, \text{ power on/off}\}$$

A bank-to-turn autopilot provides this incidence angle α together with the bank angle ϕ_{BV}. Because the symmetry plane coincides with the load factor plane, the transformation of the forces to body axes is like in Eq. (9.30) with the angle of attack α assuming the role of the total angle of attack α'

$$C_A = -C_L \sin \alpha + C_D \cos \alpha$$
$$C_N = C_L \cos \alpha + C_D \sin \alpha \tag{9.33}$$

When you read the newer literature,[5] you will find the aerodynamic coefficient defined in the positive direction of the body axes, C_X and C_Z. They are in the opposite direction of C_A and C_N and are obtained from lift and drag coefficients by

$$C_X = C_L \sin \alpha - C_D \cos \alpha$$
$$C_Z = -C_L \cos \alpha - C_D \sin \alpha \tag{9.34}$$

In either case, the aerodynamic force vector for Newton's equation is

$$[f_a]^B = \bar{q}S \begin{bmatrix} -C_A \\ 0 \\ -C_N \end{bmatrix} = \bar{q}S \begin{bmatrix} C_X \\ 0 \\ C_Z \end{bmatrix} \tag{9.35}$$

The trimmed aerodynamic tables of aircraft are usually built as functions of Mach and alpha. Several sets may be required for power on/off, or different configurations, like flaps in/out or gear in/out. Sometimes skin-friction corrections with altitude are also included.

The generic cruise missile simulation CADAC CRUISE5 can serve as an example. Its C_L and C_D tables are given as functions of Mach and angle of attack for three c.m. locations. With the turbojet engine providing continued thrust, we model only power-on drag. Let us now turn to the propulsive forces.

9.2.2 Propulsion

Most of our needs for modeling thrust forces have already been covered in Sec. 8.2.4. The equations for rockets and combined-cycle airbreathing engines apply here as well. I will only expand on turbojet propulsion because it plays such an important role in cruise missiles and aircraft.

Review the section on turbojet propulsion in Chapter 8. The thrust formula

$$F = \dot{m}_a(V_e - V) \tag{9.36}$$

(with V as the flight velocity, V_e the exhaust velocity, and $\dot{m}_a$ the airflow rate) is commonly replaced by tables for thrust, fuel flow, and dynamic transients. We will work through an example that is used for cruise missiles and aircraft.

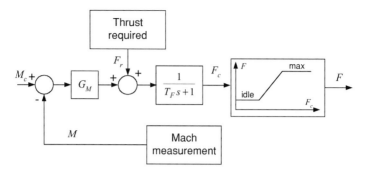

Fig. 9.6 Mach hold control loop.

Cruise missiles have to maintain Mach number under maneuvers and environmental effects. Particularly challenging are the terrain-following and obstacle-avoidance flight patterns. We will model a Mach-number hold system suitable for cruise missiles and aircraft under such conditions.

The thrust required F_r to maintain a certain Mach number is equal to the drag force projected onto the centerline of the turbine. If the turbine axis is parallel to the body 1 axis, we require that

$$F_r = \frac{\bar{q}SC_D}{\cos \alpha} \tag{9.37}$$

This value is used in the Mach hold control loop of Fig. 9.6. The commanded Mach number M_c is compared with the measured value M, and the difference is sent through a gain G_M that changes units to Newtons. The demanded thrust F_c is realized by the turbojet after spool-up or spool-down delays, characterized by the first-order lag time constant T_F. However, it may exceed the maximum possible thrust or drop below the idle thrust. Limiting tables restrict this excess demand. They are, in general, functions of Mach number and altitude. In simulations, the achieved thrust F is added to the right-hand side of Newton's dynamic equations.

The time constant T_F and the gain G_M are possibly a function of power setting. As an example, the spool-up time constant for the Falcon turbojet engine F100-PW-200 is between $T_F = 0.2 \rightarrow 1.0$ s. The gain can be calculated from a simple transfer function. We complete the control loop of Fig. 9.6 by the vehicle transfer functions, represented by the vehicle mass m^B, an integrator, and the conversion to Mach number by the sonic speed V_s (see Fig. 9.7). The closed-loop transfer function is of second degree, characterized by the natural frequency ω_n and damping ζ. We

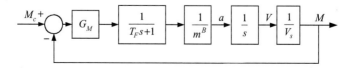

Fig. 9.7 Mach control loop.

can eliminate ω_n and solve for the gain

$$G_M = \frac{m^B V_s}{4 T_F \zeta^2}$$

With $T_F = 0.2$ s given and $\zeta = 0.707$ selected, the value for the Falcon is $G_M = 3.4 \times 10^6$.

Once the thrust F of the turbojet is given, we only have to determine the mass of the vehicle from the fuel consumed and the initial gross weight. The specific fuel consumption b_F serves this purpose. It is usually given in tabular form as a function of Mach number and altitude. Multiplied by the thrust, it provides the fuel flow, which, integrated over time, supplies the expended fuel.

You can find the details of the implementation in the five-DoF cruise missile simulation CADAC CRUISE5 and the six-DoF simulation CADAC FALCON6. The engine decks are provided in English units, just to keep your unit conversion skills sharp.

9.2.3 Autopilot

When the Wright brothers flew their first contraption, they controlled the airframe with stick and rudder. Manually they counteracted wind gusts and steered the plane on its course. By the end of World War II, the flying machines became so sophisticated that the pilot needed some help from electronic instruments. For simple flight conditions, like steady cruise over long distances, the electronics would take over completely, and with hands off, the automatic pilot would control the aircraft. Today, autopilots are found in every aircraft. Missiles, lacking the human touch, cannot fly without them. We are even now conceptualizing combat aircraft that relegate the pilot to a ground controller.

There are many good references on control theory in general and autopilots in particular. A good introduction to classical and modern control is the textbook by Dorf and Bishop,[9] which uses the popular MATLAB[10] software package. Pamadi[7] treats flight controllers from a classical viewpoint, whereas Steven and Lewis[6] approach them from a modern angle. One of my favorite books on advanced topics is by Stengel[11] addressing the stochastic effects of control. If you are fluent in German, you should consider the standard text by Brockhaus.[12]

Autopilots stabilize airframes, improve control response, convert guidance signals to actuator commands, and maintain constant flight parameters. They compare the commanded inputs with the measured states and shape the error signal for execution by the actuators. Figure 9.8 shows the position of the autopilot for a piloted

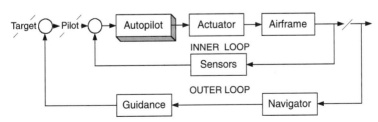

Fig. 9.8   Autopilot for aircraft (inner loop) and guidance loop for missiles.

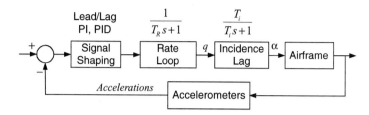

Fig. 9.9 Pseudo-five-DoF pitch plane acceleration controller for missiles (similar for yaw plane).

aircraft (inner loop only) and for a missile (inner loop with guiding outer loop). For an aircraft the pilot sets the course; for a missile the goal is to reach a certain target state, be it for intercept, rendezvous, or specific end conditions.

The signal of the feedback loop determines the type of autopilot. A flight-path controller operates on flight-path-angle measurements. In the horizontal plane it is also referred to as heading autopilot. To hold altitude, height measurements are used in the altitude controller. A bank-angle controller executes constant turns. For missiles the acceleration controllers are particularly important because the commands from the guidance system are expressed in body accelerations.

9.2.3.1 Acceleration controller.
Let us start with the heart of a missile autopilot. It consists of body rate and acceleration feedback with lead/lag shaping filters as depicted in Fig. 9.9. For five-DoF modeling we simply represent the rate feedback loop by a first-order transfer function with the time constant T_R, which is representative of the full rate autopilot response. If the flight conditions are changing dramatically, you should consider making the value of T_R a function of dynamic pressure.

How can we generate the angle of attack from the output of the rate loop? We make use of the incidence-lag relationship. Consider Fig. 9.10 with its lift force L and thrust F, and apply Newton's second law in the direction normal to the velocity vector v_B^E:

$$mV\dot{\gamma} = F\alpha + \bar{q}SC_{L_\alpha}\alpha \tag{9.38}$$

where we approximated the lift coefficient by its slope $C_L = C_{L_\alpha}\alpha$ and assumed

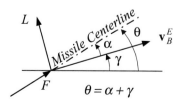

Fig. 9.10 Incidence angle lag.

small incidence angles. Taking the time derivative of the angular relationship

$$\dot{\alpha} = \dot{\theta} - \dot{\gamma} = q - \dot{\gamma} \tag{9.39}$$

and substituting it into Eq. (9.38) yields the incidence-lag differential equation

$$\dot{\alpha} = q - \frac{F + \bar{q}SC_{L_\alpha}}{mV}\alpha \tag{9.40}$$

with the time constant

$$T_i = \frac{mV}{F + \bar{q}SC_{L_\alpha}} \tag{9.41}$$

and the Laplace transfer function of angle of attack wrt pitch-rate response

$$\frac{\alpha(s)}{q(s)} = \frac{T_i}{T_i s + 1} \tag{9.42}$$

The value of the incidence-lag time constant T_i decreases with increasing lift and thrust, reflecting the improved responsiveness of the airframe. For accurate modeling the lift slope coefficient C_{L_α} should be made a function of Mach number and possibly of angle of attack.

The airframe block in Fig. 9.9 represents Newton's equations with the aerodynamic and thrust tables providing the specific forces. Accelerometers, nowadays located in the IMU, measure the accelerations. To keep the simulation simple, higher-order sensor dynamics are neglected.

As an example, I use the acceleration feedback autopilot of the CADAC SRAAM5 simulation. Its position in the logic flow is shown in Fig. 9.11. The guidance module sends the pitch and yaw acceleration commands a_{Nc} and a_{Lc}, respectively, to the autopilot, and the measured acceleration $[\hat{f}_{sp}]^B$ comes from the INS. Rocket thrust F is needed for the T_i calculation. The output, incidence angles α and β, is transmitted to the aerodynamic tables.

The block diagrams, Figs. 9.12 and 9.13, show in greater detail the actual implementation of the pitch and yaw loops. They are very similar, but watch out for the signs (they have been the nemesis of many student projects). We are following the sign conventions of missile aerodynamics. Normal acceleration is positive up and lateral acceleration positive to the right. Pitch rate q, positive up, produces positive α, but yaw rate r, positive to the right, generates negative β.

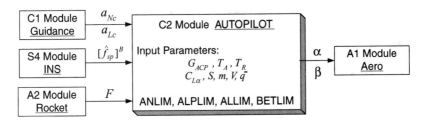

Fig. 9.11   Acceleration autopilot of the SRAAM5 five-DoF simulation.

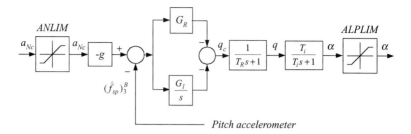

Fig. 9.12   Pitch acceleration loop.

The acceleration command, by convention in gs, is converted—after limiting—to the units meters/seconds squared. The error signal is fed through a proportional gain G_R and, for tracking accuracy, through an integrator with gain G_I. After the rate-loop and the incidence-lag transfer functions, the incidence angle is limited before being sent to the aerodynamic tables.

This autopilot must control the vehicle throughout its expansive flight envelope. For air-to-air missiles the excursions are from subsonic launch to triple the sonic speed at motor burnout; incidence angles may reach 50 deg; and the dynamic pressure can change by a factor of 20. Gain scheduling of G_R and G_I provides this flexibility, given the representative time constant T_R of the body-rate feedback loop.

The rate loop transfer function is based on the simplified moment equation about the c.m.

$$I\dot{q} = \bar{q}SlC_{m_\delta}\delta \tag{9.43}$$

with I the moment of inertia, l the moment reference length, δ the control fin deflection, and C_{m_δ} the control moment derivative. Figure 9.14 depicts the rate feedback loop for the pitch plane. You should be able to produce the equivalent yaw loop. G_A is the rate loop gain that converts the error signal into a control deflection, followed by the control limiter, the control effectiveness term, and the inertial integrator. The two negative signs are introduced to abide by the aerodynamic convention that a positive control deflection generates a negative pitch rate, but because they cancel, they are of no consequence.

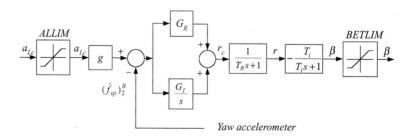

Fig. 9.13   Yaw acceleration loop.

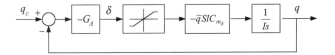

Fig. 9.14   Rate feedback loop for pitch control.

The closed-loop transfer function is of first order with the gain of one and the time constant

$$TR = \frac{I}{G_A \bar{q} S I C_{m\delta}} \tag{9.44}$$

Note that the time constant is inversely proportional to the dynamic pressure and the open loop gain G_A. Its values should be taken from response curves of six-DoF simulations and expressed as a function of $\bar{q}$. The CADAC SRAAM5 simulation uses the simple linear relationship $T_R = 0.22 - 2 \times 10^{-7}\bar{q}$ based on the six-DoF CADAC SRAAM6 data.

I recommend that you use the following root locus technique to calculate the G_R and G_I gain scheduling. But first, we have to complete the acceleration feedback loop of Fig. 9.9 by providing the transfer function for the airframe.

Refer back to Eq. (9.38) and recognize that the acceleration normal to the velocity vector is $a = V\dot{\gamma}$. Then, with the definition of the incidence-lag time constant T_I of Eq. (9.41)

$$a = \frac{T + \bar{q} S C_{L_\alpha}}{m} \alpha = \frac{V}{T_i} \alpha \tag{9.45}$$

and the complete acceleration loop is shown in Fig. 9.15. The open-loop transfer function for the root locus procedure is

$$\frac{a(s)}{\varepsilon(s)} = G_{ACP} \frac{(s + 1/T_A)}{s(s + 1/T_R)(s + 1/T_i)} \tag{9.46}$$

with the root locus gain $G_{ACP} = G_R V/(T_R T_i)$ and lead time constant $T_A = G_R/G_I$. The root locus emanates from the three poles and terminates at the zero and at infinity along two vertical asymptotes.

I picked the gain $G_{ACP} = 12.2$ with the closed-loop roots $-2.7525 + 2.7689i$, $-2.7525 - 2.7689i$, -0.6670 for best performance resulting in $G_R = 0.0055$ and $G_I = 0.0046$. This type of analysis, applied throughout the flight envelope, gave me the gain schedule $G_{ACP} = (0.002\bar{q})^{0.575}$ and constant $T_A = 1.2$ (see CADAC SRAAM5 simulation).

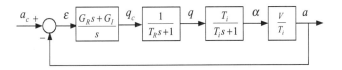

Fig. 9.15   Acceleration feedback loop for pitch control.

If the rate loop time constant T_R is not known, a simplified analysis can guide us to select appropriate values. We first develop the transfer function $\delta(s)/a_c(s)$, then impose the initial value theorem for step input, and finally obtain the relationship

$$\frac{\delta_0}{a_{c(\text{step})}} = \frac{G_A G_{\text{ACP}} T_R T_i}{V} \tag{9.47}$$

The initial fin deflection δ_0, which is also the maximum deflection, is proportional to an acceleration step input $a_{c(\text{step})}$, related by the gains G_A and G_{ACP} and the time constants T_R, T_I. Given the maximum control fin deflection and the desired maximum acceleration capability, Eq. (9.47) can be evaluated. However, G_A and G_{ACP} are also not known prior to the root locus analysis. Therefore, we have to employ an iterative design technique: Assume a value for T_R, conduct the root locus analysis, and verify that the desired acceleration can be achieved. Substituting into Eq. (9.47) the expressions for T_R, T_i, Eqs. (9.44) and (9.41), and solving for a_c yields

$$a_{c(\text{max})} = G_{\text{ACP}} \frac{\bar{q} S l C_{m_\delta} \left(T + \bar{q} S C_{L_\alpha} \right)}{m I} \delta_{(\text{max})} \tag{9.48}$$

Note that the dependence on the rate gain G_A cancels. Given the maximum control deflection, the achievable acceleration increases with increasing dynamic pressure, thrust, aerodynamic lift slope, and control derivative; and decreases with increasing mass properties. A high value of the root locus gain G_{ACP} is also desirable, but must be balanced against the stability requirements.

The acceleration feedback autopilot with inner-rate-loop stabilization finds widespread application in missiles. Its feedback signals are readily obtained from the onboard IMU, and its command signal is directly supplied by the guidance law. On the other hand, angle-of-attack feedback autopilots are also sometimes employed, particularly for high angle-of-attack maneuvers when tight incidence angle control is required. In aircraft angle-of-attack, sensors may be available, but for missiles the feedback signal must be synthesized from IMU measurements and may, therefore, lack accuracy.

Congratulations, you have persevered through the labyrinth of autopilot design for skid-to-turn missiles. But what if you have to model a cruise missile or a bank-to-turn hypersonic vehicle? I will lead you through the steps to modify what you have learned and combine it with a bank-to-turn controller.

9.2.3.2 Bank-to-turn autopilot.   As you know, the yawing degree of freedom is neglected in five-DoF aircraft simulations by enforcing zero sideslip. This corresponds to the assumption of perfectly coordinated bank-to-turn maneuvers. Let us make use of this simplification and build a basic bank-to-turn autopilot.

We maintain the acceleration feedback channel in the pitch plane. However, this body-fixed plane is now rotated about the velocity vector through the bank angle ϕ_{BV} wrt the vertical plane (see Fig. 9.16). The lateral acceleration is, given the normal load a_N,

$$a_L = a_N \sin \phi_{\text{BV}} \tag{9.49}$$

When the bank angle is 90 deg, the total normal load contributes to the lateral maneuver.

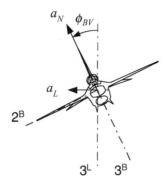

Fig. 9.16 Banking aircraft.

To maintain the same maneuver direction under negative angle of attack, the aircraft must bank in the opposite direction. We see that we must build into the banking logic a switching function that is dependent on the normal load factor sign. Figure 9.17 exhibits the autopilot embedded in a bank-to-turn vehicle. The pitch acceleration feedback autopilot is carried over unaltered from Fig. 9.15. The lateral acceleration command, beyond a small threshold, is divided by the normal load factor, given the correct sign, and limited in magnitude. Now we represent the roll degree of freedom by a simple first-order transfer function (remember our pseudo-five-DoF approach). You should get the value of its time constant T_ϕ from a six-DoF simulation, possibly making it a function of dynamic pressure. Most likely, the six-DoF simulation has a closed-loop roll transfer function of second order. If it is optimally damped ($\zeta = 0.7$), then the time constant of the first-order approximation is

$$T_\phi = \frac{1}{0.707\omega_n}$$

where ω_n is the natural frequency. This converts for a roll autopilot with position and rate feedback to

$$T_\phi = \frac{2I_p}{\bar{q}Sl} \frac{1}{K_p\{C_{l_\delta} - [l/(2V)]C_{l_p}\}} \tag{9.50}$$

with I_p the roll moment of inertia, K_p the inner roll rate gain, C_{l_δ} and C_{l_p} the roll control and damping derivatives, respectively. The details are not as important as the facts that the roll time constant increases with moment of inertia and decreases with dynamic pressure.

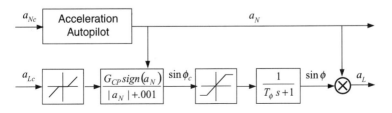

Fig. 9.17 Bank-to-turn autopilot.

Finally, to establish the lateral acceleration a_L, the achieved bank angle is multiplied by a_N.

For the bank-to-turn autopilot the acceleration command a_N must be given in the normal load factor plane of the aircraft. More frequently, the guidance command will be expressed in the vertical plane. This poses no problem for the skid-to-turn vehicle. However, for bank-to-turn implementations we have to convert it first by dividing it by the bank angle $\cos \phi_{BV}$.

I have to confess to a simplification that I have glossed over so far. The roll DoF in a six-DoF simulation is with respect to the body 1 axis. On the other hand, the bank angle in a five-DoF simulation is about the velocity vector. The two axes differ by the angle of attack. Strictly speaking, the first-order lag in Fig. 9.17 is the representation of the roll loop response, but we interpret it as the bank angle DoF. The effect is negligible, particularly because the angle of attack of bank-to-turn vehicles is usually less than 10 deg.

You can find the details of the implementation of Fig. 9.17 in the cruise missile simulation CADAC CRUISE5. The gain G_{CP} could have the value one but is usually increased to two or three to counteract the flattening of the $\sin \phi_c$ curve. To prevent division by zero, I added the small bias of 0.001 to the denominator. The threshold prevents roll oscillations under noisy lateral acceleration commands. A typical value of the threshold is 0.0174 g.

This bank-to-turn autopilot applies to aircraft, cruise missiles, hypersonic planar configured vehicles, and ramjet-powered missiles. If the vehicle cruises for any length of time, it may have to maintain altitude. We now modify the pitch channel to introduce altitude feedback.

9.2.3.3 Altitude hold autopilot.   There is little difference between a five- and six-DoF altitude hold autopilot. Two feedback loops are wrapped around the acceleration autopilot (see Fig. 9.18) with two gains G_H and G_V determining the dynamic response. To prevent large error signals driving the acceleration loop, I inserted an altitude rate limiter *HDTLIM*. The measured signals are altitude rate from the INS and altitude from an altimeter. According to our pseudo-five-DoF approach, the acceleration autopilot provides the angle of attack that is needed in the aerodynamic table look-up routines. The airframe block provides the remaining dynamics and measurements.

Let us address the need for the gravity bias term. When the vehicle flies straight and level, the altitude error signal is zero. However, to counteract gravity the airframe must generate a 1-g load factor. This commanded input is provided by the gravity bias term.

We have to distinguish between the implementations for skid-to-turn and bank-to-turn vehicle. While the altitude controller operates in the vertical plane, the

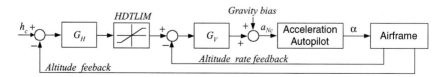

Fig. 9.18   Altitude hold autopilot.

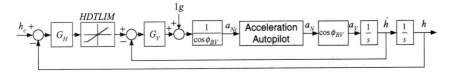

Fig. 9.19   Altitude hold autopilot for bank-to-turn and skid-to-turn vehicles.

pitch plane of the vehicle can be banked through the angle ϕ_{BV}. The adjustments through $\cos \phi_{BV}$ are shown in Fig. 9.19, as well as the two integrations that convert the vertical acceleration a_V into altitude rate and altitude. If $\phi_{BV} \equiv 0$, the vehicle executes skid-to-turn maneuvers and $\cos \phi_{BV} = 1$.

To determine the gains G_H and G_V, we start first with the design of the acceleration autopilot followed by a root locus analysis of the inner and outer altitude loops. Figure 9.20 shows the root locus patterns. The complete altitude hold loop has five closed-loop poles with a dominant oscillatory complex conjugate pair. For the CRUISE5 concept I selected the constant values

$$G_H = 0.5, \quad G_V = 1, \quad \text{HDTLIM} = 20 \text{ m/s}$$

They give good performance also for terrain following and obstacle avoidance flights.

We will stop here. You may require other autopilot functions for your specific vehicles, but I hope that these three examples, illustrating the idiosyncrasy of pseudo-five-DoF simulations, will enable you to devise your own designs. On the CADAC Web site you will find several other options like flight-path-angle hold, thrust vector pointing for reentry vehicles, yaw and pitch rate hold, etc. Look up the FORTRAN code and see if any one suits your applications.

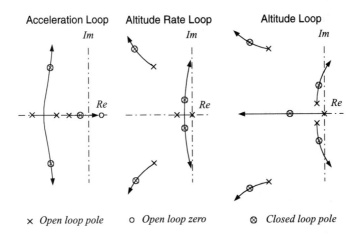

Fig. 9.20   Root locus analysis from inner acceleration loop to outer altitude controller.

9.2.4 Guidance

First, let us attempt a definition: Guidance is the logic that issues steering commands to the vehicle to accomplish certain flight objectives. For a missile the objective may be to hit a target. Given the vehicle and target states—position, velocity, and angles—the guidance algorithm generates the autopilot commands that steer the missile to the target.

For an aircraft on final approach, the objective is the descent on a glide slope leading to the touchdown point on the runway. If the pilot is at the controls, he issues the steering commands directly to the fin actuators, based on the situational awareness presented to him by the cockpit instruments. He fulfills the function of the guidance logic and the autopilot. Alternatively, if a stability augmentation system is engaged, he sends via his stick guidance signals to the autopilot. In the case of an automatic landing system, the electronics take over completely and provide the guidance logic for a hands-off touchdown.

Dealing with unmanned vehicles, like missiles and projectiles, the guidance logic is the most important function for ensuring mission success. Therefore, it is imbued with particular stature and is called the *guidance law*. You have probably heard about the proportional navigation law, which more appropriately should be called proportional guidance law. Some of its prefixes indicate performance improvements, like augmented, higher order, etc. Other guidance laws have names like line guidance, parabolic guidance, squint angle guidance, and others.

The open literature does not cover all aspects of guidance because some of the tricks of this trade are either classified or proprietary to industry. However, a few good references have been published just recently. *Advances in Missile Guidance Theory* by Ben-Asher[13] addresses guidance from a modern control aspect. An easier text to read is the third edition by Zarchan *Tactical and Strategic Missile Guidance.*[14] Practical guidance aspects are provided in *Modern Navigation Guidance, and Control Processing* by Lin.[15] Also many of us owe much insight into optimal guidance to Bryson and Ho and their classic *Applied Optimal Control.*[16]

For our simplified five-DoF representation of missiles and aircraft, I will limit the discussion to the basic, but all-pervasive proportional navigation law for missiles, and to line guidance, suitable for way-point guidance or landing approaches. Some of the more advanced schemes are introduced in conjunction with the full six-DoF simulations (see Chapter 10).

Although we live in this chapter in the pseudo-five-DoF world, with its special autopilot provisions, the guidance loop (outer feedback loop) is affected little by these simplifications. This fact is useful in two ways. During the conceptual phase of a vehicle design, detailed guidance studies can be conducted without detailed knowledge of aerodynamic and autopilot specifications. Alternately, if a full six-DoF simulation must be simplified, e.g., shortening run time by simplifying aerodynamics and reducing the autopilot bandwidth, the guidance loop can be transferred directly to the pseudo-five-DoF implementation.

9.2.4.1 Proportional navigation.

Proportional navigation (PN) is as old as the seafaring mariners who knew that a collision will occur if another vessel maintained its beam position. Pirates used that principle to intercept their bounty. Today's missiles use PN to intercept targets.

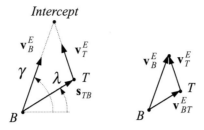

Fig. 9.21   Engagement triangle.

The earliest applications go back to the end of World War II and were reported by the Naval Research Lab[17] and later extended by the Massachusetts Institute of Technology in a formerly classified report.[18] Ever since, PN has been the premier guidance law, particularly for air-to-air missiles. I do not know of any interceptor that does not employ some form of this guidance scheme.

The displacement vector of the target c.m. T wrt the missile c.m. B, s_{TB}, is called the line-of-sight (LOS) vector. Its orientation will remain fixed in inertial space if the missile is on a collision course. The other important vector is the differential velocity v_{BT}^E of the missile c.m. B wrt the target frame T, which is obtained from the inertial velocities of the missile and the target

$$v_{BT}^E = v_B^E - v_T^E \tag{9.51}$$

(For a discussion of differential and relative velocities, refer back to Example 4.17). The Earth serves here as an inertial frame. As long as the relative velocity vector is pointing at the target, i.e., parallel to the LOS, an intercept will occur (assuming constant velocities). This engagement triangle is shown in Fig. 9.21. Although it is valid in three dimensions, let us consider it from a two-dimensional standpoint. The flight-path angle γ and the LOS angle λ are defined wrt an inertial datum. These angles must remain constant for the intercept. If the target maneuvers evasively, the missile velocity vector must be adjusted according to the change in LOS angle. For instance, if the target speeds up, λ increases, and, therefore, γ must be increased in order to maintain the engagement triangle. In other words, the time rate of change of γ must be made proportional to the time rate of change of λ, i.e.,

$$\dot\gamma = N\dot\lambda \tag{9.52}$$

with the navigation ratio N as the proportionality constant. Equation (9.52) is the famous PN relationship. The navigation ratio determines the lead of the missile velocity vector wrt the target. For most missiles it is given the value between two and four. Modern optimal control assigns it the value three.[16]

Let us look at Fig. 9.22 and follow an engagement that starts outside the engagement triangle. Assume that missile and target velocities are constant. Initially, the missile B_0 flies directly toward the target T_0, and the glide-path angle γ_0 equals the LOS angle λ_0. The advancing target turns the LOS and generates a $\dot\lambda$, which, magnified by N, turns the missile by $\dot\gamma$. Now, $\gamma_1 > \lambda_1$, and the missile's flight-path

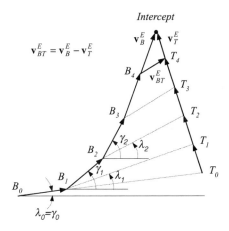

Fig. 9.22 Engagement starting from arbitrary initial conditions.

angle starts to lead the LOS angle. Eventually, the engagement triangle does not change shape and $\dot{\lambda} = 0$, $\dot{\gamma} = 0$ until intercept.

In applications the LOS rate $\dot{\lambda}$ is measured by the missile seeker and converted through the PN law into the acceleration command for the autopilot. The missing link is the relationship between $\dot{\gamma}$ and acceleration a. We derive it from the fact that the acceleration normal to the missile velocity vector $\boldsymbol{v}_B^E$ is (with $V = |v_B^E|$)

$$a = V\dot{\gamma} \tag{9.53}$$

and we obtain with Eq. (9.52) the steering command for the autopilot

$$a = NV\dot{\lambda} \tag{9.54}$$

This relationship applies only to planar engagements. The extension to the three-dimensional case follows by similitude. The angular velocity vector of the LOS frame O wrt the inertial Earth frame E is ω^{OE}, and the angular velocity vector of the velocity vector $\boldsymbol{v}_B^E$ wrt Earth is ω^{VE}. Following Eq. (9.52), we formulate in three dimensions

$$\omega^{VE} = N\omega^{OE} \tag{9.55}$$

Furthermore, taking our cue from Eq. (9.53), the acceleration $\boldsymbol{a}$ normal to the missile velocity is proportional to the cross product of ω^{VE} and the unit vector $\boldsymbol{u}_v$ of $\boldsymbol{v}_B^E$:

$$\boldsymbol{a} = V\Omega^{VE}\boldsymbol{u}_v \tag{9.56}$$

Substituting Eq. (9.55) into the last relationship yields the acceleration command

$$\boldsymbol{a} = NV\Omega^{OE}\boldsymbol{u}_v - \boldsymbol{g} \tag{9.57}$$

with the added gravity bias term $\boldsymbol{g}$. Without this term and no signal from the seeker, zero acceleration would be commanded, resulting in a ballistic trajectory.

The g-bias term counteracts the sagging tendency of the trajectory under seeker control.

This form of PN has been given the special designation pure PN[19] and is characterized by the fact that the acceleration command is normal to the inertial missile velocity vector v_B^E. Another form, the so-called true PN,[19] generates its acceleration command normal to the relative velocity vector v_{BT}^E. It has the same form as pure PN, portrayed in Eq. (9.57), except that u_v is taken as the unit vector of v_{BT}^E and $V = |v_{BT}^E|$.

Let us pause, sit back, and look at the effect that a rotating round Earth can have on the PN implementation. Certainly, the angular velocity of the Earth is negligible wrt the rapidly changing LOS rate of a missile engagement. Therefore, the preceding equations are equally valid for the rotating Earth case.

For practical applications the inertial LOS rates ω^{OE} and the unit velocity vector are given in body axes, whereas the gravity bias is expressed in local-level axes (or geographic axes for round-rotating Earth). The acceleration command to the autopilot must be delivered in body axes:

$$[a]^B = NV[\Omega^{OE}]^B[u_v]^B - [T]^{BL}[g]^L \qquad (9.58)$$

Full information awareness is assumed. The INS provides $[T]^{BL}$. The seeker delivers $[\omega^{OE}]^B$ and, in addition, $[u_v]^B$ for true PN. Some simplifications are possible. If the acceleration levels of the engagements are high, like in close-in combat, gravity can be neglected, and therefore $[T]^{BL}$ is not needed. Infrared imaging seekers do not deliver full target-missile kinematics as radio frequency seekers do. Therefore, if the relative LOS vector cannot be constructed, pure PN is implemented and $[u_v]^B$ calculated from INS data.

9.2.4.2 CADAC implementation of PN.

For five-DoF CADAC implementations the guidance law is programmed in the C1 Guidance Module (see Fig. 9.23) and provided with the necessary input from the Seeker Module S1 (see Fig. 9.30) and the INS Module S4. The two components of the acceleration signal are converted to units in gs and sent as normal acceleration command a_{Nc} and lateral acceleration command a_{Lc} to the autopilot (see Autopilot Sec. 9.2.3.1). The third component parallel to the body 1 axis is not used.

The key design parameter is the navigation ratio N. It usually is a constant but could be made a function of the closing speed between target and missile.

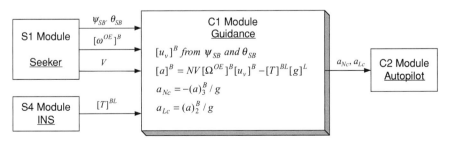

Fig. 9.23   Computer implementation of PN guidance law.

To determine its value, start with the number three and conduct sensitivity studies throughout the engagement envelopes. You will have to compromise and emphasize the more probable intercepts.

We have succeeded in modeling the PN guidance law, but only in its simplest form. The more complex form of thrust-corrected PN is reserved for Sec. 10.2.5.1. Let us now turn to a yet unpublished guidance law, which I have used for waypoint guidance and landing approaches of airplanes.

9.2.4.3 Line guidance, scalar case.
In some applications it is desirable that the target point is approached from a certain direction. For instance, an aircraft coming in for a landing must, while descending along the glide slope, line up with the the the runway and approach an initial point (IP) at the beginning of the runway before touchdown. The IP is the target point, and the heading and depression angles define the approach line. A second example describes the delivery of submunitions against tanks traveling on a highway. The cruise missile, which carries the submunitions, must approach in the direction of the highway and start dispensing the submunitions at the head of the tank column. Again, an approach line and IP must be defined. Other applications include the delivery of hard target warheads along a steep trajectory, reconnaissance flights along railroad tracks, and waypoint guidance with desired headings.

The task of line guidance is to align the vehicle toward the line and keep the velocity vector aligned until the vehicle flies through the IP. We call this line the line of attack (LOA) in contrast to the LOS, which is the line between the vehicle c.m. and the IP. Figure 9.24 depicts the geometry for the landing approach of an aircraft.

The objective of line guidance is to drive the LOS toward the LOA. As they merge, the aircraft will fly through the IP along the desired approach line. Let us first derive the governing law for horizontal steering, which can be visualized as the horizontal projection of Fig. 9.24.

In addition to the LOS coordinate system $]^O$, we define the coordinate system $]^F$ associated with the LOA of the final approach (see Fig. 9.25). The velocity of the aircraft B wrt the Earth frame E is v_B^E, and its two components in the 2 direction of the $]^F$ and $]^O$ coordinates are $(v_B^E)_2^F$ and $(v_B^E)_2^O$. To accomplish the objective of line guidance, both components must be driven to zero by the acceleration

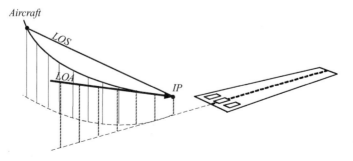

Fig. 9.24   Line guidance with LOA and LOS.

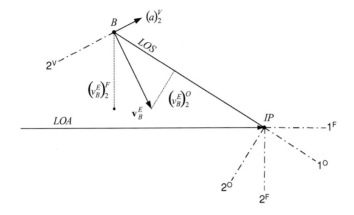

Fig. 9.25 Line guidance engagement in horizontal plane.

command, normal to the velocity vector. From these considerations we formulate the horizontal line guidance law

$$(a)_2^V = K\left[-\left(v_B^E\right)_2^O + G\left(v_B^E\right)_2^F\right] \tag{9.59}$$

The velocity vector is turned towards the *IP* if the acceleration command $(a)_2^V$ reduces the $(v_B^E)_2^O$ component. The aircraft c.m. *B* is moved faster toward the LOA if $(a)_2^V$ increases the $(v_B^E)_2^F$ component. The *G* gain balances the two terms, and the *K* gain adjusts for the different units. When v_B^E is on the LOA, both components are zero. Gain *K* is the guidance gain and is usually selected to be constant with values between one and three. The bias gain *G*, on the other hand, needs to decrease as the aircraft approaches the *IP* point, or else, dynamic overshoot can prevent the aircraft from flying through the *IP*. I found an exponential decrease suitable for all applications

$$G = \left(1 - \exp\left(-\frac{|s_{TB}|}{d}\right)\right) \tag{9.60}$$

where $|s_{TB}|$ is the distance of the vehicle to the *IP* and *d* is the norm distance at which the gain has the value $G = 0.633$. A typical value is $d = 1000$ m.

The development of the line guidance law for the vertical plane is similar. Applying Eq. (9.59) to the third directions of the coordinate systems $]^V$, $]^O$, and $]^F$ yields

$$(a)_3^V = K\left[-\left(v_B^E\right)_3^O + G\left(v_B^E\right)_3^F\right] - g\cos\theta_{VL} \tag{9.61}$$

with the gravity bias term $g\cos\theta_{VL}$. Note that $(a)_3^V$ is positive down and normal to the velocity vector v_B^E. For small angles of attack the normal load factor command is approximately $a_{Nc} = -(a)_3^V$; otherwise, we use $a_{Nc} = -(a)_3^V\cos\alpha$. In the horizontal plane the lateral acceleration command is $a_{Lc} = (a)_2^V$.

Most line guidance applications have a fixed *IP* point (or target) from which emanates a fixed LOA. As an extension, only slow target movements compared

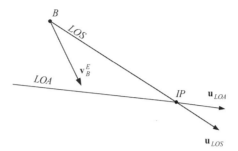

Fig. 9.26   Three-dimensional engagement.

to the missile speed can be accommodated. If $G = 0$, we have the special case of point guidance, also called pursuit guidance, with the velocity vector of the vehicle pointed at the target, irrespective of the approach direction.

9.2.4.4 Line guidance, vector case.
If you are theoretically inclined and want to see some vector and tensor mechanics in action, you will like this general derivation of line guidance. We will start with the two conditions for line guidance, derive the vector form, convert it to tensors, and eventually express it in matrices for programming. Equations (9.59) and (9.61) will be fully recovered.

The principle of line guidance is stated as follows (see Fig. 9.26): If the vehicle B is positioned on the LOA and the velocity vector v_B^E points at the target, then the two vector products of the unit vectors u_{LOS} and u_{LOA} are zero:

$$\varepsilon_P = V_B^E u_{LOS} = 0 \quad \text{and} \quad \varepsilon_L = V_B^E u_{LOA} = 0 \tag{9.62}$$

ε_P is the so-called *point attack error*, employed to correct LOS errors, and ε_L is the *line attack error*, used to bias the velocity vector towards the LOA. If both errors are zero, the vehicle flies on the LOA toward the target. We combine them with the variable gain G for the total error vector

$$\varepsilon = \varepsilon_P - G\varepsilon_L \tag{9.63}$$

ε_L opposes ε_P until the vehicle has reached the LOA. For this reason the $G\varepsilon_L$ term is also called the *line bias* term.

The commanded acceleration is the vector product of the total error ε and the velocity vector $V u_v$ multiplied by a constant gain K (note E is the skew-symmetric form of ε):

$$a = KVEu_v \tag{9.64}$$

The acceleration vector acts normal to the velocity vector and is zero if the total error is zero, or if ε is parallel to u_v, which never occurs. Substituting Eq. (9.63) and then Eq. (9.62) into the last equation yields

$$a = KV(E_P u_v - GE_L u_v) = KV(U_v U_{LOS} u_v - GU_v U_{LOA} u_v) \tag{9.65}$$

The two terms on the right-hand side are vector triple products (see Sec. 2.2.5.1).

They convert to

$$a = KV[u_{LOS} - u_v\bar{u}_v u_{LOS} - G(u_{LOA} - u_v\bar{u}_v u_{LOA})] \tag{9.66}$$

We abbreviate the two terms by introducing the point guidance vector c_P and the line guidance vector c_L

$$c_P = u_{LOS} - u_v\bar{u}_v u_{LOS}; \quad c_L = u_{LOA} - u_v\bar{u}_v u_{LOA} \tag{9.67}$$

and, combining it with the gravity bias term g, we obtain the three-dimensional line guidance law

$$a = KV(c_P - Gc_L) - g \tag{9.68}$$

The point guidance vector generates an acceleration command that drives the missile toward the intercept point, biased toward the LOA by the line guidance vector and its gain.

The two control vectors have a geometrical interpretation. Rewriting Eq. (9.67), we recognize the plane projection tensor $N_v = E - u_v\bar{u}_v$ of Sec. 2.3.5:

$$c_P = (E - u_v\bar{u}_v)u_{LOS} = N_v u_{LOS}; \quad c_L = (E - u_v\bar{u}_v)u_{LOA} = N_v u_{LOA} \tag{9.69}$$

This plane, normal to the velocity vector, is called the *latax plane* after the British designation for lateral acceleration of a missile. It contains the two control vectors, which are just the projections of the unit LOS and LOA vectors, respectively. Because both control vectors lie in the latax plane, then so does the acceleration command.

We are now ready to derive the component Eqs. (9.59) and (9.61) from the vector Eq. (9.68). To carry out our objective, we have to employ some geometry. Figure 9.27 displays the engagement of missile B against target T. For clarity, it depicts only the formation of the point guidance vector c_P (the projection of the unit LOS vector u_{LOS} on the latax plane). The line guidance vector would be formed by a similar projection of the unit LOA vector.

To develop the component equations, we use an alternate route to form the point guidance vector. Instead of u_{LOS}, we project the unit velocity vector u_v on the LOS

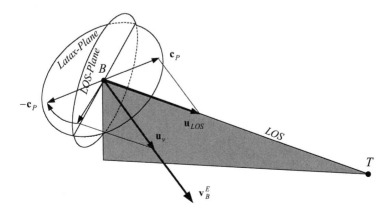

Fig. 9.27   Point guidance vector [c_P].

plane and rotate the projection through the rotation tensor R^{VO} of the velocity frame V wrt the LOS frame O. The resultant vector is in the opposite direction of the point guidance vector. Expressed in mathematical terms, we summarize

$$c_P = -R^{VO} N_O u_v \tag{9.70}$$

where N_O is the planar projection tensor onto the LOS plane. This relationship is valid in any coordinate system. In particular, c_P should be expressed in velocity coordinates $]^V$ and u_v in LOS coordinates $]^O$

$$[c_P]^V = -[R^{VO}]^V [T]^{VO} [N_O]^O [u_v]^O = -[N_O]^O [u_v]^O \tag{9.71}$$

where $[R^{VO}]^V [T]^{VO} = [E]$ [see Eq. (4.6)].

We build the line guidance vector by a similar process

$$c_L = -R^{VF} N_F u_v \tag{9.72}$$

where N_F is the planar projection tensor onto the plane normal to the LOA. Its component form is in velocity coordinates $]^V$ and LOA coordinates $]^F$

$$[c_L]^V = -[R^{VF}]^V [T]^{VF} [N_F]^F [u_v]^F = -[N_F]^F [u_v]^F \tag{9.73}$$

Substituting Eqs. (9.71) and (9.73) into Eq. (9.68) and multiplying out the matrices yields the line guidance law in component form

$$[a]^v = KV \left(-[N_O]^O [u_v]^O + G[N_F]^F [u_v]^F \right) - [T]^{VL} [g]^L$$

$$= K \left(- \begin{bmatrix} 0 & 0 & 0 \\ 0 & 1 & 0 \\ 0 & 0 & 1 \end{bmatrix} \begin{bmatrix} (v_B^E)_1^O \\ (v_B^E)_2^O \\ (v_B^E)_3^O \end{bmatrix} + G \begin{bmatrix} 0 & 0 & 0 \\ 0 & 1 & 0 \\ 0 & 0 & 1 \end{bmatrix} \begin{bmatrix} (v_B^E)_1^F \\ (v_B^E)_2^F \\ (v_B^E)_3^F \end{bmatrix} \right)_{VL}$$

$$- \begin{bmatrix} \cos\theta_{VL}\cos\psi_{VL} & \cos\theta_{VL}\sin\psi_{VL} & -\sin\theta_{VL} \\ -\sin\psi_{VL} & \cos\psi_{VL} & 0 \\ \sin\theta_{VL}\cos\psi_{VL} & \sin\theta_{VL}\sin\psi_{VL} & \cos\theta_{VL} \end{bmatrix} \begin{bmatrix} 0 \\ 0 \\ g \end{bmatrix}$$

$$= \begin{bmatrix} 0 \\ K\left[-(v_B^E)_2^O + G(v_B^E)_2^F \right] \\ K\left[-(v_B^E)_3^O + G(v_B^E)_3^F \right] \end{bmatrix} - g \begin{bmatrix} -\sin\theta_{VL} \\ 0 \\ \cos\theta_{VL} \end{bmatrix}$$

These are Eqs. (9.59) and (9.61) except for the gravity component $g\sin\theta_{VL}$, which was neglected earlier in Eq. (9.59) thanks to the small angle assumption for θ_{VL}.

9.2.4.5   CADAC implementation of line guidance.

For the five-DoF CADAC implementation, the guidance law is programmed in the C1 Guidance Module and provided with the necessary input from the INS Module S4, possibly updated by seeker information (see Fig. 9.28). Two components of the acceleration signal are converted to units in gs and sent as normal acceleration command a_{Nc} and lateral acceleration command a_{Lc} to the autopilot (see Autopilot Sec. 9.2.3.1).

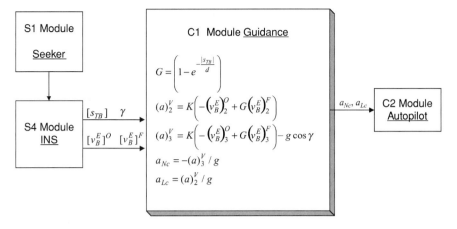

Fig. 9.28   Computer implementation of line guidance law.

The key input parameters are the guidance gain K, the bias gain G, and the representative norm-distance d. The autopilot will convert the lateral acceleration commands to bank-angle commands if the vehicle is an aircraft or cruise missile.

You can find the line guidance law used in the CADAC CRUISE5 simulation for waypoint guidance and approach to an IP. Given these examples, you should be able to adopt line guidance to your particular needs.

9.2.5 Sensors

Sensors measure the states of a vehicle relative to a reference frame, process the data, and send the information to the cockpit display or the guidance computer for further action. They execute a distinct navigation task, contributing to the situational awareness of the vehicle. We restrict ourselves to sensors that establish the LOS to a point, be it an IP, aimpoint, or target. They are often referred to as *seekers* because they seek out the location of an object. At the risk of disappointing you, I will not discuss such navigation aids as star trackers, altimeters, angle of attack vanes, etc., but concentrate on microwave radar and electro-optical (EO) sensors found in aircraft and homing missiles.

Radar was conceived by the British during World War II to track incoming German aircraft. Ever since, it has been improved in power, accuracy, and flexibility. EO seekers also had their crude beginnings during WWII in German guided missiles. They proved their deadly accuracy in the Gulf War. As we search the literature, we find many books on radar and optics, but few that relate to the task of sensing the LOS of an object. My favorite introductory text is by Hovanessian.[20] He treats radar and EO sensors from an application-oriented standpoint. In addition, I can recommend two handbooks that should be on your shelf as you wrestle with the modeling task of sensors: the infrared handbook by the company ERIM[21] and the radar handbook by Skolnik.[22]

The main distinction between radar and EO systems is their operational frequency in the electromagnetic spectrum. The radar bands we consider are the X, Ku, and Ka bands. The EO wavelengths span the visible spectrum $0.38-0.76$ μm

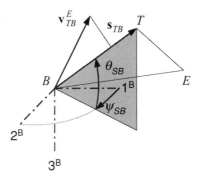

Fig. 9.29   Kinematic seeker geometry.

and the infrared spectrum $3-14\ \mu$m. Radars are active sensors, i.e., they emit the energy that they receive, whereas most EO systems are passive systems, relying on natural energy sources. An exception is the laser radar or ladar, which is an active sensor.

An aircraft or missile sensor has two distinct tasks: to acquire the target during the acquisition phase and to track the target during the tracking phase. Each phase makes different demands on the sensor. The acquisition phase requires the sensor to detect a target rapidly in a large search area, followed by accurate tracking possibly under adverse conditions.

Another distinguishing feature of sensors is the mounting type of the sensing element. A dish antenna would be mounted on a gimbal, but an array of wave guides can be mounted on a plate and bolted to the aircraft. EO seekers are usually placed on gimbals; however, advanced focal plane arrays lend themselves to body-fixed arrangements. Accordingly, we distinguish between gimbaled and strap-down sensors. Before we embark on these detailed models, let us take the kinematic approach of an ideal LOS pointing continually at the target—a simple but useful model.

9.2.5.1 Kinematic seeker.   The main purpose of a seeker is to establish (acquire) and to maintain (track) the LOS to the target. We make use of this fact by defining the LOS displacement vector s_{TB} of the center of the target T wrt the c.m. of the body B. With the help of the differential velocity v_{TB}^E of the center of the target T wrt the vehicle frame B, we can calculate the inertial LOS rate ω^{OE} of the LOS frame O wrt the inertial Earth frame E and send it to the guidance processor. The orientation of the LOS is measured relative to the body by the two angles ψ_{SB}, θ_{SB}, the seeker azimuth and elevation angles, respectively. Figure 9.29 depicts the geometric situation.

We can derive from purely kinematic considerations the most important variable, i.e., the inertial LOS rates ω^{OE}. We form the LOS vector s_{TB} from the two displacement vectors of the vehicle s_{BE} and the target s_{TE}, where E is an arbitrary reference point on the Earth

$$s_{TB} = s_{TE} - s_{BE} \tag{9.74}$$

then generate the unit LOS vector

$$\boldsymbol{u}_{TB} = \frac{\boldsymbol{s}_{TB}}{|\boldsymbol{s}_{TB}|} \tag{9.75}$$

We need also the differential velocity vector $\boldsymbol{v}_{TB}^E$, which we obtain from the inertial velocities of the vehicle and the target $\boldsymbol{v}_B^E$ and $\boldsymbol{v}_T^E$, respectively:

$$\boldsymbol{v}_{TB}^E = \boldsymbol{v}_T^E - \boldsymbol{v}_B^E \tag{9.76}$$

Now, the cross product of the two vectors will provide us with the LOS rate $\boldsymbol{\omega}^{OE}$

$$\boldsymbol{\omega}^{OE} = \frac{1}{|\boldsymbol{s}_{TB}|} \boldsymbol{U}_{TB} \boldsymbol{v}_{TB}^E \tag{9.77}$$

This equation has been derived without reference to a coordinate system and is therefore valid in any coordinate system. We express the LOS rate in body coordinates and the right-hand side in local-level coordinates

$$[\omega^{OE}]^B = \frac{1}{|\boldsymbol{s}_{TB}|} [T]^{BL} [U_{TB}]^L [v_{TB}^E]^L \tag{9.78}$$

The onboard INS supplies the transformation matrix $[T]^{BL}$.

For display purposes you can also calculate the seeker angles ψ_{SB} and θ_{SB} from the unit LOS vector expressed in body axes

$$[u_{TB}]^B = [T]^{BL} [u_{TB}]^L \tag{9.79}$$

from which you get the seeker angles as polar angles

$$\psi_{SB} = \arctan\left[\frac{(u_{TB})_2^B}{(u_{TB})_1^B}\right], \quad \theta_{SB} = \arctan\left\{\frac{-(u_{TB})_3^B}{\sqrt{[(u_{TB})_1^B]^2 + [(u_{TB})_2^B]^2}}\right\} \tag{9.80}$$

Also of interest is the *closing speed*, which is the relative velocity projected on the LOS

$$V_c = [\overline{u_{TB}}]^L [v_{TB}^E]^L \tag{9.81}$$

It is negative if the target is closing on the vehicle. The time it takes the vehicle to intercept the target is calculated from the absolute value of the closing speed and the distance to the target. With the assumption of constant relative velocity and nonmaneuvering target, the time-to-go is

$$t_{\text{go}} = \frac{|\boldsymbol{s}_{TB}|}{|V_c|} \tag{9.82}$$

This simple kinematic model serves both radar and EO-type seekers. It represents their ideal performance, unencumbered by field-of-view limits, gimbal dynamics, tracker errors, environmental and target effects. It is useful for simple simulations with emphasis on other than seeker phenomena. I use it to calculate synthetic LOS rates in midcourse based on INS information and to debug guidance simulations before incorporating the actual seeker model. Therefore, you will

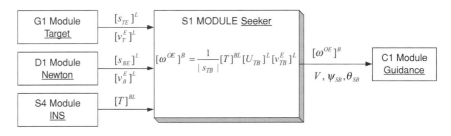

Fig. 9.30   Computer implementation of a simple kinematic seeker.

find it as a kinematic option in most of the CADAC simulations that incorporate seekers.

Figure 9.30 shows the implementation of the kinematic seeker in the S1 Module of a five-DoF CADAC simulation. The target and dynamics modules provide the kinematic variables to calculate the inertial LOS rates, converted to body axes by the transformation matrix obtained from the INS. Pay particular attention to the absolute velocity V that is sent to the guidance module. When I discussed the PN implementation, I distinguished between pure PN with $V = |v_B^E|$ and true PN with $V = |v_{TB}^E|$. The INS can provide the vehicle velocity v_B^E, but not the differential velocity v_{TB}^E; nor can a seeker measure the latter directly. Radar seekers, however, can supply the closing speed V_c of Eq. (9.81).

Although the kinematic sensor model serves mostly ideal seeker representations, it can be adapted to situations that are more realistic. For instance, for acquisition calculations you can use the ideal LOS range-to-go $|s_{TB}|$; for field-of-view limits the seeker angles ψ_{SB} and θ_{SB} give you the values to establish break-lock conditions; you can add noise to these angles or to the LOS rates to corrupt the true LOS, or you can even apply uncertainties to the target vector v_T^E to represent target measurement errors.

More realistic sensor models require detailed gimbal dynamics, system noise, target signature, environmental effects, and countermeasures. Most of these error sources depend on the operational frequency band. We will first treat the common dynamic effects and then address radar and EO sensors separately.

9.2.5.2   Dynamic seeker.

Tracking of a target necessitates the persistent pointing of the sensor's beam at the target. The difference between the true LOS and the beam is the tracking error, which is exploited by the tracking mechanism to keep the target in the field of view. Mechanical or electronic gimbals serve this purpose. In the case of mechanical gimbals, the sensor is isolated from the body by a gimbaled platform and is therefore called a *gimbaled seeker*. On the other hand, if the beam is steered electronically, the antenna is mounted on the vehicle body and is referred to as a *strap-down seeker*. Radar seekers can be of both types, whereas EO seekers are generally gimbaled. I will concentrate here on the more pervasive gimbaled seekers.

A voltage proportional to the tracking error torques the gimbals of the seeker. This voltage is proportional to the inertial LOS rate. You may remember that the PN law requires just this inertial LOS rate as input. Because the torquers are not

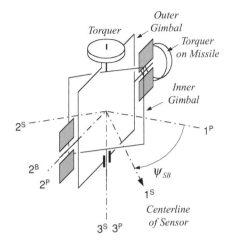

Fig. 9.31   Seeker platform with two gimbals.

fixed in the inertial frame but rather mounted on the body or the gimbals, rate gyros compensate for the body rates.

A typical gimbal arrangement is shown in Fig. 9.31. A motor, mounted on the vehicle body, torques the outer gimbal or pitch gimbal. Its associated coordinate system is $]^P$ with the 2^P axis coinciding with the vehicle's 2^B axis and the 1^P axis displaced by the pitch seeker angle θ_{SB} from the 1^B axis (not shown). The inner gimbal, or yaw gimbal, is torqued about the 3^P axis through the angle ψ_{SB} to arrive at the inner gimbal coordinate system $]^S$. Lastly, the sensor's beam is aligned with the 1^S axis.

Of interest is the TM $[T]^{SB}$ of the inner gimbal wrt the body coordinates. We acquire it by the sequence of transformations

$$]^S \xleftarrow{\psi_{SB}}]^P \xleftarrow{\theta_{SB}}]^B$$

depicted in Fig. 9.32

$$[T]^{PB} = \begin{bmatrix} \cos\theta_{SB} & 0 & -\sin\theta_{SB} \\ 0 & 1 & 0 \\ \sin\theta_{SB} & 0 & \cos\theta_{SB} \end{bmatrix}$$

$$[T]^{SP} = \begin{bmatrix} \cos\psi_{SB} & \sin\psi_{SB} & 0 \\ -\sin\psi_{SB} & \cos\psi_{SB} & 0 \\ 0 & 0 & 1 \end{bmatrix}$$

and the multiplication of the two matrices

$$[T]^{SB} = [T]^{SP}[T]^{PB} = \begin{bmatrix} \cos\psi_{SB}\cos\theta_{SB} & \sin\psi_{SB} & -\cos\psi_{SB}\sin\theta_{SB} \\ -\sin\psi_{SB}\cos\theta_{SB} & \cos\psi_{SB} & \sin\psi_{SB}\sin\theta_{SB} \\ \sin\theta_{SB} & 0 & \cos\theta_{SB} \end{bmatrix}$$

$$(9.83)$$

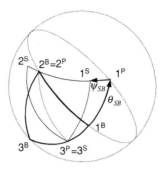

Fig. 9.32 Seeker and body axes.

The angles ψ_{SB} and θ_{SB} are called *seeker gimbal angles*. You must limit them in your simulation to model the seeker gimbal limits and, for more realism, limit their time derivatives to represent the gimbal rate limits. Typical values for limiting gimbal angle and rate are 63 deg and 400 deg/s, respectively.

We now build the block diagram that is the basis for our gimbaled seeker model and consider several error sources and dynamic effects. Target *scintillation* is caused by spatial fluctuations of the reflected energy. If the wavelength is in the radio frequency (RF) spectrum, scatterers that are distributed over the target reflect different amounts of energy as a function of target attitude and aspect angles. For EO sensors energy in the visible or infrared spectrum is emitted or reflected from the target surfaces, giving rise to a similar phenomenon. We model scintillation as a random shift of the aimpoint O from the center of the target T. The stochastic model is a first-order Gaussian–Markov process applied independently to the three coordinates of the aimpoint displacement $[s_{OT}]^T$ in target axes. The standard deviations could be made a function of the aspect angles θ_{OT} and ψ_{OT} of the LOS wrt the target axes. Scintillation shifts the true LOS vector s_{TB} to a point at the apparent aimpoint O (see Fig. 9.33). We express the corrupted LOS $[s_{OB}]$ in the local-level coordinate system $]^L$ and therefore need to include the transformation matrix $[T]^{TL}$

$$[s_{OB}]^L = [\bar{T}]^{TL}[s_{OT}]^T + [s_{TB}]^L$$

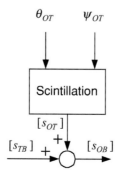

Fig. 9.33 Scintillation.

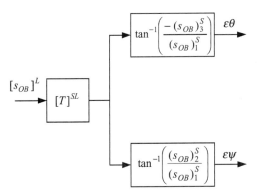

Fig. 9.34   Tracking errors.

The tracking error is the deviation between the apparent LOS and the centerline of the sensor beam and is measured by the tracker in the inner gimbal coordinates. We transform $[s_{OB}]^L$ to the inner gimbal coordinates using Eq. (9.83) and the TM of missile wrt local-level coordinates $[T]^{BL}$ (see Fig. 9.34)

$$[s_{OB}]^S = [T]^{SB}[T]^{BL}[s_{OB}]^L$$

The pitch and yaw tracking errors $\varepsilon\theta$ and $\varepsilon\psi$ are obtained from the components as shown in Fig. 9.34. Because the tracking errors are small angles, we do not expect any trouble from the arc tangent function.

The tracking errors are corrupted by noise and radome errors. Processing of the incoming signal intoduces Gaussian-distributed noise with its standard deviation possibly a function of signal strength. The beam, penetrating the radome, is deflected by the material properties of the radome such that the larger the seeker gimbal angles the greater the deviations. A linear relationship is commonly assumed with a typical value of 1% of the pitch and yaw gimbal angles. Signal processing introduces also a time lag that is modeled by a first-order transfer function with time constant T_S and gain G_S. The output of the tracker torques the gimbals in order to zero the tracking error. This is in effect the inertial LOS rate $\dot{\lambda}_q$ that drives the PN guidance law (see Fig. 9.35).

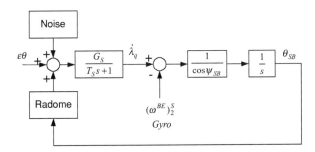

Fig. 9.35   Pitch tracker loop.

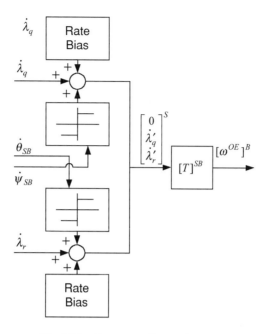

Fig. 9.36   Cross coupling and output.

The torquing signal must be compensated by the vehicle body rates, as measured by rate gyros, mounted on the inner gimbal. For the pitch loop it is the variable $(\omega^{BE})_2^S$ as shown in Fig. 9.35. Furthermore, because the pitch torquer is not located on the inner gimbal but is mounted on the vehicle body, the torquing signal is divided by $\cos \psi_{SB}$. Finally, the integrating effect of the torquing moment produces the pitch gimbal angle θ_{SB}. A similar loop exists for the yaw channel.

A possible rate bias and coulomb friction cross coupling can further corrupt the inertial LOS rate. The rate bias is caused by friction in the torquers about their respective axes, whereas coulomb friction is dry stiction, a function only of the direction of the angular rate, which couples the two gimbals. Figure 9.36 shows the signal flow of both the pitch and yaw LOS rates. Although the LOS rates are wrt the inertial frame, they are still expressed in inner gimbal coordinates. A final coordinate transformation $[T]^{SB}$ brings them into the form used by the guidance law, i.e., the angular velocity of the LOS frame O with respect to the Earth frame E (inertial frame), expressed in body axes, namely $[\omega^{OE}]^B$.

To sum up, Figs. 9.33–9.36 are combined and presented in Fig. 9.37. The tracker loops are now closed, feeding back the gimbal angles to form the TM $[T]^{SB}$, which multiplied by $[T]^{BL}$ yields $[T]^{SL}$.

Figure 9.37 represents a fairly detailed seeker model, suitable for five- and six-DoF simulations. With the appropriate error values it models missile seekers, both for air-to-air active radar and air-to-ground EO seekers. It can also be used for aircraft acquisition and tracking radar. If the gimbals are reversed, i.e., the outer gimbal rotates about the vehicles vertical axes (outer yaw gimbal), then Eq. (9.83) must be changed, but the block diagram remains essentially intact. The

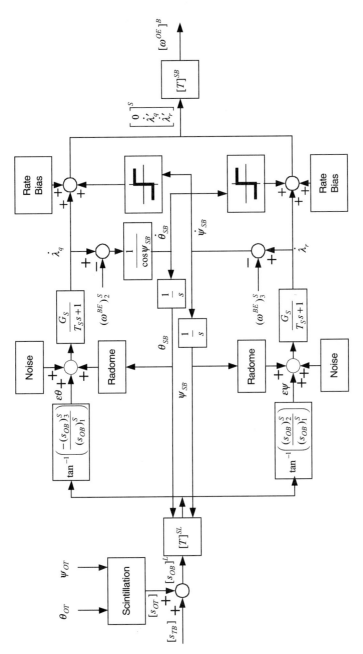

Fig. 9.37   Dynamic seeker block diagram.

only modification is in the tracker loop. The $1/\cos\psi_{SB}$ term is removed and the $1/\cos\theta_{SB}$ term instead inserted into the yaw channel. One feature that I have left out in the model is the gimbal dynamics. However, they are of such high bandwidth that they have little effect on the LOS rates and can therefore be neglected.

Primary challenges in seeker simulations are the building of the acquisition tables and the generation of error statistics. They will depend on the operational frequency of the sensor. Therefore, we have to discuss the physical implication of microwave and optical systems separately. First, let us treat radar seekers, followed by the EO systems.

9.2.5.3 Radar.

The modeling task of radar can be extremely tedious if the signal processing details are the focus. Fortunately, for our simple five-DoF simulations we can adopt a top-level viewpoint and proceed with an error model that corrupts the true LOS. The error sources are target glint, clutter, radome diffraction, and, for gimbaled systems, servo noise, coulomb friction, and rate gyro bias. Electronic countermeasures can degrade both the acquisition and the tracking performance.

The acquisition capability of a given radar sensor is essentially a function of the radar cross section of the target σ, the radar scan time T_s (scan duration), and the search area Ω (in steradians). The required level of the signal-to-noise ratio S/N for target detection can vary as a function of ground clutter, atmospheric backscattering, or environmental conditions. For modeling purposes we consider a functional relationship of the acquisition range R, derived from the radar range equation

$$R = \sqrt[4]{K\frac{\sigma T_s}{(\text{S/N})\Omega}} \tag{9.84}$$

where K represents the sensor specific constant that contains such terms as average power, aperture, receiver noise temperature, and system losses. To evaluate Eq. (9.84), S/N is expressed in natural units and not in decibels.

The radar cross section of the target, as presented to the seeker, is a function of the target attitude. We can therefore include the target aspect angle in the range calculations, possibly using tables that are functions of azimuth and elevation angles.

A more sophisticated model looks at the individual scattering points that intercept the transmitted electromagnetic energy and reflect portions of it toward the transmitting antenna. The total reflected energy, expressed in radar cross section, is in the form of a Rayleigh distribution (see Sec. 10.3.1.2). From one radar scan to the next, the radar cross section fluctuates according to the random draw from this distribution. This phenomenon is called *scintillation*. Given an acceptable false alarm rate, the *probability of detection* becomes a function of the S/N ratio, which is a function of the radar cross section. The greater the S/N ratio is, the higher the probability of detection. Tables relate these three parameters (false alarm rate, S/N, and probability of detection) for a particular radar and target configuration. If you want to bring your model to such a level of sophistication, consult the reference by Hovanessian and Ahn.[23]

These discussions are equally valid for aircraft and missile radars. In addition, the aircraft radar may have the capability to keep track of several targets while scanning the search area for new opportunities. With this search-while-track capability the radar computer maintains multiple target tracks that are displayed in the cockpit. The pilot assigns a missile to a target, thus transferring the target state to the missile guidance computer. After launch the missile steers toward the target, possibly receiving (via data link) further target updates. When the missile comes into acquisition range, it turns on its seeker and initiates the search mode. Assuming constant frame time T_s, the acquisition range is according to Eq. (9.84) inversely proportional to the fourth root of the search area. Therefore, the better the midcourse accuracy, the smaller the volume that needs to be searched, and the longer the range at which the target is acquired.

Once target lock has occurred, the seeker must continue tracking the target under adverse conditions like countermeasures, clutter, and atmospheric backscattering. If the noise should increase, the radar may be unable to retain track and may break lock. Hopefully, enough time remains to reacquire the target, or, if the countermeasures emanate from the target, the seeker can switch over to the home-on-jam mode.

A radar sensor measures four quantities: range, range rate, azimuth, and elevation angles of the LOS. These measurements are converted into the displacement s_{TB} of the target relative to the missile and the velocity v_T^B of the target c.m. relative to the missile frame. The angular noise sources, depicted in Fig. 9.35, are of Gaussian nature with the same standard distribution for the pitch and yaw channels:

$$\sigma_\theta = \frac{\theta_{BW}}{2\sqrt{S/N}} \qquad (9.85)$$

where θ_{BW} is the antenna beam width in radians and the S/N ratio is decreasing with range according to the radar range equation.[20] Range and range-rate measurements are also random with the following standard deviations

$$\sigma_R = \frac{c\tau}{4\sqrt{S/N}} \qquad (9.86)$$

$$\sigma_{\dot{R}} = \frac{\lambda \Delta f_s}{4\sqrt{S/N}} \qquad (9.87)$$

where c is the speed of light, τ the pulse length, λ the wavelength of the carrier, and Δf_s the Doppler filter bandwidth. As the missile approaches the target, the signal strength increases and so does the S/N ratio, resulting in improved measurements in all channels. Yet, disturbances like countermeasures, clutter, and backscattering can drive up the noise level, and scintillation becomes more pronounced toward the target.

Clutter is the unwanted energy return from scatterers on the ground, which may enter the seeker through the main or side lobes. Particularly the altitude return through the side lobes can interfere with the target detection process. It is, in general, a difficult undertaking to model clutter accurately. We must be satisfied here with an approximation based on the modified radar range equation. The ground clutter S/N is a function of equivalent clutter radar cross section σ_c, radar scan time T_s (scan duration), and the search area Ω:

$$(S/N)_c = K' \frac{T_s \sigma_c}{\Omega R^3} \qquad (9.88)$$

Note that ground clutter decreases with the third power of range (as long as the radar beam is smaller than the clutter background.)

Atmospheric backscattering is particularly noticeable in heavy rain. It is a function of the frequency band. The higher the frequency is, the stronger the effect becomes. Again we use the modified radar range equation, introduce the equivalent rain scattering cross section σ_s, and formulate the backscatter S/N

$$(S/N)_s = K'' \frac{T_s \sigma_s}{\Omega R^2} \tag{9.89}$$

which decreases with the second power of range and, therefore, not as fast as ground clutter.

Fortunately, Doppler radar can detect targets in clutter and backscattering provided these targets are moving like airplanes and cruise missiles. In effect, the clutter rejection of modern airborne radars is so good that the acquisition range is solely determined by system noise and can therefore be calculated by Eq. (9.84) alone.

For air-to-ground radar against stationary targets, clutter and backscattering are restricting the acquisition range severely. As an example, let the required system $(S/N)_t$ for target detection be 10 dB. For a given radar against moving targets (no clutter or backscattering corruption), the acquisition range is, let us say, 10 km. If the target is stationary, it can be corrupted by ground clutter $(S/N)_c = 12$ dB and possibly by rain scattering of $(S/N)_s = 20$ dB. Under these circumstances, the acquisition range moves closer to the target where the reflected energy from the target is at the higher level of S/N of

$$S/N = (S/N)_t + (S/N)_c + (S/N)_s = 42 \, \text{dB} \tag{9.90}$$

At that range the detectable signal is still 10 dB above the accumulative noise, as required. For our example, the acquisition range has been reduced from 10 to about 1.2 km by the ground clutter and rain scattering.

Once target acquisition has occurred, clutter and backscattering effects are lowered significantly by predictive filtering. Therefore, system noise characterized by Eqs. (9.85–9.87) adequately models the radar tracking noise.

For a pseudo-five-DoF simulation you can start with a kinematic seeker and check out your code thoroughly. If you model an air-to-air missile, you can analyze its kinematic performance adequately and establish launch zones. However, if your interest is in miss distance, you need to upgrade your model to a dynamic seeker and include the major noise sources. For radar seekers you can follow the preceding format and generate the numerical values from the seeker's specifications. If you design a new missile and the seeker does not exist, you should query the experts— good luck!—and establish several levels of error models. As a result of your analysis, you may be able to define these specifications and thus guide the radar developer in the design process.

Radar sensors, although they exhibit robust performance, may not be accurate enough for missiles with small warheads. Furthermore, their cost can be prohibitive for low-cost solutions. A viable alternative is the EO sensor operating in the visible or infrared spectrum.

9.2.5.4 EO sensors.
The technology leaps in focal plane arrays have made the EO sensor an excellent candidate for missile seekers. Detector costs have

plummeted, whereas array sizes have increased. In military applications the infrared (IR) spectrum is preferred because it opens the envelope to adverse weather and night operations. We are particularly interested in the 8.5- to 12.5-μm wave band, where mercury-cadmium-telluride detectors operate at temperatures of 70 to 80 K. Besides passive sensors receiving the thermal energy from emitting targets or reflected natural energy, there are also active sensors under development that emit and receive IR energy in radar fashion. They combine a laser emitter with radar processing techniques and are therefore called *ladars*. Modern CO_2-based ladars operate in the 10.6-μm wavelength, an area in the spectrum where atmospheric attenuation is at a relative minimum.

We will concentrate here on the modeling of passive IR sensors, either used as hot-spot trackers or as imaging seekers. As in radar, our ambition is not in the detailed modeling of the processing algorithms—I leave this to the experts—but our interest is in a top-level representation of the errors that corrupt the LOS between the sensor and the target. The active ladar sensor, on the other hand, can be treated like a radar, and you can refer back to the preceding section for details.

Some of the important error sources of passive IR sensors are atmospheric attenuation (water vapor, mist, fog, rain, clouds), ground clutter, processing delays, and, of course, countermeasures. In addition, we have to model the dynamic errors like spectral target scintillation, radome diffraction, gimbal friction, cross coupling, and rate gyro errors. The dynamic errors were addressed in the section on dynamic seekers. In the following we look at the physical properties of the passive IR sensor and how they affect the acquisition and tracking performance.

IR sensors measure the heat energy and calculate the temperature gradients to produce a TV-like image at night as well as during the day. For a given sensor the acquisition range is a function of the radiation intensity of the target J_t, the S/N, the number of detectors n, and the dwell time calculated from the frame time T_f over the search area Ω (in steradian)

$$R = \sqrt{K \frac{J_t}{(\text{S/N})}} \sqrt[4]{\frac{nT_f}{\Omega}} \tag{9.91}$$

Notice the similarities with the radar equation (9.84). The radar cross section has been replaced by the radiation intensity of the target J_t, and the scan and frame time are synonymous. However, the detection range is inversely proportional to the square root of S/N, whereas the fourth root applies to radars. The difference is based on the fact that the emitted energy has to travel the distance twice for radars but only once for passive IR sensors. K represents the sensor specific constant that contains such terms as aperture, focal length, detector detectivity, and losses. Equation (9.91) is valid for point sources against a clear background and without atmospheric attenuation. It describes the acquisition performance of a hot-spot sensor under ideal conditions quite well.

If the target is embedded in a background with variable spectral radiation emittance, like a vehicle traveling over land, the noise level of the system is increased, and the acquisition range is decreased likewise. This deteriorating effect depends on many variables, e.g., terrain type, sun angle, and seasonal changes. For simple simulations we just increase the threshold S/N by the background conditions $(\text{S/N})_c$.

Atmospheric attenuation is expressed as a loss per kilometer in decibels. It is a function of temperature, visibility, and humidity, as well as the spectral band of the sensor, and is formulated as an incremental signal-to-noise ratio $\Delta(S/N)_a$. The threshold S/N is then

$$S/N = (S/N)_t + (S/N)_c + \Delta(S/N)_a R \qquad (9.92)$$

This equation is similar to Eq. (9.90) but warrants further explanations. The threshold S/N establishes the acquisition range through Eq. (9.91), i.e., as the missile approaches the target, the signal strength in the detector increases to a level so that the S/N for target detection is reached. Without ground clutter and atmospheric attenuation the sensor specifications require, for target detection to occur, that the signal must be above the system noise by a certain factor. This is expressed by the sensor $(S/N)_t$. Ground clutter raises this factor and is additive because we use logarithmic units. Furthermore, the atmospheric attenuation increases this factor even more; however, it is not constant but is a function of acquisition range.

To implement Eq. (9.92) in your simulation, keep a running account of this threshold S/N and calculate the acquisition range from Eq. (9.91) (do not forget to convert from decibels to natural units: $x = 10^{dB/10}$). As the missile approaches, the target and its LOS range become equal to the acquisition range and target acquisition occurs.

Once the seeker starts to track the target, the uncertainties are dominated by dynamic errors and not signal processing phenomena. Just consider that the beam width of an IR sensor in the 10-μm wave band and with an aperture of 10 cm is 0.1 mr, small enough to be overwhelmed by dynamic errors.

So far, we have limited our discussion to targets that are essentially *point emitters*—far removed targets and objects with a strong radiating heat source fall into this category. Vintage IR seekers, like those of the Stinger and Sidewinder missiles, can only track such point sources. One of their drawbacks is that they are very susceptible to flare countermeasures. With the introduction of IR focal plane arrays, it has become feasible to image the target and to correlate the image with stored templates. If a match is found, the sensor locks on to the target and guides the missile to intercept. Sophisticated processing does not only acquire the target, but also classifies it and selects a particular vulnerable aimpoint. Turn with me now to a top-level discussion of these imaging seekers.

The image of such a seeker is either produced by a line scanner or a staring array. In both cases we consider the number of pixels on target: the more pixels, the higher the resolution of the target. Processing the temperature gradients from the pixels forms the image.

As the missile approaches the target and the threshold S/N is exceeded, the seeker starts to image the area where the target is expected to be located. The processor compares the temperature gradients with a prestored template of the target. When a match is found, the difference between the predicted and actual target location is used to improve the navigation solution. This imaging/update cycle repeats until the target fills the array completely.

Modeling of the acquisition phase consists of two parts. First, we calculate the threshold S/N from Eq. (9.92) and the associated acquisition range, Eq. (9.91). This procedure represents a deterministic approach. An alternate stochastic model is based on curves of the probability of acquisition vs range-to-target with the target size and the atmospheric conditions as parameters. These curves, calculated or measured, approximate parabolas with vertices at the probability of one and decrease with range. With p, the parabola parameter, the probability of acquisition is

$$P_{acq} = 1 - \frac{R^2}{4p} \tag{9.93}$$

Developing the tables of $p = f$ {target size, atmospheric conditions} can involve time-consuming tests and calculations. So, be forewarned! As a simplified model, I have used a linear curve fit of p as a function of target size at fixed weather conditions. To determine the occurrence of the in-range event of a particular computer run, draw a number from a uniform distribution. If P_{acq} is greater than that number, the seeker starts imaging the scene.

To ensure that the target is contained in the scene, the pixels must cover an area large enough to account for the pointing uncertainty of the sensor's centerline. This uncertainty is primarily determined by the midcourse navigation accuracy. With the INS position error given by its standard deviation σ_{INS} and the targeting error by σ_{Tar}, the pointing error is (in units of length)

$$\sigma_p = \sqrt{\sigma_{INS}^2 + \sigma_{Tar}^2} \tag{9.94}$$

The second effect to be modeled is the target acquisition time, consisting of the template imaging and matching process. Before launch the three-dimensional target template is stored onboard the missile processor. It consists of high-contrast facets in the form of a wire frame model. Once the sensor is within acquisition range, the three-dimensional template is readied for correlation by projecting it into the plane normal to the LOS. The pixels of the focal plane must cover this two-dimensional picture and the uncertainty area surrounding it. The time to image and process the data is directly proportional to the number of the pixels such engaged.

Each pixel has an instantaneous field-of-view of ε_i, given in radians. A typical value is 0.75 mr. We calculate the number N_a of pixels involved in the search process by covering three standard deviations or 99.7% of the pointing error (see Fig. 9.38):

$$N_a = \left(2\frac{3\sigma_p}{\varepsilon_i R}\right)^2 \tag{9.95}$$

If we designate each pixel's imaging time as Δt_i and its processing time as Δt_p,

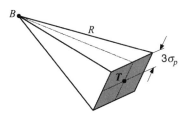

Fig. 9.38   Pixel on target uncertainty.

then the duration of the acquisition T_a is

$$T_a = N_a(\Delta t_i + \Delta t_p) \tag{9.96}$$

In your simulation tracking of the target should begin at the time the missile enters the acquisition range and acquisition time period T_a has elapsed. At this instant the first navigation update is sent to the INS and both the target location and INS navigation errors are reduced to the sensor's uncertainties.

After the first update the error basket has been reduced significantly, particularly by the elimination of the targeting error. Before acquisition the navigation solution was carried out in an absolute frame of reference. After acquisition the missile guides relative to the target, thus making the absolute targeting error irrelevant.

During tracking, the size and dynamics of the target determines the numbers of pixels engaged in the imaging and correlation process. For a stationary target we take three times the linear size of the target l_T. The number of active pixels is then

$$N_t = \left(\frac{3l_T}{\varepsilon_i R} \right)^2$$

and the duration of imaging and processing is

$$T_t = N_t(\Delta t_i + \Delta t_p)$$

T_t is significantly smaller than T_a, and, therefore, the update rate during tracking is faster than the acquisition time. Furthermore, most imaging seekers take advantage of the fact that imaging of the next frame can occur during processing of the preceding image. Because imaging is faster than processing, the update rate is determined by the processing of the pixels only. A 20-Hz update rate is the current state of the art. For a maneuvering target all pixels may be required to keep the target in the field of view. Then T_t may not be much smaller than T_a.

The tracking accuracy of imaging seekers is not determined by the beam width of the pixels, but by the template matching process. During mission planning, photography is used to build a three-dimensional wire frame model of the target. If the aspect angles and the range at which the picture was taken are known imprecisely, an error will creep into the tracking performance. Moreover, during target tracking the aspect angles and the range are corrupted by the INS errors. Both phenomena, prelaunch and in-flight distortions, contribute primarily to the tracking errors.

The sensor measures the azimuth and elevation angles of the LOS to the target aimpoint. These angles are taken relative to the missile body. For gimbaled seekers they are the gimbal angles. The measurements are corrupted by the correlation process, consisting of the mission planning and tracking errors and the dynamic errors of the gimbals. For a well-designed and fabricated seeker the dominant errors are not caused by the gimbals but by the template matching process.

We model the mission planning and tracking angular distortions by ε_m and ε_t, respectively, and the range errors as ΔR_m and ΔR_t. The measurement errors in the azimuth and elevation plane can then be formulated by

$$\varepsilon_{az} = K_{\varepsilon,az}(\varepsilon_m + \varepsilon_t) + K_{R,az}(\Delta R_m + \Delta R_t)$$

$$\varepsilon_{el} = K_{\varepsilon,el}(\varepsilon_m + \varepsilon_t) + K_{R,el}(\Delta R_m + \Delta R_t) \tag{9.97}$$

where K are constants for a particular target, obtained from extensive testing and analysis. In your simulation you can keep the values of ε_m and ΔR_m fixed, whereas ε_t and ΔR_t are provided directly by the INS error model. If you execute Monte Carlo runs, you could interpret the values of ε_m and ΔR_m as standard deviations of a random Gaussian draw.

I have led you from simple kinematic seeker formulations to fairly complex imaging sensors and discussed both radar and IR implementations. As long as you pursue top-level system simulations, you should have enough information to model the seeker for your particular application—by the way, you can include these seeker models also in your six-DoF simulations. However, I caution you, if you should embark on building a specific simulation for the development of a seeker you must consult the seeker specialist and learn the finer points of seeker modeling.

9.3   Simulations

Modeling and simulation are closely related subjects. So far, in this chapter, I concerned myself with the modeling of kinematics, dynamics, aerodynamics, propulsion, autopilots, guidance, and seekers. I derived the principal equations and indicated their validity, applicability, and limitations. To advance to the next stage of building a simulation, we have to proceed from theory to praxis. As often, the praxis is much more complex than the theory leads us to believe. Building a simulation is a tedious, time-consuming process, whose reward lies only in the final accomplishment. Hopefully, you have a sample simulation as a baseline, and your job is to modify it for a new application.

In this section we focus on the CADAC simulation environment. You should have read by now Appendix B, explaining the CADAC architecture and be familiar with its basic modular structure. By necessity, I had to be selective and chose as a prime example the simple air-to-air missile AIM5. The detailed description should enable you to build the simulation by yourself. However, just in case you do not have the time or patience, you can download it from the CADAC Web site. A more sophisticated version of an air-to-air missile, the SRAAM5 simulation, is documented briefly and can also be found on the CADAC Web site. For the cruise missile enthusiast I provide the CRUISE5 model with turbojet propulsion and GPS guidance. If you want to gain proficiency in five-DoF modeling, you should conduct the appended projects that will introduce you to the FALCON5 aircraft and the AGM5 air-to-ground missile.

The modular structure of CADAC allows us to deal with each subsystem of the vehicle separately. I have taken advantage of this characteristic already when I discussed the implementation of individual modules in the preceding sections of this chapter. One key feature is the control of the interfaces between the modules. Only because of their strict enforcement is it possible to exchange modules among simulations and across organizations. Utility programs, provided on the CADAC Web site, help you to maintain these interfaces and integrate other modules with minimal effort.

9.3.1   AIM5 Air Intercept Missile

This air-to-air missile example incorporates the rudimentary models of aerodynamics, propulsion, autopilot, guidance, and seeker of the preceding sections. It

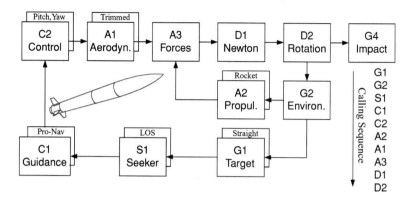

Fig. 9.39   CADAC pseudo-five-DoF air-to-air simulation: AIM5.

represents an air intercept missile with a solid, single-burn rocket motor, an acceleration hold autopilot, simple proportional navigation, and a kinematic seeker.

Figure 9.39 depicts the modules of the AIM5 simulation. Each module in CADAC is identified by a two-character code, which is its subroutine name. For clarification a title is added. After the initialization the integration loop starts with the Target Module G1 and ends with the Kinematic Module D2. The integration loop continues until the missile has reached the closet point to the target. Then the Impact Subroutine G4 is executed to stop the run and to display miss distance information. The sequence of execution is: G1, G2, S1, C1, C2, A2, A1, A3, D1, D2, and finally G4. I will describe these modules in the same order.

The Target Module G1 of an air-to-air missile can be very complex owing to the fact that airborne targets are highly maneuverable and take evasive actions. The SRAAM5 simulation has all of the details, but here we keep the model simple and limit ourselves to targets that fly straight at constant speed. The Target Module, see Fig. 9.40, consists of two subroutines: the *initialization subroutine*, which prepares the state variable $[s_{TE}]^L$ for integration, and the actual *target subroutine*, which calculates the position of the target by integrating the velocity $[v_T^E]^L$. Both vectors are sent to the S1 seeker and D1 Newton Modules for further processing. Because we assume constant target velocity, the three components of the vector $[v_T^E]^L$ are provided just as input.

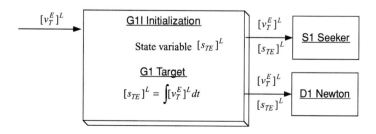

Fig. 9.40   G1 Target Module.

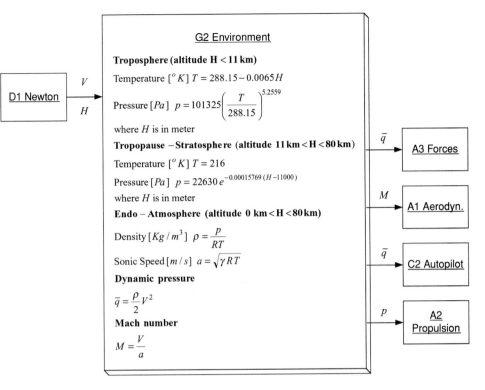

Fig. 9.41   G2 Environmental Module.

The Environmental Module G2 calculates the air density, pressure, temperature, and speed of sound. We make use of the ISO 1962 atmosphere (see Sec. 8.2.1). Accordingly, the temperature and pressure calculations are separated into two altitude layers: troposphere (<11 km) and tropopause through stratosphere ($11 \rightarrow 80$ km). The density calculations, based on the perfect gas law, are the same in both layers.

Figure 9.41 provides the atmospheric equations in SI units. In addition, the Mach number and dynamic pressure calculations are given. These parameters are used in the A3, C2, and A1 Modules. For backpressure calculations the atmospheric pressure is also needed in the A2 propulsion module. We encounter some problems with the input variable velocity V and altitude H. Following the calling sequence, they are established in D1 *after* the computations in G2. However, this is not too serious because the time lag is only the length of the integration interval. For the first time computation the initial values of V and H come from the input file.

The Seeker Module S1 was discussed in the preceding section. We use the simple kinematic seeker as depicted in Fig. 9.30. In our case we do not model the INS module, i.e., we assume perfect knowledge of the transformation matrix $[T]^{BL}$, which is calculated in the D2 kinematics module.

The Guidance Module C1 has also been detailed earlier in Fig. 9.23, and again the INS is assumed perfect.

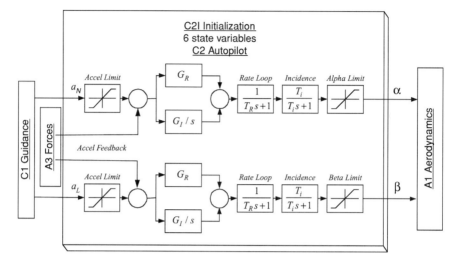

Fig. 9.42　C2 Autopilot Module.

The Autopilot Module C2 fashions the acceleration feedback autopilot (Sec. 9.2.3.1) with integral and proportional feedforward. Six state variables are initialized in C2I. They correspond to the Laplace transform variable s in Fig. 9.42: the integrator of the PI compensator and the first-order transfer functions of the rate loop and incidence angle lag dynamics.

The acceleration commands come from the guidance module and are expressed in normal acceleration (positive up) and lateral acceleration (positive right). In an actual autopilot the feedback signals come from accelerometers, either body mounted or as part of the INS. For our simple simulation we use the idealized acceleration, calculated in the A3 Forces Module.

The output of our Autopilot Module is in terms of incidence angles. As you may recall from our discussion in Sec. 9.1.1, pseudo-five-DoF simulations use autopilot transfer functions to model the attitude dynamics. The incidence angles are the output, ready to be used in the aerodynamic tables of the A1 Aerodynamic Module. However, be aware that in the real world the autopilot output are pitch yaw and roll control commands to the control surfaces, which in turn, through the surface deflections, generate moments and produce incidence angles that generate the aerodynamic forces.

The Aerodynamic Module A1 determines the aerodynamic coefficients from the trimmed aerodynamic tables, provided in the A1 Initializing Module. Because our missile has tetragonal symmetry, we chose the model of Sec. 9.2.1.1. The lift and drag coefficients are a function of Mach number and total angle of attack α'. Drag is also affected by the exhaust plume of the rocket. We must therefore carry two tables power-on and power-off drag.

The two incidence angles, as depicted in Fig. 9.43, arrive from the autopilot and are converted into aeroballistic incidence angles α' and ϕ'. With α', Mach number, and the power flag, the tables are interpolated for C_L and C_D. Before the body coefficients C_A, C_Y, and C_N can be sent to the A3 Forces Module, we

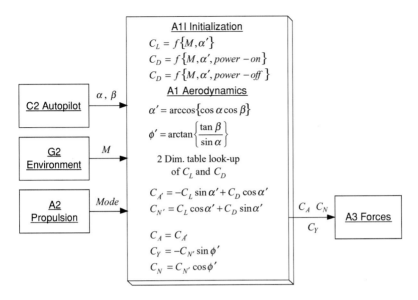

Fig. 9.43   A1 Aerodynamic Module.

have to transform C_L and C_D to aeroballistic axes and then through ϕ' to body axes.

The A3 Forces Module converts the coefficients into aerodynamic forces, combines them with the thrust from the Propulsion Module A2, and sends them to the Module D1 to be used in Newton's equations (see Fig. 9.44). It also prepares the specific force $[f_{\text{sp}}]^B$, the ideal measurement of the accelerometers, to be available for the autopilot. The A3 Modules perform an additional task. The transformation

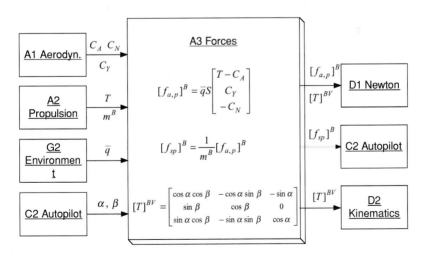

Fig. 9.44   A3 Forces Module.

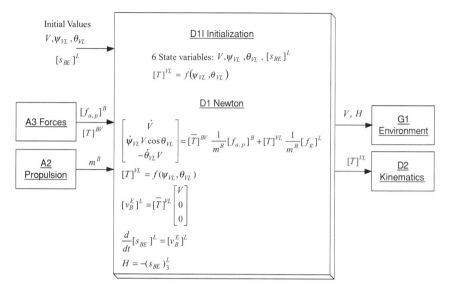

Fig. 9.45　D1 Newton Module.

matrix $[T]^{BV}$, derived in Eq. (9.8), is calculated from the incidence angles, provided by the autopilot and is sent to the D1 Newton Module and the D2 Kinematics Module.

The D1 Newton Module is the center of the simulation. Here, the state variables of the vehicle (velocity and displacement) are integrated, starting from the initial conditions. The equations of motion, Eq. (9.27) for flat Earth, are programmed (see Fig. 9.45). They solve for the vehicle speed V and the flight-path angles ψ_{VL}, θ_{VL}. At this point the $[T]^{VL}$ transformation matrix Eq. (9.21) is also required for the conversion of the gravity acceleration to velocity coordinates. However, for $[T]^{VL}$ to be available for the first integration, it must be initialized using the initial flight-path angles. After converting the velocity vector to local-level coordinates, the location of the c.m. B of the vehicle wrt an arbitrary Earth reference point E is obtained by integration. Speed V and altitude H are sent to the environmental module and the transformation matrix $[T]^{VL}$ to the Kinematic Module.

The D2 Kinematic Module, Fig. 9.46, is intrinsic to the pseudo-five-DoF implementation. Instead of integrating Euler's equations, as in full six-DoF simulations, we derive the body rates $[\omega^{BE}]^B$ from kinematic relationships, as outlined in Sec. 9.1.5. From Eq. (9.26) we determine the body rates, and Eq. (9.22) provides us with the direction cosine matrix. Figure 9.46 reflects these equations and the extraction of the Euler angles $\psi_{BL}, \theta_{BL}, \phi_{BL}$ from the direction cosine matrix.

The G4 Stop Subroutine, Fig. 9.47, is not a module as CADAC defines it. It is just a subroutine, called by SUBROUTINE STAGE3, and without the initialization capability of a module. Its purpose is to stop the run and print out intercept time and miss distance.

To get an accurate calculation, a linear interpolation is carried out between the last two integration steps. When the closing speed between the missile and the

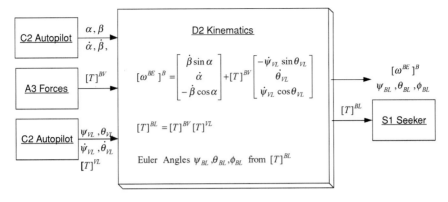

Fig. 9.46   D2 Kinematic Module.

target aircraft switches sign, the run is terminated, and the two displacement vectors of the missile $[s_{B\bar{B}}]^L$ and the target $[s_{T\bar{T}}]^L$ between the last two integrations are calculated. Refer to Fig. 9.47 for the equations, which were derived in Problem 2.9. The integration time step Δt and the run time of the next to the last integration step $t_{\bar{B}}$ are provided by the CADAC executive routine.

We have circled the whole loop of Fig. 9.39 and defined the major features of each module. I have left off some minor calculations that would have only cluttered the figures. You can find these details in the source code of the AIM5 simulation, stored on the CADAC Web site.

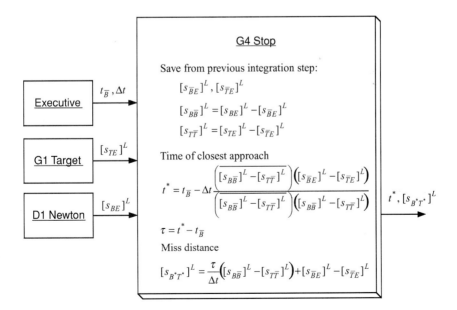

Fig. 9.47   G4 end-of-run subroutine.

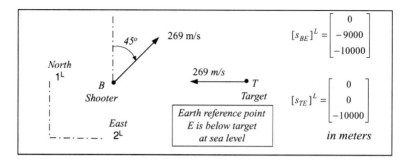

Fig. 9.48   Horizontal engagement (initial conditions).

9.3.1.1   Horizontal engagement.   Let us put this simulation to work and build two test cases. The first scenario is a coaltitude engagement at 10 km altitude. Both target and shooter fly at constant speed. The initial situation is given in Fig. 9.48. We study the sensitivity of the engagement to the navigation gain N using the values 2, 3, and 4 and comparing the flight time, Mach number, and miss distance at intercept.

Figure 9.49 depicts the engagement for the three cases. We conclude that the higher the navigation gain the tighter the engagement and the smaller the miss

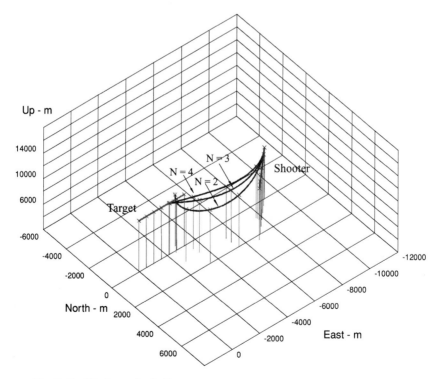

Fig. 9.49   Horizontal missile engagements with navigation gain $N = 2, 3, 4$.

Table 9.1   Summary of horizontal engagement

Navigation gain	Time of flight, s	Mach number at intercept	Miss distance, m
2		No intercept occurred	
3	8.05	2.5	2.59
4	7.69	2.8	1.31

distance. The missile with $N = 2$ fails completely to make the intercept. The details of the simulation can be found on the CADAC Web site. The INPUT.ASC file contains the parameters of the AIM5 missile like launch mass, reference area, autopilot parameters, initial conditions, etc., and aerodynamic tables and thrust characteristics are given in Modules A1 and A2.

Table 9.1 summarizes the engagement conditions. Based on this study we would select $N = 4$, wondering if we should not have increased the gain even further. However, we must realize that the simulation is an idealized model of the engagement and does not include any noise sources. The effect of these real-world conditions requires a compromise between accuracy and stability. Higher gains lead to tighter but also less stable intercepts. The navigation gain of four is a good compromise.

To investigate further the reason for the failed run with $N = 2$, we plot the sideslip angle and the lateral acceleration for the three cases in Fig. 9.50. Note that higher navigation gains demand more lateral acceleration earlier, making the intercept more benign. With $N = 2$ the missile is so sluggish that it misses the target completely and saturates its lateral channel after flyby.

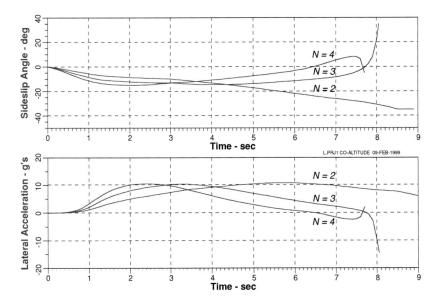

Fig. 9.50   Missile response with navigation gains $N = 2, 3, 4$ for horizontal engagement.

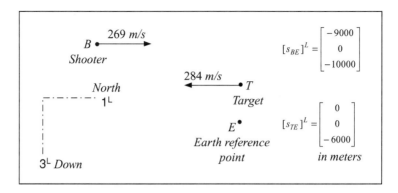

Fig. 9.51   Vertical engagement (initial conditions).

9.3.1.2 Vertical engagement.   The second scenario is a shoot-down engagement in the vertical plane. The initial situation is given in Fig. 9.51. Both aircraft cruise at 0.9 Mach, but, because of different altitudes, they do not have the same velocity. As in the horizontal case, we study the sensitivity of the missile performance to the navigation gains $N = 2, 3, 4$ and compare flight time, Mach number, and miss distance at intercept.

Figure 9.52 shows the vertical engagement of the three cases. In this shoot-down scenario the missile intercepts the target with all three navigation gains (see Table 9.2). The smaller gain values lead to trajectories with less depression because the acceleration command is reduced. This response of the missile acceleration is confirmed by Fig. 9.53. Observe how the angle of attack and the resultant normal acceleration depress down the missile quicker with higher gains. With $N = 2$ the missile almost overshoots the target, save for a final pull-down effort that exhausts its maneuvering capability. Higher gains exhibit the opposite trend. After the initial downward acceleration the missile has to pull up for the intercept.

This sample simulation of an air-intercept missile highlights several points. Any aerospace vehicle simulation should exhibit a modular structure to mirror the vehicle's subsystems. With the interfaces clearly defined, several designers can craft the code in parallel, shortening the development time significantly. As these simulations multiply, a library can be maintained and modules exchanged with other applications. I have conserved many resources by reuse of modules and exchange with other organizations.

Table 9.2   Summary of vertical engagement

Navigation gain	Time of flight, s	Mach number at intercept	Miss distance, m
2	8.23	2.3	18.4
3	7.78	2.9	6.3
4	7.67	2.9	0.8

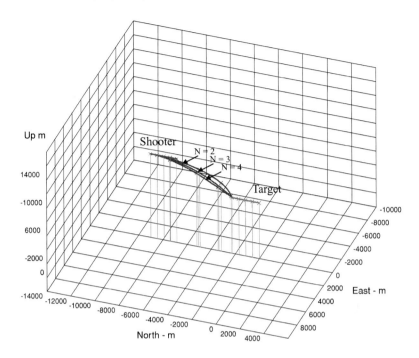

Fig. 9.52   Vertical missile engagements with navigation gain N = 2, 3, 4.

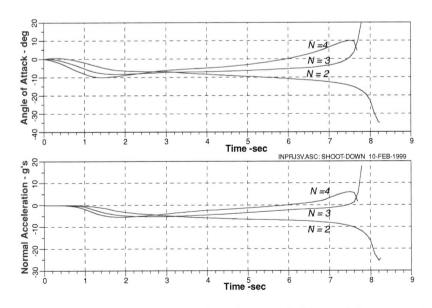

Fig. 9.53   Missile response with navigation gains N = 2, 3, 4 for vertical engagement.

If you do not have your own modular simulation, you can adopt CADAC as your simulation environment. More than 20 years have been invested in its development and perfection. Although it abides by some stringent FORTRAN conventions, the reward is a code that can easily be modified and shared. Appendix B provides an introduction. More details can be found in the program documentation, made available on the CADAC Web site. The vehicle's modules are embedded in the executive routine CADX.FOR. It handles input/output and integration of state variables. Modules and CADX.FOR compiled together require at run time two input files CADIN.ASC and HEAD.ASC. The first defines the input and the second the output. To inspect these files, download them from the CADAC AIM5 folder on the Web site.

The output, as defined by HEAD.ASC, is written to a TRAJ.BIN file. Now, all of the postprocessing options of CADAC are available. With KPLOT you can display two- and three-dimensional trajectories, as I have done in Figs. 9.49, 9.50, 9.52, and 9.53. Other options allow you to build ASCII files of selected variables, make strip charts, plot footprints, and launch envelopes. If you run multiple stochastic trajectories, you can display histograms, bivariate distributions (error ellipses), and the mean traces of Monte Carlo replications.

After these detailed explanations of a simple pseudo-five-DoF simulation, you should be able to decipher on your own the source code of the more complex short-range air-to-air missile SRAAM5. I will therefore be brief and refer you to the CADAC Web site for the details.

9.3.2 SRAAM5—Short-Range Air-to-Air Missile

SRAAM5 is a good example of how far a five-DoF simulation can grow in sophistication and still maintain the simple pseudo structure of the dynamic equations of motion. Figure 9.54 shows the structure with the shaded modules indicating new

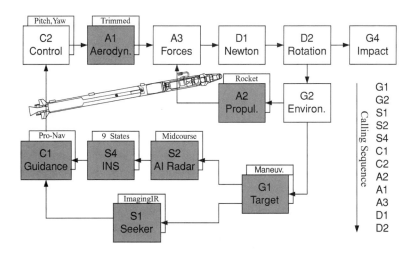

Fig. 9.54   CADAC SRAAM5 pseudo-five-DoF simulation.

elements. The A1 and A2 Modules are specific to the SRAAM concept, while the guidance loop is expanded by a midcourse and a terminal mode. In midcourse the shooter aircraft sends target data over data link to the missile INS navigator, which in turn delivers the kinematic LOS rates to the pro-nav guidance law. In the terminal phase the LOS rates are provided by an imaging infrared seeker. A brief description of the new modules follows.

The *G1 Target Module* implements realistic engagement scenarios are implemented. They are known by special designations, like Pre-Merge (shooter centered), One Circle Fight, Two Circle Fight, Lufbery Circle, Target Centered Engagement, Chase Circle, Head-On Circle, and Twin Circle. You just need to set the flag MTARG and thus invoke the preprogrammed initial conditions for target and shooter aircraft.

S1 Seeker Module uses imaging IR sensors as the current state of the art for short-range air-to-air missiles. Although only generic data are used, the roll/pitch gimbals and the coordinate systems are quite realistic. The description of the seeker is at an advanced level and is therefore deferred to Sec. 10.2.6.

The *S2 Air Intercept Radar* is a simple kinematic model of an acquisition and tracking radar, located in the shooter aircraft. It is used to acquire and track the target and transmit that information to the missile at launch and during an optional midcourse phase.

S4 INS Module is used in midcourse only. The nine error state equations (three positions, three velocities, and three tilts) bring realism to the fly-out accuracy. The derivation of the error equations is postponed until Sec. 10.2.4.

The *C1 Guidance Module* provides a midcourse and terminal guidance phase. In midcourse a simple pro-nav law is implemented (see Sec. 9.2.4.1), whereas the terminal phase relies on an advanced formulation of the pro-nav law, described in Sec. 10.2.5.1.

The *A1 Aerodynamics Module* uses trimmed aerodynamic lift and drag coefficients of a generic short-range air-to-air missile expressed in tables as a function of Mach and angle of attack, for power on/off, and three c.m. locations. The length of the missile is 2.95 m, and its diameter 0.1524 m.

The *A2 Propulsion Module* gives the thrust of a single-pulse rocket motor as a table of thrust vs time with backpressure corrections. Vehicle mass and c.m. shifts are also updated. Launch mass is 91.7 kg, and the motor fuel is 35.3 kg.

To your chagrin, the discussion of several features are postponed until the next chapter because of their advanced nature. Actually, I built the six-DoF version of SRAAM6 first and then converted the inner loop (equations of motion, aerodynamics and autopilot) to the five-DoF model. The outer (guidance) loop transferred with only minor modifications to the SRAAM5 simulation. If resources permit, I recommend this approach from six-DoF to five-DoF modeling because in the process we can validate our simplified model with the six-DoF truth model.

The AIM5 and SRAAM5 simulations are representative of highly maneuverable missiles with tetragonal symmetry executing skid-to-turn maneuvers. Cruise missiles, requiring long range performance, are designed with high-aspect-ratio wings for efficient cruise. They are steered like airplanes by bank-to-turn maneuvers. We turn now to the CRUISE5 simulation and all of the attributes of a typical cruise missile.

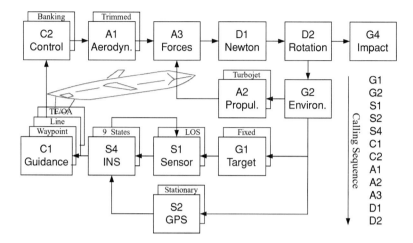

Fig. 9.55   CADAC CRUISE5 pseudo-five-DoF simulation.

9.3.3 CRUISE5—Cruise Missile

Cruise missiles fly like autonomously controlled airplanes, but do not return, unless used as reconnaissance vehicles. The CRUISE5 simulation is representative of these types of vehicles. Figure 9.55 shows the modular structure and the calling sequence. During the cruise phase, the missile navigates with GPS/INS from waypoint to waypoint until it reaches the terminal area. Now the sensor, pointed by the INS at the target, becomes active and provides the refined navigation solution to the line guidance law. The autopilot converts the acceleration commands into bank-to-turn steering signals, which the airframe executes until impact.

I will be brief with the description of the modules in the hope that you have gained enough savvy to explore on your own the source code provided at the CADAC Web site.

The *G1 Target Module* models a simple point target, possibly moving along a straight line.

The *G2 Environmental Module* uses the ARDC 1959 standard atmosphere, and an option is provided for the tabular input of a special atmosphere. Wind and gust effects are modeled in a simplified manner.

The *A1 Aerodynamic Module* provides the lift and drag coefficients are provided as functions of Mach and angle of attack for three c.m. locations. With an aspect ratio of 20, best lift-over-drag value is about 10. Span is 4.3 m and reference area $= 0.929$ m^2. Refer to Sec. 9.2.1.2 for more detail.

The *A2 Propulsion Module* gives the thrust available (max thrust) and idle thrust (min thrust) as provided by tables as a function of altitude and Mach number. Given the thrust, the fuel flow rate is obtained from tables, which depend on thrust and altitude. A Mach hold controller maintains the cruise speed. The modeling of the turbojet is discussed in Sec. 9.2.2. Takeoff gross mass is 1019 kg, and maximum fuel mass is 195 kg.

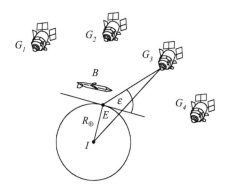

Fig. 9.56   GPS constellation.

The *S2 GPS Module* consists of four GPS satellites and a 17 states Kalman filter that updates the INS navigation solution. Figure 9.56 shows the satellite constellation. Because the CRUISE5 simulation uses the flat-Earth assumption, the Earth is nonrotating, and the satellites have to be assumed fixed. Although unrealistic, this simplification is acceptable for short duration flights and for generic navigation error studies. However, it is unsuitable for flight-test support.

My discussion of GPS and its filter is very abbreviated because it would lead us too far into an advanced topic that is well covered in the literature.[24] For those of you who have the drive and motivation, a solid foundation in GPS navigation together with the following diagrams should be sufficient to help you interpret the FORTRAN code of the S2 Module.

Based on the satellite and cruise-missile geometry of Fig. 9.56, there are five dilution of precision calculations that are summarized in Fig. 9.57. The pseudo-range measurements are corrupted by ionospheric and tropospheric effects, and additional errors are introduced by path delays, receiver noise and resolution, receiver dynamic noise, clock bias, and clock frequency uncertainties. These errors corrupt the measurements that are input to the Kalman filter as shown in Fig. 9.58.

The Kalman filter consists of 17 states that model nine INS errors and eight instrument uncertainties (three INS positions, three INS velocities, three INS tilts, three INS accelerometer biases, three INS gyro biases, one clock bias, one clock frequency error). Figure 9.59 shows the linear dynamic F matrix with the specific force coupling terms f_1, f_2, and f_3, the direction cosine matrix $[T]^{BL}$, and the clock time constant τ_c.

Finally, Fig. 9.60 combines all the elements into the sequence of the filtering process. As in any Kalman filter, there is an extrapolation and update phase. The update of the states is sent to the INS, its sensors, and the GPS clock. A typical update rate is 1 Hz. Although the equations have the appearance of a linear filter, in effect the measurements are nonlinear, and therefore we are modeling an extended Kalman filter that linearizes the observation matrix at every integration step.

Unless you are a control engineer, these diagrams may overwhelm you. Get a good grip of Kalman filtering through books like that by Maybeck.[25] Then with

For Every GPS Satellite

Azimuth α,
Elevation ε

$$\delta = \frac{\pi}{2} + \varepsilon$$

$$DGE = R_\oplus \cos \delta + \sqrt{(R_\oplus \cos \delta)^2 + DGI^2 - R_\oplus^2}$$

$$[s_{GE}]^L = \text{MATCAR} (DGE, \alpha, \varepsilon)$$

$$[u_{GE}]^L = \frac{[s_{GE}]^L}{|s_{GE}|}$$

Dilution of Precision Calculations

$$[H_{GPS}]^{4 \times 4} = \begin{bmatrix} [\overline{u}_{G_1 E}]^L & 1 \\ [\overline{u}_{G_2 E}]^L & 1 \\ [\overline{u}_{G_3 E}]^L & 1 \\ [\overline{u}_{G_4 E}]^L & 1 \end{bmatrix}$$

$$[DOP]^{4 \times 4} = \left([\overline{H}_{GPS}][H_{GPS}] \right)^{-1}$$

Time $\quad TDOP = \sqrt{DOP(4,4)}$

Vertical $\quad VDOP = \sqrt{DOP(3,3)}$

Horizontal $\quad HDOP = \sqrt{DOP(1,1) + DOP(2,2)}$

Position $\quad PDOP = \sqrt{\sum_{i=1}^{3} DOP(i,i)}$

Geometric $\quad PDOP = \sqrt{\sum_{i=1}^{4} DOP(i,i)}$

Fig. 9.57 Dilution of precision (DOP) calculations.

perseverance, you should be able to back out the equations from the FORTRAN code. After a series of test cases, you may feel like you have mastered the subject.

The *S1 Sensor Module* can detect and image a landmark or target and update, just like GPS, the INS navigation solution. Although the sensor is a simple kinematic model, it has a full set of errors that corrupt the true LOS. The measurement is sent to the INS in raw form without filtering to keep the implementation lucid. Several fixes can occur on one image improving the update consecutively.

To understand the update process, we first define *miss distance* as the vectorial sum of guidance and navigation errors (see Fig. 9.61). The *navigation error* is the

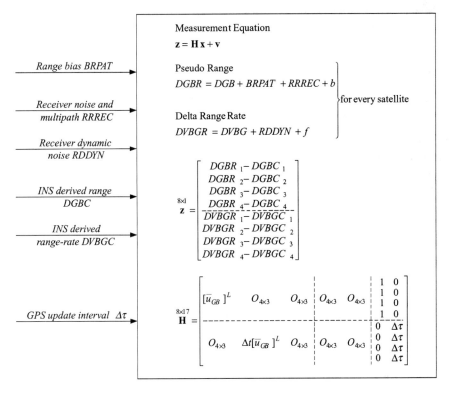

Fig. 9.58   Measurement errors and observation matrix.

distance between the true and estimated position of the target, and the *guidance error* encompasses the remaining vehicle errors, originating from the guidance law, the autopilot, air-frame maneuver limits, and environmental effects.

Figure 9.62 depicts the two steps of seeker measurement and navigation update. In preparation of the measurement, the INS points the seeker at $\hat{T}$, but the seeker pointing error instead misplaces the beam at point A, the aimpoint. The scene is imaged and the measured update determined. The measured update corrects the navigation error except for the inherent pointing error.

Before the navigation update the scene is correlated with the stored template, but unfortunately a correlation error creeps in that corrupts the measured update. Now the processor updates the estimated target location from $\hat{T}^-$ to $\hat{T}^+$. The original navigation error is reduced by the measured update, except for the correlation error. The vehicle executes a correction and flies towards the new target point $\hat{T}^+$.

The *S4 INS Module* is tightly integrated into the GPS and sensor loop. It stabilizes the GPS algorithms and points the sensor beam at the target. Its navigation solution is updated either by GPS in absolute coordinates or by the sensor in target-relative coordinates. It presents the single interface to the guidance processor. As in the SRAAM5 simulation, the nine error state equations (three positions, three

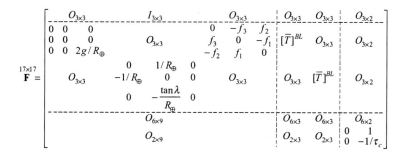

Filter Equation
$\dot{x} = Fx + w$

$$17\times1 \quad x = \begin{bmatrix} \overset{3\times1}{\varepsilon s} & \text{position error} \\ \overset{3\times1}{\varepsilon v} & \text{velocity error} \\ \overset{3\times1}{\varepsilon r} & \text{tilt error} \\ \hline \overset{3\times1}{\varepsilon a} & \text{accel. bias} \\ \overset{3\times1}{\varepsilon g} & \text{gyro. bias} \\ \hline b & \text{clock bias} \\ f & \text{clock frequency} \end{bmatrix}$$

Fig. 9.59   Seventeen states Kalman filter.

velocities, and three tilts) bring realism to the fly-out accuracy. The derivation of the error equations can be found in Sec. 10.2.4.

The *C1 Guidance Module* converts the navigation information from the INS to pitch acceleration and bank-angle commands that are executed by the control module. Three options are provided: waypoint guidance, terrain following/obstacle avoidance (TF/OA), and terminal line guidance. For course planning waypoints are identified on the map, and the cruise missile follows the sequence along preset headings, employing the lateral line guidance law as outlined in Sec. 9.2.4. In the vertical plane either the altitude hold loop of the Control Module C2 is engaged, or the TF/OA guidance law is invoked. As the missile approaches the terminal area, line guidance steers the missile in both planes to the target.

Obstacles are generated by two stochastic functions. An exponential distribution determines the waiting distance for the next obstacle to occur. The height of the obstacles is determined by a Rayleigh distribution. Terrain roughness is provided by the Newton Module D1. The onboard sensor can track terrain and obstacles either in a look-down or look-ahead mode. The look-ahead mode assumes that a terrain database of the area has been loaded in the guidance processor.

The *C2 Control Module* provides several autopilot options. Although a skid-to-turn mode is also given, we make use only of the bank-to-turn option. The key features are the acceleration controllers and the altitude hold autopilot. Both types are discussed in Sec. 9.2.3. The heading and flight-path-angle controllers are of lesser significance.

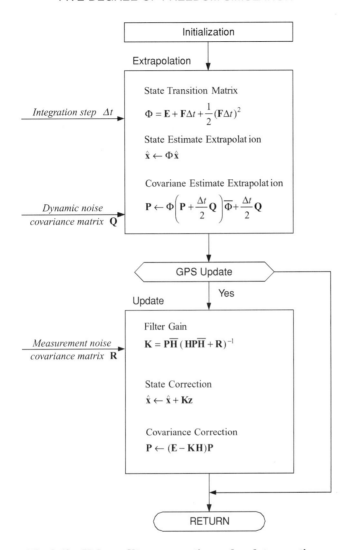

Fig. 9.60   Kalman filter propagation and update equations.

The *A3 Forces Module* is where the aerodynamic and propulsive forces combines and sends them as specific forces to the INS and Newton Modules. The TM $[T]^{BV}$ is also calculated from Eq. (9.11).

The *D1 Newton Module's* primary purpose is to solve the translational equations of motion. It also contains a subroutine that generates stochastic terrain, compatible for a look-down or look-ahead terrain sensor. The stochastic model is a second-order autocorrelation function driven by white Gaussian noise. The parameters of Table 9.3 describe the roughness of the terrain at three typical locations.

Table 9.3   Typical terrain of three levels of roughness

Location	Terrain type	Correlation length, m	Factor of second correlation length	Roughness std deviation, m	Slope std deviation
Ellsworth, KS	Smooth	536.4	0.726	9.14	0.02
Green River, UT	Medium	609.6	0.444	30.48	0.075
Black Top Mt, NM	Rough	914.4	0.444	152.4	0.25

The *D2 Rotation Module* replaces the Euler equations of six-DoF simulations. Its main output is the body rate $[\omega^{BE}]^B$ for the INS module and the direction cosine matrix $[T]^{BL}$ for various transformation tasks.

The *G4 Impact Module* computes the waypoint and target miss distances are computed. As defined in Fig. 9.61, the navigation and guidance errors that make up the miss distance are also recorded. The target, specified by a point, is a surface, which is oriented relative to Earth by two angles. The equations for the miss distance in the target plane are taken from Problem 2.8.

We have completed the round through the modules of Fig. 9.55. It is now up to you to exercise the test cases provided at the CADAC Web site. Particularly interesting is the study of maneuver requirements for various terrain roughness. Figure 9.63 shows a typical trajectory plot of the CRUISE5 vehicle, hugging a medium-rough terrain at an average clearance of 50 m while avoiding obstacles placed on the surface. Well, CRUISE5 did hit an obstacle, placed at a very inopportune location, just as it descended into a valley. Assuming it survived intact, it pops up, and then dives into the target under line guidance.

Pseudo-five-DoF simulations take an important place in the arsenal of an aerospace engineer. They are easier to build than six-DoF simulations, and, in contrast to three-DoFs, model attitude dynamics, although simplified. During the preliminary design phase, they serve to establish the performance of the vehicle. With the maturing of the concept, however, more information becomes available, especially in aerodynamics and autopilot design. Now the construction of a six-DoF model can begin. The next chapter provides the details.

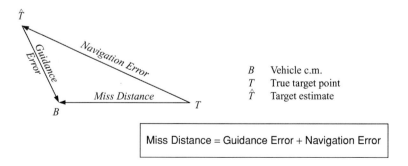

B	Vehicle c.m.
T	True target point
$\hat{T}$	Target estimate

Miss Distance = Guidance Error + Navigation Error

Fig. 9.61   Miss distance definition.

1. Seeker Measurement

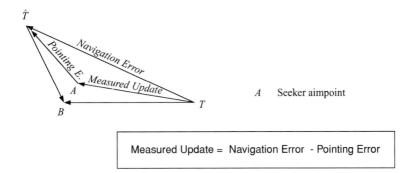

A   Seeker aimpoint

Measured Update = Navigation Error - Pointing Error

2. Navigation Update

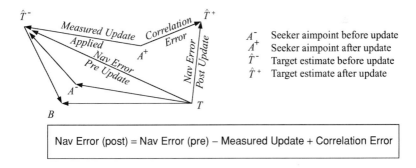

A^- Seeker aimpoint before update
A^+ Seeker aimpoint after update
$\hat{T}^-$ Target estimate before update
$\hat{T}^+$ Target estimate after update

Nav Error (post) = Nav Error (pre) − Measured Update + Correlation Error

Fig. 9.62   Seeker measurement and navigation update.

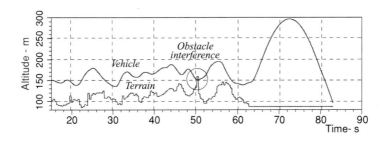

Fig. 9.63   Terrain-following obstacle avoidance.

References

[1]Hemsh, M. J., *Tactical Missile Aerodynamics: General Topics*, Vol. 1, *Tactical Missile Aerodynamics: Prediction Methodology*, Vol. 2, Progress in Aeronautics and Astronautics, AIAA, Washington, DC, 1992.

[2]Blake, W., "Missile DATCOM, User's Manual," Air Force Research Lab., WL-TR-93-3043, Wright Patterson AFB, OH, 1993 (distribution unlimited; for computer code call 937255-6764).

[3]Gentry, A. E., Smyth, D. N., and Oliver, W. R., "The Mark IV Supersonic Hypersonic Arbitrary Body Program," Air Force Flight Dynamics Lab., AFFDL-TR-73-159, Vols. I, II, and III, Wright–Patterson AFB, OH, 1973.

[4]Hoak, D. E., et al., "The USAF Stability and Control DATCOM," Air Force Research Lab. TR-83-3048, Wright Patterson AFB, OH.

[5]Roskam, J., *Airplane Flight Dynamics and Automatic Flight Control*, Parts I and II, Roskam Aviation and Engineering, Lawrence, KS, 1979.

[6]Stevens, B. L., and Lewis, F. L., *Aircraft Control and Simulation*, Wiley, New York, 1992.

[7]Pamadi, B. N., *Performance, Stability, Dynamics, and Control of Airplanes*, AIAA Education Series, AIAA, Reston, VA, 1998.

[8]Etkin, Bernard, *Dynamics of Atmospheric Flight*, Wiley, New York, 1972.

[9]Dorf, R. C., and Bishop, R. H., *Modern Control Systems*, Addison Wesley, Longman, Reading, MA, 1998.

[10]*MATLAB*,®Math Works, Inc., Natick, MA, 1998.

[11]Stengel, R. B., *Optimal Control and Estimation*, Dover, New York, 1994.

[12]Brockhaus, R., *Flugregelung*, Springer-Verlag, Berlin, 1994.

[13]Ben-Asher, J. Z., *Advances in Missile Guidance Theory*, Vol. 180, Progress in Astronautics and Aeronautics, AIAA, Reston, VA, 1998.

[14]Zarchan, P., *Tactical and Strategic Missile Guidance*, Vol. 176, Progress in Astronautics and Aeronautics, AIAA, Reston, VA, 1998.

[15]Lin, C.-F., *Modern Navigation Guidance, and Control Processing*, Prentice–Hall, Upper Saddle River, NJ, 1991.

[16]Bryson, A. E., Jr, and Ho, Y., *Applied Optimal Control*, Hemisphere, New York, 1997.

[17]Newell, H. E., Jr., "Guided Missile Kinematics," U.S. Naval Research Lab., Washington, DC, 22 May 1945.

[18]Gallagher, J. M., Jr., and Trembath, N. W., "Study of Proportional Navigation Systems with Linearized Kinematics," Dynamic Analysis and Control Lab., Massachusetts Inst. of Technology, Cambridge, MA, Rept. 75, 15 April 1953.

[19]Guelman, M., "The Closed-Form Solution of True Proportional Navigation," *IEEE Transactions on Aerospace and Electronic Systems*, Vol. AES-12, No. 4, 1976, pp. 472–482.

[20]Hovanessian, S. A., *Introduction to Sensor Systems*, Artech House, 1988.

[21]Wolfe, W. L., and Zissis, G. J. (ed.), *The Infrared Handbook*, ERIM, Ann Arbor, MI, 1989.

[22]Skolnik, M. I., *Radar Handbook*, 2nd ed., McGraw–Hill, New York, 1989.

[23]Hovanessian, S. A., and Ahn, H. H., "Calculation of Probability Detection with Target Scintillation," *IEEE Transactions on Aerospace and Electronic Systems*, March 1973, p. 300.

[24]Janisczek, P. M. (ed.), "Global Positioning System," *Navigation Journal*, Vols. 1–4, Inst. of Navigation, Alexandria, VA, 1980, 1984, 1986, 1993.

[25]Maybeck, P. S., *Stochastic Models, Estimation, and Control*, Vol. 1, Academic Press, New York, 1979.

Problems

9.1   AIM5 engagement studies.   Investigate the performance of the AIM5 air-to-air missile as modeled by the CADAC AIM5 simulation. Performance is defined in terms of launch envelope and miss distance. The missile's fly-out time is limited by the onboard battery life of 30 s.

Task 1:   Download the AIM5 project from the CADAC Web site and run the test cases INHORI.ASC, INHOR3.ASC, and INVERT3.ASC. Get familiar with the AIM5 documentation of Sec. 9.3.1.

Task 2:   Analyze the performance of the AIM5 missile using the CADAC SWEEP methodology. Set up input files for the following engagement codes taken from Tables P9.1 and P9.2:

Input file	Codes
INLOW.ASC	L1/T1
INMED.ASC	L2/T2
INHIGH.ASC	L3/T3
INMIX.ASC	L1/T2

Run the four SWEEP cases by setting up fans with varying launch ranges DHTB and angles AZTLX. Contour the launch envelopes, limited by miss distance not greater than 5 m and flight time not greater than 30 s.

Task 3:   Document your results. Provide your four input files and four contour plots.

9.2   Automatic approach to LAX with Falcon aircraft.   Using CRUISE5 as a template, build a pseudo-five-DoF simulation of an aircraft with the following module structure (you have to change only the shaded Modules A1 and A2).

Task 1:   Familiarize yourself with the CRUISE5 simulation, run the test cases, and plot selected parameters.

Task 2:   Strip CRUISE5 of unnecessary modules and run test cases INCRUISE.ASC, INCLIMB.ASC, and INIP.ASC for verification.

Task 3:   Download from the CADAC Web site the FALCON5 aerodynamics and propulsion data decks and create the new Modules A1 and A2.

Task 4:   Adopt the test cases INCRUISE.ASC, INCLIMB.ASC, and INIP.ASC to the FALCON5 simulation and execute them for verification. (You may have to tweak the autopilot.)

Table P9.1   Missile launch conditions

Launch conditions	Altitude, m	Mach	Heading, deg
L1	1,000	0.6	0
L2	3,000	0.75	0
L3	10,000	0.9	0

Table P9.2   Target aircraft conditions at time of missile launch

Target conditions	Altitude, m	Mach	Heading deg
T1	1000	0.6	180
T2	3000	0.75	180
T3	5000	0.9	90

Task 5:   Guide your FALCON5 to an automatic landing approach at L.A. International Airport, Runway 076 deg at 34°27′N, 118°26′W (start of runway). Begin over Santa Catalina Island, 33°50′N, 118°025′W, heading 300 deg, speed 164 m/s, altitude 3000 m. Approach IP (500 m before start of runway) at heading 76 deg and glide slope−10 deg with Mach 0.3 and flaps out (20% increase in drag, 60% increase in lift).

(a) Build the input file INLAX.ASC. Consider inserting an additional waypoint. Bring the aircraft down in altitude steps at 20 km from the IP with extended flaps.

(b) Display the three-dimensional trajectory in KPLOT-PITA with time traces of altitude, Mach number, flight-path angle, and heading angle.

9.3   AGM5 air-to-ground missile.
Convert the AIM5 air-to-air missile model into an unpropelled air-to-ground missile simulation and determine its performance. (You have to modify only the shaded modules.)

Task 1:   Familiarize yourself with the AIM5 simulation, run the test cases, and plot selected parameters. You will benefit from first executing Problem 9.1.

Task 2:   Download from the CADAC Web site the AGM5 aerodeck and replace the A1 module. Implement the miss distance calculations of Problem 2.8 in a new G4 subroutine.

Task 3:   Build the trajectory input file INTRAJ.ASC. Launch: heading = 10 deg (from North), altitude = 3000 m, speed = 300 m/s. Stationary target: Distance from launch point = 9000 m at azimuth = − 10 deg (from North), altitude = 100 m, target plane tilted up 70 deg. Record miss distance and plot altitude, Mach number, and flight-path angle vs time.

Task 4:   Conduct an autopilot gain sensitivity study for the trajectory of Task 3. Build the input file INMULTI.ASC for the three gains GACP = 4,40,400. Record

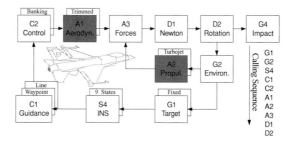

miss distance and plot angle of attack vs time of the three traces. What gain yields the smallest miss distance?

Task 5: Determine the footprint for the launch condition of Task 3 on a horizontal target plane at 100 m altitude. Build the input file INPRINT.ASC and set it up for a SWEEP run. The footprint is limited by miss distance less than 5 m. Plot the footprint.

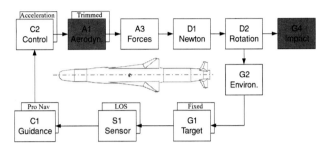

Task 6: Compile your results in a brief report. Include the input files IN-TRAJ.ASC, INMULTI.ASC, and INPRINT.ASC, the plots from Tasks 3, 4, and 5, and the miss distances of Tasks 3 and 4.

10
Six-Degree-of-Freedom Simulation

We have assembled all of the gear we need to embark on the ultimate trip of full six degrees of freedom. In Chapter 5 you learned to model the three translational degrees of freedom, and Chapter 6 taught you the secrets of three rotational degrees of freedom. Newton and Euler should have become your favorite companions.

You also have become familiar with simple simulations. Chapter 8 instructed you how to build a three-DoF simulation and under what circumstances it can satisfy your needs. Then you moved up one notch and added two more degrees of freedom in Chapter 9 to arrive at that peculiar form of pseudo-five-DoF simulations. Both have important roles to play during the development cycle of an aerospace vehicle. As an engineer, you have to adjust your tools to the task at hand and the resources available. Start simple! It is better to have a reliable three-DoF simulation ready in time than a nonvalidated and tardy six-DoF simulation.

Eventually, the time will come when you have to strike out and go for maximum fidelity. But count your cost, and let your sponsor know the extensive resources he has to set aside. You will need lots of data: full aerodynamic and thrust tables—trimmed data will no longer suffice; a complete flight control design; mass parameters that include moments of inertia; and other subsystems, like sensors and guidance logic. You also will need lots of time. I have yet to finish a simulation, completely verified and validated, within the projected period. From experience, I know that once a simulation runs reasonably well it takes another third of the development time to produce a reliable product.

The challenge is great, but very rewarding. I will help you with the details. We develop together two types of equations of motion. For in-atmosphere, near-Earth flight we assume the flat Earth to be the inertial frame. Tactical missiles and aircraft fall into this category, and I will add spinning missiles and Magnus rotors to the collection. Quaternions will serve us to compute the attitude angles.

Other types of equations are required for hypersonic and orbital vehicles. They are based on an inertial frame aligned with the solar ecliptic and convecting with the center of the Earth—the inertial frame J2000 you encountered in Chapter 3. We shall study the elliptical geometry of Earth, the gravitational field, and introduce the geodetic coordinate system. Then I shall formulate the equations of motion relative to the inertial and geographic frames. The attitude angles will be computed from the direction cosine matrix equations.

Next, we have to flesh out the right-hand side of the equations of motion. I present aerodynamic models for airplanes, hypersonic vehicles, and missiles. The autopilot for six-DoF simulations requires an entirely new treatment with such options as rate damping loop, roll autopilot, heading angle, and flight-path-angle controller, acceleration autopilot, and altitude hold autopilot. Actuators are now also part of the control loop.

All of the seeker and guidance options of five-DoF simulations can directly be integrated into your six-DoF. I shall just expand your horizons. The modeling of INS errors becomes now important; extensions of proportional navigation and the advanced guidance law will improve the intercept performance of your missile, and a gimbaled IR sensor provides you with a modern missile seeker.

This chapter will not satisfy all of your needs. Its purpose is mainly to furnish enough examples so that you can develop the specialized components for your own simulation. It should also help you decipher six-DoF simulations when you are asked to evaluate or to modify them. For that reason, I refer you also to the simulation examples on the CADAC Web site, which enrich this chapter: the FALCON6 aircraft simulation, the GHAME6 hypersonic vehicle, and the SRAAM6 short-range air-to-air missile.

10.1   Six-DoF Equations of Motion

Before we embark on our journey, it would be to your advantage to stop by at Chapter 5 and review Newton's law and Chapter 6 on Euler's law. Otherwise, starting from these general principles the derivation here is self-contained.

As always, I shall first formulate the equations in an invariant tensor form first, followed by the all-important decision of coordinate systems. The form of the MOI tensor will lead us into two different directions. For missiles with tetragonal symmetry, the equations are so simple that we write them in scalar form. For aircraft with planar symmetry however, we have to live with an inverted moment of inertia matrix.

The selection of the inertial reference frame is always an important decision that you have to make before you begin your modeling task. For near-Earth aerospace vehicles, including low-Earth orbiting satellites, the Earth-centered, ecliptic-oriented frame serves as inertial reference (J2000). It can be combined with a spherical or elliptical Earth shape. However, frequently the trajectory of a vehicle is so slow compared to orbital speed and so close to the Earth that the centrifugal and Coriolis accelerations can be neglected. In these cases we choose the Earth as the inertial frame and unwrap the Earth's surface into a plane tangential to the launch point. We speak now of the flat-Earth approach.

The choice is yours to make. I will first discuss the easier flat-Earth case, applicable to tactical missiles and aircraft and add two special cases: spinning missile and Magnus rotor. Then I will lure you over to the much more difficult modeling task of the elliptical Earth. We will make excursions into geodesy and study the shape and gravitational field of the Earth before we derive the equations of motions. The inertial frame is the Earth-centered, ecliptic-oriented frame. To provide you with the full spectrum of formulations, I will not only use the inertial frame as a reference frame but also the Earth frame. You will encounter both options in six-DoF simulations.

10.1.1   Flat-Earth Equations of Motion

Unless you work exclusively with space applications, you will encounter the flat-Earth equations of motion more often in six-DoF simulations. We first derive the general form of the equations of motion and then specialize them for missiles and aircraft based on the characteristic of the MOI tensor. Spinning missiles and Magnus rotors will each require a fresh approach.

Let us first derive the translational equations of a flight vehicle subjected to the aerodynamic and thrust forces $f_{a,p}$ and the gravitational acceleration g. Newton's second law wrt the inertial frame I states that the time rate of change of linear momentum equals the externally applied forces:

$$mD^I v_B^I = f_{a,p} + mg$$

where m is the vehicle mass and v_B^I is the velocity of the missile c.m. B wrt the inertial reference frame I. The flat-Earth assumption allows us to declare the Earth frame E as the inertial frame. Therefore, Newton's law becomes

$$mD^E v_B^E = f_{a,p} + mg$$

To model the aerodynamic forces, we need access to the incidence angles. They are calculated from the velocity vector as perceived from the vehicle as reference frame. Its time rate of change $D^B v_B^E$ is therefore referred to the vehicle B. This desired shift of reference frame is accomplished through the Euler transformation

$$mD^B v_B^E + m\Omega^{BE} v_B^E = f_{a,p} + mg \qquad (10.1)$$

These are the translational equations of a vehicle over a flat Earth. The second term on the left-hand side is identified as the *tangential acceleration* term. The equation is valid in any coordinate system.

Now comes the important event of picking the coordinate system. Actually, I have already preempted the decision by transforming the rotational derivative to the body frame. To generate the ordinary time derivative, the rotational derivative must be expressed in the coordinate system $]^B$ associated with the frame B. Expressing all terms in $]^B$ yields

$$m\left[\frac{dv_B^E}{dt}\right]^B + m[\Omega^{BE}]^B[v_B^E]^B = [f_{a,p}]^B + m[g]^B$$

This selection suits us well because it models $[v_B^E]^B$, the linear velocity of vehicle wrt Earth, as needed for incidence angle calculations; $[\Omega^{BE}]^B$, the angular velocity of vehicle wrt Earth; and $[f_{a,p}]^B$, the aerodynamic and propulsive forces, all expressed in the preferred coordinate system $]^B$. Only the last term is better formulated in the local-level coordinate system $]^L$ so that g takes on the simple form $[\bar{g}]^L = [0 \ 0 \ g]$. Alas, the direction cosine matrix $[T]^{BL}$ appears, which we will derive later from Euler's equations. The translational equations in matrix form, suitable for programming, are then

$$m\left[\frac{dv_B^E}{dt}\right]^B + m[\Omega^{BE}]^B[v_B^E]^B = [f_{a,p}]^B + m[T]^{BL}[g]^L \qquad (10.2)$$

Much labor is involved in modeling the forces $[f_{a,p}]^B$. In the next section I will give you a taste for the complexity of six-DoF aerodynamics, both for missiles and aircraft. Here we proceed blissfully and code the equations directly with program languages like FORTRAN or C. You can also choose the simulation environment

CADAC, which provides the utility subroutines that enable vector state variable integration and matrix manipulations. Written out in coordinate form, we get

$$
m \left\{ \begin{bmatrix} du/dt \\ dv/dt \\ dw/dt \end{bmatrix}^B + \begin{bmatrix} 0 & -r & q \\ r & 0 & -p \\ -q & p & 0 \end{bmatrix}^B \begin{bmatrix} u \\ v \\ w \end{bmatrix}^B \right\}
$$

$$
= \begin{bmatrix} f_{a,p_1} \\ f_{a,p_2} \\ f_{a,p_3} \end{bmatrix}^B + \begin{bmatrix} t_{11} & t_{12} & t_{13} \\ t_{21} & t_{22} & t_{23} \\ t_{31} & t_{32} & t_{33} \end{bmatrix}^{BL} \begin{bmatrix} 0 \\ 0 \\ mg \end{bmatrix}^L
$$

These equations are simple enough to be expressed in scalar form:

$$
\frac{du}{dt} = rv - qw + \frac{f_{a,p_1}}{m} + t_{13}g
$$

$$
\frac{dv}{dt} = pw - ru + \frac{f_{a,p_2}}{m} + t_{23}g \qquad (10.3)
$$

$$
\frac{dw}{dt} = qu - pv + \frac{f_{a,p_3}}{m} + t_{33}g
$$

The integration of these differential equations yields the velocity vector that must be integrated once more to obtain the location of the missile c.m. B wrt an Earth reference point E

$$
[D^E s_{BE}] = [v_B^E] \qquad (10.4)
$$

The integration is best carried out in the local-level coordinate system. Therefore, given $[v_B^E]^B$, we program the second set of differential equations as

$$
\left[\frac{ds_{BE}}{dt} \right]^L = [\bar{T}]^{BL} [v_B^E]^B \qquad (10.5)
$$

and in coordinate form

$$
\begin{bmatrix} (ds_{BE}/dt)_1 \\ (ds_{BE}/dt)_2 \\ (ds_{BE}/dt)_3 \end{bmatrix}^L = \overline{\begin{bmatrix} t_{11} & t_{12} & t_{13} \\ t_{21} & t_{22} & t_{23} \\ t_{31} & t_{32} & t_{33} \end{bmatrix}}^{BL} \begin{bmatrix} u \\ v \\ w \end{bmatrix}^B \qquad (10.6)
$$

Here we have the six first-order differential equations that govern the translational motions of a vehicle with the Earth as the inertial reference frame. Equations (10.3) are nonlinear and coupled by the body rates p, q, and r and the direction cosine matrix $[T]^{BL}$ with the attitude equations. The nonlinearity enters through the incidence angles in the aerodynamic force and moment calculations [for instance, see Eqs. (10.61) and (10.62)]. Equations (10.6) are linear differential equations, again coupled by $[T]^{BL}$ with the attitude equations.

The rotational degrees of freedom are governed by Euler's law that states that the time rate of change of angular momentum equals the externally applied moments.

To conform to the translational equations, we pick E as the inertial frame:

$$D^E l_B^{BE} = m_B \qquad (10.7)$$

where $l_B^{BE} = I_B^B \omega^{BE}$ is the angular momentum of body B wrt frame E referred to the c.m. B, I_B^B is the MOI of missile body B referred to the c.m., and m_B the aerodynamic and thrust moments referred to the c.m.

Just as in the case of the translational equations, we take the perspective of the vehicle body frame B and use therefore the Euler's transformation to transfer the rotational derivative to the body frame

$$D^B l_B^{BE} + \Omega^{BE} l_B^{BE} = m_B$$

Let us focus on the rotational derivative, expanding the angular momentum vector, and applying the chain rule

$$D^B l_B^{BE} = D^B \left(I_B^B \omega^{BE} \right) = D^B I_B^B \omega^{BE} + I_B^B D^B \omega^{BE} = I_B^B D^B \omega^{BE}$$

For a rigid body $D^B I_B^B$ is zero, a simplification that motivates the transformation to the body frame. The equation reduces to

$$I_B^B D^B \omega^{BE} + \Omega^{BE} I_B^B \omega^{BE} = m_B \qquad (10.8)$$

These are the rotational equations of a vehicle with respect to a flat Earth. The second term on the left-hand side is the apparent gyroscopic effect, arising from the choice of the body as reference frame.

As a coordinate system, we pick again the body coordinates because they express the MOI tensor in constant form, yield the ordinary time derivative of the body rates, and express the aerodynamic moments in their preferred form:

$$[I_B^B]^B \left[\frac{d\omega^{BE}}{dt} \right]^B + [\Omega^{BE}]^B [I_B^B]^B [\omega^{BE}]^B = [m_B]^B \qquad (10.9)$$

For an arbitrary $[I_B^B]^B$ we premultiply the equation by its inverse and solve for the time derivative

$$\left[\frac{d\omega^{BE}}{dt} \right]^B = \left([I_B^B]^B \right)^{-1} \left(-[\Omega^{BE}]^B [I_B^B]^B [\omega^{BE}]^B + [m_B]^B \right) \qquad (10.10)$$

These three first-order nonlinear differential equations couple with the translational equations (10.2) only through the aerodynamic moments $[m_B]^B$. The form of the MOI tensor $[I_B^B]^B$ establishes a major demarcation. Missiles exhibit pure diagonal forms, resulting in simple equations, whereas aircraft have one off-diagonal element that leads to more complex solutions. We reserve therefore in the following sections different treatments for missiles and aircraft. Here we complete that part of the treatment of the attitude kinematics, which is common to both.

Given the body rates $[\overline{\omega^{BE}}]^B = [p \quad q \quad r]$ from Eq. (10.10), a second set of differential equations provides the body attitudes. We discussed the three possibilities in Chapter 4: Euler angle, quaternion, or direction cosine equations. Here we pursue the quaternion approach. Go back to Sec. 4.3.3 and review quaternions, or, like many other impatient engineers, just program the following equations.

The four-quaternion elements q_0, q_1, q_2, and q_3 are related to the three body-rate components by four linear differential equations:

$$
\begin{Bmatrix} \dot{q}_0 \\ \dot{q}_1 \\ \dot{q}_2 \\ \dot{q}_3 \end{Bmatrix} = \frac{1}{2} \begin{bmatrix} 0 & -p & -q & -r \\ p & 0 & r & -q \\ q & -r & 0 & p \\ r & q & -p & 0 \end{bmatrix} \begin{Bmatrix} q_0 \\ q_1 \\ q_2 \\ q_3 \end{Bmatrix}
\tag{10.11}
$$

To solve the equations, they must be initialized. Unfortunately, quaternions are hard to visualize; therefore, it is more likely that the initial Euler angles are given. The conversions are

$$
q_0 = \cos\left(\frac{\psi}{2}\right)\cos\left(\frac{\theta}{2}\right)\cos\left(\frac{\phi}{2}\right) + \sin\left(\frac{\psi}{2}\right)\sin\left(\frac{\theta}{2}\right)\sin\left(\frac{\phi}{2}\right)
$$

$$
q_1 = \cos\left(\frac{\psi}{2}\right)\cos\left(\frac{\theta}{2}\right)\sin\left(\frac{\phi}{2}\right) - \sin\left(\frac{\psi}{2}\right)\sin\left(\frac{\theta}{2}\right)\cos\left(\frac{\phi}{2}\right)
$$

$$
q_2 = \cos\left(\frac{\psi}{2}\right)\sin\left(\frac{\theta}{2}\right)\cos\left(\frac{\phi}{2}\right) + \sin\left(\frac{\psi}{2}\right)\cos\left(\frac{\theta}{2}\right)\sin\left(\frac{\phi}{2}\right)
$$

$$
q_3 = \sin\left(\frac{\psi}{2}\right)\cos\left(\frac{\theta}{2}\right)\cos\left(\frac{\phi}{2}\right) - \cos\left(\frac{\psi}{2}\right)\sin\left(\frac{\theta}{2}\right)\sin\left(\frac{\phi}{2}\right)
\tag{10.12}
$$

For example, an aircraft's Euler angles are zero if it starts flying north, straight, and level. Its quaternions are also zero with the exception of the scalar component $q_0 = 1$.

If the differential equations (10.11) could be solved exactly, we would not have to worry about maintaining the orthonormality of the quaternion. However, the use of numerical integration schemes introduces errors caused by finite word length and discretization of time. We can maintain the unit norm of the quaternion by a proven trick.[1] Let the orthonormality error λ be

$$
\lambda = 1 - \left(q_0^2 + q_1^2 + q_2^2 + q_3^2\right)
$$

and add to the right side of Eq. (10.11) the factor $k\lambda\{q\}$, where $\{q\}$ is the four-element quaternion vector. The constant k is chosen such that $k\Delta t \le 1$, with Δt the integration interval. Its value is not crucial. I have tried several k and found the effect quite acceptable. For the CADAC SRAAM6 simulation, after some testing I picked $k = 0.5$ at $\Delta t = 0.001$.

Given the quaternions, the Euler angles are derived from the following relationships:

$$
\tan\psi = \frac{2(q_1 q_2 + q_0 q_3)}{q_0^2 + q_1^2 - q_2^2 - q_3^2}, \qquad \sin\theta = -2(q_1 q_3 - q_0 q_2),
$$

$$
\tan\phi = \frac{2(q_2 q_3 + q_0 q_1)}{q_0^2 - q_1^2 - q_2^2 - q_3^2}
\tag{10.13}
$$

The first equation has singularities at $\psi = \pm 90$ deg, and the last equation at $\phi = \pm 90$ deg. These singularities are not serious because they occur during off-line calculations and not inside the differential equations. They can easily be bypassed by programming around them.

The direction cosine matrix $[T]^{BL}$ could be calculated from the Euler angles. However, because of the singularities, it is better to use the quaternion relationship directly:

$$[T]^{BL} = \begin{bmatrix} q_0^2 + q_1^2 - q_2^2 - q_3^2 & 2(q_1 q_2 + q_0 q_3) & 2(q_1 q_3 - q_0 q_2) \\ 2(q_1 q_2 - q_0 q_3) & q_0^2 - q_1^2 + q_2^2 - q_3^2 & 2(q_2 q_3 + q_0 q_1) \\ 2(q_1 q_3 + q_0 q_2) & 2(q_2 q_3 - q_0 q_1) & q_0^2 - q_1^2 - q_2^2 + q_3^2 \end{bmatrix}$$

(10.14)

In summary, the three translational degrees of freedom are governed by the six first-order differential equations (10.2) and (10.6). The three rotational degrees of freedom are calculated from the three first-order differential equations (10.10) and the four kinematic differential equations (10.11).

With the general equations of motion of a vehicle over a flat Earth in place, we now turn to special expressions of Eq. (10.10). The form of the MOI tensor will determine whether we call them missile or aircraft equations.

10.1.1.1 Missile equations.

Missiles have simple MOI tensors. Because of symmetry, the body axes coincide with the principal axes, and the tensor exhibits only diagonal elements. A particular important case is the missile with tetragonal symmetry, which manifests two equal principal MOI. (We exclude here the cruise missiles with their aircraft-like symmetries and attribute them to the aircraft equations in the next section.)

Equation (10.10) with a diagonal MOI tensor $[I_B^B]^B$ is in component form

$$\begin{bmatrix} dp/dt \\ dq/dt \\ dr/dt \end{bmatrix}^B = \begin{bmatrix} I_1^{-1} & 0 & 0 \\ 0 & I_2^{-1} & 0 \\ 0 & 0 & I_3^{-1} \end{bmatrix}^B \left(- \begin{bmatrix} 0 & -r & q \\ r & 0 & -p \\ -q & p & 0 \end{bmatrix}^B \right.$$

$$\left. \times \begin{bmatrix} I_1 & 0 & 0 \\ 0 & I_2 & 0 \\ 0 & 0 & I_3 \end{bmatrix}^B \begin{bmatrix} p \\ q \\ r \end{bmatrix}^B + \begin{bmatrix} m_{B_1} \\ m_{B_2} \\ m_{B_3} \end{bmatrix}^B \right)$$

(10.15)

and it is easily expressed in three scalar equations:

$$\frac{dp}{dt} = I_1^{-1} \left[(I_2 - I_3) qr + m_{B_1} \right]$$

$$\frac{dq}{dt} = I_2^{-1} \left[(I_3 - I_1) pr + m_{B_2} \right]$$

(10.16)

$$\frac{dr}{dt} = I_3^{-1} \left[(I_1 - I_2) pq + m_{B_3} \right]$$

These first-order differential equations are nonlinear and coupled only through the aerodynamic moments m_{B_1}, m_{B_2}, and m_{B_3} to the translational equations. Their integration yields the body rates p, q, and r. Let me point out again that the term *body rates* in flight mechanics refers to the angular velocity of the body wrt the inertial frame expressed in body axes, in short $[\overline{\omega^{BE}}]^B = [p \ q \ r]$.

Many missiles exhibit tetragonal symmetry, i.e., $I_3 = I_2$, and therefore possess rotational equations that are even simpler:

$$\frac{dp}{dt} = I_1^{-1} m_{B_1}$$

$$\frac{dq}{dt} = I_2^{-1}\left[(I_2 - I_1)pr + m_{B_2}\right]$$

$$\frac{dr}{dt} = I_2^{-1}\left[(I_1 - I_2)pq + m_{B_3}\right]$$

We are now in a position to summarize the equations that form the core of a missile's six-DoF simulation. They are displayed in Fig. 10.1. The translational degrees of freedom, represented by the velocity components u, v, and w, are solved by Newton's equation (10.3); and the rotational DoF, expressed in body rates p, q, and r, are governed by Euler's equation (10.16). The kinematic equations block calculates body attitudes in the form of Euler angles ψ, θ, and ϕ and the direction cosine matrix $[T]^{BL}$ from body rates using the quaternion methodology [Eqs. (10.11), (10.13), and (10.14)]. For aerodynamic table look-ups the incidence angles α' and ϕ', peculiar to missiles, are displayed [see Eqs. (3.22) and (3.23)]. Finally, the aerodynamic coefficients and propulsive forces are combined in the forces and moments block. Notice the special missile features of expressing the force coefficients as C_A, C_Y, and C_N in body axes. Refer to Sec. 10.2.1.3, in which I treat the modeling of missile aerodynamics in greater depth.

As I mentioned, these blocks form just the basic loop of a six-DoF simulation. In a full-up missile model they are joined by an autopilot and actuator to form the inner loop. An outer loop, supplying the navigation and guidance functions, completes the missile simulation. Modeling details of these additional components will be presented in Secs. 10.2.2, 10.2.4, and 10.2.5.

10.1.1.2 Aircraft equations. An aircraft, in contrast to a missile, does not have all of its principle axes aligned with the body axes. The 1^B axis may be parallel to what is called the *water line* (an expression borrowed from naval architecture), aligned with the zero-lift attitude, or just poke through the nose of the radome. Whatever the case may be, it is not, in general, the principal MOI axes, nor is the 3^B axis. However, thanks to the planar symmetry wrt the 1^B and 3^B axes, the 2^B axis is indeed a principal axis. Such a configuration gives rise to a MOI tensor in body coordinates of the form

$$\left[I_B^B\right]^B = \begin{bmatrix} I_{11} & 0 & I_{13} \\ 0 & I_2 & 0 \\ I_{31} & 0 & I_{33} \end{bmatrix}$$

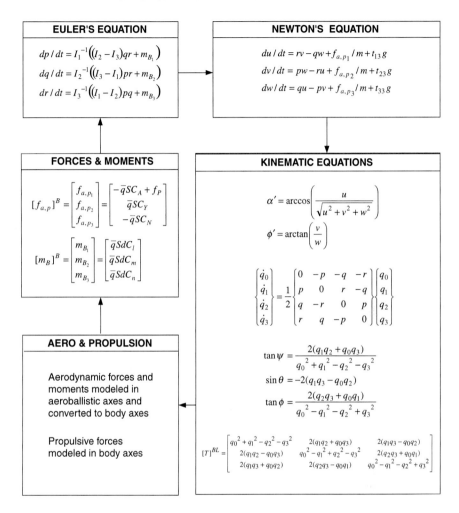

Fig. 10.1   Summary of kinematic and dynamic equations of missiles.

with product of inertia components at the 1,3 and 3,1 locations and $I_{13} = I_{31}$. On the diagonal we have the axial moments of inertia I_{11} and I_{33} and the principal moment of inertia I_2.

We can program Euler's equation (10.10) directly, tolerating the inversion of the moment of inertia matrix, or, because there are still four zero elements in $[I_B^B]^B$, it is possible to solve for the body rate components explicitly and write down the scalar equations.

Before proceeding, I want to expand the angular momentum vector with additional rotary devices like propellers or turbojets, which are an important feature of airplanes. Let the rotary frame be B_R and its c.m. B_R. The angular momentum is designated $l_{B_R}^{B_R E}$ and the angular momentum of the remaining rigid aircraft cell is as previously $l_B^{BE} = I_B^B \omega^{BE}$. Combining both according to Example 6.6, the total

MOI referred to the common c.m. C becomes

$$l_C^{\Sigma BE} = l_B^{BE} + l_{B_R}^{B_R}$$

and Euler's law

$$D^E l_B^{BE} + D^E l_{B_R}^{B_R} = m_C$$

As before, we transform the rotational derivative to the reference frame associated with the aircraft cell B:

$$D^B l_B^{BE} + \Omega^{BE} l_B^{BE} + D^B l_{B_R}^{B_R E} + \Omega^{BE} l_{B_R}^{B_R E} = m_C$$

Now, let us simplify the solution by neglecting changes in rotary motion $D^B l_{B_R}^{B_R E} = 0$, i.e., assuming that the propellers or turbojets spin at constant speed. With the additional stipulation that the aerodynamic and propulsive moments can be referred to the cell's c.m. B (rather than C), we obtain the tensor equation for the attitude motions, without incurring any significant errors

$$I_B^B D^B \omega^{BE} + \Omega^{BE}(I_B^B \omega^{BE} + l_{B_R}^{B_R E}) = m_B$$

Expressed in aircraft cell coordinates $]^B$

$$[I_B^B]^B \left[\frac{d\omega^{BE}}{dt}\right]^B + [\Omega^{BE}]^B([I_B^B]^B[\omega^{BE}]^B + [l_{B_R}^{B_R E}]^B) = [m_B]^B$$

and solved for the body-rate derivative

$$\left[\frac{d\omega^{BE}}{dt}\right]^B = ([I_B^B]^B)^{-1}[-[\Omega^{BE}]^B([I_B^B]^B[\omega^{BE}]^B + [l_{B_R}^{B_R E}]^B) + [m_B]^B] \quad (10.17)$$

Compare this equation with Eq. (10.10) and note the coupling of the rotary angular momentum $[l_{B_R}^{B_R E}]^B$ with the aircraft body rates $[\Omega^{BE}]^B$. This vector product models the gyroscopic effect that the rotary engine imparts on the aircraft.

Before we express the component equations, let us make the reasonable assumption that the spin axes of all rotary parts are parallel to the aircraft longitudinal axes; therefore, $[l_{B_R}^{B_R E}]^B = [l_R \ 0 \ 0]$. Now with $I_{13} = I_{31}$, we express Eq. (10.17) in coordinates

$$\begin{bmatrix} dp/dt \\ dq/dt \\ dr/dt \end{bmatrix}^B = \left(\begin{bmatrix} I_{11} & 0 & I_{13} \\ 0 & I_2 & 0 \\ I_{13} & 0 & I_{33} \end{bmatrix}^B\right)^{-1} \left(-\begin{bmatrix} 0 & -r & q \\ r & 0 & -p \\ -q & p & 0 \end{bmatrix}^B\right.$$

$$\times \left(\begin{bmatrix} I_{11} & 0 & I_{13} \\ 0 & I_2 & 0 \\ I_{13} & 0 & I_{33} \end{bmatrix}^B \begin{bmatrix} p \\ q \\ r \end{bmatrix}^B + \begin{bmatrix} l_R \\ 0 \\ 0 \end{bmatrix}^B\right) + \left.\begin{bmatrix} m_{B_1} \\ m_{B_2} \\ m_{B_3} \end{bmatrix}^B\right)$$

Because the MOI matrix has off-diagonal elements, solving for the body-rate derivatives is somewhat convoluted. I will spare you the details and just provide the result:

$$\frac{dp}{dt} = \frac{1}{I_{11}I_{33} - I_{13}^2}\{[(I_2I_{33} - I_{33}^2 - I_{13}^2)r$$
$$- I_{13}(I_{33} + I_{11} - I_2)p - I_{13}l_R]q + I_{33}m_{B_1} - I_{13}m_{B_3}\}$$

$$\frac{dq}{dt} = \frac{1}{I_2}\{[(I_{33} - I_{11})p - l_R]r + I_{13}(p^2 - r^2) + m_{B_2}\} \qquad (10.18)$$

$$\frac{dr}{dt} = \frac{1}{I_{11}I_{33} - I_{13}^2}\{[(-I_{11}I_2 + I_{11}^2 + I_{13}^2)p$$
$$- I_{13}(I_{33} + I_{11} - I_2)r + I_{11}l_R]q + I_{11}m_{B_3} - I_{13}m_{B_1}\}$$

As you can see, the scalar equations of motions of an aircraft are much more complicated than those for missiles, Eq. (10.16). You have a choice now, either to program these scalar equations or the matrix equations (10.17). With today's computer power at you fingertips, I recommend you use the simpler, less error prone matrix formulation.

Comparing Eqs. (10.16) and (10.18) provides us with some insight into the differences of missiles and aircraft dynamics. Let us first contrast the roll equations. The product of inertia I_{13} couples the pitch and yaw rates to the roll degrees of freedom of the aircraft. The greater its value, the stronger the coupling. If $I_{13} = 0$, the missile roll equation is recovered. The pitch and yaw equations exhibit similar trends. Other phenomena caused by I_{13} are the coupling of the aerodynamic yawing moment m_{B_3} into the roll axis and the aerodynamic rolling moment m_{B_1} into the yaw axis. Finally, the rotary angular momentum l_R couples through the pitch rate into both the roll and yaw derivative equations, while the yaw rate connects it with the pitch derivative.

Could we have drawn the same conclusions from the matrix equation (10.17)? The answer is yes, in a general sense; it shows the coupling of the body rates $[\omega^{BE}]^B$ through the MOI tensor $[I_B^B]^B$ into the body-rate derivatives, as well as the coupling of the aerodynamic moments $[m_B]^B$ through the inverse $([I_B^B]^B)^{-1}$. Furthermore, the vector product of $[\Omega^{BE}]^B$ couples the rotary angular momentum $[l_{B_k}^{B_k}]^B$. However, by comparing Eqs. (10.17) with (10.10) we would not have discovered the much simpler missile equations, unless we had scrutinized the simplifying effect of the diagonal moment of inertia matrix.

To sum up, I have assembled the important aircraft equations in Fig. 10.2, just as I did with the missile equations in Fig. 10.1. Euler's equations are copied from Eq. (10.18), and Newton's equations are adopted unchanged from the missile model. The incidence angles are now angle of attack α and sideslip angle β, and the aerodynamic forces are modeled in body axes directly.

The missile and aircraft six-DoF equations are the bread and butter of simulation engineers. They were derived with the assumption that the Earth is the inertial frame and that the curved Earth can be unfurled into a flat plane. How these equations are integrated with other subsystems, such as autopilots, actuators and navigation systems will be described later. Before we leave this section, however, I would like

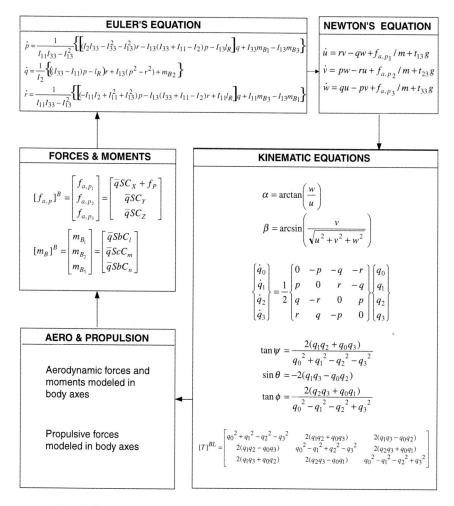

Fig. 10.2   Summary of kinematic and dynamic equations of aircraft.

to derive the especially interesting equations of motions for spinning missiles and
Magnus rotors.

10.1.1.3 Spinning missile equations.   As you stroll through an armament
museum and visit the tactical missile section, you will notice missiles with canard
or tail control. Tail-controlled missiles, like the Advanced Medium-Range Air-to-
Air Missile, have full three axes stabilization and tight control over the three body
rates, roll, pitch, and yaw. Canard-controlled missiles on the other hand exhibit
a nasty coupling of pitch and yaw control into the rolling moment. Because this
rolling moment is difficult to suppress, the airframe is left to roll freely, or, like in
the case of the Sidewinder air-to-air missile, the roll rate is dampened by so-called
rolerons. A third type of missile is intentionally rolled for stability. A good example

is the Rolling Airframe Missile, which uses angular momentum for gyroscopic stability.

The first type of missile can be modeled by the equations summarized in Fig. 10.1. Spin-stabilized missiles, on the other hand, require a special set of six-DoF equations that decouple the fast roll rate from the pitch and yaw dynamics. A so-called nonspinning reference frame B' is introduced that slips like a sheath over the spinning body. Its justification comes from the aerodynamic modeling of spinning symmetrical bodies that gives rise to such terms as the Magnus lift and moment coefficients.

The translational equations are transferable from the flat-Earth equations (10.1) and (10.4) by substituting frame B' for B.

$$m D^{B'} v_B^E + m \Omega^{B'E} v_B^E = f_{a,p} + mg$$

with its matrix version expressed in nonspinning body axes $]^{B'}$

$$m \left[\frac{dv_B^E}{dt} \right]^{B'} + m [\Omega^{B'E}]^{B'} [v_B^E]^{B'} = [f_{a,p}]^{B'} + m[T]^{B'L} [g]^L \qquad (10.19)$$

The $]^{B'}$ coordinates, because they do not follow the rolling of the missile, are obtained from the local-level coordinates by only two rotations, namely, yaw and pitch, thus, the second axis of $]^{B'}$ always lies in the horizontal plane.

The position of the missile follows from the integration of the second set of differential equations (10.4), expressed in $]^{B'}$ coordinates.

$$\left[\frac{ds_{BE}}{dt} \right]^L = [\bar{T}]^{B'L} [v_B^E]^{B'} \qquad (10.20)$$

These translational equations of spinning missiles have the same appearance as those of nonspinning missiles. The only difference consists in applying the nonspinning reference frame B' and its associated coordinate system $]^{B'}$. Major differences exist, however, in the attitude equations.

The derivation of the attitude equations starts with Eq. (10.7), applied to the nonspinning body frame B'

$$D^{B'} l_B^{BE} + \Omega^{B'E} l_B^{BE} = m_B \qquad (10.21)$$

We divide the angular velocity ω^{BE} of the missile body B wrt the reference frame E into two parts:

$$\omega^{BE} = \omega^{BB'} + \omega^{B'E}$$

The first part is the spin rate, and the second part is the angular velocity of the nonspinning shell wrt Earth. Similarly, the angular momentum is split into two parts:

$$l_B^{BE} = I_B^B \omega^{BE} = I_B^B \omega^{BB'} + I_B^B \omega^{B'E}$$

As called for by Eq. (10.21), we apply the rotational derivative wrt to B'

$$D^{B'} l_B^{BE} = D^{B'} \left(I_B^B \omega^{BB'} \right) + D^{B'} \left(I_B^B \omega^{B'E} \right)$$

$$= I_B^B D^{B'} \omega^{BB'} + I_B^B D^{B'} \omega^{B'E}$$

where we made use of $D^{B'}I_B^B = 0$, based on the assumption that the missile has rotational symmetry. Thus, the attitude equation of motion is

$$I_B^B D^{B'} \omega^{BB'} + I_B^B D^{B'} \omega^{B'E} + \Omega^{B'E} I_B^B \omega^{BB'} + \Omega^{B'E} I_B^B \omega^{B'E} = m_B$$

The first term models the change in angular momentum caused by the variable spin rate of the missile. For a missile with constant spin rate, this term is zero. The second term represents the change of angular momentum of the nonspinning body shell as it is subjected to yaw and pitch rate changes. The gyroscopic coupling moment is expressed by the third term, which usually dominates the weaker coupling of the fourth term.

The attitude equation is expressed in nonspinning body coordinates:

$$[I_B^B]^{B'} \left[\frac{d\omega^{BB'}}{dt} \right]^{B'} + [I_B^B]^{B'} \left[\frac{d\omega^{B'E}}{dt} \right]^{B'} + [\Omega^{B'E}]^{B'} [I_B^B]^{B'} \left([\omega^{BB'}]^{B'} \right.$$

$$\left. + [\omega^{B'E}]^{B'} \right) = [m_B]^{B'} \tag{10.22}$$

Carefully distinguish between the letters B and B'. The point B is always the c.m. of the missile, and the coordinate system is always the nonrotating $]^{B'}$; but the frame may be the body fixed frame B, as in I_B^B, or the nonrolling frame B', as in $\omega^{B'E}$.

Rolling missile airframes usually possess rotational symmetry relative to the 1^B axis. The MOI tensor therefore has a particular simple diagonal form, which is the same in the $]^B$ and $]^{B'}$ coordinate systems. With the two cross-principal moments of inertia being equal $I_3 = I_2$, we have

$$[I_B^B]^{B'} = \begin{bmatrix} I_1 & 0 & 0 \\ 0 & I_2 & 0 \\ 0 & 0 & I_2 \end{bmatrix}^{B'}, \quad [\omega^{BB'}]^{B'} = \begin{bmatrix} \omega \\ 0 \\ 0 \end{bmatrix}^{B'}, \quad [\omega^{B'E}]^{B'} = \begin{bmatrix} p' \\ q' \\ r' \end{bmatrix}^{B'}$$

I also included the components of the spin angular velocity $[\omega^{BB'}]^B$ and the angular velocity of the nonrolling frame wrt Earth $[\omega^{BE}]^{B'}$. Substituting these matrices into Eq. (10.22) yields

$$\begin{bmatrix} I_1(d\omega/dt) \\ 0 \\ 0 \end{bmatrix}^{B'} + \begin{bmatrix} I_1 dp'/dt \\ I_2 dq'/dt \\ I_2 dr'/dt \end{bmatrix}^{B'} + \begin{bmatrix} 0 & -r' & q' \\ r' & 0 & -p' \\ -q' & p' & 0 \end{bmatrix}^{B'} \begin{bmatrix} I_1 & 0 & 0 \\ 0 & I_2 & 0 \\ 0 & 0 & I_2 \end{bmatrix}^{B'}$$

$$\times \left(\begin{bmatrix} \omega \\ 0 \\ 0 \end{bmatrix}^{B'} + \begin{bmatrix} p' \\ q' \\ r' \end{bmatrix}^{B'} \right) = \begin{bmatrix} m_{B_1} \\ m_{B_2} \\ m_{B_3} \end{bmatrix}^{B'}$$

Matrix multiplications deliver the three scalar differential equations

$$I_1 \left(\frac{d\omega}{dt} + \frac{dp'}{dt} \right) = m_{B_1}$$

$$I_2 \left(\frac{dq'}{dt} \right) + I_1 \omega r' - p' r' (I_2 - I_1) = m_{B_2} \qquad (10.23)$$

$$I_2 \left(\frac{dr'}{dt} \right) - I_1 \omega q' + p' q' (I_2 - I_1) = m_{B_3}$$

By definition, the rolling motion of the nonrolling airframe is zero; therefore $dp'/dt = p' = 0$, and the equations simplify to

$$\frac{d\omega}{dt} = I_1^{-1} m_{B_1}$$

$$\frac{dq'}{dt} = I_2^{-1} \left(-I_1 \omega r' + m_{B_2} \right) \qquad (10.24)$$

$$\frac{dr'}{dt} = I_2^{-1} \left(I_1 \omega q' + m_{B_3} \right)$$

These are the Euler equations for spinning missiles. Compare them with the missile equations (10.16). They differ in the gyroscopic terms. Because of an overriding spin angular velocity ω, the gyroscopic moments $I_1 \omega r'$ and $I_1 \omega q'$ dominate the cross coupling. The yaw rate r' couples into the pitch degrees of freedom and, inversely, the pitch rate affects the yaw rate equation. This coupling gives rise to a nutational mode, which effects the missile dynamics profoundly. The programming of these equations follows the same layout as provided in Fig. 10.1, except that the body rates p, q, and r are replaced by the primed variables q' and r', while p' and ϕ are zero.

10.1.1.4 Magnus rotor.
If you liked the derivation of the spinning missile equations, you will really enjoy this excursion into the dynamics of bodies that spin about a horizontal axis. They are my favorite subject because I spent five years studying their characteristics, culminating in my dissertation.[2]

Spinning bodies were first investigated by Heinrich Gustav Magnus in 1852. He demonstrated experimentally that a body rotating in an airstream experiences a force that acts substantially normal to the airflow. An autorotating flight vehicle, designed to develop this Magnus force efficiently and to employ it as the major lift force in free flight, is called a Magnus rotor. It consists of a center body with driving vanes and endplates.

Figure 10.3 depicts a Magnus rotor spinning about a near horizontal axis and gliding with the velocity v_B^E. The spin axes 2^S and the velocity vector form a plane containing the roll axis 1^S normal to the spin axis. The yaw axis 3^L is parallel to the gravitational acceleration.

The dynamics of Magnus rotors can best be explained in terms of a horizontally spinning gyroscope that is subjected to aerodynamic forces. The dynamic motions of the glide trajectory are rolling $\dot\phi$, yawing $\dot\psi$, and sideslipping β. They exhibit three modes: nutation, precession, and undulation. The *nutation* and *precession*

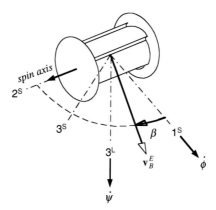

Fig. 10.3   Magnus rotor in descending flight.

modes are those of a gyroscope with the modification that the nutation is aerodynamically dampened. New is the *undulation* mode. It arises from the interaction of aerodynamic moments—generated by β—and gyroscopic precession. In general, this mode can be oscillatory or aperiodic.

You can understand the fun I had working on this project, developing the theory, building the computer simulation, and flight testing some specimens. For our current purpose we limit the discussion to the derivation of the equations of motion, the starting point of any six-DoF simulation of Magnus rotors.

As always, the translational equation is based on Newton's second law

$$m D^E v_B^E = f_a + f_g$$

and the rotational equation derives from Euler's law

$$D^E l_B^{BE} = m_B$$

with aerodynamic forces and moments f_a, m_B and the gravitational force $f_g = mg$.

Just as I used a nonspinning frame B' for the spinning missile, in the same vein I introduce here a frame that is attached to the Magnus rotor but does not follow its spin. It is called the *stability* frame and is defined by the spin axis and the projection of the velocity vector on the symmetry plane of the rotor. The associated stability coordinate system has its 2^S axis aligned and in the direction of the spin vector, its 1^S axes parallel and in the direction of the projection of the velocity vector on the symmetry plane. The third axis 3^S completes the triad. The transformation matrix $[T]^{SL}$ of the stability wrt the local-level axes follows the conventional transformation of Euler angles: yaw ψ_{SL}, pitch θ_{SL}, and roll ϕ_{SL} (see Sec. 3.2.2.4).

The first step is to transform Newton's equation to the stability frame and express the terms in stability coordinates

$$m \left[\frac{d v_B^E}{dt} \right]^S + m [\Omega^{SE}]^S [v_B^E]^S = [f_a]^S + m [T]^{SL} [g]^L \tag{10.25}$$

Focus on the meaning of $[\Omega^{SE}]^S$. It is the skew-symmetric form of the angular velocity vector of the stability frame wrt Earth, which embodies the rolling and yawing rates, but excludes the spinning motion. The choice of the stability coordinates is dictated by the method of the aerodynamic force measurement. In the wind tunnel the sting supporting the model is centered at the spin axis, allowing the Magnus rotor to rotate freely. The force measurements, taken by the sting, are in stability axes. By the way, the term *stability axes* is borrowed from airplane dynamics. There and here, the 1^S axis is also defined as the projection of the velocity vector on the symmetry plane of the aircraft.

From the velocity vector $[v_B^E]^S$ and the transformation matrix $[T]^{SL}$, we can get by integration the position of the Magnus rotor $[s_{BE}]^L$, i.e., the displacement of the c.m. B with respect to an Earth reference point E, expressed in local-level coordinates.

$$\left[\frac{ds_{BE}}{dt}\right]^L = [\bar{T}]^{SL}[v_B^E]^S \tag{10.26}$$

Equations (10.25) and (10.26) are the translational equations of Magnus rotors. The angular velocity $[\Omega^{SE}]^{B'}$ and the transformation matrix $[T]^{SL}$ are still to be supplied by the rotational equations.

Similar to spinning missiles, the more exciting part is the derivation of the attitude equation. Applying Euler's law and transforming the rotational derivative to the stability frame, we obtain

$$D^S l_B^{BE} + \Omega^{SE} l_B^{BE} = m_B \tag{10.27}$$

We divide the angular velocity ω^{BE} of the missile body B wrt the reference frame E into two parts:

$$\omega^{BE} = \omega^{BS} + \omega^{SE}$$

The first part is the spin rate, and the second one is the angular velocity of the nonspinning stability frame wrt Earth. Similarly, the angular momentum is split into two parts

$$l_B^{BE} = I_B^B \omega^{BE} = I_B^B \omega^{BS} + I_B^B \omega^{SE} \tag{10.28}$$

We form the rotational derivative as required by Eq. (10.27)

$$D^S l_B^{BE} = D^S\left(I_B^B \omega^{BS}\right) + D^S\left(I_B^B \omega^{SE}\right)$$
$$= I_B^B D^S \omega^{BS} + I_B^B D^S \omega^{SE} \tag{10.29}$$

where we made use of $D^S I_B^B = 0$ because the Magnus rotor's MOI about the spin axis is independent of spin attitude. This simplifying assumption is true at least approximately because the effect of the driving vanes is insignificant. Substituting Eqs. (10.28) and (10.29) into Eq. (10.27) yields the atitude equation of motion

$$I_B^B D^S \omega^{BS} + I_B^B D^S \omega^{SE} + \Omega^{SE} I_B^B \omega^{BS} + \Omega^{SE} I_B^B \omega^{SE} = m_B \tag{10.30}$$

The total rate of change of angular momentum consists of the spin contribution—first term—and the stability frame rotations—second term. The primary gyroscopic coupling stems from the third term, which dominates the fourth term.

Just as the stability coordinates were used for the translational equations, they are also best suited for the attitude equation. The aerodynamic moment $[m_B]^S$ is given in that form, the angular velocity $[\omega^{SE}]^S$ is needed by the translational equations, and the rotational time derivatives become the ordinary derivatives. For these reasons we express the rotational equations in stability coordinates

$$[I_B^B]^S \left[\frac{d\omega^{BS}}{dt}\right]^S + [I_B^B]^S \left[\frac{d\omega^{SE}}{dt}\right]^S + [\Omega^{SE}]^S [I_B^B]^S ([\omega^{BS}]^S + [\omega^{SE}]^S) = [m_B]^S$$

(10.31)

Another important fact is that the MOI tensor is constant in stability coordinates and has a particularly simple diagonal form (note because of symmetry $I_3 = I_1$)

$$[I_B^B]^S = \begin{bmatrix} I_1 & 0 & 0 \\ 0 & I_2 & 0 \\ 0 & 0 & I_1 \end{bmatrix}^S, \quad [\omega^{BS}]^{S'} = \begin{bmatrix} 0 \\ \omega \\ 0 \end{bmatrix}^S, \quad [\omega^{SE}]^S = \begin{bmatrix} p' \\ q' \\ r' \end{bmatrix}^S$$

with the angular velocities also included. Substituting the matrices into Eq. (10.31) yields

$$\begin{bmatrix} 0 \\ I_2(d\omega/dt) \\ 0 \end{bmatrix}^S + \begin{bmatrix} I_1(dp'/dt) \\ I_2(dq'/dt) \\ I_1(dr'/dt) \end{bmatrix}^S + \begin{bmatrix} 0 & -r' & q' \\ r' & 0 & -p' \\ -q' & p' & 0 \end{bmatrix}^S \begin{bmatrix} I_1 & 0 & 0 \\ 0 & I_2 & 0 \\ 0 & 0 & I_1 \end{bmatrix}^S$$

$$\times \left(\begin{bmatrix} 0 \\ \omega \\ 0 \end{bmatrix}^S + \begin{bmatrix} p' \\ q' \\ r' \end{bmatrix}^S \right) = \begin{bmatrix} m_{B_1} \\ m_{B_2} \\ m_{B_3} \end{bmatrix}^S$$

and the three scalar equations

$$I_1\left(\frac{dp'}{dt}\right) - I_2\omega r' - q'r'(I_2 - I_1) = m_{B_1}$$

$$I_2\left(\frac{d\omega}{dt} + \frac{dq'}{dt}\right) = m_{B_2}$$

$$I_1\left(\frac{dr'}{dt}\right) + I_2\omega p' + p'q'(I_2 - I_1) = m_{B_3}$$

Although the pitch rate of the stability frame is not zero—magnitude of the undulating flight-path rate, it is much smaller than the spin rate, and therefore the terms containing q' can be neglected. With this simplification and solving for the derivatives yields the simple attitude equations of Magnus rotors:

$$\frac{dp'}{dt} = \frac{\left(I_2 \omega r' + m_{B_1}\right)}{I_1}$$

$$\frac{d\omega}{dt} = \frac{m_{B_2}}{I_2} \qquad (10.32)$$

$$\frac{dr'}{dt} = \frac{\left(-I_2 \omega p' + m_{B_3}\right)}{I_1}$$

The first and last equations govern the gyroscopic coupling, which is a function of the spin MOI ratio I_2/I_1, the spin rate ω, and the angular coupling rates r' and p'. A positive yaw rate r' induces a positive change in roll rate p', whereas the roll rate has the opposite effect on the yaw rate. This purely gyroscopic behavior is modified by the aerodynamic moments m_{B_1} and m_{B_3}. The second equation simply describes the spin history. In a typical flight the Magnus rotor will reach a steady-state spin rate, indicating that the torque of the driving vanes equals the skin friction on the cylinder and the endplates.

To build a complete six-DoF simulation of Magnus rotors, you would have to program Eqs. (10.25), (10.26), and (10.32), supplemented by quaternion equations for the calculation of the transformation matrix $[T]^{SL}$. Figure 10.2, with the proper substitutions for Newton's and Euler's equations, can serve as a blueprint.

You may have noticed already the great similarity between the equations of a spinning missile and a Magnus rotor. Both use a nonspinning frame that slips over the body, which we called either the nonspinning body frame B' or the stability frame S. Associated with the frames are the nonspinning body or stability coordinate systems. They serve as the axes for recording the aerodynamic forces and moments in wind tunnels. The gyroscopic coupling is evident in both cases and gives rise to nutation about the velocity vector of the missile and normal to the velocity vector for the Magnus rotor. In either case the development of a full six-DoF simulation would follow the same pattern.

The four models of missile, aircraft, spinning missile, and Magnus rotor cover every conceivable aerospace vehicle that you may have to simulate. You can program them in matrix form or, in some simple instances, as scalar equations. If guidance and control systems are part of the vehicle, you will have to supplement your simulation with the appropriate code.

One simplification is common to all: the equations of motion were derived with the Earth as inertial reference frame and its curvature unfurled into a flat surface. This assumption is acceptable as long as the vehicle travels at moderate speeds (less than Mach 5) and at low altitudes (less than 30 km). At higher speeds and altitudes, we have to abandon the flat-Earth assumption and consider Earth as a spheroid moving with respect to the sun's ecliptic.

10.1.2 Elliptical Earth

As we travel into the higher regions of the atmosphere, or leave it altogether for the wide expanse of space, our speed gets faster and our view of the globe broader. It becomes evident that the Earth is round and rotating with respect to the sun. As travelers, we experience such effects as centrifugal and Coriolis accelerations and are looking for another immutable reference frame somewhere out there.

This situation motivates us to study the shape of the Earth in more detail. First, I shall give a brief overview of geodesy, the geodetic coordinate system, and the gravitational attraction of the Earth. Then we are prepared to formulate the six-DoF equations of motions in an inertial and geographic format. For the kinematic equations we shall pursue the direction cosine approach rather than use quaternions.

Some of this material should be familiar to you from Chapters 8 and 9. Here, however, we expand our horizon to an elliptical Earth and deal with the mechanics of developing a simulation model for hypersonic vehicles.

10.1.2.1   Geodesy.

Geodesy can be defined as the science concerned with the precise positioning of points on the surface of Earth and the determination of the exact size and shape of Earth. It also involves the study of the variations of Earth's gravitational attraction and the application of these variations to the precise measurements of Earth's shape.

The science of geodesy is as old as the early Greeks. Whereas Homer lived on a flat Earth, Pythagoras, the mathematician, envisioned the Earth to be fashioned according to a perfect sphere. Of the same opinion was Aristotle, the dominant scientific authority until the Middle Ages.

One of the great pursuits of the ages—and continuing today—is the accurate measurement of the Earth's circumference. It amazes me to read that Eratosthenes, a Greek scholar experimenting in Egypt, deducted from measurements of sun angles that the circumference at the equator is 25,000 miles. Today we peg it at 24,899 miles. However, when Ptolemy came on the scene, he convinced the majority of the scholars that the number should be revised downward to 18,000 miles. It still stood at that value when Columbus made his plans to sail to India. Fortunately so, because otherwise Columbus may have been unwilling to travel such a great distance and America might have been discovered by Russian cosmonauts peeking down from space!

The Pythagorean spherical concept offers a simple surface, which is mathematically easy to deal with. Many astronomical and navigational computations use it to represent Earth. Although the sphere is a close approximation to the true figure of Earth and satisfactory for many purposes, for the geodesist interested in measuring long distances—spanning continents and oceans—a more exact configuration is necessary.

Because Earth is in fact slightly flattened at the poles and bulges somewhat at the equator, the geometrical figure used in geodesy is an ellipsoid of revolution or *spheroid*. Two dimensions define a spheroid; the semimajor axis and the flattening parameter. The flattening parameter is defined as the fractional change between the semimajor axis a and semiminor axes b:

$$f = \frac{a - b}{a}$$

Some references also call it the *ellipticity* of the spheroid and invite some confusion with the term *eccentricity*, which is defined as

$$e = \sqrt{\frac{a^2 - b^2}{a^2}}$$

Thanks to the World Geodetic Systems (WGS) Committee, chartered in 1960, the numerical precision of the reference ellipsoid has been refined ever since. Today, we use the WGS 84 (Ref. 3) system with the values $a = 6,378,137.0$ m and $e = 0.08181919$.

For completeness' sake, however, let me point out that the real shape of Earth is arbitrary and is only successively approximated by an alternating series of elliptical and pear-shaped surfaces. So far, our classification is based on geometric considerations only (geometrical geodesy). The WGS 84 ellipsoid is that first approximation beyond the sphere that fits the actual surface of the Earth best (at sea level). Alternately, the actual shape of the Earth is also referred to as *geoid*. It is defined as that surface to which the oceans would conform over the entire Earth if free to adjust to the combined effect of Earth's mass attraction and the centrifugal force of Earth's rotation. Notice the geoid is defined with the aid of physical forces. Therefore, this branch of geodesy is also called *physical geodesy*.

Do not be intimidated by geodesy. You will never learn all of the tricks of its trade unless you make it your life's passion. If that is the case, I recommend you start with the book by Torge.[4] Fortunately, for six-DoF simulations we are satisfied with the WGS 84 ellipsoidal approximation and do not distinguish between geometrical and physical geodesy.

Briefly review Secs. 3.1 and 3.2. There we define the inertial frame I and the associated inertial coordinate system $]^I$. The inertial frame's attitude is fixed in the ecliptic, but the center of Earth is also one of its points. In effect, the inertial frame convects with Earth's motion around the sun but does not follow Earth's rotation. The inertial coordinate axes are aligned with the vernal equinox and Earth's spin axis.

Earth's frame E, the associated coordinate system $]^E$, and the geographic coordinate system $]^G$ are also defined in Sec. 3.2. The transformation matrix $[T]^{GE}$ is a function of longitude and latitude [Eq. (3.13)]. Because in Chapter 3 Earth was presumed to be spherical, latitude was measured from the equator to the geocentric displacement vector of the vehicle. However, for an elliptical Earth model we now have to refine the definition of latitude, while longitude remains the same.

We call the latitude of the WGS 84 model the *geodetic latitude* λ_d, and the earlier definition is relabeled the *geocentric latitude* λ_c. Figure 10.4 depicts the geometrical situation. To define the geodetic latitude, project the vehicle c.m. B onto the local tangent plane at B_0 and continue the plumb line until it crosses the

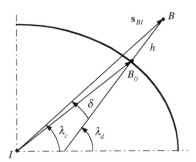

Fig. 10.4   Geocentric and geodetic latitudes.

plane through the equator. The angle between the plumb line and the equatorial plane is the geodetic latitude λ_d. This is the latitude that is used in cartography. If you want to go back to a round Earth, simply use the geocentric latitude λ_c.

Let us review and revise carefully the definition of the geographic coordinate system. We rename the former geographic system and call it the *geocentric coordinate system* while maintaining the same symbol $]^G$. Its 1^G and 2^G axes are still parallel to the plane normal to s_{BI} and point north and east, respectively, while the 3^G axis is parallel and points in the opposite direction of s_{BI}.

A new entity, the geodetic coordinate system, is required. Its 1^D and 2^D axes are parallel to the plane normal to the plumb line and point north and east, respectively. The 3^D axis is parallel to the plumb line, pointing downward. The geodetic and geocentric coordinate systems are related by the deflection angle δ, as indicated in Fig. 10.4, and defined by

$$\delta = \lambda_d - \lambda_c \tag{10.33}$$

Its value depends on the geodetic latitude λ_d, the flattening parameter f, and the altitude of the vehicle h. Britting[5] has given an approximation

$$\delta = f\,\sin(2\lambda_d)\left(1 - \frac{f}{2} - \frac{h}{R_0}\right) \tag{10.34}$$

where R_0 is the distance between the points B_0 and I, which can be calculated from the Earth's major axis a, its flattening parameter, and the geodetic latitude of the vehicle

$$R_0 = a\left\{1 - \frac{f}{2}[1-\cos(2\lambda_d)] + \frac{5f^2}{16}[1-\cos(4\lambda_d)]\right\} \tag{10.35}$$

The accuracy of R_0 is better than one meter, and δ is within 4.5 arcsec at 30 km altitude. The transformation matrix of the geodetic wrt geocentric coordinate systems is governed by this deflection angle δ

$$[T]^{DG} = \begin{bmatrix} \cos\delta & 0 & \sin\delta \\ 0 & 1 & 0 \\ -\sin\delta & 0 & \cos\delta \end{bmatrix} \tag{10.36}$$

For spherical Earth $\delta = 0$, the geocentric and geodetic coordinate systems coincide.

Given the longitude l (now also called geodetic longitude) and the geodetic latitude λ_d, the transformation matrix of the geodetic wrt the Earth coordinates is adapted from Eq. (3.13)

$$[T]^{DE} = \begin{bmatrix} -\sin\lambda_d\cos l & -\sin\lambda_d\sin l & \cos\lambda_d \\ -\sin l & \cos l & 0 \\ -\cos\lambda_d\cos l & -\cos\lambda_d\sin l & -\sin\lambda_d \end{bmatrix} \tag{10.37}$$

Now regard Fig. 10.5. We can extend Eq. (10.37) to inertial coordinates by replacing the geodetic longitude with the celestial longitude l_i, which is measured from the 1^I axis. The celestial longitude is composed of the celestial longitude of Greenwich l_G and the geodetic longitude of the vehicle $l_i = l_G + l$. However,

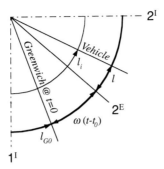

Fig. 10.5   Celestial longitude.

because the Greenwich meridian rotates with the Earth rate $\omega_\oplus$ during the fly-out time of the vehicle, it increases from its initial location l_{G0} by $\omega_\oplus(t - t_0)$ becoming $l_G = l_{G0} + \omega_\oplus(t - t_0)$. The celestial longitude of the vehicle at time t is then

$$l_i = l_{G0} + \omega_\oplus(t - t_0) + l \tag{10.38}$$

Substituting l_i for l in Eq. (10.37) yields the transformation matrix of the geodetic wrt the inertial coordinate systems

$$[T]^{DI} = \begin{bmatrix} -\sin \lambda_d \cos l_i & -\sin \lambda_d \sin l_i & \cos \lambda_d \\ -\sin l_i & \cos l_i & 0 \\ -\cos \lambda_d \cos l_i & -\cos \lambda_d \sin l_i & -\sin \lambda_d \end{bmatrix} \tag{10.39}$$

Both transformation matrices will be useful when we build our six-DoF simulations.

Another important kinematic problem is the determination of the inertial position of the vehicle $[s_{BI}]^I$ in inertial coordinates, given the vehicle's longitude l, latitude λ_d, altitude h, and initial Greenwich celestial longitude l_{G0}. We start with the displacement vector of the vehicle in geocentric coordinates $[\overline{s}_{BI}]^G = [0 \ 0 \ -(R_0+h)]$, using the approximation $R_0+h \approx |s_{BI}|$, which is according to Britting[5] less than 1 m for endo-atmospheric flight. R_0 is given by Eq. (10.35). Then we transform from geocentric to inertial coordinates and obtain the result

$$[s_{BI}]^I = [\bar{T}]^{DI}[T]^{DG}[s_{BI}]^G \tag{10.40}$$

where the two transformation matrices are given by Eqs. (10.39) and (10.36), respectively.

The inverse problem is the determination of the vehicle's longitude, latitude, and altitude, given the inertial displacement vector $[s_{BI}]^I$. Unfortunately, a straight calculation is not possible because the deflection angle itself is a function of latitude. Therefore, we take recourse to an iterative approach that proceeds as follows:

1) Initialize the geodetic latitude with the geocentric value $\lambda_d = \lambda_c = \arcsin [(s_{BI})_3^I/|s_{BI}|]$.

2) Start an iterative loop in which you calculate the radius to the ellipse R_0 with Eq. (10.35), the altitude h from the approximation $h = |s_{BI}| - R_0$, and the deflection δ using Eq. (10.34).

3) Now update the approximation $\lambda_d \leftarrow \lambda_d + \delta$ and stop the iterative loop if the improvement drops below a certain value, say 10^{-7} radians.

4) Finally, calculate the geodetic longitude

$$l = \arcsin\left\{(s_{BI})_2^I / \sqrt{\left[(s_{BI})_1^I\right]^2 + \left[(s_{BI})_2^I\right]^2}\right\} - l_{G0} - \omega_\oplus(t - t_0)$$

In most situations it takes fewer than 10 iterations for convergence. Yet, it is always advisable that you guard against excessive numbers of iterations.

Let us wrap up our adventure in geodesy with an excursion to the modeling of the gravitational field and the resulting gravitational acceleration that acts on the flight vehicle. As we have discussed in Sec. 8.2.2, the force of attraction F between two bodies M and m, separated by the r distance, is according to Newton

$$F = G\frac{Mm}{r^2}$$

where G is the universal gravitational constant. If m represents the flight vehicle and M the Earth, then the acceleration on the vehicle is

$$g = \frac{F}{m} = \frac{GM}{r^2}$$

If the Earth were a perfect sphere of uniform density, g would sweep out a gravitational field consisting of concentric spheres. Yet, Earth being a geoid, we have to call the mathematicians for help. They invented the spherical harmonics that successively improve the representation of the gravitational field. The first term represents the sphere, and, together with the second term, the desired model of an elliptical gravitational field is reached.

The WGS folks have worked since 1960 to improve this elliptical gravitational model. In 1984 they determined that the value of GM is 3.986005×10^{14} m^3/s^2 and that the second-degree zonal gravitational coefficient is $\bar{C}_{2,0} = -4.841668 \times 10^{-4}$. The acceleration acting on a vehicle caused by the gravitational reference ellipsoid is given in geocentric coordinates as

$$[g]^G = \frac{GM}{|s_{BI}|^2}\begin{bmatrix} -3\sqrt{5}\bar{C}_{2,0}\left(\dfrac{a}{|s_{BI}|}\right)^2 \sin\lambda_c \cos\lambda_c \\ 0 \\ 1 + \dfrac{3}{2}\sqrt{5}\bar{C}_{2,0}\left(\dfrac{a}{|s_{BI}|}\right)^2 (3\sin^2\lambda_c - 1) \end{bmatrix} \tag{10.41}$$

with all variables already defined. The main contribution is in the 3^G direction. If you neglect the ellipticity, i.e., $\bar{C}_{2,0} = 0$, you recover the spherical value of $g = GM/|s_{BI}|^2$. Notice, because of symmetry, there is no gravitational component in the east direction.

Hitherto, we have labored through the preliminaries of modeling an elliptical Earth. You will use most of these equations to build the utility routines for your six-DoF simulation. In particular, you will have to program a subroutine that calculates $[T]^{GI}$, given l, λ_d, h (called CADTGI84 in CADAC); another one that determines l, λ_d, h, given $[s_{BI}]^I$ (called CADGEO84 in CADAC); and the inverse that provides $[s_{BI}]^I$, given l, λ_d, h (called CADINE84 in CADAC). You find these subroutines in CADAC utility file UTL3.FOR on the CADAC Web site. Finally, the gravitational acceleration is given by Eq. (10.41) (in the CADAC GHAME6 simulation, Module G2, subroutine G2GRAV). Now we are prepared to derive the equations of motions.

10.1.2.2 Equations of motion relative to the inertial frame.

In the preceding flat-Earth approach we had the choice of modeling the equations of motion either in local-level, presumed inertial, or body coordinates. Now we have five possibilities: inertial, Earth, geographic, geodetic, or body coordinates. As bewildering as these choices can be, surveying the literature I find these options: ENDOSIM[6] integrates Newton's equation in inertial coordinates, Euler's equation in body coordinates and uses quaternions, requiring the introduction of inertial Euler angles. Another U.S. Army simulation[7] solves Newton's equation in geocentric coordinates, while Euler's equation is again given in body coordinates.

There is little doubt that Euler's equation should be integrated in body coordinates because of the inherent invariance of the MOI tensor. However, for Newton's equation we have a choice. I prefer the inertial approach, integrating both the acceleration and velocity equations in inertial coordinates. The other option is the integration of the geographic acceleration and velocity. We will describe it later. We also have to decide whether we use quaternions or the direction cosine matrix. Because quaternions are closely related to Euler angles, the introduction of inertial Euler angles becomes necessary. I propose that we use the direction cosine matrix and avoid confusing angle definitions.

The derivation of the equations of motion is easy and short. Newton's law provides

$$mD^I v_B^I = f_{a,p} + mg$$

The left side is best expressed in inertial coordinates, but on the right-hand side the forces are formulated in body coordinates and the gravitational acceleration in geocentric coordinates:

$$m[D^I v_B^I]^I = [\bar{T}]^{BI}[f_{a,p}]^B + m[\bar{T}]^{GI}[g]^G$$

This leads to the first set of differential equations:

$$\left[\frac{dv_B^I}{dt}\right]^I = \frac{1}{m}[\bar{T}]^{BI}[f_{a,p}]^B + [\bar{T}]^{GI}[g]^G \tag{10.42}$$

The $[T]^{BI}$ transformation is the direction cosine matrix, and the other transformation $[\bar{T}]^{GI}$ is obtained from Eqs. (10.36) and (10.39):

$$[\bar{T}]^{GI} = [\bar{T}]^{DI}[T]^{DG}$$

Given the inertial velocity $[v_B^I]^I$, the inertial displacement vector of the vehicle $[s_{BI}]^I$ is obtained from the second set of differential equations:

$$\left[\frac{ds_{BI}}{dt}\right]^I = [v_B^I]^I \tag{10.43}$$

Equations (10.42) and (10.43) are the translational differential equations, which couple with the rotational equations through the direction cosine matrix $[T]^{BI}$ and the forces $[f_{a,p}]^B$. A precarious issue is the initialization of the state variables. We will deal with it after we have derived the attitude equation.

The attitude equation is obtained as before from Euler's law

$$D^I l_B^{BI} = m_B$$

Express the angular momentum as the MOI tensor and the inertial body rates $l_B^{BI} = I_B^B \omega^{BI}$; transform the rotational derivative to the body frame; and finally express the tensors in body axes

$$\left[\frac{d\omega^{BI}}{dt}\right]^B = \left([I_B^B]^B\right)^{-1}\left(-[\Omega^{BI}]^B[I_B^B]^B[\omega^{BI}]^B + [m_B]^B\right) \tag{10.44}$$

This differential equation is just like the one for the flat-Earth case [Eq. (10.10)], except that the Earth frame E is replaced by the inertial frame I.

Now I must address the initialization of the state variables. So far, we attained nine first-order differential equations, and therefore nine state variables must be initialized. They are divided into three pairs of three each: $[v_B^I]^I$ from Newton's equation (10.42), $[s_{BI}]^I$ from the velocity integration Eq. (10.43), and $[\omega^{BI}]^B$ from Euler's equation (10.44). The initialization of the direction cosine matrix will be addressed in the next section.

When you build a simulation, you should make it user friendly. Particularly the input should be easy to visualize. Therefore, you want to replace $[v_B^I]^I$ by the geographic velocity vector $[v_B^E]^D$, expressed in geodetic coordinates. We generate such a relationship by first reviewing the definition of these velocities $v_B^I \equiv D^I s_{BI}$ and $v_B^E \equiv D^E s_{BE}$, where $s_{BI} = s_{BE} + s_{EI}$ (E is any fixed point on the Earth and I is Earth's center), and therefore

$$D^I s_{BI} = D^E s_{BE} + D^E s_{EI} + \Omega^{EI} s_{BI}$$

with $D^E s_{EI}$ equals zero because s_{EI} is constant in the Earth frame. Thus

$$v_B^I = v_B^E + \Omega^{BE} s_{BI} \tag{10.45}$$

and expressed in the appropriate coordinate systems

$$[v_B^I]^I = [\bar{T}]^{DI}[v_B^E]^D + [\Omega^{EI}]^I[s_{BI}]^I \tag{10.46}$$

Furthermore, because $[T]^{DI}$ is only a function of longitude and latitude, we can summarize: Given the geodetic longitude, latitude, and altitude, and the geographic velocity vector $[v_B^E]^D$, the inertial velocity vector $[v_B^I]^I$ can be initialized. A convenient alternative to $[v_B^E]^D$ is the initialization by the geographic speed $|v_B^E|$ and the heading angle χ and the flight-path angle γ, both relative to the local tangent plane. All you need is a conversion routine as, for instance, the CADAC subroutine MATCAR.

As far as the angular velocity $[\omega^{BI}]^B$ is concerned, it is usually initialized by zero values. Only in extreme cases, like close-in air-to-air combat, may the missile have to be given the angular rates of the launch aircraft. Then the aircraft's initial geographic angular rate $[\omega^{BE}]^D$ has to be converted to the inertial body rate in body coordinates

$$[\omega^{BI}]^B = [T]^{BD}[\omega^{BE}]^D + [\omega^{EI}]^I$$

where $[\omega^{EI}]^I$ is the Earth's rotation.

Now, there is an alternate way to initialize the inertial velocity. It is based on Euler and incidence angles. From these angles and the geographic speed the geographic velocity vector can be derived. Express the geographic velocity vector in body and relative wind coordinates $]^W$

$$[v_B^E]^B = [\bar{T}]^{WB}[v_B^E]^W = [\bar{T}]^{WB}\begin{bmatrix} |v_B^E| \\ 0 \\ 0 \end{bmatrix}$$

With the transformation matrix Eq. (3.18) introduce the incidence angles

$$[v_B^E]^B = |v_B^E|\begin{bmatrix} \cos\alpha\,\cos\beta \\ \sin\beta \\ \sin\alpha\,\cos\beta \end{bmatrix}$$

and transform the velocity to geodetic coordinates

$$[v_B^E]^D = [\bar{T}]^{BD}[v_B^E]^B$$

The transformation matrix $[T]^{BD}$ contains the Euler angles referred to the geodetic coordinate axes and follows the format of Eq. (3.14). Once $[v_B^E]^D$ is computed, the determination of $[v_B^I]^I$ follows the preceding outline.

10.1.2.3 Determination of direction cosine matrix.

Finally, we solve the basic kinematic problem, namely, given the inertial body rates in body coordinates calculate the direction cosine matrix $[T]^{BI}$. We have dealt with this issue before in Chapter 4 and presented three possible approaches. For the flat Earth, six-DoF simulations of the preceding section, I recommended the quaternion solution for two reasons: 1) only four linear differential equations need be solved and 2) the Euler angles are readily computed from the direction cosine matrix $[T]^{BL}$. Here, the question arises what are the Euler angles associated with $[T]^{BI}$? They would be related to the inertial coordinate system and, therefore, void of any flight mechanical meaning.

Hence, let us calculate $[T]^{BI}$ directly from $[\omega^{BI}]^B$, using the direction cosine differential equations (4.66)

$$\left[\frac{dT}{dt}\right]^{BI} = [\overline{\Omega^{BI}}]^B [T]^{BI} \tag{10.47}$$

where $[\Omega^{BI}]^B$ is the skew-symmetric form of $[\omega^{BI}]^B$, provided by Euler's equation.

The initialization of $[T]^{BI}$ is of immediate interest. Most likely, the simulation is initialized by the standard Euler angles, referred to as the geodetic coordinate system. Accordingly, the transformation matrix $[T]^{BD}$ can be calculated. Yet, we also need to compute $[T]^{DI}$ to complete the initialization process of $[T]^{BI} = [T]^{BD}[T]^{DI}$. This is easily done: Given the initial geodetic longitude, latitude, and altitude, $[T]^{DI}$ is obtained from Eq. (10.39).

Once the vehicle is airborne and $[T]^{BI}$ is being updated, we have to solve the inverse problem, namely, what are the Euler angles relative to geodetic coordinates? "No problem," you say, let us just calculate $[T]^{DI}$ from the current longitude, latitude, and altitude and solve for $[T]^{BD} = [T]^{BI}[\bar{T}]^{DI}$. The Euler angles are then

$$\theta = \arcsin(-t_{13})$$

$$\psi = \arccos\left(\frac{t_{11}}{\cos\theta}\right)\operatorname{sgn}(t_{12}) \tag{10.48}$$

$$\phi = \arccos\left(\frac{t_{33}}{\cos\theta}\right)\operatorname{sgn}(t_{23})$$

where the t_{ij} are the elements of $[T]^{BD}$. Correct, just make sure to program around $\cos\theta = 0$ to avoid unwanted divide by zero run-time errors.

All would be well if we had a perfect computer to solve the differential equations. However, finite integration steps and limited word length cause the orthogonality of the direction cosine matrix to be slowly lost. Consequently, we have to reestablish orthogonality after every integration step. The following procedure, patterned after Savage,[8] works quite well.

Simply orthogonalize $[T]^{BI}$ by the following recursive algorithm

$$[T(n+1)]^{BI} = [T(n)]^{BI} + \tfrac{1}{2}\{[E] - [T(n)]^{BI}[\overline{T(n)}]^{BI}\}[T(n)]^{BI} \tag{10.49}$$

where $[T(n+1)]^{BI}$ is the new orthogonalized matrix, based on the current direction cosine matrix $[T(n)]^{BI}$ attained by integration. Notice, the value in the parentheses is a zero matrix if $[T(n)]^{BI}$ is orthogonal. To show that indeed the orthogonality has improved, I invite you on a side tour, which you are free to bypass.

Let us expand the last term

$$[\delta T]^{BI} = \tfrac{1}{2}\{[E] - [T(n)]^{BI}[\overline{T(n)}]^{BI}\}[T(n)]^{BI}$$

using the base vector format of the transformation matrix [see Eq. (3.5)]

$$[T]^{BI} = \begin{bmatrix} \bar{b}_1 \\ \bar{b}_2 \\ \bar{b}_3 \end{bmatrix} ; \quad [T]^{BI}[\bar{T}]^{BI} = \begin{bmatrix} \bar{b}_1\bar{b}_1 & \bar{b}_1\bar{b}_2 & \bar{b}_1\bar{b}_3 \\ \bar{b}_2\bar{b}_1 & \bar{b}_2\bar{b}_2 & \bar{b}_2\bar{b}_3 \\ \bar{b}_3\bar{b}_1 & \bar{b}_3\bar{b}_2 & \bar{b}_3\bar{b}_3 \end{bmatrix}$$

where b_i is the (3×1) column base vector and $\bar{b}_i$ its transpose, a (1×3) row vector,

$$[\delta T]^{BI} \equiv \begin{bmatrix} \delta \bar{b}_1 \\ \delta \bar{b}_2 \\ \delta \bar{b}_3 \end{bmatrix} = \frac{1}{2} \begin{bmatrix} (1 - \bar{b}_1 b_1)\bar{b}_1 - \bar{b}_1 b_2 \bar{b}_2 - \bar{b}_1 b_3 \bar{b}_3 \\ -\bar{b}_2 b_1 \bar{b}_1 + (1 - \bar{b}_2 b_2)\bar{b}_2 - \bar{b}_2 b_3 \bar{b}_3 \\ -\bar{b}_3 b_1 \bar{b}_1 - \bar{b}_3 b_2 \bar{b}_2 + (1 - \bar{b}_3 b_3)\bar{b}_3 \end{bmatrix}$$

$$= \frac{1}{2} \begin{bmatrix} -\bar{b}_1 b_2 \bar{b}_2 - \bar{b}_1 b_3 \bar{b}_3 \\ -\bar{b}_2 b_1 \bar{b}_1 - \bar{b}_2 b_3 \bar{b}_3 \\ -\bar{b}_3 b_1 \bar{b}_1 - \bar{b}_3 b_2 \bar{b}_2 \end{bmatrix}$$

Substituting into the recursive relationship and writing out the lines yields

$$\bar{b}_1(n+1) = \bar{b}_1(n) - \tfrac{1}{2}[\bar{b}_1(n)b_2(n)\bar{b}_2(n) + \bar{b}_1(n)b_3(n)\bar{b}_3(n)]$$

$$\bar{b}_2(n+1) = \bar{b}_2(n) - \tfrac{1}{2}[\bar{b}_2(n)b_1(n)\bar{b}_1(n) + \bar{b}_2(n)b_3(n)\bar{b}_3(n)]$$

$$\bar{b}_3(n+1) = \bar{b}_3(n) - \tfrac{1}{2}[\bar{b}_3(n)b_1(n)\bar{b}_1(n) + \bar{b}_3(n)b_2(n)\bar{b}_2(n)]$$

After these preliminaries we want to check if the orthogonality error decreases as a result of the recursive algorithm. Before the update the error of any two base vectors has the scalar value $\varepsilon_{ik} = \bar{b}_i(n)b_k(n)$. Let us say this error is of order $\mathcal{O}(\varepsilon)$. Now we want to show that after the update the error $\bar{b}_i(n+1)b_k(n+1)$ has decreased by an order of magnitude to $\mathcal{O}(\varepsilon^2)$. Multiply any two base vectors

$$\bar{b}_1(n+1)b_2(n+1) = \bar{b}_1(n)b_2(n) - \tfrac{1}{2}\bar{b}_1(n)b_1(n)\bar{b}_1(n)b_2(n)$$

$$- \tfrac{1}{2}\bar{b}_1(n)b_1(n)\bar{b}_1(n)b_2(n) + \mathcal{O}(\varepsilon^2) + \mathcal{O}(\varepsilon^3)$$

and recognize that the first three terms on the right-hand side cancel. Hence, the orthogonality error of the updated base vectors was reduced to $\mathcal{O}(\varepsilon^2)$. Because this conclusion holds for base vectors one and two, it is true for any two combinations and thus also for the direction cosine matrix $[T]^{BI}$.

You can use Eqs. (10.47–10.49) to solve the basic kinematic problem in your simulation. They will deliver the direction cosine matrix $[T]^{BI}$ and the Euler angles ψ, θ, and ϕ wrt the geodetic coordinates. Besides the inertial body rates $[\omega^{BI}]^B$, you will also need the geodetic longitude, latitude, and altitude of the vehicle. For an example you may refer to CADAC GHAME6 simulation, Module G3.

In summary, Fig. 10.6 combines the most important parts of a six-DoF simulation for a hypersonic vehicle over an elliptical Earth. The CADAC GHAME6 model is patterned after this diagram. Newton's equations are given by Eqs. (10.42) and (10.43) and Euler's equations by Eq. (10.44). The computation of the direction cosine matrix follows Eqs. (10.47) and (10.48). For the representation of the aerodynamic forces, I chose the lift and drag coefficients, defined in stability coordinates.

The initialization of the state variables is also indicated. The geographic velocity is given in geodetic variables, speed $|v_B^E|$, heading χ, flight-path angle γ, geodetic longitude l, latitude λ_d, and altitude h. They are converted to the inertial velocity vector $[v_B^I]^I$ and the vehicle's initial position $[s_{BI}]^I$. If not zero, the initial geographic angular rate $[\omega^{BE}]^D$, given in geodetic coordinates, is converted to inertial

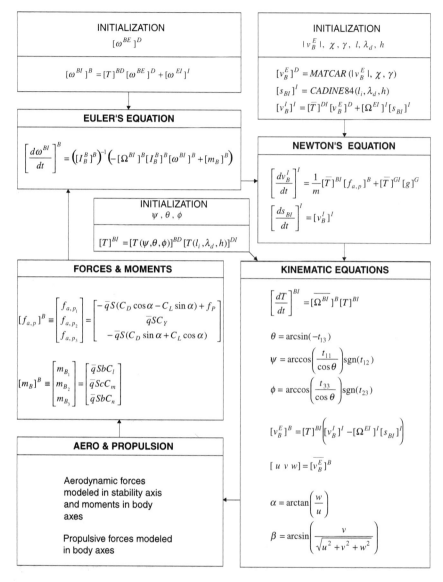

Fig. 10.6 Summary of kinematic and dynamic equations of a hypersonic vehicle over an elliptical Earth.

body rates $[\omega^{BI}]^I$. Finally, the direction cosine matrix is initialized by the vehicle's Euler angles.

10.1.2.4 Equations of motion relative to the Earth frame.

The geographic modeling of the equations of motion over an elliptical Earth focuses on the integration of the geographic acceleration $D^E v_B^E$ relative to the Earth rather than the inertial acceleration $D^I v_B^I$. Make no mistake, the J2000 is still the inertial frame for Newton's and Euler's laws. We just introduce the Earth as a convenient reference frame. I will derive the relationship between the two velocities and their accelerations before presenting the translational equations. As before, the attitude equation holds no surprises.

There are two reasons for focusing on the geographic velocity v_B^E: 1) the atmosphere is rotating with the Earth, and the aerodynamic forces are a function of the geographic velocity; 2) the vehicle's position and velocity are traced more conveniently over the Earth than in the inertial frame.

By definition, the inertial velocity is $v_B^I \equiv D^I s_{BI}$, where s_{BI} is the location of the vehicle c.m. B wrt the center of the Earth I. To introduce the geographic velocity, we change the reference frame from I to E

$$D^I s_{BI} = D^E s_{BI} + \Omega^{EI} s_{BI}$$

Now, introduce a reference point on the Earth E (any) as an intermediary between B and I:

$$s_{BI} = s_{BE} + s_{EI} \tag{10.50}$$

and substitute it into the preceding equation

$$D^I s_{BI} = D^E s_{BE} + D^E s_{EI} + \Omega^{EI} s_{BI}$$

$D^E s_{EI} = 0$ because s_{EI} is constant in the Earth frame. Using the definition of geographic velocity $v_B^E \equiv D^E s_{BE}$, we get the desired relationship between the inertial and geographic velocities:

$$v_B^I = v_B^E + \Omega^{EI} s_{BI} \tag{10.51}$$

The translational differential equation is derived from Newton's second law

$$m D^I v_B^I = f_{a,p} + mg$$

It must be modified to bring out the geographic velocity. We first deal with the rotational derivative and transform it to the frame E

$$D^I v_B^I = D^E v_B^I + \Omega^{EI} v_B^I \tag{10.52}$$

then we modify the first term on the right-hand side by substituting Eq. (10.51)

$$D^E v_B^I = D^E v_B^E + D^E \left(\Omega^{EI} s_{BI} \right)$$

and subject the last term to the chain rule

$$D^E \left(\Omega^{EI} s_{BI} \right) = D^E \Omega^{EI} s_{BI} + \Omega^{EI} D^E s_{BI}$$

The first term on the right-hand side is zero because the Earth's angular velocity is constant. The second term can be expanded:

$$\Omega^{EI} D^E s_{BI} = \Omega^{EI} \left(D^E s_{BE} + D^E s_{EI} \right) = \Omega^{EI} D^E s_{BE} = \Omega^{EI} v_B^E$$

Collecting terms

$$D^E v_B^I = D^E v_B^E + \Omega^{EI} v_B^E$$

substituting them into Eq. (10.52) yields

$$D^I v_B^I = D^E v_B^E + \Omega^{EI} v_B^E + \Omega^{EI} v_B^I$$

and finally replacing the last v_B^I by Eq. (10.51) we get a relationship between the inertial and gravitational accelerations

$$D^I v_B^I = D^E v_B^E + 2\Omega^{EI} v_B^E + \Omega^{EI} \Omega^{EI} s_{BI} \qquad (10.53)$$

We identify the Coriolis term $2\Omega^{EI} v_B^E$ and the centrifugal term $\Omega^{EI} \Omega^{EI} s_{BI}$ that we first encountered in Sec. 5.3.1. Replacing the inertial acceleration in Newton's equation by Eq. (10.53), we obtain the translational equation

$$D^E v_B^E = \frac{1}{m} f_{a,p} + g - 2\Omega^{EI} v_B^E - \Omega^{EI} \Omega^{EI} s_{BI} \qquad (10.54)$$

It is exclusively a function of the geographic velocity and acceleration, although it governs the translational equations of motion relative to the inertial frame. It is also still in the invariant tensor form and therefore holds in any coordinate system. In particular, if we chose a coordinate system that is associated with the Earth's reference frame E, the rotational time derivative will turn into the ordinary time derivative. We have three choices; the Earth $]^E$, the geocentric $]^G$, or the geodetic $]^D$ coordinate systems. Most likely, we want the geographic velocity computed in geodetic coordinates. In this case we program

$$\left[\frac{dv_B^E}{dt} \right]^D = \frac{1}{m} [\bar{T}]^{BD} [f_{a,p}]^B + [T]^{DG} [g]^G$$

$$- 2[T]^{DI} [\Omega^{EI}]^I [\bar{T}]^{DI} [v_B^E]^D - [T]^{DI} [\Omega^{EI}]^I [\Omega^{EI}]^I [s_{BI}]^I \qquad (10.55)$$

Each term on the right-hand side is expressed in the most suitable coordinate system: the forces in body coordinates, Earth's gravitational acceleration in geocentric coordinates, Earth's angular velocity in inertial coordinates, and the geographic velocity in geodetic coordinates. The vehicle's displacement is referenced to the center of Earth $[s_{BI}]^I$ and is shown in inertial coordinates. It could as easily, with the help of Eq. (10.50), be expressed in geodetic Earth-referenced position $[s_{BE}]^D$ (point E can be any reference point fixed in the Earth frame)

$$[s_{BI}]^I = [\bar{T}]^{DI} [s_{BE}]^D + [s_{EI}]^I$$

The transformation matrices $[T]^{DG}$ and $[T]^{DI}$ have already been given, where $[T]^{BD}$ is obtained from the direction cosine matrix calculations.

An alternate form of Eq. (10.55) is also useful. It stems from the requirement of calculating the incidence angles α and β. They are derived from the geographic velocity expressed in body axes $[v_B^B]^B$. We could of course obtain it from $[v_B^E]^B = [\bar{T}]^{BD} [v_B^E]^D$, but some aerospace professionals prefer to formulate the geographic acceleration in body coordinates proper. In that case we take the perspective of the geographic velocity from the vehicle's body frame and make an Euler transformation to body frame B

$$D^E v_B^E = D^B v_B^E + \Omega^{BE} v_B^E$$

Substitution into Eq. (10.54) yields

$$D^B v_B^E + \Omega^{BE} v_B^E = \frac{1}{m} f_{a,p} + g - 2\Omega^{EI} v_B^E - \Omega^{EI}\Omega^{EI} s_{BI}$$

Expressed in body coordinates the rotational derivative will convert to the ordinary time derivative, and Newton's equations are programmed as

$$\left[\frac{dv_B^E}{dt}\right]^D = -[\Omega^{BE}]^B [v_B^E]^B + \frac{1}{m}[f_{a,p}]^B + [T]^{BG}[g]^G$$

$$- 2[T]^{BI}[\Omega^{EI}]^I[\bar{T}]^{BI}[v_B^E]^B - [T]^{BI}\big([\Omega^{EI}]^I[\Omega^{EI}]^I[s_{BI}]^I\big) \quad (10.56)$$

where $[\Omega^{BE}]^B = [\Omega^{BI}]^B - [T]^{BI}[\Omega^{EI}]^I$, $[\Omega^{BI}]^B$ is obtained from Euler's equation, and $[T]^{BI}$ from the direction cosine equation. Equation (10.56) is also found in this form in the book by Stevens and Lewis [Eqs. (1.3–1.6)].[9]

Having obtained the geographic velocity v_B^E defined as $v_B^E \equiv D^E s_{BI}$ from integrating Newton's equation, a second differential equation solves for the inertial position s_{BI}:

$$D^I s_{BI} = D^E s_{BI} + \Omega^{EI} s_{BI}$$

Note that the inertial point I in $D^E s_{BI}$ is located at the center of Earth and belongs also to the rotating Earth frame E. For this reason the geographic velocity can be defined two ways: $v_B^E = D^E s_{BE} = D^E s_{BI}$. However, let me caution you that the point E in this context can only be the center of the Earth and not just any point of the Earth frame, as was the case earlier. With this stipulation we formulate the second translational equation:

$$D^I s_{BI} = v_B^E + \Omega^{EI} s_{BI} \quad (10.57)$$

For programming purposes we have to distinguish whether the geographic velocity is given in geodetic or body coordinates. Accordingly, we have two options:

$$\left[\frac{ds_{BI}}{dt}\right]^I = [\bar{T}]^{BG}[v_B^E]^G + [\Omega^{EI}]^I[s_{BI}]^I \quad (10.58)$$

$$\left[\frac{ds_{BI}}{dt}\right]^I = [\bar{T}]^{BI}[v_B^E]^B + [\Omega^{EI}]^I[s_{BI}]^I \quad (10.59)$$

Equations (10.55) and (10.58) constitute the translational equations with emphasis on geodetic coordinates, whereas Eqs. (10.56) and (10.59) focus on the body coordinate system. Both give the same results. It is up to your customer to express her preference. The initialization of the geographic velocity v_B^E may be the deciding factor. $[v_B^E]^D$ is easier to visualize than $[v_B^E]^B$. Yet in any case, you will need to calculate both: $[v_B^E]^B$ to derive the incidence angles and $[v_B^E]^D$ to plot the velocity components over the Earth.

The derivation of the attitude equation harbors no surprises. Euler's law, referred to the inertial frame I and expressed in body axes, leads to the same rotational equations (10.44), already used and repeated here for completeness:

$$\left[\frac{d\omega^{BI}}{dt}\right]^{B} = \left([I_{B}^{B}]^{B}\right)^{-1}\left(-[\Omega^{BI}]^{B}[I_{B}^{B}]^{B}[\omega^{BI}]^{B} + [m_{B}]^{B}\right) \qquad (10.60)$$

The direction cosine matrix is as already calculated from the inertial body rates $[\omega^{BI}]^{B}$, Eqs. (10.47–10.49).

Let us list the state variables that are to be integrated: the geographic velocity $[v_{B}^{E}]^{D}$ or $[v_{B}^{E}]^{B}$; the inertial position $[s_{BI}]^{I}$; the inertial body rates $[\omega^{BI}]^{B}$; and the direction cosine matrix. Now, the second formulation of the six-DoF equations of motions over an elliptical Earth is complete.

Here are your choices. You can model your hypersonic vehicle relative to the inertial or the geographic frames over an elliptical rotating Earth and exercise several methods of initialization. I recommend the inertial approach because of its simplicity and I would combine it with the direction cosine matrix computations. If your customer requests a spherical Earth instead of its elliptical shape, you can accommodate him easily by setting the flattening parameter to zero and the semimajor axis of the ellipsoid to the mean Earth radius.

As you build your simulation, let the CADAC GHAME6 code be your guide. There you will also find other interesting modeling applications for aerodynamics and autopilots, which are the subject of the following sections.

10.2 Subsystem Models

The equations of motions are the most important part of a simulation, but by no means the most complicated ones. Their models are already in mathematical form and easily programmed. However, an aerospace vehicle consists of many more subsystems encompassing numerous technical disciplines.

The aerodynamicist provides the mathematical model of the aerodynamic forces and moments that act on the vehicle; thrust tables are generated by the propulsion expert; the control engineer designs the flight control system; and INS, sensors, and seekers are modeled by the appropriate experts.

We first discuss the modeling of the aerodynamics of aircraft, hypersonic vehicles, and missiles. The approach differs from the five-DoF implementation because now the aerodynamic forces and moments are untrimmed and depend on the control surface deflections. I shall acquaint you with the aerodynamic models of a typical aircraft, like the Fighting Falcon, a NASA hypersonic vehicle, and an air-to-air missile. Each model represents a different approach. The first two, with their planar airframes, use the Cartesian incidence angles of angle of attack and sideslip angle. The missile is modeled with the polar incidence angles of total angle of attack and aerodynamic roll angle.

For propulsion you can reach back to Secs. 8.2.4 and 9.2.2, where I discussed rocket motors, subsonic turbojets, ramjets, and scramjets. Those models can also be used in six-DoF simulations. If you want to build a detailed engine model however, you need to consult with propulsion experts and the references that they suggest.

In six-DoF simulations autopilots and flight control systems are represented by their full-up designs. No simplifications are required as in the pseudo-five-DoF simulations. When introducing you to six-DoF autopilots, I provide fairly simple structures. They are based on pole placement techniques that have the advantage of adapting to varying flight regimes.

The actuators execute the autopilot commands and turn them into control surface deflections. For aircraft we deal with aileron, elevator, and rudder actuators and in some cases with elevons and rudder. Missiles most likely have four control surfaces with four actuators. Their autopilot commands roll, pitch, and yaw, and the signals are divided up for the four surface actuators. Besides aerodynamic control, I also give you a model of thrust vector control for agile missiles.

Then we take on the subsystems of navigation and guidance. The INS is an integral part of any modern aerospace vehicle. I derive the equations of space stabilized and local-level systems and provide you with a nine-state error formulation.

Earlier you were introduced to proportional navigation. I expand on that discussion and derive the compensated PN and advanced guidance law (AGL). Whereas PN was formulated in inertial LOS rates, AGL uses differential velocity and displacement, expressed in inertial coordinates, as navigation input.

Finally, I augment the seeker discussion of Sec. 9.2.5 with an imaging IR seeker. We will investigate the intricate modeling of focal plane arrays, and mechanical and virtual gimbals. We also address noise and biases that corrupt the signals.

These subsystems are intended to serve as examples for modeling important components of aerospace vehicles. Although complete in themselves, they do not represent the most sophisticated models that you will encounter in six-DoF simulations. Regard them as instructional material that will get you started, guiding you from concepts to mathematical models and code implementation. You will find most of the subsystems, discussed in this section, coded up in the six-DoF simulations FALCON6, GHAME6, and SRAAM6, accessible on the CADAC Web site.

10.2.1 Aerodynamics

In the preceding section we concentrated on the left side of Newton's and Euler's equations. Now we move over to the right-hand sides and discuss the aerodynamic forces and moments. The modeling of aerodynamics is more of an art than a science because the physics of the airflow over a vehicle is so complicated that even today's supercomputers have to bow to the supremacy of the wind tunnel. However, the art of modeling aerodynamics is well understood, and we make use of the extensive background material available to us.

Before we proceed, I encourage you to review the five-DoF aerodynamics section 9.2.1 and the Taylor-series development in Sec. 7.3. For six-DoF modeling the book by Stevens and Lewis[9] provides additional up-to-date insight into the linearization process.

Let us revisit the functional dependencies of the aerodynamic forces and moments of Sec. 9.2.1:

aero forces and moments

$$
= f \left\{ \underbrace{M, \text{Re}}_{\substack{\text{flow} \\ \text{characteristics}}}, \underbrace{\alpha, \beta, \dot{\alpha}, \dot{\beta}}_{\substack{\text{incidence angles} \\ \text{and rates}}}, \underbrace{p, q, r}_{\text{body rates}}, \underbrace{\delta p, \delta q, \delta r}_{\substack{\text{control surface} \\ \text{deflections}}}, \text{c.g., power, shape, scale} \right\}
$$

where the flow characteristics are determined by Mach number M = velocity/sonic speed and Reynolds number Re = inertia forces/frictional forces. The incidence angles, here in Cartesian form, are the main dependencies. Their derivatives are usually of secondary importance and sometimes combined with the body rates, although their physical origins are quite distinct. The control surfaces are the main moment generators and are crucial for controlled flight. If the vehicle burns much fuel, the c.g. is bound to shift and has to be accounted for. Equally important is the effect of the exhaust plume on the vehicle drag. Hence, power-on/off effects must be included. Finally, the geometric descriptors are shape and scale, but are constant for a given vehicle.

You can imagine there are many techniques to simplify the functional form. The two primary methods are 1) the brute force modeling by tables and 2) the expansion in Taylor series. Diversity arises from the mix of the two approaches. Rather than being all inclusive, I will select three types of vehicles and furnish their aerodynamic models. The FALCON6 represents the aircraft model, NASA's GHAME6 vehicle exemplifies the hypersonic flight regime, and tactical missiles are embodied by the SRAAM6 concept.

10.2.1.1   Aircraft.

Aircraft aerodynamicists, like Lanchester,[10] have pioneered the modeling of aerodynamics. His system of aerodynamic coefficients has served us well over the years. Even rocketeers, after first using ballistic coefficients, have joined up and have used since the early 1960s[11] the same framework.

The six-DoF aerodynamic model of an aircraft can be as simple as the linear terms of the Taylor-series expansion or as complicated as tables with five independent variables. You will find the simple representations in simulations that emphasize autopilot design studies, whereas flight simulators require high-fidelity aerodynamic response for pilot training. We will take the middle road and essentially follow Stevens and Lewis.[9]

Today, the formatting of force coefficients is mostly in body coordinates with the positive sense following the direction of the body axes

$$[\overline{f_a}]^B = \bar{q}S[C_X \quad C_Y \quad C_Z]$$

where $\bar{q}$ is the dynamic pressure and S the reference area. Gone are the times of using lift and drag coefficients, although we shall encounter this archaic form shortly at the hypersonic vehicle. The format of the moments, referenced to the c.g. B, is noncontroversial and has always been in body coordinates

$$[\overline{m_B}]^B = \bar{q}S[C_l \times b \quad C_m \times c \quad C_n \times b]$$

with the reference lengths b as span and c as chord.

Figure 10.7 should help you visualize the positive directions of the force and moment coefficients. The controls of an aircraft are aileron δa, elevator δe, and rudder δr. Their positive deflections are defined by the following equations. Whereas a positive aileron deflection causes a positive rolling moment, elevator and rudder deflections are defined as positive if they produce positive increments of lift and side force.

Aileron:

$$+\delta a \quad \rightarrow \quad +\Delta LL \qquad \text{rolling moment}$$

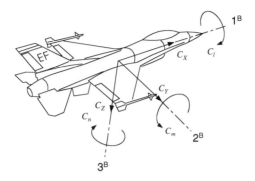

Fig. 10.7   Aircraft force and moment coefficients.

Elevator:

$$+\delta e \quad \rightarrow \quad +\Delta L \qquad \text{lift force}$$

Rudder:

$$+\delta r \quad \rightarrow \quad +\Delta Y \qquad \text{side force}$$

Now let us expand the aerodynamic force coefficients into a Taylor series but only with respect to the body rates p, q, r and the control deflections aileron, elevator, and rudder, δa, δe, δr. The coefficients remain in general a function of Mach M, angle of attack α, and sideslip angle β. With some variations the following model has been used for the F16 (Ref. 9). Please refer to Eq. (10.61). The axial force coefficient C_X essentially represents drag. Its dependency on Mach number and angle of attack is in tabular form. In addition, the pitch-rate elevator deflection is also tabular because of its nonlinear behavior. Only one Taylor-series expansion term relative to pitch rate is included. It models the drag effect, caused by the increased local angle of attack on the tail, as a result of the pitch rate. Because the side force C_Y remains usually small, it contains only tables of Mach numbers with the exception of the damping derivatives C_{Y_r} and C_{Y_p}, which are also a function of α; otherwise, only the linear terms in the Taylor series of β, δa, δr are taken into account. The normal force coefficient C_Z is similar to C_X, except that its dependency on δe can be linearized:

$$C_X = C_{X_o}(M, \alpha, \delta e) + \frac{c}{2V} C_{X_q}(M, \alpha) q$$

$$C_Y = C_{Y_\beta}(M)\beta + C_{Y_{\delta a}}(M)\delta a + C_{Y_{\delta r}}(M)\delta r$$

$$+ \frac{b}{2V} \left[C_{Y_r}(M, \alpha) r + C_{Y_p}(M, \alpha) p \right] \qquad (10.61)$$

$$C_Z = C_{Z_o}(M, \alpha, \beta) + C_{Z_{\delta e}}(M, \alpha)\delta e + \frac{c}{2V} C_{Z_q}(M, \alpha) q$$

The rolling and yawing moment coefficients C_l and C_n of Eq. (10.62) are modeled in the same fashion with respect to their nonlinear behavior in β. Their dependencies on aileron and rudder deflections have been linearized. In contrast, the pitching moment C_m is not a function of β, but the elevator deflection is included in the C_{m_0} table. An important effect has the center of gravity (c.g.) location on the pitching and yawing moments. Just consider the shift of fuel or cargo during flight. Furthermore, in wind-tunnel testing the moment center of the wind-tunnel model was probably not placed at the c.g. location of the full-up aircraft. In either case, be it dynamic or static, a moment arm $x_{cgR} - x_{cg}$ appears, which couples C_Z into C_m and C_Y into C_n. The fixed reference point is at x_{cgR}, whereas the actual c.g is at x_{cg}.

$$C_l = C_{l_0}(M, \alpha, \beta) + C_{l_{\delta a}}(M, \alpha, \beta)\delta a + C_{l_{\delta r}}(M, \alpha, \beta)\delta r$$

$$+ \frac{b}{2V}\left[C_{l_r}(M, \alpha)r + C_{l_p}(M, \alpha)p\right]$$

$$C_m = C_{m_0}(M, \alpha, \delta e) + \frac{c}{2V}C_{m_q}(M, \alpha)q + \frac{C_Z}{c}(x_{cgR} - x_{cg}) \quad (10.62)$$

$$C_n = C_{n_0}(M, \alpha, \beta) + C_{n_{\delta a}}(M, \alpha, \beta)\delta a + C_{n_{\delta r}}(M, \alpha, \beta)\delta r$$

$$- \frac{C_Y}{b}(x_{cgR} - x_{cg}) + \frac{b}{2V}\left[C_{n_r}(M, \alpha)r + C_{n_p}(M, \alpha)p\right]$$

Stevens and Lewis[9] provide numerical values for the coefficients at the single Mach number of 0.6. You can refer to the aerodynamic tables and mass properties in Appendix A of their book or can go to the CADAC FALCON6 simulation, Module A1.

I had difficulties locating six-DoF aerodynamics of modern aircraft in the open literature. Many of the models are either classified or considered proprietary by the manufacturer. If you have access to flight simulators, ask the software engineer to show you the aerodynamic model. You will be surprised by its sophistication. Here, we are contented with a middle-of-the-road approach.

10.2.1.2 Hypersonic vehicle.

You may be surprised, but hypersonic vehicles are modeled just like airplanes. The framework of Lanchester[10] applies even here. However, to gather the aerodynamic data is a gargantuan task. The greatest challenge is probably the SSTO vehicle. It takes off at low subsonic speeds, transitions through the sonic barrier to the supersonic regime, and, after about Mach 5, enters the hypersonic region. Eventually, the air density becomes so small, at about 100 km, that the aerodynamic effects are essentially nil.

It would be nice to have the resources to gather the data in wind tunnels. It takes three different facilities (subsonic, supersonic, and hypersonic) to cover the flight regimes. Yet, even then one has to make compromises. It is difficult to match the wind-tunnel's Reynolds number with free flight, and the Mach number tops out at about 10. You may take recourse to the highly touted computational fluid dynamics Mafia. However, their supercomputers devour your budget as well, and they will still send you back to the transonic tunnel to fill in the gaps.

There is a third option available to you. Since the dawn of the computer age and the maturing of the FORTRAN language, aerodynamicists have condensed their knowledge in semi-empirical codes. We already have met the Digital DATCOM

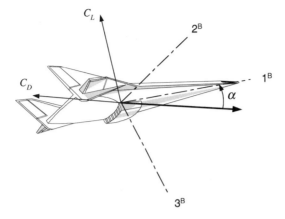

Fig. 10.8   Hypersonic vehicle lift and drag coefficients.

for aircraft and the Missile DATCOM programs. Both operate from dimensioned vehicle data using analytical formulas and interpolated wind-tunnel databases. In the early 1970s, so-called panel methods were developed, which are particularly useful in the supersonic and hypersonic regimes. Best known is the U.S. Air Force Supersonic/Hypersonic Arbitrary Body Program (S/HABP),[12] which is very popular with industry. It was upgraded in the late 1980s by McDonnell Douglas, St. Louis,[13] and is based on the impact pressure calculation method, which is most accurate at hypersonic speeds. Its capabilities, including viscous effects, extend down into the supersonic regime using so-called embedded flowfields.

Another semi-empirical code, the so-called PAN AIR method,[14] is NASA's contribution to supersonic and hypersonic aerodynamic prediction. It solves the linear partial differential flow equations numerically by approximating the configuration's surfaces by panels. PAN AIR is a higher-order panel method that is limited to inviscid flow.

NASA's Generic Hypersonic Aerodynamics Model Example (GHAME) serves us as a test case.[15] It is a hypersonic vehicle that takes off horizontally under turbojet propulsion. Once at Mach 2 the ramjet propels the vehicle to Mach 6, at which the scramjet takes over until the limits of the atmosphere are reached. The aerodynamic model was developed with the S/HABP code and from experimental data based on the space shuttle and the experimental vehicles X-24B and C.

Contrary to the FALCON6 model, GHAME aerodynamics uses lift and drag as force components, as depicted in Fig. 10.8. Not shown are the side force and the moment conventions because they remain unchanged. The NASP-like vehicle is controlled by rudder and two elevons: δv_l, left and δv_r, right. The elevons function as ailerons δa and elevators δe according to the relationships

$$\delta a = \frac{(\delta v_l - \delta v_r)}{2} \tag{10.63}$$

$$\delta e = \frac{(\delta v_l + \delta v_r)}{2} \tag{10.64}$$

The aerodynamic forces are given in stability axes (see Sec. 3.2.2.5)

$$[\overline{f_a}]^S = \bar{q}S[-C_D \quad C_Y \quad -C_L]$$

whereas the moments are in body axes as before

$$[\overline{m_B}]^B = \bar{q}S[C_l \times b \quad C_m \times c \quad C_n \times b]$$

NASA formulated the aerodynamics by simple Taylor-series expansion, including only the linear terms and making all derivatives tabular functions of Mach number and angle of attack:

$$C_L = C_{L_0}(M, \alpha) + C_{L_\alpha}(M, \alpha)\alpha + C_{L_{\delta e}}(M, \alpha)\delta e + C_{L_q}(M, \alpha)q\frac{c}{2V}$$

$$C_D = C_{D_0}(M, \alpha) + C_{D_\alpha}(M, \alpha)\alpha$$

$$C_Y = C_{Y_0}(M, \alpha) + C_{Y_\beta}(M, \alpha)\beta + C_{Y_{\delta a}}(M, \alpha)\delta a + C_{Y_{\delta r}}(M, \alpha)\delta r \quad (10.65)$$

$$+ C_{Y_p}(M, \alpha)p\frac{b}{2V} + C_{Y_r}(M, \alpha)r\frac{b}{2V}$$

Yet, by no means do these equations represent a linear model. Look at the lift slope derivative $C_{L_\alpha}(M, \alpha)$. It is a tabular function of α and is multiplied again by α. The same is true for the drag derivative $C_{D_\alpha}(M, \alpha)$. The other derivatives are linear in terms of the remaining states, but still nonlinear in Mach and α. The effect of body rates on the lift and side forces, i.e., the C_{L_q}, C_{Y_p}, and C_{Y_r} derivatives, can often be neglected.

The moment coefficients follow a similar pattern:

$$C_m = C_{m_0}(M, \alpha) + C_{m_\alpha}(M, \alpha)\alpha + C_{m_{\delta e}}(M, \alpha)\delta e + C_{m_q}(M, \alpha)q\frac{c}{2V}$$

$$C_l = C_{l_0}(M, \alpha) + C_{l_\beta}(M, \alpha)\beta + C_{l_{\delta a}}(M, \alpha)\delta a + C_{l_{\delta r}}(M, \alpha)\delta r$$

$$+ C_{l_p}(M, \alpha)p\frac{b}{2V} + C_{l_r}(M, \alpha)r\frac{b}{2V} \quad (10.66)$$

$$C_n = C_{n_0}(M, \alpha) + C_{n_\beta}(M, \alpha)\beta + C_{n_{\delta r}}(M, \alpha)\delta r + C_{n_{\delta a}}(M, \alpha)\delta a$$

$$+ C_{n_p}(M, \alpha)p\frac{b}{2V} + C_{n_r}(M, \alpha)r\frac{b}{2V}$$

All four terms in the pitching moment coefficient C_m are important. Yet, the rolling and yawing moment coefficients C_l and C_n usually have negligible trim coefficients C_{l_0} and C_{n_0}. In effect, Fig. 7.1 predicts that they are zero. Notice the strong cross coupling between the rolling and yawing moments. It occurs through the famous Dutch-roll derivative C_{l_β} and the body rates p, q and the controls $\delta a, \delta r$.

We are fortunate that NASA has provided us with complete tables. If you want to get a feel for their numerical values, you can turn to the CADAC GHAME6 simulation and consult Module A1. Now, finished with these aircraft-type vehicles, we turn to missiles with tetragonal symmetry.

10.2.1.3 Missiles.
Most missiles have simple geometrical shapes. In rocketry the four fin, axis-symmetrical configuration, first fashioned by Wernher von Braun as the V2, are prevalent still today. Air-to-air missiles like ASRAAM and AMRAAM exhibit this tetragonal symmetry, as well as our CADAC SRAAM model.

The tetragonal symmetry fosters the expression of the aerodynamics in aeroballistic axes. Refer back to Fig. 3.18 for the definition of the total angle of attack α' and aerodynamic roll angle ϕ'. Both replace the standard incidence angles α and β. The aerodynamic forces and moment are expressed in coefficient form in aeroballistic coordinates $]^R$

$$[\overline{f_a}]^R = \bar{q}S[-C_A \quad C_Y' \quad -C_N']$$

$$[\overline{m_B}]^R = \bar{q}Sl[C_l \quad C_m' \quad C_n']$$

where $\bar{q}$ is the dynamic pressure, S the reference area (cross section), l the reference length (diameter), and the prime refers to the aeroballistic axes. Figure 10.9 depicts the positive sense of these coefficients. Pay particular attention to the positive direction of the aerodynamic roll angle ϕ'. It is taken positive from the aeroballistic axes to the body axes. Also notice that the axial force and the rolling moment coefficients are invariant under this transformation.

The functional dependency of a given configuration is reduced to the following form:

$$
\begin{array}{ll}
C_A = f(M, \delta'q, \delta'r, \text{power}) & C_l = f(M, \alpha', \phi', p, \delta p) \\
C_Y' = f(M, \alpha', \phi', \delta'r) & C_m' = f(M, \alpha', \phi', \text{c.g.}, q', \delta'q) \\
C_N' = f(M, \alpha', \phi', \delta'q) & C_n' = f(M, \alpha', \phi', \text{c.g.}, r', \delta'r)
\end{array}
$$

where the Reynolds and incidence rate dependencies were neglected.

The body rates p, q', r' and the control commands $\delta p, \delta q', \delta r'$ are expressed in aeroballistic coordinates $]^R$ and obtained from body coordinates by the

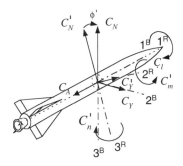

Fig. 10.9   Force and moment coefficients in aeroballistic axes.

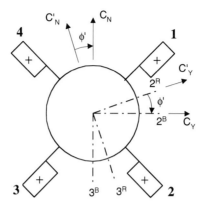

Fig. 10.10   Positive deflection of control fins viewed from rear.

transformations

$$
\begin{bmatrix} p \\ q' \\ r' \end{bmatrix}^{R} =
\begin{bmatrix} 1 & 0 & 0 \\ 0 & \cos\phi' & -\sin\phi' \\ 0 & \sin\phi' & \cos\phi' \end{bmatrix}
\begin{bmatrix} p \\ q \\ r \end{bmatrix}^{B}
$$

$$
\begin{bmatrix} \delta p \\ \delta q' \\ \delta r' \end{bmatrix}^{R} =
\begin{bmatrix} 1 & 0 & 0 \\ 0 & \cos\phi' & -\sin\phi' \\ 0 & \sin\phi' & \cos\phi' \end{bmatrix}
\begin{bmatrix} \delta p \\ \delta q \\ \delta r \end{bmatrix}^{B}
\qquad (10.67)
$$

The control commands require further explanations. The starting point is Fig. 10.10 with the convention of positive surface deflections. For reinforcement I have also included again the positive direction of the aerodynamic roll angle ϕ'. Because there are four fins, but only three attitude degrees of freedom, we combine them mathematically to form the three control commands

$$
\delta p = \tfrac{1}{4}(-\delta 1 - \delta 2 + \delta 3 + \delta 4)
$$

$$
\delta q = \tfrac{1}{4}(\delta 1 + \delta 2 + \delta 3 + \delta 4) \qquad (10.68)
$$

$$
\delta r = \tfrac{1}{4}(-\delta 1 + \delta 2 - \delta 3 + \delta 4)
$$

A fourth combination is also possible, but it produces only drag and no control moment. Look at Fig. 10.10 and work out the three relationships. Four positive fin deflections create a positive pitch command; the first two fins up and the last two fins down generate a positive roll command; and alternating the fin deflections causes a positive yaw command. This convention is in agreement with aircraft practices and is summarized in the following equations.

Roll:

$$
+\delta p \quad \rightarrow \quad +\Delta L \qquad \text{rolling moment}
$$

Pitch:

$$+\delta q \quad \rightarrow \quad +\Delta N \qquad \text{normal force}$$

Yaw:

$$+\delta r \quad \rightarrow \quad +\Delta Y \qquad \text{side force}$$

Now we are ready to expand the coefficients. Please refer to Eq. (10.69). The axial force coefficient C_A consists of the base and skin friction drag C_{Ao} and the power on/off effect ΔC_A. Furthermore, it has a first-order term in α' and a second-order three-dimensional table look-up in $\delta_{\text{eff}} = (|\delta q| + |\delta r|)/2$ (the effective fin deflection). You can see that great care has been taken to model the axial force because its fidelity determines the fly-out range of the missile. The side force and normal force coefficients C_Y' and C_N' are dependent on the roll orientation of the missile relative to the load factor plane, described by the angle ϕ' ($\phi' = 0$ when the missile maneuvers in the cruciform configuration):

$$C_A = C_{Ao}(M) + \Delta C_{A(\text{power})}(M) + C_{A_{\alpha'}}(M)\alpha' + C_{A_{\delta^2 \text{eff}}}\left(M, \alpha', \delta_{\text{eff}}^2\right)\delta_{\text{eff}}^2$$

$$C_Y' = \Delta C_{Y,\phi'}'(M, \alpha') \sin 4\phi' + C_{Y_{\delta r}}'(M, \alpha')\delta'r \qquad (10.69)$$

$$C_N' = C_{No}'(M, \alpha') + \Delta C_{N,\phi'}'(M, \alpha') \sin^2 2\phi' + C_{N_{\delta q}}'(M, \alpha')\delta'q$$

Inspecting Eq. (10.69), you see that this dependency is modeled by sine functions. Both terms, $\sin 4\phi'$ and $\sin^2 \phi'$, are periodic every 90 deg, corresponding to the tetragonal symmetry of the missile. However, the increment in side force reaches its maximum at 22.5 deg and its minimum at 67.5 deg. The normal force effect is always positive and reaches its peak every 45 deg. Realize that in aeroballistic axes the side force component is always small because, by definition, the missile responds in the load factor plane. Any perturbations caused by rotational asymmetries are of second order and are many times just neglected. The other terms in Eq. (10.69) are the control effectiveness derivatives $C_{Y_{\delta r}}'$ and $C_{N_{\delta q}}'$.

The moment coefficients are modeled in a similar fashion. They consist of primary terms, damping derivatives, control effectiveness, and c.g. adjustments. The primary roll term $C_{l,\phi_{\alpha^2}'}(M, \alpha')\alpha'^2 \sin 4\phi'$ of Eq. (10.70) is an attempt to model the roll coupling caused by vortices impinging on the tail surfaces at high angles of attack. It is dependent on the roll orientation of the missile, $\sin 4\phi'$, which we encountered already in the C_Y' coefficient:

$$C_l = C_{l,\phi_{\alpha^2}'}(M, \alpha')\alpha'^2 \sin 4\phi' + C_{l_p}(M, \alpha')\frac{pl}{2V} + C_{l_{\delta p}}(M, \alpha')\delta p$$

$$C_m' = C_m'(M, \alpha') + \Delta C_{m,\phi'}'(M, \alpha') \sin^2 2\phi' + C_{m_q}'(M)\frac{q'l}{2V}$$

$$\qquad (10.70)$$

$$+ C_{m_{\delta q}}'(M, \alpha')\delta'q - \frac{C_N'}{l}(x_{\text{cg},R} - x_{\text{cg}})$$

$$C_n' = \Delta C_{n,\phi'}'(M, \alpha') \sin 4\phi' + C_{n_r}'(M)\frac{r'l}{2V} + C_{n_{\delta r}}'(M, \alpha')\delta'r - \frac{C_Y'}{l}(x_{\text{cg},R} - x_{\text{cg}})$$

This yaw–roll coupling is 90-deg periodic in ϕ', causing a positive or negative rolling moment, depending on the orientation of the fins relative to the load factor plane. The other two terms of the rolling coefficient bear no surprises.

The pitching moment coefficient, similarly affected by the orientation of the load factor plane, possesses a perturbation term $\Delta C'_{m,\phi'}(M, \alpha')\sin^2 2\phi'$. It vanishes at 90-deg increments between maxima. The yawing moment coefficient C'_n repeats the pattern of C_l.

To adjust for weight shifts during flight, the pitching and yawing moment must be corrected for the shift in c.g. Just like the aircraft equations (10.62), the adjustment is made with the moment arm $(x_{\text{cg},R} - x_{\text{cg}})$, where $x_{\text{cg},R}$ is the reference location (at launch or from wind-tunnel tests.) and x_{cg} the true location.

I have presented to you just one missile model, which is quite useful for preliminary performance studies. It has been adopted for many air-to-ground and air-to-air simulations and linked with the Digital DATCOM aerodynamic prediction program. If you want to see it in action, you should refer to the CADAC SRAAM6 simulation, study the Module A1, and run the test cases.

10.2.1.4 Summary. The three aerodynamic models, although limited in sophistication, should give you an appreciation of the six-DoF treatment of aerodynamics. There is a distinct difference between aircraft and missiles, corresponding to their respective symmetries. Planar vehicles are modeled with the Cartesian incidence angles α and β, whereas tetragonal missiles lend themselves to the treatment with polar incidence angles α' and ϕ'. However, you will find many missile simulations that use also the Cartesian incidence angle scheme. Next time you get your hands on a six-DoF simulation, see if you recognize some of the modeling features of this section.

10.2.2 Autopilot

The aerodynamics, propulsion, and mass properties, activated by Newton's and Euler's equations, model the vehicle dynamics. They define the plant to be controlled. An aircraft needs rate damping for stability augmentation, roll control for coordinated turns and flight-path-angle tracking. For long duration flights, altitude and heading hold autopilots are indispensable for a weary crew. Missiles, flying in the atmosphere, are primarily controlled by rate and acceleration feedback loops. A roll controller maintains their orientation. For longer duration flights the trajectory can be shaped by a flight-path tracker or altitude hold autopilot. Outside the atmosphere missile attitude is controlled by angular feedback loops.

The feedback signals come from sensors that are located throughout the vehicle. The INS, an indispensable component in any modern aerospace vehicle, provides rate and acceleration feedback for all three axes, directly from its accelerometers and gyros, and attitude angles from its navigation computer. If wind is not a factor, it can also calculate vertical and horizontal flight-path angles. The air data computer provides Mach number and dynamic pressure. Using incidence angles as feedback signals is problematic because it is difficult to accurately measure or compute the angle of attack or sideslip angle. The alpha vanes that you see on the outside of airplanes are only used as warning devices and not as autopilot sensors.

The coverage in this section picks up from Sec. 9.2.3 where I discussed autopilots of pseudo-five-DoF simulations. As you may recall, in five-DoF simulations autopilots do not issue control commands to the control surfaces but model closed-loop response, delivering angle of attack to the trimmed aerodynamic tables. In the real, nonpseudo world the autopilot, or the human operator, controls the aileron, elevator, and rudder directly, and only through the airframe and the sensors are the dynamic loops closed.

Most of the autopilot designs of Sec. 9.2.3 do not apply here, with the exception of the altitude hold autopilot. Its outer altitude position and rate loops can be maintained as long as the inner acceleration autopilot is replaced by its six-DoF equivalent. So let us get a new start and build control systems for aircraft and missiles that are suitable for our six-DoF simulations.

Before we begin, let me caution you, however, that my autopilot designs are simplified versions of what you actually find onboard these vehicles. Aircraft flight control systems can be very sophisticated and are beyond the scope of this discussion. If you are tasked to model an existing vehicle, you should get the autopilot specification and replicate it as faithfully as possible. Only then will the customer trust the results of your analysis. However, if your job is to study a generic system or to develop a new concept, you will find the following designs useful.

To make the autopilots serve a variety of vehicles and flight conditions, I employ the pole placement technique. You can find this method discussed by Stevens and Lewis.[9] Yet if you have some familiarity with classic and modern linear control, you should be able to follow this self-contained presentation.

10.2.2.1 Rate damping loop.
The most elementary autopilot function is the improvement of the dynamic stability of the vehicle, both in the pitch and yaw planes. This is accomplished by augmenting the aerodynamic damping derivative by rate feedback loops. The feedback sensors are the rate gyros from the INS system. We shall have to deal with the pitch and yaw planes separately. The pitch-rate autopilot is the same for both aircraft and missiles, although we may have to use different symbols, like lift slope C_{L_α} and the normal slope C_{N_α}. The first one is given in velocity axes and the second one in body axes. Missiles prefer C_{N_α}, whereas aircraft use the more traditional C_{L_α}. If the missile has tetragonal symmetry, the yaw aerodynamics behaves like those of the pitch plane, and therefore the same autopilot serves both planes. In the aircraft case, however, the lateral aerodynamics is distinctly different and requires a separate treatment.

Let us apply the pole placement technique to this simple example. We employ the open-loop transfer function of the pitch plane, close the loop with a gain, and calculate the closed-loop poles. Because the gain is a free variable, we can choose it such that the closed-loop damping achieves a desired value for all flight conditions.

From Sec. 7.4.2 we carry over the linear perturbation equations of the missile pitch plane in the two state variables $\bar{x} = [q \ \alpha]$, pitch rate and angle of attack. (Note that in this autopilot section we adopt the bolded vector convention of modern control.)

$$\begin{bmatrix} \dot{q} \\ \dot{\alpha} \end{bmatrix} = \begin{bmatrix} M_q & M_\alpha \\ 1 & -\dfrac{N_\alpha}{V} \end{bmatrix} \begin{bmatrix} q \\ \alpha \end{bmatrix} + \begin{bmatrix} M_\delta \\ -\dfrac{N_\delta}{V} \end{bmatrix} \delta$$

where $N_\alpha = (\bar{q}S/m)C_{N_\alpha}$, $N_\delta = (\bar{q}S/m)C_{N_\delta}$, $M_\alpha = (\bar{q}Sd/I_2)C_{m_\alpha}$, $M_q = [\bar{q}Sd^2/(2I_2V)]C_{m_q}$, $M_\delta = (\bar{q}Sd/I_2)C_{m_\delta}$ are the dimensional derivatives for the normal

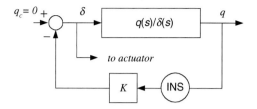

Fig. 10.11   Rate feedback loop.

force N and the pitching moment M; δ is the pitch control surface deflection; and V the flight speed. These plant equations are in the general form of

$$\dot{x} = Fx + gu$$

Now we build the transfer function of pitch rate q with respect to pitch control δ. Take the Laplace transformation of the plant equation and solve for $x(s)$

$$sx(s) = Fx(s) + gu(s)$$

$$(sI - F)x(s) = gu(s)$$

$$x(s) = \Phi(s)gu(s)$$

where $\Phi(s) = (sI - F)^{-1}$ is the resolvant matrix. The input $u(s)$ is δ, whereas the output $y(s)$ could be either q or α. The vector $\bar{h}(s) = [1 \ \ 0]$ will pick the output q from the general relationship

$$y(s) = \bar{h}\Phi(s)gu(s)$$

Therefore, the desired transfer function is

$$\frac{q(s)}{\delta(s)} = \bar{h}\Phi(s)g = \frac{M_\delta[s + N_\alpha/V - (M_\alpha/M_\delta)(N_\delta/V)]}{s^2 + (N_\alpha/V - M_q)s - M_\alpha - M_q N_\alpha/V} \quad (10.71)$$

You can verify this equation by comparing it with Roskam (Ref. 16, p. 426). The transfer function is positioned in a simple feedback loop (Fig. 10.11) and has two complex conjugate poles and a zero with the root locus pattern shown in Fig. 10.12. Selecting the closed-loop damping coefficient ζ determines the gain K.

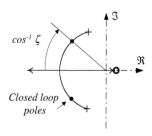

Fig. 10.12   Root locus diagram of rate autopilot.

To compute this relationship, we abbreviate the transfer function by

$$\frac{q(s)}{\delta(s)} = \frac{G(s+z)}{s^2 + as + b}$$

and build the closed-loop transfer function

$$\frac{q(s)}{q_c(s)} = \frac{G(s+z)}{s^2 + (a + KG)s + b + KGz} = \frac{G(s+z)}{s^2 + 2\zeta\omega_n s + \omega_n^2}$$

The last formulation introduces the desired closed-loop parameters ζ and ω_n. Equating terms of equal powers in the denominator yields the formula for the rate loop gain

$$K = \frac{1}{G}\left[-(a - 2\zeta^2 z) + \sqrt{(a - 2\zeta^2 z)^2 - (a^2 - 4\zeta^2 b)}\right] \qquad (10.72)$$

where the negative sign before the radical was discarded because it does not lead to a useful solution. The resulting natural frequency of the closed-loop poles, which follows from the selection of ζ, becomes

$$\omega_n = \sqrt{b + KGz}$$

No matter what the flight conditions are, within the limitations of linear approximations the damping of the closed-loop response is guaranteed to be the value that you select. However, it could come at a penalty. A high gain K, causing a high natural frequency ω_n, may drive the controls prematurely into saturation. You need to check out the control deflections for some of the more extreme flight conditions before you commit to a desired ζ. If you cannot reconcile all conditions with one value ζ, you could make it dependent on dynamic pressure.

The calculation of K in a simulation and in an actual flight control system is carried out on-line, making use of sensors and stored data. Tables provide the aerodynamic data and mass properties, while Mach number and dynamic pressure come from the air data computer. Combined, they establish the transfer function parameters. Together with the required damping ζ, the gain K can be computed.

This simple rate loop is used in the CADAC SRAAM6 simulation, Module C2, Subroutine C2RATE. It works quite well during the initial phase, as the missile leaves the launch rail, requiring stabilization. Because the SRAAM6 has tetragonal symmetry, both pitch and yaw channels have identical rate loops.

The GHAME6 simulation of the hypersonic SSTO vehicle requires a yaw stability augmentation system (SAS). Our rate loop satisfies this need, yet must be adjusted to airplane yawing conventions. A positive β generates a negative side force Y, whereas a positive α causes a positive normal force N. Accounting for this difference, the linearized perturbation equations in the yaw plane are for airplanes

$$\begin{bmatrix} \dot{r} \\ \dot{\beta} \end{bmatrix} = \begin{bmatrix} LN_r & LN_\beta \\ -1 & \dfrac{Y_\beta}{V} \end{bmatrix} \begin{bmatrix} r \\ \beta \end{bmatrix} + \begin{bmatrix} LN_\delta \\ \dfrac{Y_\delta}{V} \end{bmatrix} \delta$$

where $Y_\beta = (\bar{q}S/m)C_{Y_\beta}$, $Y_\delta = (\bar{q}S/m)C_{Y_\delta}$, $LN_\beta = (\bar{q}Sb/I_{33})C_{n_\beta}$, $LN_r = [\bar{q}Sb^2/(2I_{33}V)]C_{n_r}$, $LN_\delta = (\bar{q}Sb/I_{33})C_{n_\delta}$ are the dimensioned derivatives for the side force Y and the yawing moment LN; δ is the rudder deflection. According to the pole placement procedure, we obtain the open-loop transfer function, again confirmed by Ref. 16, p. 458:

$$\frac{r(s)}{\delta(s)} = \frac{LN_\delta[s - Y_\beta/V + (LN_\beta/LN_\delta)(Y_\delta/V)]}{s^2 - (Y_\beta/V + LN_r)s + LN_\beta + LN_r Y_\beta/V} \tag{10.73}$$

and with the abbreviated closed-loop transfer function

$$\frac{r(s)}{r_c(s)} = \frac{G(s+z)}{s^2 + (a + KG)s + b + KGz} = \frac{G(s+z)}{s^2 + 2\zeta\omega_n s + \omega_n^2}$$

We can use Eq. (10.72) again to calculate the yaw loop gain K. You can find an example for a yaw loop autopilot in the code CADAC GHAME6, Module C2, Subroutine C2YSAS.

10.2.2.2 Roll position autopilot.

The roll position autopilot has a dual purpose. Missiles, maneuvering independently in the pitch and yaw plane, use the roll autopilot to suppress any roll excursions that may be caused by aerodynamic cross coupling. Although earlier designs, like the Sidewinder air-to-air missile, incorporated only roll damping, all modern missiles possess roll position loops. If the airframe lacks tetragonal symmetry—like cruise missiles or airplanes, the roll loop is used to bank the vehicle for lateral maneuvers.

We build a dual feedback controller, utilizing the roll rate from the INS gyro and roll position from INS navigation computations. The inner rate loop augments the aerodynamic damping and the outer position loop executes the roll command. The transfer function is rather simple, consisting of the roll damping derivative $LL_p = (\bar{q}Sb/I_{11})(b/2V)C_{l_p}$ and roll control derivative $LL_{\delta a} = (\bar{q}Sb/I_{11})C_{l_{\delta a}}$ [see Eq. (7.57)]

$$\frac{p(s)}{\delta a(s)} = \frac{LL_{\delta a}}{s - LL_p}$$

Figure 10.13 depicts the two feedback loops and Fig. 10.14 the associated root locus diagram. The roll autopilot has two gains K_ϕ and K_p that require definition. They give us the added flexibility to specify not only the damping but also the natural frequency of the closed-loop poles. We build first the closed-loop transfer

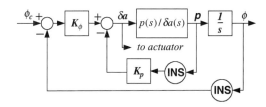

Fig. 10.13   Roll rate and position feedback loops.

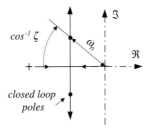

Fig. 10.14   Root locus diagram of roll autopilot.

function and set it equal to the desired form that introduces the parameters ζ and ω_n:

$$\frac{\phi(s)}{\phi_c(s)} = \frac{K_\phi LL_{\delta a}}{s^2 + (K_p LL_{\delta a} - LL_p)s + K_\phi LL_{\delta a}} = \frac{K_\phi LL_{\delta a}}{s^2 + 2\zeta\omega_n s + \omega_n^2} \qquad (10.74)$$

Comparing terms of equal power of s, we can calculate the two gains

$$K_\phi = \frac{\omega_n^2}{LL_{\delta a}} \quad \text{and} \quad K_p = \frac{2\zeta\omega_n + LL_p}{LL_{\delta a}}$$

The comments I made about rate loops apply here as well. To prevent aileron saturation in extreme flight conditions, you should schedule ζ and ω_n and calculate derivatives on-line from stored data. You will be glad to know that, thanks to the integrator in the forward loop, the steady-state error of the roll loop is zero, i.e., the roll position autopilot is a perfect tracker. An implementation of the roll autopilot is given in the CADAC GHAME6 simulation, Module C2, Subroutine C2ROLL.

10.2.2.3 Heading angle tracker.

Heading changes of an aircraft are executed by roll control. As the lift vector is banked, a horizontal force component generates a lateral acceleration that turns the velocity vector horizontally. Direct sideslip control is ineffective because of the small lateral projected area of the aircraft; also, the ensuing adverse yaw-roll coupling would be undesirable.

We build the heading angle tracker by simply wrapping a heading loop around the roll position autopilot. The schematic is shown in Fig. 10.15. To derive the transfer function from the roll angle ϕ to the heading angle χ, we set up the relationship between the lateral acceleration and the lateral force, caused by the lift vector, banked through the angle ϕ; and with the small angle assumption of ϕ, we can

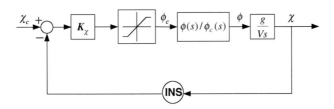

Fig. 10.15   Heading angle tracker enclosing roll autopilot.

simplify the expression

$$mV\dot{\chi} = L\sin\phi = mg\phi$$

and obtain the desired transfer function

$$\frac{\chi(s)}{\phi(s)} = \frac{g}{Vs}$$

shown in the last block of Fig. 10.15. The gain K_χ translates the heading error $\chi_c - \chi$ into the roll command ϕ_c after limits are applied.

We employ again the pole placement technique to calculate this gain. Clinching the open-loop transfer function from Fig. 10.15 and with Eq. (10.74)

$$T_o(s) = K_\chi \frac{g}{V} \frac{\omega_n^2}{s^3 + 2\zeta\omega_n s^2 + \omega_n^2 s}$$

and closing the feedback loop, while setting it equal to the generic third-order transfer function,

$$T_c(s) = \frac{K_\chi(g/V)\omega_n^2}{s^3 + 2\zeta\omega_n s^2 + \omega_n^2 s + K_\chi(g/V)\omega_n^2} = \frac{K_\chi(g/V)\omega_n^2}{(s + p_\chi)(s^2 + 2\zeta_x\omega_{n\chi}s + \omega_{n\chi}^2)}$$

with the free parameters p_χ, ζ_χ, and $\omega_{n\chi}$, which determine the modal response of the heading tracker. Equating equal powers of s yields three equations, but, including K_χ, there are four unknowns. Hence, one additional equation is required. After having studied the root locus for some time, I found that positioning p_χ on the real axes, directly under the complex poles of the roll autopilot, grants a well-behaved heading response. Thus, we set

$$p_\chi = \zeta\omega_n$$

Eliminating ζ_χ and $\omega_{n\chi}$, we calculate the gain from

$$K_\chi = \frac{V}{g}\zeta\omega_n(1 - \zeta^2) \tag{10.75}$$

The heading gain depends on the data entry ζ and ω_n of the roll autopilot and the flight speed V. If the optimal $\zeta = 0.707$ is selected, the gain is simply $K_\chi = 0.147V\omega_n/g$. Just as the roll autopilot, the heading loop is also a perfect tracker, thanks to the integrator in the forward loop. You can find the implementation of the heading tracker in the CADAC GHAME6 simulation, Module C2, subroutine C2HEAD. It operates in conjunction with subroutine C2ROLL.

10.2.2.4 Acceleration autopilot.

Highly agile air-to-air missiles rely on tight autopilot control for successful target intercept. Because they zero miss distance by lateral acceleration, you will find that all modern missiles are controlled by acceleration controllers. To improve performance, an inner rate loop is added for stability augmentation. Both acceleration and body-rate feedback are provided by the accelerometer and rate sensors of the INS.

Airplanes are less likely to be controlled by acceleration autopilots because passenger comfort is more important than maneuverability. However, for special

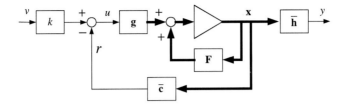

Fig. 10.16   Linear system with feedback gain $\bar{c}$ and gain k.

applications, like unmanned combat air vehicles or cruise missiles, the normal load factor plane may contain an acceleration feedback loop. Such guidance systems as terrain following and obstacle avoidance or target homing require rapid response that only an acceleration autopilot can provide.

Having used classical control for the rate and roll autopilots, I use modern pole placement techniques for designing the acceleration tracking loops. For best performance, proportional and integral (PI) techniques are applied—proportional control for quick response and integral control for zeroing steady-state errors. After a general introduction to state variable pole placement techniques, I will deal with the pitch plane acceleration autopilot and solve for the three feedback gains that satisfy the specified closed-loop response. For missiles the yaw plane implementation is the same.

Let us formulate the problem (see Fig. 10.16). Given is the linear time-variant plant $\dot{x} = F(t)x + g(t)u$. What is the controller gain c in order for the closed-loop poles to take their places at the locations p_i, $i = 1, \ldots, n$ for all times t; and what is the gain k so that the steady-state gain is one? We consider only SISO systems, with v as input and y as output.

The solution follows the same steps we used for the rate loop, except now we employ the state variable form. The closed-loop system is $\dot{x} = [F(t) - g(t)\bar{c}]x + kg(t)v$ with the closed-loop fundamental matrix $[F(t) - g(t)\bar{c}]$, whose eigenvalues must be equal to the desired closed-loop poles. Therefore, the condition for pole placement is

$$\text{Det}|Is - F(t) + g(t)\bar{c}| = \prod_{i=1}^{n}(s - p_i) \qquad (10.76)$$

with n the number of states.

The forward gain is calculated on the basis that $y = v$ when steady state has been reached, which is the case when $\dot{x} = 0$. Employing the steady-state closed-loop system equation and the output relationship $y = \bar{h}x$, we can solve for the forward gain

$$k = -\frac{1}{\bar{h}[F(t) - g(t)\bar{c}]^{-1}g(t)} \qquad (10.77)$$

Equation (10.76) is evaluated by equating terms of equal power in s. This process leads to a system of linear equations, which can be solved for the controller gain c

$$c = P^{-1}d \qquad (10.78)$$

where P is an $n \times n$ matrix and d an $n \times 1$ vector.

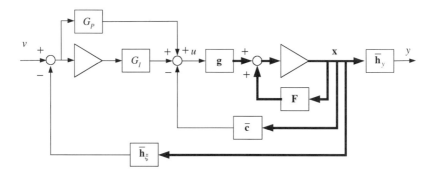

Fig. 10.17   Acceleration PI autopilot with feedback gains c and feedforward gains G_P and G_I.

For implementation, either in a simulation or onboard a flight vehicle, Eqs. (10.77) and (10.78) are calculated on-line, given the $F(t)$ matrix and $g(t)$ vector. Note that in general two matrix inversions of order n must be carried out. Fortunately, for the acceleration autopilot we can solve Eq. (10.78) algebraically, whereas later on the flight-path-angle autopilot of Sec. 10.2.2.6 requires the full matrix inversion. To adapt the controller gains to highly variable flight conditions, the desired closed-loop poles p_i can be made a function of some flight parameter, say, time, dynamic pressure, or altitude.

Let us get specific with the acceleration autopilot. Fig. 10.17 portrays the structure. The proportional and integral feedforward branches with their respective gains G_P and G_I are clearly visible. The major feedback loop is via the rate and acceleration gains $\bar{c} = [k_2 \ k_1]$. A second acceleration feedback with unit gain $\bar{h}_\xi$ is wrapped around the outside to improve performance. We derive from this schematic the relationship for the control variable

$$u = -\bar{c}x + G_I \int (v - \bar{h}_\xi x)\, dt + G_P(v - \bar{h}_\xi x)$$

Then we augment the states by introducing the scalar auxiliary variable $\xi = \int(v - \bar{h}_\xi x)\, dt$ with its state equation

$$\dot{\xi} = v - \bar{h}_\xi x$$

Substituting u into the open-loop system yields the closed-loop system, augmented by the auxiliary variable ξ:

$$\begin{bmatrix} \dot{x} \\ \dot{\xi} \end{bmatrix} = \begin{bmatrix} F - g(\bar{c} + G_P\bar{h}_\xi) & G_I g \\ -\bar{h}_\xi & 0 \end{bmatrix} \begin{bmatrix} x \\ \xi \end{bmatrix} + \begin{bmatrix} G_P g \\ 1 \end{bmatrix} v \tag{10.79}$$

which we abbreviate by the equation

$$\dot{x}' = F'(t)x' + g'(t)v$$

The eigenvalues of this closed-loop fundamental matrix F' must be equal to the desired closed-loop poles. The condition for pole placement is

$$\text{Det}[\boldsymbol{I}s - \boldsymbol{F}'(t)] = \prod_{i=1}^{n}(s - p_i)$$

and specifically

$$\text{Det}\left|\begin{bmatrix} \boldsymbol{I}_{nxn}s - \boldsymbol{F} + \boldsymbol{g}(\bar{\boldsymbol{c}} + \boldsymbol{G}_P\bar{\boldsymbol{h}}_\xi) & \vdots & -\boldsymbol{G}_I\boldsymbol{g} \\ \hdashline \boldsymbol{h}_\xi & \vdots & s \end{bmatrix}\right| = \prod_{i=1}^{n}(s - p_i) \qquad (10.80)$$

The evaluation of this equation leads to the on-line calculation of the gains $\boldsymbol{c}$, $\boldsymbol{G}_I$ and $\boldsymbol{G}_P$, given the desired closed-loop poles p_i.

Now we are ready to design the acceleration autopilot for agile missiles. From Chapter 7 we carry over the linearized plant Eq. (7.69). It is similar to the rate equation, except for the angle of attack, which is replaced by the normal acceleration a:

$$\begin{bmatrix} \dot{q} \\ \dot{a} \end{bmatrix} = \begin{bmatrix} M_q & \dfrac{M_\alpha}{N_\alpha} \\ N_\alpha & -\dfrac{N_\alpha}{V} \end{bmatrix} \begin{bmatrix} q \\ a \end{bmatrix} + \begin{bmatrix} M_\delta \\ 0 \end{bmatrix} \delta$$

To solve Eq. (10.80), we first establish the following correspondences with the plant equations:

$$\bar{\boldsymbol{x}} = [q \quad a]; \quad \boldsymbol{F} = \begin{bmatrix} M_q & \dfrac{M_\alpha}{N_\alpha} \\ N_\alpha & -\dfrac{N_\alpha}{V} \end{bmatrix}; \quad \bar{\boldsymbol{g}} = [M_\delta \quad 0];$$

$$u = \delta; \quad \bar{\boldsymbol{h}}_\xi = [0 \quad 1]; \quad \bar{\boldsymbol{c}} = [k_2 \quad k_1]$$

Then we apply them to Fig. 10.17 and display them in Fig. 10.18. Recognize the two acceleration feedback loops, the inner rate loop, and the two PI feedforward branches. The rate and acceleration signals are taken from the INS, and the δ

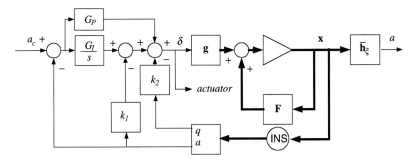

Fig. 10.18   Acceleration autopilot for missiles.

signal is sent to the pitch actuator. The gains k_1, k_2, G_I are calculated from the pole placement condition Eq. (10.80)

$$\text{Det} \left| \begin{bmatrix} s - M_q & -\frac{M_\alpha}{N_\alpha} \\ -N_\alpha & s + \frac{N_\alpha}{V} \end{bmatrix} + \begin{bmatrix} k_2 M_\delta & (k_1 + G_P)M_\delta \\ 0 & 0 \end{bmatrix} \ \vdots \ -\begin{bmatrix} G_I M_\delta \\ 0 \end{bmatrix} \right.$$
$$\left. \begin{matrix} [0 \quad 1] & \vdots & s \end{matrix} \right|$$
$$= (s - p_1)(s - p_2)(s - p_3) \tag{10.81}$$

Because the closed-loop system is of third order—two states and one integrator—three poles can be specified: one pair of conjugate complex poles and one real pole:

$$p_1 = -a + ib = -\zeta\omega_n + i\omega_n\sqrt{1 - \zeta^2}$$
$$p_2 = -a - ib = -\zeta\omega_n - i\omega_n\sqrt{1 - \zeta^2} \tag{10.82}$$
$$p_3 = -p$$

You pick the damping ζ, the natural frequency ω_n, and a pole location p (positive if stable) for the desired closed-loop response. The right-hand side of Eq. (10.81), expanded in polynomials, is

$$(s - p_1)(s - p_2)(s - p_3) = s^3 + a_m s^2 + b_m s + c_m$$

where

$$a_m = -p_1 - p_2 - p_3 = 2\zeta\omega + p$$
$$b_m = p_1 p_2 + p_1 p_3 + p_3 p_2 = \omega^2 + 2\zeta\omega p \tag{10.83}$$
$$c_m = -p_1 p_2 p_3 = \omega^2 p$$

Evaluating the left-hand determinant of Eq. (10.81) and equating terms of equal power yields the three gains

$$G_I = \frac{\omega^2 p}{N_\alpha M_\delta}$$

$$k_2 = \frac{1}{M_\delta}\left(2\zeta\omega + p + M_q - \frac{N_\alpha}{V}\right) \tag{10.84}$$

$$k_1 = \frac{1}{N_\alpha M_\delta}\left(\omega^2 + 2\zeta\omega p + M_\alpha + \frac{M_q N_\alpha}{V} - k_2\frac{M_\delta N_\alpha}{V}\right) - G_P$$

Because of the simple structure, it was possible to solve for the gains without a matrix inversion. The position feedforward gain G_P is not accessible to the pole placement technique. It must be determined based on a root locus analysis.

In a typical application the dimensioned aerodynamic derivatives are loaded into the missile computer as tables of Mach and angle of attack; the velocity is available from the onboard INS. If the missile experiences little change in dynamic pressure, it may be sufficient to specify fixed closed-loop poles. However, in the case of widely changing flight conditions, the closed-loop poles are scheduled as a function of dynamic pressure. As a rule, it is sufficient to schedule the natural frequency ω_n only while keeping the damping ζ and the real pole p location fixed. The gain G_P is usually fixed and apportions additional feedforward for increased speed of response.

As you check out your design, make sure that the gains are not too high and cause the control to saturate prematurely. Lowering the bandwidth ω_n can alleviate a potential problem. For tetragonal missiles you can use the same acceleration loop for both the pitch and yaw planes. A roll position controller will round out your autopilot design. You will find such an example in CADAC SRAAM6, Module C2 Subroutines C2PI and C2ROLL.

10.2.2.5 Altitude hold autopilot.
As mentioned in Sec. 9.2.3.3, there is little difference between an altitude hold autopilot of a pseudo-five-DoF or six-DoF implementation. Refer back to Fig. 9.19 and substitute the acceleration autopilot that we just designed. The altitude and rate measurements are provided by the INS. As before, the root locus analysis will also serve us here and will help us to determine the two gains G_V and G_H. However, I am not building an adaptive gain scheduler as before because altitude corridors are usually fixed, and a constant set of gains will suffice. The hypersonic vehicle GHAME is controlled by this autopilot. You can see the code implementation in the CADAC GHAME6 simulation, Module C2, Subroutine C2ALT.

10.2.2.6 Flight-path-angle controller.
Finally, I want to present to you one more autopilot function, the flight-path-angle controller. The aircraft relies on it for climb and descent, especially if the flight-path angle is constrained by air safety rules. You will find this design eminently interesting because it is a perfect example for the pole placement technique of state variables, as embodied by Eqs. (10.77) and (10.78).

To be specific, we replace the general variable names of Fig. 10.16 with those of our example and present it in Fig. 10.19. The commanded and achieved light path

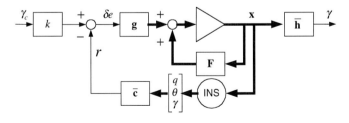

Fig. 10.19   Flight-path-angle tracking loop.

angle is γ_c and γ, respectively, supplemented by two more state variables: pitch rate q and pitch angle θ. All three states are provided by the INS. The three feedback gains c, multiplied by the state vector x, produce the scalar feedback variable r. The elevator deflection δe is sent to the actuator. We proceed to calculate the feedback gains c and the feedforward gain k.

First, we copy the plant equation $\dot{x} = Fx + g\delta e$ from Eq. (7.73)

$$
\begin{bmatrix} \dot{q} \\ \dot{\theta} \\ \dot{\gamma} \end{bmatrix} = \begin{bmatrix} M_q & M_\alpha & -M_\alpha \\ 1 & 0 & 0 \\ 0 & L_\alpha/V & -L_\alpha/V \end{bmatrix} \begin{bmatrix} q \\ \theta \\ \gamma \end{bmatrix} + \begin{bmatrix} M_{\delta e} \\ 0 \\ L_{\delta e}/V \end{bmatrix} \delta e
$$

with the dimensional derivatives $L_\alpha, L_{\delta e}, M_\alpha, M_q, M_{\delta e}$. Substituting the plant into Eq. (10.76) yields on both sides third-order polynomials in s, the same order we encountered in the acceleration autopilot. We can, therefore, carry over Eqs. (10.82) and (10.83), but, unfortunately, we cannot solve for the gains by elimination. Instead we have to use Eq. (10.78) and invert the matrix P to calculate the feedback gains:

$$
\begin{bmatrix} c_1 \\ c_2 \\ c_3 \end{bmatrix} = \begin{bmatrix} M_{\delta e} & 0 & L_{\delta e}/V \\ \hline \begin{matrix} M_{\delta e}L_\alpha/V- \\ L_{\delta e}M_\alpha/V \end{matrix} & M_{\delta e} & -M_qL_{\delta e}/V \\ \hline 0 & \begin{matrix} M_{\delta e}L_\alpha/V- \\ L_{\delta e}M_\alpha/V \end{matrix} & \begin{matrix} M_{\delta e}L_\alpha/V- \\ L_{\delta e}M_\alpha/V \end{matrix} \end{bmatrix}^{-1} \begin{bmatrix} a_m + M_q - L_\alpha/V \\ b_m + M_\alpha + L_\alpha M_q/V \\ c_m \end{bmatrix}
$$

Once we have c, we proceed to calculate the feedforward gain k from Eq. (10.77). All variables are known, and $\bar{h} = [0 \ 0 \ 1]$ picks out the state variable γ. Using k in the forward loop ensures that the steady-state gain is one, and γ will track γ_c after the transients have died down.

10.2.2.7 Summary.
We accumulated a whole assortment of autopilot designs for our six-DoF simulations. The *rate dampers* stabilize missiles off the launch rail and augment the directional stability of aircraft. *Roll autopilots* are used either to suppress the roll excursions for skid-to-turn missiles or to bank an aircraft into a turn. If this turn leads into a new direction, the *heading tracker* ensures that the new bearing is maintained. *Acceleration autopilots* are squarely in the domain of missiles. They execute guidance commands and exploit the full maneuvering capability of the missile airframe. Aircraft are less likely to employ acceleration autopilots, unless they are destined to fly low and hug the terrain while avoiding obstacles. Yet, we embedded an acceleration controller inside an *altitude hold loop* for air corridor flying. Finally, the climb and descent of an aircraft is controlled by the *flight-path-angle controller*.

The output of these autopilots are sent to actuators, which rotate control surfaces resulting in moments about the vehicle's c.m. We will now turn to the dynamic modeling of these actuating devices.

10.2.3 Actuator

In gliders and small airplanes the pilot's stick movements are sent directly by cable to the control surfaces. His muscle strength is sufficient to overcome the small control moments. In larger airplanes, however, mechanical or electrical devices must boost the human power. Moreover, if you take the man out of the loop, as for instance in missiles, and replace him by low-voltage autopilot signals, considerable amplification and power input are required to move the surfaces. The devices that deliver that boost are called *actuators*.

An actuator is a device that actualizes steering inputs to motivators. These motivators can be aileron, elevator, and rudder, or could be gimbaled nozzles of rockets. Even reaction jets are grouped into this category. We distinguish accordingly between actuators for aerodynamic control, thrust vector control (TVC), and reaction jet control systems (RCS). Hydraulics, pneumatics, or electromehanical devices can accomplish the power amplification. Power consumption, size, and cost are important selection criteria.

For six-DoF simulations we are mostly concerned with the accurate modeling of the dynamic characteristics of these devices. Needless to say that actuator companies, like Chandler Evans, invest great resources in presenting to the customer accurate performance specifications. These include mathematical models that can be used in system simulations for performance studies. The models are of high order and include all known nonlinear effects.

My purpose is less ambitious. I want to show you simple models, which nonetheless convey the salient characteristics of actuators. Most likely, you will have to model actuators for aerodynamic and thrust vector control. The more esoteric RCS are used for precision steering in exo-atmospheric vehicles, as direct force motivators. Their response is so fast that static modeling is sufficient.

10.2.3.1 Aerodynamic controls.
The most widely used method to control endo-atmospheric vehicles is through aerodynamic surfaces. By deflecting them, moments are generated about the c.m., which in turn rotate the airframe. The resulting incidence angles generate aerodynamic forces, which accelerate the vehicle in the desired direction.

We start with the description of missiles, followed by aircraft. Figure 10.20 displays the positive sense of the aerodynamic moments and the convention of positive

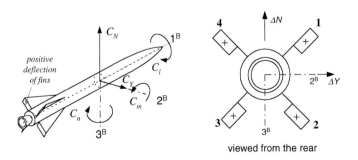

viewed from the rear

Fig. 10.20   Definition of positive aerodynamic moment and surface deflections.

surface deflections. There are only three moments, rolling C_l, pitching C_m, and yawing C_n, but four surfaces to produce them. Therefore, the four surface deflections $\delta 1$, $\delta 2$, $\delta 3$, and $\delta 4$ are combined to compute three apparent controls, called roll δp, pitch δq, and yaw δr:

$$\delta p = \tfrac{1}{4}(-\delta 1 - \delta 2 + \delta 3 + \delta 4)$$

$$\delta q = \tfrac{1}{4}(\delta 1 + \delta 2 + \delta 3 + \delta 4) \qquad (10.85)$$

$$\delta r = \tfrac{1}{4}(-\delta 1 + \delta 2 - \delta 3 + \delta 4)$$

A fourth relationship does not result in a moment, but only in a pure axial force:

$$\delta d = \tfrac{1}{4}(\delta 1 - \delta 2 - \delta 3 + \delta 4)$$

By the way, the proposal has been made to exploit this drag force for retarding a reentry missile.

Unfortunately, no consensus exists concerning the positive sense of control surfaces for missiles. We follow here the recommendation of the former North American Aviation Corporation. Another convention defines the surface deflections as positive when they contribute to a positive rolling moment. Our approach (see the following equations) has the advantage that it agrees with aircraft conventions for positive control deflections.

Roll:

$$+\delta p \quad \rightarrow \quad +\Delta C_l \qquad \text{rolling moment}$$

Pitch:

$$+\delta q \quad \rightarrow \quad +\Delta C_N \qquad \text{normal force}$$

Yaw:

$$+\delta r \quad \rightarrow \quad +\Delta C_Y \qquad \text{side force}$$

Positive roll control (aileron) produces a positive rolling moment; positive pitch control (elevator) generates a positive normal force (but a negative pitching moment); and positive yaw control (rudder) creates a positive side force (but a negative yawing moment).

The missile autopilot sends the roll, pitch, and yaw commands to the actuators. Yet, before they can be utilized, they have to be separated into individual fin commands:

$$\delta 1 = -\delta p + \delta q - \delta r$$

$$\delta 2 = -\delta p + \delta q + \delta r$$

$$\delta 3 = +\delta p + \delta q - \delta r \qquad (10.86)$$

$$\delta 4 = +\delta p + \delta q + \delta r$$

then each actuator module can convert the fin command δi_c into an actual surface deflection δi, where $i = 1, 2, 3, 4$. We represent the response of the fin actuator

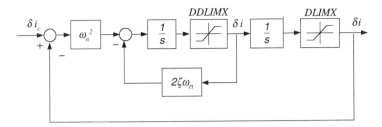

Fig. 10.21   Block diagram of fin actuator.

by a second-order transfer function

$$\frac{\delta i(s)}{\delta i_c(s)} = \frac{\omega_n^2}{s^2 + 2\zeta\omega_n s + \omega_n^2}$$

with natural frequency ω_n and damping ζ. Although the transfer function models only the linearized dynamics, we include two important nonlinearities as portrayed in Fig. 10.21. DLMIX limits the deflection of the fin, and DDLMIX restricts the maximum fin rate. Limiting the fin rate should not be neglected because it can become the source of serious performance degradation.

This actuator facsimile is the standard model of the CADAC six-DoF simulations. Its implementation calls for careful coding of the limiting feature of the derivative $\dot{\delta i}$. For details you can consult the CADAC SRAAM6 simulation, Module C4.

The same model is also used for aircraft-type vehicles, like cruise missiles, hypersonic vehicles, and, of course, the FALCON6. Some simplifications however apply. The basic control surfaces of aircraft are aileron δa, elevator δe, and rudder δr. Autopilot commands can be fed directly to the surface actuators. Their positive directions are shown in Fig. 10.22. Aircraft control conventions, like those in the following equations, are similar to missile control.

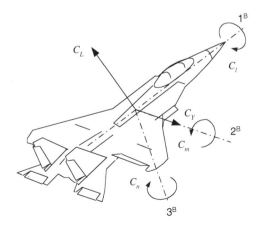

Fig. 10.22   Positive control surface deflections.

Aileron:

$$+\delta a \quad \rightarrow \quad +\Delta C_l \qquad \text{rolling moment}$$

Elevator:

$$+\delta e \quad \rightarrow \quad +\Delta C_L \qquad \text{lift force}$$

Rudder:

$$+\delta r \quad \rightarrow \quad +\Delta C_Y \qquad \text{side force}$$

Again, only the aileron deflection produces a positive rolling moment. Elevator and rudder cause negative moments, while their positive sense is defined by the positive forces they produce.

For programming purposes we can copy the code of missiles, as represented by Fig. 10.21, and insert it into our aircraft simulation. I have done so for the CADAC GHAME6 simulation, Module C4.

Aerodynamic surfaces are sometimes inadequate to control the vehicle. Greater agility may be required of a missile. A hypersonic vehicle can reach such heights that, despite its velocity, the dynamic pressure has fallen below acceptable values. For these applications thrust vector control could be the solution.

10.2.3.2 Thrust vector control.

Instead of using aerodynamics to turn the vehicle, thrust vector control employs propulsive moments to increase the incidence angle and thus maneuvers the vehicle through the ensuing aerodynamic forces. Outside the atmosphere, in the absence of aerodynamics, direct force control must be applied, using reaction control jets. We limit our discussion here to the endo-atmospheric application of thrust vector control.

A common feature is the deflection of the propulsive vector from the vehicle centerline in order to produce a moment about the vehicle's c.m. The deflection can be produced by turning the exhaust plume with jet tabs, like the original German V2, by pintel nozzles or titling the whole nozzle assembly. In either case the simulation model is the same. We use the moving nozzle as an example.

Figure 10.23 explains the geometry. The moment arm is the distance between the throat of the nozzle and the c.m. of the vehicle: $\Delta x = |x_p - x_{c.m.}|$. (As a reminder of what you learned in earlier chapters, I have drawn the body axes not

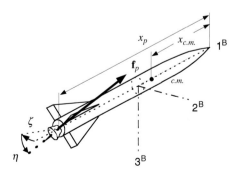

Fig. 10.23   Definition of nozzle deflection angles.

through the c.m.). The nozzle deflection is given by yaw angle ζ in the $1^B, 2^B$ plane and the pitch angle η in the displaced $1^B, 3^B$ plane. With the sequence of transformation of the nozzle coordinates wrt the body coordinates $]^N \xleftarrow{\eta}]\xleftarrow{\zeta}]^B$, you should be able to derive the following transformation matrix:

$$[T]^{NB} = \begin{bmatrix} \cos\zeta\cos\eta & \sin\zeta\cos\eta & -\sin\eta \\ -\sin\zeta & \cos\zeta & 0 \\ \cos\zeta\sin\eta & \sin\zeta\sin\eta & \cos\eta \end{bmatrix}$$

The thrust force in body axes is

$$[f_p]^B = [\bar{T}]^{NB}[f_p]^N$$

where $[\bar{f_p}]^N = [t \; 0 \; 0]$ with t as the thrust magnitude. Substituting the transformation matrix yields

$$[f_p]^B = \begin{bmatrix} \cos\eta\cos\zeta \\ \cos\eta\sin\zeta \\ -\sin\eta \end{bmatrix} t$$

The vector product of force and moment arm produces the moments about the c.m. that turn the missile:

$$[m_p]^B = \begin{bmatrix} 0 & -(f_p)_3^B & (f_p)_2^B \\ (f_p)_3^B & 0 & -(f_p)_1^B \\ -(f_p)_2^B & (f_p)_1^B & 0 \end{bmatrix} \begin{bmatrix} \Delta x \\ 0 \\ 0 \end{bmatrix} = \begin{bmatrix} 0 \\ (f_p)_3^B \\ -(f_p)_2^B \end{bmatrix} \Delta x$$

$$= \begin{bmatrix} 0 \\ -\sin\eta \\ -\cos\eta\sin\zeta \end{bmatrix} t|x_p - x_{c.m.}|$$

Either a positive ζ or η causes a negative moment. Notice also that the force components oppose the maneuver, just as it is the case with aerodynamic controls. For instance, a positive η produces a force component $(f_p)_3^B = -t\sin\eta$, which counteracts the pitch-down maneuver. This adverse control force is noticeable at the beginning of the maneuver until the aerodynamic force, produced by the incidence angle, overpowers it. To avoid this effect, aerodynamic control surfaces and reaction jets have been placed forward of the c.m. Located there, they actually aid in the maneuver.

A drawback of TVC is the lack of roll control. Twin nozzles and peripheral reaction jets have been applied to overcome this deficiency. However, they come at a cost and performance penalty.

The dynamic response of gimbaled nozzles can be patterned after the model we used for aerodynamic controls Fig. 10.21. With two gimbals we introduce the η and ζ actuators into the control loop. Typical values for their deflection and rate limiters are 10 deg and 100 deg/s, respectively. You can find applications of TVC in the CADAC SRAAM6 simulation, Module C3.

10.2.4 Inertial Navigation System

Simulations of aerospace vehicles most likely require a model of an inertial navigation system (INS). I cannot imagine a modern missile or aircraft that does not employ an INS for navigation. There are the ballistic missiles with their high-precision gimbaled platforms, the passenger planes with laser gyros, and tactical missiles with inexpensive strap-down systems.

If you are tasked to simulate an INS, you can approach it from two aspects. Either you duplicate mathematically the functioning hardware with its imperfections, or you use the analytical error equations to corrupt the true navigation states. The first approach is used for detailed INS studies, whereas the analytical method is better suited for system-level performance studies. Our focus is on system simulations where I concentrate on the error equation approach and leave the more difficult task of hardware simulations to the experts.

Sir Isaac Newton unknowingly laid the foundation for inertial navigation. His second law states that position can be determined by integrating the vehicle's acceleration twice. The acceleration is measured by an accelerometer. If the vehicle flew perfectly level, all we would have to add is a computer to carry out the integrations. However, missiles pursue targets, aircraft climb and descend, and satellites gyrate. To level the accelerometers, either they are mounted on a gimbaled platform, or a computer keeps track of the rotation between the accelerometers and inertial frames. For distinction, they are called either *platform* or *strap-down INS*.

The leveling of accelerometers requires gyroscopes. Their signals are used to either torque the platform or to determine the transformation matrix computationally. In both cases the so-called transfer alignment process will initialize them.

My treatment of INS simulations will be brief, with emphasis on error models that stood the test of performance studies. I assume that you have some familiarity with INS or are willing to acquire it by reading any of the standard reference texts. Two classics stand out, the book by Britting[17] and a report by Widnall and Grundy.[18] Britting treats a variety of INS systems from space stabilized to local-level platforms and strap-down systems. He painstakingly shows that the error equations of all of these various mechanizations can be condensed into one analytical form. More recent texts include detailed accounts by Chatfield[19] and a broader treatment by Biezad.[20]

The fundamental equations of INS navigation, based on Newton's law, calculate the velocity of the vehicle c.m. wrt the inertial frame in inertial coordinates $[v_B^I]^I$ and its position wrt an inertial reference point I, $[s_{BI}]^I$ in inertial coordinates

$$\left[\frac{dv_B^I}{dt}\right]^I = [\bar{T}]^{BI}[f_{sp}]^B + [g]^I$$

$$\left[\frac{ds_{BI}}{dt}\right]^I = [v_B^I]^I$$

with $[f_{sp}]^B$ the specific force measured by the accelerometers, and $[g]^I$ the gravitational acceleration in inertial coordinates. The accelerometers could be mounted on the vehicle or on a platform. In either case $]^B$ stands for the coordinate system

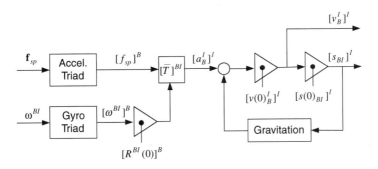

Fig. 10.24   INS principle of operation.

associated with that frame. The integration of the gyro's angular velocity of the vehicle wrt the inertial frame, expressed in inertial coordinates $[\omega^{BI}]^I$, delivers the rotation tensor of the body frame B wrt the inertial frame $[R^{BI}]^I = [\bar{T}]^{BI}$. Figure 10.24 depicts these operational equations in block diagram form. Starting with the acceleration $[a_B^I]^B$ and rate $[\omega^{BI}]^B$ measurements, the transformed specific force, combined with the gravitational acceleration, is integrated twice to attain the vehicle's position track. Six differential equations have to be solved for the basic navigation solution. Another four quaternion differential equations calculate the attitude angles.

Because the focus is not on the actual hardware, but rather on the error contribution of the INS to vehicle performance, I concentrate on their error equations. Both, the inertial and local-level systems will be discussed, for space and atmospheric vehicles, respectively. My approach is to corrupt the true values by the INS errors, thus providing the computed navigation data to the rest of the simulation. Figure 10.25 shows the process of modeling navigation parameters with uncertainties. The ε indicates perturbations, and the caret reflects computed values.

Are you ready to plunge into the details? As you will see, I shall make use of the perturbation methodology of Chapter 7 and derive the error equations of the space-stabilized navigator. To attain the error equations of terrestrial navigators, I elevate the Earth to an inertial frame and introduce the leveling feedback. A state-based formulation will round out the discussions.

10.2.4.1 Error equations for space-stabilized navigators.
The space-stabilized INS is conceptually the simplest of all navigators because Newton's

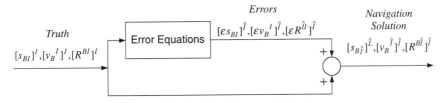

Fig. 10.25   Modeling of INS navigation parameters with uncertainties.

law assumes its most compact form when referred to the inertial frame. The error equations, based on the component perturbations of Chapter 7, form the basis for modeling the navigator of satellites or space ascent vehicles.

Newton's law states that the inertial acceleration equals the specific force f_{sp} acting on the vehicle plus the gravitational acceleration.

$$D^I v_B^I = f_{sp} + g \tag{10.87}$$

For a space-stabilized navigator we choose the inertial coordinate system, but recognize that the specific forces are most likely measured in coordinates associated with the platform or the vehicle's body.

$$\left[\frac{dv_B^I}{dt}\right]^I = [\bar{T}]^{BI}[f_{sp}]^B + [g]^I \tag{10.88}$$

The variables in this equation represent the true values, only known by God or the simulation. The values provided by the INS to the guidance processor are the so-called *computed variables*, which are corrupted by the INS errors. These errors, also called *perturbations*, are the difference between computed and true values.

Following the methodology of Chapter 7, the component perturbation of any vector x is

$$\varepsilon x = \hat{x} - R^{\hat{I}I} x \tag{10.89}$$

where $\hat{x}$ is the computed or corrupted vector and x the true variable. Perturbations of position, velocity, specific force, and gravitational acceleration can be patterned after this equation.

The rotation tensor $R^{\hat{I}I}$ takes on particular significance. It is the so-called INS tilt tensor that relates the true inertial frame I to the computed frame $\hat{I}$. Associated with the two frames are the true inertial coordinate system $]^I$ and the computed system $]\hat{I}$, respectively. All information coming from the INS is expressed in computed coordinates.

Once we introduce these coordinate systems, the component perturbations are reclaimed:

$$[\varepsilon x]^{\hat{I}} = [\hat{x}]^{\hat{I}} - [R^{\hat{I}I}]^{\hat{I}}[x]^{\hat{I}} = [\hat{x}]^{\hat{I}} - [R^{\hat{I}I}]^{\hat{I}}[T]^{\hat{I}I}[x]^I$$

and with $[R^{\hat{I}I}]^{\hat{I}}[T]^{\hat{I}I} = [E]$

$$[\varepsilon x]^{\hat{I}} = [\hat{x}]^{\hat{I}} - [x]^I \tag{10.90}$$

This equation consists of column matrices, only valid in the chosen coordinate systems. It is not an invariant tensor formulation like Eq. (10.89).

Retracing the development in Sec. 4.1.4, the tilt tensor under small perturbations consists of a unit tensor and a skew-symmetric tensor

$$[R^{\hat{I}I}]^{\hat{I}} = [E]^{\hat{I}} + [\varepsilon R^{\hat{I}I}]^{\hat{I}}$$

where the perturbation tensor of rotation [see Eq. (4.26)] is expressed by the small

Table 10.1 Perturbations of INS variables

Variable	Perturbation
Velocity	$\left[\varepsilon v_B^I\right]^I = \left[v_B^I\right]^I - \left[v_B^I\right]^I$
Position	$[\varepsilon s_{BI}]^I = [s_{BI}]^{\hat{I}} - [s_{BI}]^I$
Specific force	$[\varepsilon f_{sp}]^B = [\hat{f}_{sp}]^B - [f_{sp}]^B$
Gravitational acceleration	$[\varepsilon g]^I = [\hat{g}]^I - [g]^I$
Coordinate transformation	$[T]^{\hat{I}I} = [E]^I - [\varepsilon R^{\hat{I}I}]^I$

angle components

$$[\varepsilon R^{\hat{I}I}]^I = \begin{bmatrix} 0 & -\varepsilon\psi & \varepsilon\theta \\ \varepsilon\psi & 0 & -\varepsilon\phi \\ -\varepsilon\theta & \varepsilon\phi & 0 \end{bmatrix}$$

which can be reduced to the tilt vector $[\overline{r^{\hat{I}I}}]^I = [\varepsilon\phi \ \ \varepsilon\theta \ \ \varepsilon\psi]$.

This tilt vector represents actually the attitude perturbation, as demonstrated by this simple exercise. Apply the component perturbations Eq. (10.89) to the tilt vector and recognize that the vector product is zero:

$$\varepsilon r = r^{\hat{I}I} - R^{\hat{I}I} r^{\hat{I}I} = r^{\hat{I}I}$$

Indeed, the tilt vector is the tilt perturbation.

The tilt rotation tensor is related to the transformation matrix perturbation by

$$[\bar{T}]^{\hat{I}I} = [R^{\hat{I}I}]^I = [E]^I + [\varepsilon R^{\hat{I}I}]^I$$

and taking the transpose yields the perturbation of the coordinate transformation matrix

$$[T]^{\hat{I}I} = [E]^I - [\varepsilon R^{\hat{I}I}]^I$$

In summary, for the derivation of the error equations I have provided the necessary perturbations in Table 10.1. Yet we still need to investigate the time derivative of the velocity vector perturbation. Apply the rotational time derivative wrt the perturbed inertial frame $\hat{I}$ to the velocity perturbation

$$D^{\hat{I}} \varepsilon v_B^I = D^{\hat{I}} v_B^I - D^{\hat{I}}\left(R^{\hat{I}I} v_B^I\right) \tag{10.91}$$

The last term is expanded and $D^{\hat{I}} v_B^I$ transformed to the I frame to obtain $D^I v_B^I$:

$$D^{\hat{I}}\left(R^{\hat{I}I} v_B^I\right) = D^{\hat{I}} R^{\hat{I}I} v_B^I + R^{\hat{I}I} D^{\hat{I}} v_B^I = D^{\hat{I}} R^{\hat{I}I} v_B^I + R^{\hat{I}I} D^I v_B^I + R^{\hat{I}I} \Omega^{\hat{I}I} v_B^I$$

The first and the last term on the right-hand side can be neglected compared to the second term. Owing to the slow Schuler frequency (0.00124 rad/s), the time derivative $D^{\hat{I}} R^{\hat{I}I}$ is negligible, and the term $R^{\hat{I}I} \Omega^{\hat{I}I} v_B^I$ is small to the second order. With these simplifications Eq. (10.91) becomes

$$D^{\hat{I}} \varepsilon v_B^I = D^{\hat{I}} v_B^I - R^{\hat{I}I} D^I v_B^I$$

and expressed in $]^{\hat{I}}$ coordinates

$$\left[D^{\hat{I}}\varepsilon v_B^I\right]^{\hat{I}} = \left[D^{\hat{I}}v_B^{\hat{I}}\right]^{\hat{I}} - \left[R^{\hat{I}I}\right]^{\hat{I}}\left[T\right]^{\hat{I}I}\left[D^I v_B^I\right]^I$$

The rotational derivatives have become the ordinary time derivatives. With $[R^{\hat{I}I}]^{\hat{I}} \times [T]^{\hat{I}I} = [E]$ we have a relationship for the perturbed time derivative, which resembles Eq. (10.90):

$$\left[\frac{d\left(\varepsilon v_B^I\right)}{dt}\right]^{\hat{I}} = \left[\frac{d\left(v_B^{\hat{I}}\right)}{dt}\right]^{\hat{I}} - \left[\frac{d\left(v_B^I\right)}{dt}\right]^I$$

Now we have all of the tools available to build the INS error equations. Substituting the perturbations into Eq. (10.88)

$$\left[\frac{d\left(v_B^{\hat{I}}\right)}{dt}\right]^{\hat{I}} - \left[\frac{d\left(\varepsilon v_B^I\right)}{dt}\right]^{\hat{I}} = [\bar{T}]^{BI}\left([\hat{f}_{sp}]^B - [\varepsilon f_{sp}]^B\right) + [\hat{g}]^{\hat{I}} - [\varepsilon g]^{\hat{I}}$$

$$= [\bar{T}]^{\hat{I}I}[\bar{T}]^{B\hat{I}}\left([\hat{f}_{sp}]^B - [\varepsilon f_{sp}]^B\right) + [\hat{g}]^{\hat{I}} - [\varepsilon g]^{\hat{I}}$$

$$= \left([E]^{\hat{I}} - [\varepsilon R^{\hat{I}I}]^{\hat{I}}\right)[\bar{T}]^{B\hat{I}}\left([\hat{f}_{sp}]^B - [\varepsilon f_{sp}]^B\right) + [\hat{g}]^{\hat{I}} - [\varepsilon g]^{\hat{I}}$$

Factoring out the last terms and dropping the second-order term $[\varepsilon R^{\hat{I}I}]^{\hat{I}}[\bar{T}]^{B\hat{I}} \times [\varepsilon f_{sp}]^B$,

$$\left[\frac{d\left(v_B^{\hat{I}}\right)}{dt}\right]^{\hat{I}} - \left[\frac{d\left(\varepsilon v_B^I\right)}{dt}\right]^{\hat{I}} = \underline{[\bar{T}]^{B\hat{I}}[\hat{f}_{sp}]^B} - [\bar{T}]^{B\hat{I}}[\varepsilon f_{sp}]^B$$

$$+ [\varepsilon R^{\hat{I}I}]^{\hat{I}}[\bar{T}]^{B\hat{I}}[\hat{f}_{sp}]^B + \underline{[\hat{g}]^{\hat{I}}} - [\varepsilon g]^{\hat{I}}$$

The underlined expressions are satisfied identically because Newton's law also holds for the unperturbed state. Eliminating these terms, we have arrived at the INS error equation

$$\left[\frac{d\left(\varepsilon v_B^I\right)}{dt}\right]^{\hat{I}} = [\bar{T}]^{B\hat{I}}[\varepsilon f_{sp}]^B - [\varepsilon R^{\hat{I}I}]^{\hat{I}}[\bar{T}]^{B\hat{I}}[\hat{f}_{sp}]^B + [\varepsilon g]^{\hat{I}} \qquad (10.92)$$

It reveals to us that the velocity error is obtained by integrating the specific force error, the coupling between the specific force and the tilt, and the gravitational modeling error $[\varepsilon g]^{\hat{I}}$. The specific force error $[\varepsilon g_{sp}]^B$ is a direct result of the accelerometer uncertainties, and the tilt $[\varepsilon R^{\hat{I}I}]^{\hat{I}}$ is caused by gyro imperfections. From the INS navigation computation comes $[T]^{B\hat{I}}$, and $[\hat{f}_{sp}]^B$ is the output of the body-mounted accelerometers. We conclude from the error equation that the INS sensors play a dominant part in the INS quality.

10.2.4.2 Instrument errors. The gyros and accelerometers are either located on a gimbaled platform or mounted on the vehicle's body. High-precision navigation systems have platforms—just look at the multimillion dollar INS of the Peacekeeper ICBM. Yet, advances in instrument technology and processing capability have made it possible to replace the gimbals with mathematical models at much lower cost. These devices are called strap-down INS. As already mentioned, our error treatment applies to both; only the numerical values of the parameters reflect the different performance levels. In our discussion, however, we emphasize the strap-down implementation.

A strap-down INS, isolated from structural frequencies by vibration dampers, has two instrument clusters. Its accelerometer cluster consists of three instruments that sense the specific force along the three body axes, and the gyro cluster contains three rotary devices that measure the inertial angular velocity of the vehicle relative to the same three axes.

We model only those errors that remain after factory and prelaunch calibrations have taken place. These primary error sources for accelerometers are random bias and noise, scale factor error and misalignment. The same types of errors apply to gyros, augmented by mass unbalance for mechanical instruments.

The accelerometer error has the form

$$[\varepsilon f_{sp}]^B = [\varepsilon b_a]^B + ([S_a]^B + [M_a]^B)[f_{sp}]^B$$

consisting of the random bias and noise $[\varepsilon b_a]^B$, the diagonal scale factor error matrix $[S_a]^B$, and the misalignment matrix $[M_a]^B$. The misalignment matrix is skew symmetric, indicating the fact that a small misalignment exists between the accelerometer cluster and the vehicle axes. Within the cluster the accelerometer axes are assumed orthogonal. The output of the accelerometer cluster is the measured specific force in body coordinates

$$[\hat{f}_{sp}]^B = [f_{sp}]^B + [\varepsilon f_{sp}]^B$$

which is a combination of the true value $[f_{sp}]^B$, known only by the simulation and the instrument error $[\varepsilon f_{sp}]^B$.

The gyro error is composed of similar terms

$$[\varepsilon \omega^{BI}]^B = [\varepsilon b_g]^B + ([S_g]^B + [M_g]^B)[\omega^{BI}]^B + [U_g]^B[f_{sp}]^B$$

consisting of the random bias and noise vector $[\varepsilon b_g]^B$, the diagonal scale factor error matrix $[S_g]^B$, the skew-symmetric misalignment matrix $[M_g]^B$, and additionally the diagonal imbalance matrix $[U_g]^B$, which couples with the specific force. The misalignment again reflects only the cluster error of the otherwise orthogonal gyro triad. The output of the gyro cluster is the measured angular rate in body coordinates

$$[\hat{\omega}^{BI}]^B = [\omega^{BI}]^B + [\varepsilon \omega^{BI}]^B$$

composed of the true value $[\omega^{BI}]^B$ and the instrument error $[\varepsilon \omega^{BI}]^B$.

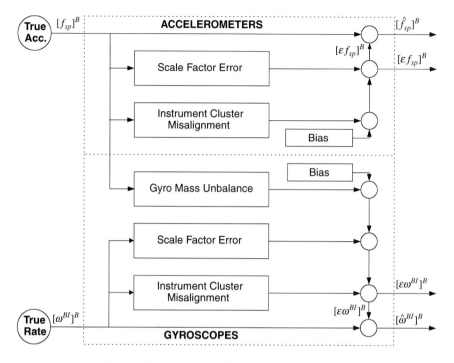

Fig. 10.26   Implementation of instrument errors.

The tilt of the INS is caused by the gyro error and grows from some initial value, unless checked by external corrections. An integrator, initialized by the uncertainty of the transfer alignment, models this process:

$$\left[\frac{d\varepsilon R^{\hat{I}I}}{dt}\right]^{\hat{I}} = [\bar{T}]^{BI}[\varepsilon\omega^{BI}]^{B} \tag{10.93}$$

Figure 10.26 summarizes the accelerometer and gyro measurement models. The true states, entering from the left and corrupted by the instrument errors, produce the measured values. Both pairs of output $[\hat{f}_{sp}]^{B}$, $[\varepsilon f_{sp}]^{B}$ and $[\omega^{BI}]^{B}$, $[\varepsilon\omega^{BI}]^{B}$ are essential for the INS error model.

10.2.4.3   Gravitational error model.   The dominant gravitational error is caused by the position error of the vehicle, i.e., the INS processor computes vehicle acceleration with an erroneous direction and magnitude of the gravitational vector. For the purposes of this derivation, we consider the gravitational field of a spherical Earth only. Any effects of higher harmonics are of lesser significance. From Newton's equation of gravitational attraction (see Sec. 8.2.2),

$$[g]^{I} = -\frac{GM}{|s_{BI}|^{3}}[s_{BI}]^{I}$$

To derive the gravitational error terms, introduce the perturbations of position and gravitational acceleration from Table 10.1:

$$[\hat{g}]^{\hat{I}} - [\varepsilon g]^{\hat{I}} = -\frac{GM}{(|s_{B\hat{I}}| - |\varepsilon s_{BI}|)^3}\left([s_{BI}]^{\hat{I}} - [\varepsilon s_{BI}]^{\hat{I}}\right)$$

If the first factor on the right-hand side is expanded into a binomial series

$$\frac{GM}{(|s_{B\hat{I}}| - |\varepsilon s_{BI}|)^3}[s_{B\hat{I}}]^{\hat{I}} = \frac{GM}{|\hat{s}_{BI}|^3}\left(1 - \frac{|\varepsilon s_{BI}|}{|\hat{s}_{BI}|}\right)^{-3} = \frac{GM}{|\hat{s}_{BI}|^3}\left(1 - 3\frac{|\varepsilon s_{BI}|}{|\hat{s}_{BI}|} + - \cdots\right)$$

and terms of second order in ε are neglected, we obtain

$$[\hat{g}]^{\hat{I}} - [\varepsilon g]^{\hat{I}} = \underline{-\frac{GM}{|s_{B\hat{I}}|^3}[s_{B\hat{I}}]^{\hat{I}}} + \frac{GM}{|s_{B\hat{I}}|^3}[\varepsilon s_{BI}]^{\hat{I}} + 3\frac{GM}{|s_{B\hat{I}}|^4}[s_{B\hat{I}}]^{\hat{I}}|\varepsilon s_{BI}|$$

where the underlined terms are satisfied identically because Newton's gravitational equation also holds for the perturbed state. The gravitational error equation is then to first-order accuracy

$$[\varepsilon g]^{\hat{I}} = -\frac{GM}{|s_{B\hat{I}}|^3}[\varepsilon s_{BI}]^{\hat{I}} - 3\frac{GM}{|s_{B\hat{I}}|^4}[s_{B\hat{I}}]^{\hat{I}}|\varepsilon s_{BI}| \qquad (10.94)$$

It exhibits the two important elements attributed to the INS navigation error. The first term on the right-hand side conveys the gravitational aberration caused by the location error $[\varepsilon s_{BI}]^{\hat{I}}$, and the second term reflects the error in the distance from the Earth's center.

All of the elements are now assembled for completing the INS error model. Figure 10.27 depicts the mathematical flow of the equations already derived. First, focus on the three integrators. They represent the three triplets of state variables: velocity error, position error, and tilt. Their initialization is carried out during the transfer alignment of the INS.

The simulation provides the true specific forces and rates, measured and corrupted by the accelerometer and gyro triads. After conversion by the tilt transformation, the specific force error is combined with the gravitational error to form the derivative of the velocity error. Like in the actual INS, two integrations lead to the position error. The major outputs of the INS model are the computed values of position $[s_{B\hat{I}}]^{\hat{I}}$, velocity $[v_B^{\hat{I}}]^{\hat{I}}$, and the direction cosine matrix $[T]^{B\hat{I}}$.

You can find this INS error model in the CADAC GHAME6 simulation, Module S4. For a hypersonic vehicle an inertial stabilized INS is quite appropriate. Moreover, the simulation builds on a legitimate inertial frame, which is a requirement for this type of model. For other simulations, based on the flat-Earth assumption, we have to proceed in a different fashion.

10.2.4.4 Error equations for terrestrial navigator.
The terrestrial navigation system uses as its main reference the local geographic frame. Although it is an inertial instrument and subject to Newton's equations, it emphasizes the local-level plane. In gimbaled systems the platform with its accelerometers is torqued to

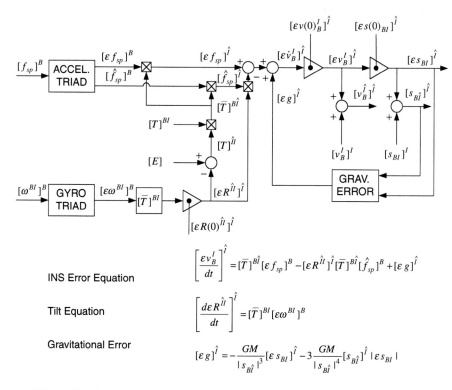

Fig. 10.27   Implementation of the error equations of the space-stabilized INS.

INS Error Equation

$$\left[\frac{\varepsilon v_B^I}{dt}\right]^{\hat{I}} = [\bar{T}]^{B\hat{I}}[\varepsilon f_{sp}]^B - [\varepsilon R^{\hat{II}}]^{\hat{I}}[\bar{T}]^{B\hat{I}}[\hat{f}_{sp}]^B + [\varepsilon g]^{\hat{I}}$$

Tilt Equation

$$\left[\frac{d\varepsilon R^{\hat{II}}}{dt}\right]^{\hat{I}} = [\bar{T}]^{BI}[\varepsilon \omega^{BI}]^B$$

Gravitational Error

$$[\varepsilon g]^{\hat{I}} = -\frac{GM}{|s_{B\hat{I}}|^3}[\varepsilon s_{BI}]^{\hat{I}} - 3\frac{GM}{|s_{B\hat{I}}|^4}[s_{B\hat{I}}]^{\hat{I}}\,|\varepsilon s_{BI}|$$

remain level, as the vehicle proceeds over the curved Earth. Double integration of the specific force renders the ground distance. For strap-down systems the onboard computer maintains the direction cosine matrix. Although the accelerometers are mounted on the vehicle, the conversion to local-level coordinates is readily made through this transformation.

Because many of the six-DoF simulations are based on the flat-Earth assumption, the terrestrial navigator is particularly well suited for this approach. The inertial frame becomes the Earth frame and the inertial coordinate system the local-level axes. However, recall that the local-level plane is not just the local tangent, but the curved surface of the Earth unwrapped into a plane. To account for this effect, we introduce the coupling of the tilt to the velocity error via the Earth's radius.

The INS error equation (10.92), already derived for the inertial-referenced INS, is now based on the Earth E as an inertial reference and the local level coordinate system $]^{\hat{L}}$, as computed by the INS processor:

$$\left[\frac{d(\varepsilon v_B^E)}{dt}\right]^{\hat{L}} = [\bar{T}]^{B\hat{L}}[\varepsilon f_{sp}]^B - [\varepsilon R^{\hat{EE}}]^{\hat{L}}[\bar{T}]^{B\hat{L}}[\hat{f}_{sp}]^B + [\varepsilon g]^{\hat{L}} \quad (10.95)$$

The tilt equation (10.93) receives an additional term because the tilt error of a terrestrial navigator grows now also as a result of the velocity error

$$
\left[\frac{d(\varepsilon R^{E\hat{E}})}{dt}\right]^{\hat{L}} = [\bar{T}]^{BL}[\varepsilon\omega^{BE}]^{B} + \begin{bmatrix} 0 & 1/R_{\oplus} & 0 \\ -1/R_{\oplus} & 0 & 0 \\ 0 & -\tan\lambda/R_{\oplus} & 0 \end{bmatrix} [\varepsilon v_{B}^{E}]^{\hat{L}} \quad (10.96)
$$

Note in particular that the second component of the velocity error couples into the third tilt component through the term $(\tan\lambda/R_{\oplus})$, which is a function of the vehicle's latitude λ. We attribute gravitational errors only to altitude uncertainties $(\varepsilon s_{BI})_{3}^{L}$ and neglect other effects. The change in magnitude of the gravitational acceleration is then based on Newton's gravitational attraction

$$
\varepsilon g = -\frac{\partial}{\partial |s_{BI}|}\left\{\frac{GM}{|s_{BI}|^{2}}\right\}(\varepsilon s_{BI})_{3}^{L} = \left(\frac{2GM}{|s_{BI}|^{3}}\right)(\varepsilon s_{BI})_{3}^{L} = \frac{2g}{|s_{BI}|}(\varepsilon s_{BI})_{3}^{L} \approx \frac{2g}{R_{\oplus}}(\varepsilon s_{BI})_{3}^{L}
$$

and in vector form replacing pro forma Earth's center I by E

$$
[\varepsilon g]^{\hat{L}} = \frac{2g}{R_{\oplus}} \begin{bmatrix} 0 \\ 0 \\ (\varepsilon s_{BE})_{3}^{\hat{L}} \end{bmatrix} \quad (10.97)
$$

With these provisions we can modify the inertial-referenced INS of Fig. 10.27 and draw the schematic of the terrestrial navigator in Fig. 10.28. The Earth frame has become the inertial frame, and the computed inertial coordinates are replaced by the computed local-level coordinates. Notice the additional leveling loop that introduces the uncertainty in the calculation of the tilt rotation tensor $[\varepsilon R^{\hat{E}E}]^{\hat{L}}$. The main output variables are the computed values of position $[s_{B\hat{E}}]^{\hat{L}}$, velocity $[v_{B}^{\hat{E}}]^{\hat{L}}$, and the direction cosine matrix $[T]^{B\hat{L}}$. They are needed as input to seeker and guidance models, as well as parameters for plotting trajectory traces. You can find an implementation of the terrestrial INS in the CADAC SRAAM6 and SRAAM5 simulations, Module S4. It is suitable for both pseudo-five- and six-DoF simulations in conjunction with the flat-Earth assumption.

The error term in Fig. 10.28 can also be reduced to a linear state variable formulation if we drop the term of the specific force perturbations $[\varepsilon f_{\text{sp}}]^{B}$ in Eq. (10.95) and reverse the vector product of the tilt skew-symmetric tensor with the specific force. Introducing the error components

$$
[\varepsilon s_{BE}]^{\hat{L}} = \begin{bmatrix} \varepsilon s_{1} \\ \varepsilon s_{2} \\ \varepsilon s_{3} \end{bmatrix}, \quad [\varepsilon v_{B}^{E}]^{\hat{L}} = \begin{bmatrix} \varepsilon v_{1} \\ \varepsilon v_{1} \\ \varepsilon v_{1} \end{bmatrix}, \quad [r^{\hat{E}E}]^{\hat{L}} = \begin{bmatrix} \varepsilon\phi \\ \varepsilon\theta \\ \varepsilon\psi \end{bmatrix}
$$

in north, east, and down coordinates, we obtain the traditional INS error equations

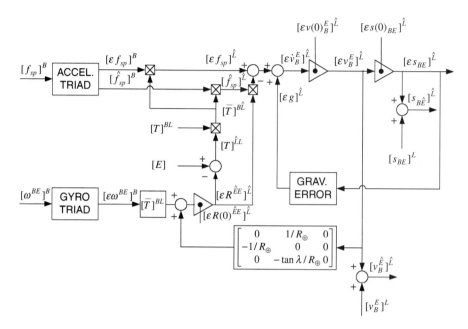

Fig. 10.28   Implementation of the error equations of the terrestrial stabilized INS.

popularized by Widnall and Grundy[18]:

$$
\begin{bmatrix}
\varepsilon \dot{s}_1 \\
\varepsilon \dot{s}_2 \\
\varepsilon \dot{s}_3 \\
\varepsilon \dot{v}_1 \\
\varepsilon \dot{v}_2 \\
\varepsilon \dot{v}_3 \\
\varepsilon \dot{\phi} \\
\varepsilon \dot{\theta} \\
\varepsilon \dot{\psi}
\end{bmatrix}
=
\left[
\begin{array}{ccc|ccc|ccc}
& O_{3\times3} & & & I_{3\times3} & & & O_{3\times3} & \\
\hline
0 & 0 & 0 & & & & 0 & -(\hat{f}_{sp})_3^{\hat{L}} & (\hat{f}_{sp})_2^{\hat{L}} \\
0 & 0 & 0 & & O_{3\times3} & & (\hat{f}_{sp})_3^{\hat{L}} & 0 & -(\hat{f}_{sp})_1^{\hat{L}} \\
0 & 0 & 2g/R_\oplus & & & & -(\hat{f}_{sp})_2^{\hat{L}} & -(\hat{f}_{sp})_1^{\hat{L}} & 0 \\
\hline
& & & 0 & 1/R_\oplus & 0 & & & \\
& O_{3\times3} & & -1/R_\oplus & 0 & 0 & & O_{3\times3} & \\
& & & 0 & -\tan\lambda/R_\oplus & 0 & & &
\end{array}
\right]
\begin{bmatrix}
\varepsilon s_1 \\
\varepsilon s_2 \\
\varepsilon s_3 \\
\varepsilon v_1 \\
\varepsilon v_2 \\
\varepsilon v_3 \\
\varepsilon \phi \\
\varepsilon \theta \\
\varepsilon \psi
\end{bmatrix}
$$

$$(10.98)$$

These are nine first-order, linear differential equations, exhibiting the two major error couplings of local-level INS systems. The tilt vector couples with the specific force into the velocity derivative channel and the velocity vector couples with the Earth's radius into the tilt derivative channel. The gravitational coupling occurs in the vertical channel between the altitude error and the vertical velocity derivative.

I found these equations particularly useful for Kalman-filter studies. To suppress errors, Kalman filters use navigation sensors to correct INS uncertainties. Embedded in the filter is a dynamic model of the INS error growth. In a typical simulation the actual error growth is modeled by Eqs. (10.95–10.97), while the Kalman filter mimics this process with the simplified state model of Eq. (10.98). If you are interested in pursuing this topic, refer to the excellent references by Maybeck[21] or Stengel.[22] As an example, the CADAC CRUISE5 simulation, Modules S1, S3, and S4 illustrate the modeling and integration of a Kalman filter between sensor measurements and INS. Although written for a five-DoF application, the code is equally pertinent for six-DoF models. To follow the code, however, a sound foundation in filtering is a prerequisite.

As a final topic, I need to address the proper initialization of INS errors. In the real world the transfer alignment process initializes the INS. Any imperfections cause initial uncertainties of the nine states of position, velocity, and tilt. Similar uncertainties should initialize the error equations in the simulation. The transfer alignment errors are known stochastically by their covariance matrix P_0, a 9 × 9 matrix of the variances and covariances of the nine-state vector εx_0. For a particular simulation run we have to extract an initial error vector from the Gaussian distribution, represented by this covariance matrix.

The Cholesky decomposition will help us in this process by taking the square root of the covariance matrix $\sqrt{P_0}$ (you can use the subroutine MATCHO in the CADAC UTL3.FOR file). Combining it with a random Gaussian (9 × 1) vector, having unit standard deviation g_{auss}, yields the initial INS error state

$$x_0 = \sqrt{P_0}\, g_{auss} \qquad (10.99)$$

A word of caution is appropriate at this point. The initial covariance matrix is not diagonal because the transfer alignment process intentionally couples states to reduce instrument and initialization errors. Therefore, a realistic simulation should not be initialized by stochastically independent states, but by correlated errors, as represented by a covariance matrix with off-diagonal covariances.

The modeling of INS is an important part of a high-fidelity six-DoF simulations. After staying with me through this much-abbreviated tour, you should understand the error equations that are found in the CADAC simulation examples. Perhaps you have even gained a general appreciation for the modeling of INS systems. However, to become better rooted in this subject, you should study in detail the recommended references.

10.2.5 Guidance

With the INS model complete we have the first element of the guidance loop that wraps around the autopilot, actuator, and airframe dynamics (see Fig. 10.29). It fulfills the navigation function by delivering the position and velocity vectors of the vehicle. Given these states, it is the responsibility of the guidance processor to guide the vehicle according to the given flight objectives, by issuing commands to the autopilot.

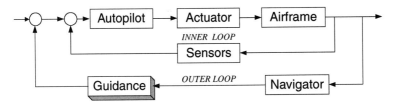

Fig. 10.29   Guidance loop wraps around the inner autopilot loop.

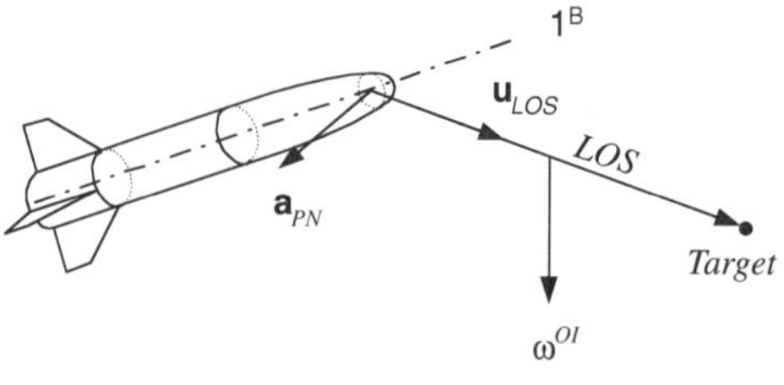

Fig. 10.30   Classical proportional navigation.

The INS provides position and velocity relative to an inertial or Earth-fixed reference. If the target states are known—moving or stationary, the navigation solution can also be expressed relative to the target reference. Then, the onboard processor can also calculate inertial LOS rates in a format familiar from LOS seekers. We already encountered such seekers in Sec. 9.2.5, and in the next section you will meet an IR seeker. Their LOS rate output in the pitch and yaw channels is in the required format for the classical proportional navigation law.

In Sec. 9.2.4 we already discussed PN and line guidance. They are equally valid for six-DoF simulations because the outer guidance loop is little affected by the inner loop, as long as the autopilot is well behaved. In this section I extend the classical proportional navigation law by correcting for longitudinal missile accelerations and formulate the so-called *compensated* PN law. Another brief excursion into modern control will expose you to the derivation of the *advanced* guidance law for missiles with strap-down seekers.

10.2.5.1 ⟋ Compensated proportional navigation.

Let us briefly review proportional navigation. We choose the guidance law option, which calculates the acceleration command normal to the LOS to the target. Therefore, in Eq. (9.57) the unit vector u_v specifically becomes the unit LOS vector u_{LOS}

$$a_{PN} = NV\Omega^{OI}u_{LOS} - g \qquad (10.100)$$

where N is the navigation gain, V the closing speed, Ω^{OI} the inertial angular velocity of the LOS wrt the inertial frame, and g the gravity bias. Figure 10.30 shows the vectors that construct this classical guidance law (disregarding g). The guidance command a_{PN} is normal to the LOS and lies in the so-called LOS plane.

You can visualize the engagement by fixing the target and flying the missile along its 1^B axis. The LOS rotates with the angular velocity of ω^{OI} in the direction shown, and the vector product with u_{LOS} produces the direction of the command a_{PN}. Now, if the missile is thrusting and therefore accelerating along its 1^B axis with a_m, a parasitic acceleration component appears in the LOS plane that should not contribute to the homing guidance. In effect it introduces intercept errors and should therefore be compensated. If that error is corrected, the PN law receives the prefix "compensated." We proceed deriving this compensation.

Refer to Fig. 10.31 for the geometric details. The missile's longitudinal acceleration a_m is projected into the LOS plane with the projection tensor P_{LOS}

$$a_{mo} = P_{LOS}a_m$$

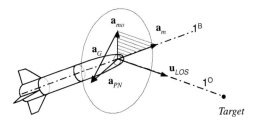

Fig. 10.31 Compensated proportional navigation.

and subtracted from the PN command a_{PN} to obtain the augmented command

$$a_G = a_{PN} - a_{mo} = a_{PN} - P_{LOS} a_m$$

$$= NV \Omega^{OI} u_{LOS} - P_{LOS} a_m$$

If we reintroduce the gravity bias, we have the form of the compensated PN law

$$a_G = NV \Omega^{OI} u_{LOS} - P_{LOS} a_m - g \qquad (10.101)$$

It consists of the basic PN term and the compensation for missile acceleration (or deceleration) and the gravity bias. Notice that I derived it in the invariant tensor form, maintaining its validity in any coordinate system.

To coordinate the law for computer implementation, we proceed in two steps. First, express the guidance command a_G in LOS coordinates, then, convert the two components from the LOS plane to body coordinates.

Let us begin with the transformation matrix $[T]^{OB}$ of the LOS coordinates relative to the body coordinates through the azimuth and elevation angles ψ_{OB}, θ_{OB}, respectively.

$$[T]^{OB} = \begin{bmatrix} \cos\theta_{OB}\cos\psi_{OB} & \cos\theta_{OB}\sin\psi_{OB} & -\sin\theta_{OB} \\ -\sin\psi_{OB} & \cos\psi_{OB} & 0 \\ \sin\theta_{OB}\cos\psi_{OB} & \sin\theta_{OB}\sin\psi_{OB} & \cos\theta_{OB} \end{bmatrix} \qquad (10.102)$$

With the missile acceleration vector given in body coordinates $[\overline{a_m}]^B = [a_m \; 0 \; 0]$, the guidance command is expressed in LOS axes (dropping again g for the time being)

$$[a_G]^O = [a_{PN}]^O - [P_{LOS}]^O [T]^{OB} [a_m]^B = [a_{PN}]^O - \begin{bmatrix} 0 & 0 & 0 \\ 0 & 1 & 0 \\ 0 & 0 & 1 \end{bmatrix} [T]^{OB} \begin{bmatrix} a_m \\ 0 \\ 0 \end{bmatrix}$$

$$= [a_{PN}]^O - a_m \begin{bmatrix} 0 \\ -\sin\psi_{OB} \\ \sin\theta_{OB}\cos\psi_{OB} \end{bmatrix}$$

Now, we focus on the two components normal to the LOS, which are the commands for the autopilot. But because its accelerometers are body mounted, the commands must be converted to body coordinates. Using the (1,1) minor matrix of Eq. (10.102), we relate the commands

$$
\begin{bmatrix} (a_G)_2^O \\ (a_G)_3^O \end{bmatrix} = \begin{bmatrix} \cos\psi_{OB} & 0 \\ \sin\theta_{OB}\sin\psi_{OB} & \cos\theta_{OB} \end{bmatrix} \begin{bmatrix} (a_G)_2^B \\ (a_G)_3^B \end{bmatrix}
$$

and solve for the body coordinates and combine them with the PN and missile accelerations

$$
\begin{bmatrix} (a_G)_2^B \\ (a_G)_3^B \end{bmatrix} = \begin{bmatrix} (a_{PN})_2^O \sec\psi_{OB} + a_m \tan\psi_{OB} \\ -(a_{PN})_2^O \tan\theta_{OB}\tan\psi_{OB} + (a_{PN})_3^O \sec\theta_{OB} - a_m \tan\theta_{OB}\sec\psi_{OB} \end{bmatrix}
$$

The component along the missile 1^B axis was discarded because it contributes nothing to the target intercept.

To sum up, you first calculate the two components of the PN command $(a_{PN})_2^O$ and $(a_{PN})_3^O$ based on the inertial LOS rate received from the seeker. Then, you combine them with the missile acceleration and bring back the gravity bias

$$
\begin{bmatrix} (a_G)_2^B \\ (a_G)_3^B \end{bmatrix} = \begin{bmatrix} (a_{PN})_2^O \sec\psi_{OB} + a_m \tan\psi_{OB} \\ -(a_{PN})_2^O \tan\theta_{OB}\tan\psi_{OB} + (a_{PN})_3^O \sec\theta_{OB} - a_m \tan\theta_{OB}\sec\psi_{OB} \end{bmatrix}
$$
$$
- \begin{bmatrix} (g)_2^B \\ (g)_3^B \end{bmatrix} \tag{10.103}
$$

You can find this implementation in the CADAC SRAAM6 simulation, Module C1. If you experiment with it, you will find that compensated PN provides some improved intercept performance during close engagements.

Compensated PN plays an essential role in many air-to-air missiles. It converts the inertial LOS rates into acceleration commands and steers the missile into the target. Some recent missile concepts, however, are equipped with strap-down seekers that deliver the target/missile dynamics in Cartesian rather than polar coordinates. For this application, the advanced guidance law, derived from optimal control, is in the right format.

10.2.5.2 Advanced guidance law.
Proportional navigation, born in the waning days of World War II, is revalidated by modern control. We will employ optimal control techniques to derive the advanced guidance law in Cartesian coordinates. It can be shown that it encompasses the classical PN law, expressed in polar coordinates. The guidance command is a function of relative position and velocity and the time-to-go until intercept t_{go}. Of particular importance for the performance of this advanced guidance law (AGL) is the accurate calculation of t_{go}. It depends on the missile and the (unknown) target motions. I will derive the so-called circle time-to-go, based on circular engagements, which performs quite well in close-in engagements.

Before we can formulate the optimality problem, we investigate the engagement geometry. Figure 10.32 shows the missile B flying an intercept against target aircraft T. For this derivation we use the Earth E as the inertial reference frame, as it is common practice for air-to-air missiles, and the local-level coordinate system

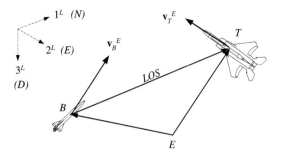

Fig. 10.32   Engagement geometry.

$]^L$. The displacement of the target wrt the missile is the difference between their displacements relative to the Earth reference point E

$$s_{TB} = s_{TE} - s_{BE}$$

Taking the rotational derivative wrt the inertial reference frame E

$$D^E s_{TB} = D^E s_{TE} - D^E s_{BE}$$

produces the differential velocity v_{TB}^E from the two relative velocities of Fig. 10.32

$$v_{TB}^E = v_T^E - v_B^E$$

Refer back to Example 4.6 to renew your understanding of differential and relative velocities. A second derivative yields the differential accelerations a_{TB}^E

$$D^E v_{TB}^E = D^E v_T^E - D^E v_B^E$$

$$a_{TB}^E = a_T^E - a_B^E$$

composed of the relative accelerations. If we pick the $]^L$ coordinate system, the rotational derivative becomes the ordinary time derivative, and we have

$$\left[\frac{d(v_{TB}^E)}{dt}\right]^L = [a_T^E]^L - [a_B^E]^L \tag{10.104}$$

Now we apply Newton's law to the missile, with f_{sp} the specific force and g the Earth's gravitational attraction, and express it in $]^L$ coordinates

$$[a_B^E]^L = [f_{sp}]^L + [g]^L$$

Substituting into Eq. (10.104) yields

$$\left[\frac{d(v_{TB}^E)}{dt}\right]^L = [a_T^E]^L - ([f_{sp}]^L + [g]^L) \tag{10.105}$$

We neglect the target and gravity accelerations and interpret the specific forces as an instantaneous response to the acceleration command $u \equiv -[f_{sp}]^L$. Furthermore, we introduce some simplifications in nomenclature: $\Delta s \equiv [s_{TB}]^L$; $\Delta v \equiv [v_{TB}^E]^L$; $\bar{x} = [\Delta s \ \ \Delta v]$. Then the state-space formulation is

$$\dot{x} = Fx + Gu; \quad F = \begin{bmatrix} 0_{3\times3} & I_{3\times3} \\ 0_{3\times3} & 0_{3\times3} \end{bmatrix}; \quad G = \begin{bmatrix} 0_{3\times3} \\ I_{3\times3} \end{bmatrix} \quad (10.106)$$

Note the very simple form of the dynamic equations. It is amazing how useful they are in applications.

We have made all preparations for the formulation of the optimal guidance problem. For you to follow the derivation, you should be familiar with the foundations of optimal control or consult Stengel[22] when you get lost.

Example 10.1 Optimality Problem

Problem. Find the control $u(x, t)$ that minimizes the cost function

$$J = \frac{1}{2}\bar{x}(t_f)Sx(t_f) + \frac{1}{2}\int_{t_0}^{t_f} \bar{u}Ru \, dt$$

subject to the dynamic constraint

$$\dot{x} = Fx + Gu; \qquad x(t_0) = x_0$$

The performance index combines two important criteria for a successful intercept. It minimizes the miss distance and limits the control power. The first term includes the weighting matrix S, which selects the square of terminal miss $\Delta s(t_f)$ from the state vector and weighs equally every component by s_s.

$$\frac{1}{2}\bar{x}(t_f)Sx(t_f) = \frac{1}{2}[\Delta s(t_f) \ \ \Delta v(t_f)] \begin{bmatrix} s_s I_{3\times3} & 0_{3\times3} \\ 0_{3\times3} & 0_{3\times3} \end{bmatrix} \begin{bmatrix} \Delta s(t_f) \\ \Delta v(t_f) \end{bmatrix}$$

$$= \frac{s_s}{2}\overline{\Delta s(t_f)}\Delta s(t_f)$$

The square of the controls are integrated over the engagement time, weighted by the 3×3 matrix $R = rI_{3\times3}$ and minimized. Apportioning values for s_s and r emphasizes either the reduction in terminal miss or in control.

Solution. Introduce the Hamiltonian

$$H = \frac{1}{2}\bar{u}Ru + \bar{\lambda}(Fx + Gu)$$

The optimal solution for $u(x, t)$ consists of the adjoint equation in the costate λ

$$\dot{\lambda} = -\left[\frac{\partial H}{\partial x}\right]; \quad \lambda(t_f) = \left[\frac{\partial \phi}{\partial x}\right] \quad (10.107)$$

$$\dot{\lambda} = -\bar{F}\lambda; \quad \lambda(t_f) = Sx(t_f)$$

and the optimality condition

$$\left(\frac{\partial H}{\partial u} = 0\right)$$

which provides the optimal guidance solution

$$u = -R^{-1}\bar{G}\lambda \tag{10.108}$$

we just have to eliminate λ.

For the elimination of λ, you need to substitute Eq. (10.108) into Eq. (10.106) and combine it with Eq. (10.107) to get the state equations augmented by the costate

$$\begin{bmatrix} \dot{x} \\ \dot{\lambda} \end{bmatrix} = \begin{bmatrix} F & -GR^{-1}\bar{G} \\ 0_{6\times6} & -\bar{F} \end{bmatrix} \begin{bmatrix} x \\ \lambda \end{bmatrix} \tag{10.109}$$

These linear differential equations can be solved using the state transition matrix $\Phi(t, t_0)$, which can be expressed in the fundamental matrix A of Eq. (10.109):

$$\Phi(t, t_0) = e^{A(t-t_0)} = I_{12\times12} + (t - t_0)A + \frac{(t - t_0)^2}{2!}A^2 + \frac{(t - t_0)^3}{3!}A^3$$

The state transition matrix is partitioned for x and λ (assuming $t_0 = 0$):

$$\begin{bmatrix} x(t) \\ \lambda(t) \end{bmatrix} = \begin{bmatrix} \Phi_{11} & \Phi_{12} \\ \Phi_{21} & \Phi_{22} \end{bmatrix} \begin{bmatrix} x_0 \\ \lambda_0 \end{bmatrix}; \quad \begin{bmatrix} \Phi_{11} & \vdots & \Phi_{12} \\ \hline \Phi_{21} & \vdots & \Phi_{22} \end{bmatrix} = \begin{bmatrix} I_{3\times3} & I_{3\times3}t & \vdots & \frac{t^3}{6r}I_{3\times3} & -\frac{t^2}{2r}I_{3\times3} \\ 0_{3\times3} & I_{3\times3} & \vdots & \frac{t^2}{2r}I_{3\times3} & -\frac{t}{r}I_{3\times3} \\ \hline 0_{3\times3} & 0_{3\times3} & \vdots & I_{3\times3} & 0_{3\times3} \\ 0_{3\times3} & 0_{3\times3} & \vdots & -I_{3\times3}t & I_{3\times3} \end{bmatrix}$$

$$\tag{10.110}$$

Notice that Φ_{21} is zero. The state transition matrix can also be used to solve the differential equations from any time t to final intercept time t_f. This gives us the opportunity to introduce the time-to-go parameter $t_{go} = t_f - t$

$$\Phi(t_f, t) = \Phi(t_f - t) = \Phi(t_{go})$$

Applied to Eq. (10.110), we obtain

$$\begin{bmatrix} x_f \\ \lambda_f \end{bmatrix} = \begin{bmatrix} \Phi_{11}(t_{go}) & \Phi_{12}(t_{go}) \\ 0_{6\times6} & \Phi_{22}(t_{go}) \end{bmatrix} \begin{bmatrix} x(t) \\ \lambda(t) \end{bmatrix}$$

From the first equation, after premultiplying by S, we get one expression

$$Sx_f = S\Phi_{11}(t_{go})x + S\Phi_{12}(t_{go})\lambda$$

From the second equation we derive $\lambda_f = \Phi_{22}(t_{go})\lambda$, which is equal to Sx_f according to Eq. (10.107):

$$Sx_f = \Phi_{22}(t_{go})\lambda$$

Combining both, we can eliminate x_f and solve for λ in terms of x:

$$\lambda = [\Phi_{22}(t_{go}) - S\Phi_{12}(t_{go})]^{-1}S\Phi_{11}(t_{go})x \tag{10.111}$$

Now we can replace λ in Eq. (10.108) and obtain the optimal solution solely as a function of the state x.

The optimal solution of guidance law consists of a gain multiplied by the state

$$u = K(t_{go})x \qquad (10.112)$$

where

$$K(t_{go}) = -R^{-1}G^T[\Phi_{22}(t_{go}) - S\Phi_{12}(t_{go})]^{-1}S\Phi_{11}(t_{go}) \qquad (10.113)$$

The guidance gain $K(t_{go})$ can be computed onboard the missile. It is a function of time-to-go, introduced through the state transition matrices. The other matrices are constant. To implement the guidance law, the differential position and velocity $\bar{x} = [\Delta s \ \Delta v]$; $\Delta s \equiv [s_{TB}]^L$; $\Delta v \equiv [v_{TB}^E]^L$ must be available to the missile guidance processor. This is by no means easily accomplished. IR seekers provide only LOS rates, which are insufficient for reconstructing the full state. RF seekers, on the other hand, measure range and range rate. They possess sufficient intelligence for the full state, but require a rather sophisticated Kalman filter. Possibly, the ultimate solution is the so-called third-party targeting. A surveillance platform downlinks accurate target position and velocity information to the missile. Onboard the missile the differential position and velocity are then formed in support of the guidance law.

The gain includes a matrix inversion, which can be executed algebraically. First we substitute the values from Eq. (10.110)

$$(\Phi_{22} - S\Phi_{12}) = \begin{bmatrix} I_{3\times3} & 0_{3\times3} \\ -I_{3\times3}t_{go} & I_{3\times3} \end{bmatrix} - \begin{bmatrix} I_{3\times3} & 0_{3\times3} \\ 0_{3\times3} & 0_{3\times3} \end{bmatrix} \begin{bmatrix} \dfrac{t_{go}^3}{6r}I_{3\times3} & -\dfrac{t_{go}^2}{2r}I_{3\times3} \\ \dfrac{t_{go}^2}{2r}I_{3\times3} & -\dfrac{t_{go}}{r}I_{3\times3} \end{bmatrix}$$

$$= \begin{bmatrix} \left(1 - \dfrac{t_{go}^3}{6r}\right)I_{3\times3} & \dfrac{t_{go}^2}{2r}I_{3\times3} \\ -I_{3\times3}t_{go} & I_{3\times3} \end{bmatrix}$$

then we take the inverse

$$(\Phi_{22} - S\Phi_{12})^{-1} = \dfrac{1}{1 + t_{go}^3/(3r)} \begin{bmatrix} I_{3\times3} & -\dfrac{t_{go}^2}{2r}I_{3\times3} \\ I_{3\times3}t_{go} & \left(1 - \dfrac{t_{go}^3}{6r}\right)I_{3\times3} \end{bmatrix}$$

and the guidance gain turns into

$$K(t_{go}) = -\dfrac{3t_{go}}{3r + t_{go}^3}[I_{3\times3} \quad I_{3\times3}t_{go}]$$

Substituting into Eq. (10.112) and replacing the state by the differential position and velocity results in the control equation

$$u = -\dfrac{3t_{go}}{3r + t_{go}^3}[I_{3\times3} \quad I_{3\times3}t_{go}]x$$

and in velocity and displacement components

$$u = -\frac{3t_{go}}{3r + t_{go}^3}(\Delta s + t_{go}\Delta v) \qquad (10.114)$$

If the control vector is not weighted, i.e., if unlimited control power is available, then $r = 0$, and Eq. (10.114) simplifies to the form of the actual AGL

$$u = -\left(\frac{N}{t_{go}^2}\Delta s + \frac{N}{t_{go}}\Delta v\right) \qquad (10.115)$$

with $N = 3$.

AGL is the PN law in Cartesian coordinates. It identifies the optimal PN gain as three. However, several air-to-air missiles operate with a higher gain—around four—to tighten up the guidance loop and to reduce the miss distance. Sometimes, the gain may even be scheduled as a function of the closing speed. In all cases, the selection must be based on extensive engagement studies that require full six-DoF fidelity and realistic noise models.

The implementation of AGL in a simulation takes the differential position and velocity from computed target and INS data $[\hat{s}_{TB}]^L$, $[\hat{v}_{TB}^E]^L$ and sends the acceleration command in body axes to the autopilot.

$$[a_c]^B = -[T]^{BL}\left(\frac{N}{t_{go}^2}[\hat{s}_{TB}]^L + \frac{N}{t_{go}}[\hat{v}_{TB}^E]^L\right) \qquad (10.116)$$

What remains to be discussed is the calculation of time-to-go.

The quality of the guidance depends on the accuracy of the time-to-go estimate. If the future trajectory were precisely known, t_{go} could be calculated error free. However, because the target is uncooperative, the missile trajectory cannot be predicted with certainty. Therefore, certain assumptions are made about both the missile and the target. The most obvious one being the straight-line extrapolation from the current conditions. Yet close-in combat is fought in circles, and these are the stressing conditions that the guidance law must excel under. We shall therefore base our t_{go} estimate on circular engagements.

Figure 10.33 shows the geometry of a particular circular engagement. We assume that the differential velocity v_{BT}^E of the missile B wrt the target T is tangential to a circle that contains both points with their displacement vector s_{TB}. If the differential missile velocity remains on the circle and is constant, we can calculate the time until intercept from it and the length of the arc ϕR:

$$t_{go} = \frac{\phi R}{|v_{BT}^E|}$$

You can calculate the arc length by first determining the angle δ from the scalar product of the two vectors $[v_{BT}^E]^L$ and $[s_{TB}]^L$:

$$\cos\delta = \frac{[s_{TB}]^L [v_{BT}^E]^L}{|s_{TB}||v_{BT}^E|}$$

then expressing

$$\varepsilon = 90 \text{ deg} - \delta$$

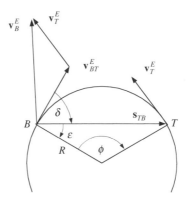

Fig. 10.33   Circle engagement.

and

$$\phi = 180 \deg - 2\varepsilon = 2\delta$$

with

$$R = \frac{|s_{TB}|}{2 \cos \varepsilon}$$

yields the arc length. The onboard guidance processor continually updates the time to go based on the v_{BT}^{E} and s_{TB} intelligence. If the missile does not follow a circle, but executes a head-on straight-line engagement, $\phi = 0$ and $R = \infty$, and the calculation breaks down. This happens rarely in air-to-air engagements. If it occurs, it lasts only for a few integration steps, and you can program around it. I have had very good results for close engagements, and no problems with long fly-out trajectories.

10.2.5.3 Summary. As a faithful follower of my exposition, you should have a good grasp of the most important guidance laws for missiles: proportional navigation, compensated proportional navigation, and the advanced guidance law. Yet be forewarned; many variants of these basic schemes will pop up in missile simulations. I hope that you will recognize them as such and become emboldened to build your own.

10.2.6   IR Seeker

In Sec. 9.2.5 you read about the basic principles of IR sensors. You may have applied them to the gimbaled model that was introduced there. If you have working code, you can drop it straight into your six-DoF simulation and start flying.

In this section we will develop another IR seeker, based on the current state of the art. Its mechanical arrangement consists of an outer roll and inner pitch gimbal. The focal plane array incorporates 128×128 elements with a resolution of 0.5 mrad and a total field of view of 3.6 deg. In clear atmosphere the detection range is 12 km. A particularly interesting feature is the so-called *virtual* gimbals,

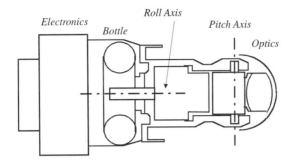

Fig. 10.34   IR seeker gimbal schematic.

which mechanize standard yaw and pitch gimbals in software. We deal therefore with two transformations, the physical gimbals of roll and pitch and the computer gimbals of yaw and pitch.

Unfortunately, the modeling task gets more complicated. Seeker kinematics is described by several transformations, and the focal plane geometry requires careful considerations. Adding uncertainties and errors poses further challenges.

10.2.6.1   Seeker kinematics.
The main parts of an IR seeker are shown in Fig. 10.34. The optics, supported by the inner pitch gimbal, carries the focal plane array and is isolated from the support by the roll gimbal. To cool the detectors, the bottle contains pressurized nitrogen, which, when vented, cools and supplies the 70–80 K operating environment. The base of the sensor houses the electronics for the gimbals and the processor of the optical data. For an air-to-air application, the image processing can include target classification, aimpoint selection, and flare rejection. The main output of the sensor is the inertial LOS rate—target relative to the missile—expressed by the angular velocity vector of the LOS frame relative to the inertial frame.

Let us assail the coordinate transformations first. The mechanical roll and pitch gimbals orient the seeker head and its optics to the body frame. The associated head coordinates have their 1^H axis aligned parallel to the optical axis and their 2^H axis with the gimbal pitch axis. As you would expect, the seeker roll axis coincides with the body 1^B axis. We build the transformation matrix $[T]^{HB}$ starting with the body axis, transforming it first through the roll gimbal angle ϕ_{HB}, and then through the pitch gimbal angle θ_{HB} to reach the head axes, i.e.,

$$]^H \xleftarrow{\theta_{HB}}] \xleftarrow{\phi_{HB}}]^B$$

Figure 10.35 should help you visualize the two transformation angles. You can spot the two centerlines of the seeker and the missile, 1^H and 1^B, respectively, separated by the angle θ_{HB}. The gimbal transformation matrix (TM) is

$$[T]^{HB} = \begin{bmatrix} \cos\theta_{HB} & \sin\theta_{HB}\ \sin\phi_{HB} & -\sin\theta_{HB}\ \cos\phi_{HB} \\ 0 & \cos\phi_{HB} & \sin\phi_{HB} \\ \sin\theta_{HB} & -\cos\theta_{HB}\ \sin\phi_{HB} & \cos\theta_{HB}\ \cos\phi_{HB} \end{bmatrix} \quad (10.117)$$

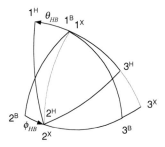

Fig. 10.35   Head wrt body transformation $[T]^{HB}$.

Notice the difference of this roll/pitch TM and that of the pitch/yaw gimbals, Eq. (9.83). Both represent mechanizations with different limitations. The pitch/yaw implementation is limited in yaw to less than 90 deg—typically 63 deg—because of mechanical constraints. Our roll/pitch gimbal arrangement does not suffer these constraints, but has a singularity straight ahead, when the sensor points in the forward direction at any roll angle.

Because of this singularity, the processor implements a virtual gimbal transformation, using the standard sequence of yaw and pitch (azimuth and elevation) from body coordinates to the so-called pointing coordinates. We have met this sequence several times. Please refer to Sec. 3.2.2 and in particular to Eq. (3.25) for details. The sequence of transformations is from body coordinates through the yaw angle ψ_{PB} and the pitch angle θ_{PB} to the pointing coordinates, i.e., $]^P \xleftarrow{\theta_{PB}}] \xleftarrow{\psi_{PB}}]^B$. Its TM is

$$[T]^{PB} = \begin{bmatrix} \cos\theta_{PB}\cos\psi_{PB} & \cos\theta_{PB}\sin\psi_{PB} & -\sin\theta_{PB} \\ -\sin\psi_{PB} & \cos\psi_{PB} & 0 \\ \sin\theta_{PB}\cos\psi_{PB} & \sin\theta_{PB}\sin\psi_{PB} & \cos\theta_{PB} \end{bmatrix} \quad (10.118)$$

To compare the three coordinate systems, Fig. 10.36 delineates the body, head, and pointing axes with the angles of transformation, ϕ_{HB}, θ_{HB}, and ψ_{HB}, θ_{HB}. You need to distinguish three different pitch planes. First, start with the missile pitch plane, subtended by 1^B and 3^B, and roll through ϕ_{HB} to the seeker pitch plane 1^H,

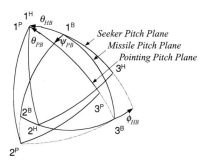

Fig. 10.36   Body, head, and pointing axes.

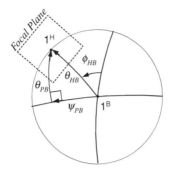

Fig. 10.37   Spherical triangle and focal plane.

3^H; second, start again with the missile pitch plane, but now yaw through ψ_{PB} to reach the pointing pitch plane 1^P, 3^P.

It is particularly informative to look down the nose of the missile and study the spherical triangle (Fig. 10.37) that encompasses the angles of both transformations. In it we see both the centerline of the missile 1^B and that of the seeker head 1^H. By Napier's rule we can relate the pitch gimbal angle θ_{HB} to the virtual gimbal angles ψ_{PB}, θ_{PB}

$$\theta_{HB} = \arccos(\cos\theta_{PB}\cos\psi_{PB}) \tag{10.119}$$

as well as the roll gimbal angle

$$\phi_{HB} = \arctan\left(\frac{\sin\psi_{PB}}{\tan\theta_{PB}}\right) \tag{10.120}$$

When $\psi_{PB} = 0$, then $\phi_{HB} = 0$; and when $\theta_{HB} = 0$, then $\phi_{HB} = 90$ deg. Can you visualize the focal plane array centered and perpendicular to 1^H?

10.2.6.2 Seeker model.   Having conquered the kinematics of the seeker, we can dig deeper and develop the mathematical seeker model. I shall break it into little bites as I have done before with the seeker in Sec. 9.2.5. But first, let us take the bird's eye view of Fig. 10.38.

An IR seeker has two major functions: imaging and tracking. They are indicated by the bar on bottom of Fig. 10.38. The detectors of the focal plane array record the target image. By processing the gray shades of the image, the aimpoint is selected and its dislocation in the focal plane recorded as the error angle.

To be realistic, we have to include in our model some of the main error sources. They are identified in Fig. 10.38 by the shaded blocks. At the detector level a dynamic phenomenon called *jitter* can destabilize the image, and biases can be introduced by the optics. The finite number of elements affects the processing of the pitch and yaw error angles, and processing of the data introduces biases.

The tracking loop of the seeker consists of the mechanical gimbals, the LOS rate estimator, and several computational transformations. Some of the mechanical errors we consider are gimbal noise and biases and the instrument errors of

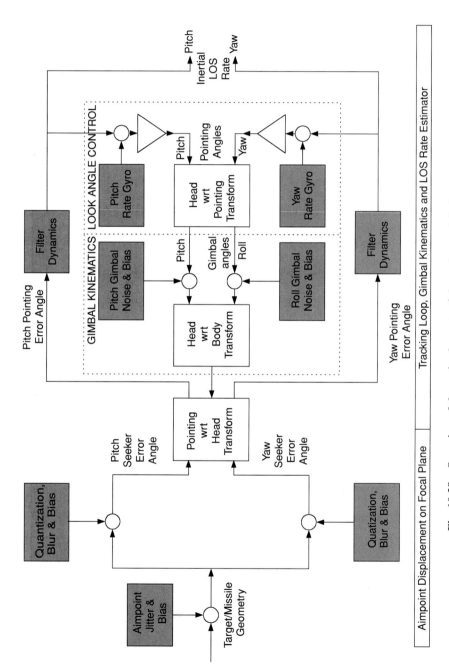

Fig. 10.38   Overview of the main elements of the seeker and their errors.

the rate gyros. The dominant time lags are caused by the filtering of the LOS signals.

Let us follow the signal flow from left to right in Fig. 10.38. The model begins with the true missile-to-target geometry s_{BT}. Corrupting the target by jitter and bias creates the actual missile-to-aimpoint displacement s_{AB} in head coordinates. Then the displacement is converted into pitch and yaw error angles and again diluted by quantization, blur, and bias uncertainties. Now the signals enter the computer-generated world of pointing coordinates and, after filtering, produce the inertial LOS rates in pointing coordinates as seeker output.

With the output generated, you may think that the seeker description is complete. But the all important seeker feedback loop still needs our attention. Through this loop the LOS rates, referred to the inertial frame, are transferred to the body frame by subtracting out the body rates. Then the pointing angles are converted into gimbal angle commands, which, after the gimbal noise and bias are added, establish the transformation matrix of the head to the pointing coordinates $[T]^{HB}$. At this point the tracker loop is closed, and the transformation matrix $[T]^{HB}$ is available for converting the error angles from head to pointing axes.

After this overview let us explore the seeker model in detail. Again, we start at the left side of the schematic and discuss first the aimpoint model shown in Fig. 10.39. The true displacement vector of the missile c.m. B wrt the target centroid T is given by the simulation in local-level coordinates $[s_{BT}]^L$. It is transformed to the head axes and converted to the actual aimpoint $[s_{AB}]^H$ by introducing the displacement vector $[s_{AT}]^H$ that models the corruption of the true target point T by jitter and bias

$$[s_{AB}]^H = [s_{AT}]^H - [s_{BT}]^H$$

I lumped these errors together into the displacement vector $[s_{AT}]^H$ and applied Gaussian noise and bias distributions to randomize the effects. Seeker specialists do not like such a top-level representation of a very complicated imaging process. Yet, if you are building a system simulation, you have to treat each subsystem with equal emphasis and have to compromise fidelity. Through cajoling, you may be able to convince your seeker expert to distill the imaging errors into this random vector $[s_{AT}]^H$.

The displacement on the focal plane is now converted into angles (see Fig. 10.40) and split into pitch and yaw error angle εY and εZ, respectively. At this point we

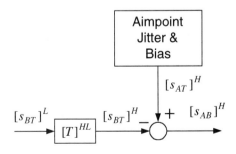

Fig. 10.39   Aimpoint model.

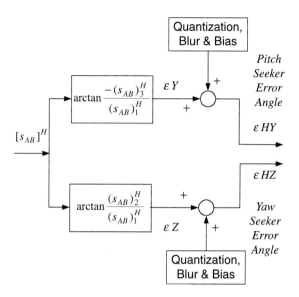

Fig. 10.40   Seeker error angles.

introduce the processing errors of quantization, blur, and bias, which you have to extract from the seeker's specification. I use random draws from a Gaussian distribution in radians to model their stochastic nature. The output of the image processor is εHY and εHZ.

Let us have a closer look at the focal plane in Fig. 10.41. Its center is the point H, which is also the piercing point of the 1^H axis (see Fig. 10.37). The 2^H and 3^H axes complete the head coordinate system. On the focal plane is also located the aimpoint A, as determined by the image processor. The true displacement error, expressed in radians, is composed of the two error angles

$$[\overline{\varepsilon AH}]^H = [0 \quad \varepsilon HZ \quad -\varepsilon HY] \tag{10.121}$$

With perfect gimbals and processing this displacement would be the desired LOS rate output of the seeker. However, the tracker loop introduces additional errors caused by the rate gyros, the virtual gimbals (pointing axes), and the mechanical

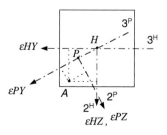

Fig. 10.41   Frontal view of focal plane array.

gimbals. Therefore, through the feedback loops the pointing axis 1^P is displaced from the center of the focal plane 1^H. The actual displacement, available for output processing, is the distance between the aimpoint A and the piercing point P of the 1^P axis $[\varepsilon AP]^H$. Our next task is to model this displacement.

The triangle of the three points A-H-P in Fig. 10.41 relates the measured value of $[\varepsilon AP]^H$ to the true displacement $[\varepsilon AH]^H$ and the pointing error $[\varepsilon PH]^H$

$$[\varepsilon AP]^H = [\varepsilon AH]^H - [\varepsilon PH]^H \tag{10.122}$$

The ε should remind us that the values are small and expressed in radians. Although the true value is given by Eq. (10.121), the pointing error, which is a function of the mechanical and virtual gimbals, still needs derivation.

The two transformations, Eqs. (10.117) and (10.118), are combined to form the pointing to head coordinate transformation

$$[T]^{PH} = [T]^{PB}[\bar{T}]^{HB} \tag{10.123}$$

Now here comes the juggling! Associated with the pointing axis 1^P and the head axis 1^H are the unit vectors

$$\overline{[p_1]}^P = [1 \quad 0 \quad 0] \qquad [\overline{h_1}]^H = [1 \quad 0 \quad 0]$$

Their tip displacement is the pointing error

$$[\varepsilon PH]^H = [\bar{T}]^{PH}[p_1]^P - [h_1]^H \tag{10.124}$$

If the two gimbal transformations were perfect, $[T]^{PH}$ would be unity and the pointing error vanished. With Eqs. (10.121) and (10.124) substituted into Eq. (10.122), we have succeeded in modeling the measured aimpoint displacement on the focal plane array.

We transition now to the tracking loop by converting to pointing coordinates

$$[\varepsilon AP]^P = [T]^{PH}[\varepsilon AP]^H \tag{10.125}$$

and separating the measurements into pitch and yaw pointing error angles

$$\varepsilon PY = -(\varepsilon AP)_3^P \quad \text{and} \quad \varepsilon PZ = -(\varepsilon AP)_2^P$$

These angle errors are the input to the tracker loops. In Fig. 10.42 I give you the entire tracking model so that you can follow in one graph the filtering, the body-rate compensation, the pointing angle generation, conversion to gimbal angle commands, and the gimbal mechanization.

The filters smooth the imaging and processing noise that come with the error angles before the LOS rates are sent out to the guidance computer. However, the filtering causes time delays of the output signals. In effect, they are the dominant lags of the seeker. We use second-order transfer functions to model either a simple bandwidth filter or a sophisticated Kalman filter.

The output of the inertial LOS rates in pitch and yaw, $\dot{\sigma}_{PY}$ and $\dot{\sigma}_{PZ}$, respectively, are also the starting point of the feedback loops. Body rates from the INS rate gyros are converted to pointing coordinates and subtracted from the inertial LOS rates to establish the body-referenced LOS rates. Their integration leads to

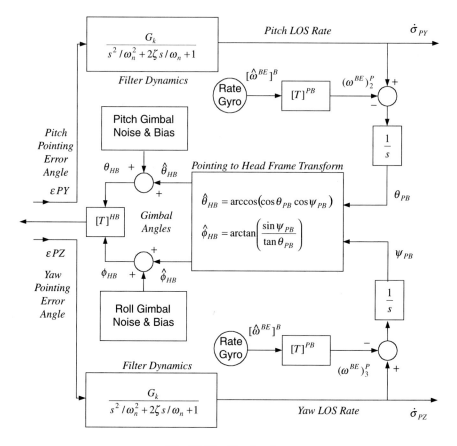

Fig. 10.42   Tracker loop.

the virtual gimbal angles ψ_{PB}, θ_{PB}, which are transformed by Eqs. (10.119) and (10.120) into the controls $\hat{\phi}_{HB}$, $\hat{\theta}_{HB}$ of the mechanical gimbals. I have elected not to model the gimbal dynamics explicitly. State-of-the art technology delivers highly accurate mechanical devices, whose bandwidth is significantly wider than the filter response. Therefore, I account only for random noise and biases in the roll and pitch gimbals. The actual gimbal angles are ϕ_{HB}, θ_{HB}. They determine the orientation of the focal plane array relative to the body frame, which is expressed mathematically by the transformation matrix $[T]^{HB}$.

The tracker loop is now complete. The TM $[T]^{HB}$ is combined with the pointing transformation $[T]^{PB}$ according to Eq. (10.123) to form the pointing wrt head TM $[T]^{PH}$, which is used to convert the seeker error angles from head to pointing coordinates [see Eq. (10.125)].

Thanks for staying with me through this tour de force. You have mastered a difficult modeling task of a modern IR seeker, applying geometric, kinematic, and dynamic tools. The conversion to code is fairly easy with the provided equations. Yet, if you prefer, you can also use the code in the CADAC SRAAM6 simulation, Module S1. It is fully integrated into the short-range air-to-air missile. Its input file provides you with typical seeker parameters that you can use as a starting point for your simulation.

With this highlight we conclude the section of subsystem modeling of six-DoF simulations. You should have a good grasp of aerodynamics, autopilots, actuators, INS, guidance, and seekers. Supplemented by relevant material from the five-DoF discussion (Sec. 7.2), you should be able to model aircraft and missiles with six-DoF fidelity. If you need more detail, you can consult the experts. Assuredly, there are many more components of aerospace vehicles that we were unable to address. Unfortunately, the open literature is of little help. You will find contractor reports a proficient source of information, provided you can get access to them.

I hope you gained an appreciation of the modeling of aerospace systems. The best way to become proficient is for you to develop your own simulations. But first learn by example. Most of the examples I used have their counterpart in the software provided on the CADAC Web site. I referred to them at the end of each section.

One more topic needs elaboration, which so far I have just alluded to: stochastic error sources for GPS, INS, seekers, and sensors and their effect on the performance of aerospace vehicles. With your curiosity aroused let us discuss these chance events.

10.3 Monte Carlo Analysis

If you know system engineers, you have heard them talk about Monte Carlo. They do not mean the picturesque city, nestled in the foothills of the Alps on the shores of the azure-blue Mediterranean, but they are referring to chance events; events that are too complex to model in all of their minute details.

We have already encountered some of them in the performance of INS and sensors. The complete error sources of INS sensors are so difficult to model that for system level analyses stochastic models like random walk, random bias, and white noise describe these uncertainties. Similarly, the IR sensor of Sec. 10.2.6 has a host of errors that model the imperfection of the electro/mechanical apparatus: aimpoint, jitter, bias, quantization, and blur of the focal plane array; rate gyros, gimbal noise, and bias of the mechanical assembly. Other uncertainties that affect the performance of an aerospace vehicle are airframe misalignments, erratic engine performance, and environmental effects of winds, gusts, and nonstandard atmospheric conditions. We already touched on nonstandard atmospheres in Chapter 8. In this section I will show you how to model winds and air turbulence, but first I must explain random events and their characterization.

What a throng of possible errors! It is the responsibility of the system engineer to define them mathematically and model them in the vehicle simulation. The stochastic nature of these errors produces random results. If the simulation were linear and the error sources were all Gaussian, the outcome would also be Gaussian. Then, the output covariances could be calculated directly from the input covariances. A single run of a so-called covariance analysis would suffice. However, the world of aerospace vehicles can seldom be fully linearized. Our highly nonlinear five- and six-DoF simulations are witnesses to that fact. Therefore, many computer runs have to be executed, each time with a new draw from the random input. All output is then collected and analyzed.

The more runs you make, the closer to the truth you get; just like the more you frequent the casino of Monte Carlo the closer to bankruptcy you get—after all, the casino has to make a profit for Prince Rainier of Monaco. The Monte Carlo technique has its root in the Manhattan Project. It involved the simulation of the probabilistic phenomenon of neutron diffusion of fissionable material. Hammersley and Handscomb (H&H)[23] give a readable summary of the 1964 status, and Zarchan[24] emphasizes guidance applications. As H&H point out, the Monte Carlo method plays an important role in experimental mathematics. It addresses problems in statistical mechanics, nuclear physics, and even genetics, which are otherwise impossible to solve. The specific Monte Carlo technique that applies to our problems is called the *direct simulation method*. Quoting from H&H, ". . . direct simulation of a probabilistic problem is the simplest form of the Monte Carlo method. It possesses little theoretical interest and so will not occupy us much in this book, but it remains one of the principal forms of Monte Carlo practice because it arises so often in various operational research problems." Could it be, because of its lack of mathematical sophistication, that it is so often called the brute force method?

There are three elements that you need to focus on: the validation of the simulation, the input parameters, and the postprocessing of the output. If you read this chapter from the beginning and exercised some of the six-DoF examples, you have a good grasp of a typical simulation. As you build your own model, make sure that the level of detail is tailored to the particular problem. For trajectory studies you should concern yourself with aerodynamics, propulsion, winds, and nonstandard atmospheres; for targeting studies you have to add navigation and guidance uncertainties with realistic stochastic error models. Foremost, however, allow plenty of time and resources to verify your work. Test your simulation under various conditions, as you would test a prototype aircraft. Have other experts review your brainchild, and do not let fatherhood pride keep you from accepting corrections.

The input parameters must be accurate, but it is sometimes difficult to pick good values for random initializations and stochastic parameters. Because their statistical distributions assume infinite sample size, you are hard pressed to find sufficient data to support your choices. That deficiency is particularly evident for new concepts that have little test data for backup.

Output data are plentiful when you make Monte Carlo runs. As a rule, the more replications you execute the better the results, but, oh horror!, the more data you have to analyze. Hopefully, your simulation environment has some or most of these chores automated. CADAC Studio provides you with a host of statistical analysis tools, which you can tailor to your needs.

To apply these methods, you have to know the bare essentials of statistics, probability, and random numbers. I assume that you have made their acquaintance so that I can concentrate here on the key elements of the Monte Carlo technique. Books by Gelb,[25] Maybeck,[26] or Stengel[27] can help you overcome potential deficiencies.

The direct simulation technique of the Monte Carlo methodology addresses primarily questions of accuracy. How precise can an aircraft navigate over water, how close will the space shuttle come to the space station, or where will the missile hit the target? I will introduce you to some of the key concepts in accuracy analysis, like the circular error probable (CEP), error ellipses, and the practice of delivery accuracy investigations. Then, using CADAC-generated diagrams, I will demonstrate with practical examples the usefulness of the Monte Carlo method in establishing the performance of an aerospace vehicle.

10.3.1 Accuracy Analysis

Let us assume the aerospace vehicle design is well established, be it as a concept or as hardware, and a validated simulation is available. You, as system engineer, have to answer questions on performance and accuracy. With your powerful PC at your beckoning, you load the simulation, provide the input data, run repetitive runs, and then sit there, overwhelmed by the output. You probably have two questions: 1) with the diversity of random input, what is the most likely statistical model of the output? and 2) how many runs are necessary for the output to be statistically significant?

Indeed, you may have a diverse array of input distributions. INS errors usually behave according to Gaussian statistics (normal distribution) and may be correlated in time (called Gauss–Markov processes). Similarly, seeker biases and noise behave mostly according to Gaussian and uniform distributions. More complex are the models of wind gusts that buffet your vehicle. Spectral densities with names like von Kármán and Dryden have a long history in aircraft analysis. It gets even more complicated, however, for a terrain-following and obstacle-avoiding cruise missile. The terrain is modeled, unless you have actual data, by a second-order autocorrelation function, driven by white Gaussian noise, and you have to select three parameters to characterize the particular terrain roughness. Obstacles are generated by two stochastic functions: an exponential distribution that determines the distance to the next obstacle and a Rayleigh distribution that randomizes obstacle height. To get more insight, I recommend you scrutinize some of your favorite six-DoF models for their stochastic prowess or look at the CADAC simulations CRUISE5 and SRAAM6.

With that many random variables taken from different types of distribution, modified and filtered by linear and nonlinear dynamics, the question is what is the most likely distribution of the output parameters. The famous *central limit* theorem provides us with the answer. It asserts that the sum of n independent random variables has an approximate Gaussian distribution if n is large. This is good news indeed for our sophisticated six-DoF simulations. The more noise sources it contains, the more Gaussian-like the output will be. But what is enough? H&H state, "In practical cases, more often than not, $n = 10$ is a reasonable number, while $n = 25$ is effectively infinite." Any respectable six-DoF simulation easily meets this condition. What a relief! We can use the well-established Gaussian statistical techniques to analyze the output.

With the output being Gaussian distributed, we can also answer the question of how many replications are necessary for a Monte Carlo analysis. It is based on the calculation of confidence intervals and their relationship to the standard deviation. Zarchan[24] provides a graph that relates the number of sample runs to confidence intervals. For instance, if 50 Monte Carlo runs produced a unit standard deviation estimate, we would have a 95% confidence that the actual standard deviation is between 0.85 and 1.28. Increasing the run number to 200 would give us, at the same confidence level, an interval between 0.91 and 1.12. In my Monte Carlo studies I never use less than 30 runs (at 95%, $0.80 < \sigma < 1.43$) and increase them to 100 runs (at 95%, $0.89 < \sigma < 1.18$) if accurate estimates are essential.

Are you now anxious to analyze your output? We will discuss first the statistical parameters of one-dimensional distributions, like altitude uncertainty or range error, the so-called *univariate* distributions, and then the two-dimensional distributions, like ground navigation error or miss distance on a surface target, the so-called *bivariate* distribution.

10.3.1.1   Univariate Gaussian distribution.

The simplest case is the one-dimensional output. Because we can assume the output parameters to be Gaussian distributed, our model is the univariate density function of the random variable x with mean μ_x and standard deviation σ_x:

$$p(x) = \frac{1}{\sigma_x \sqrt{2\pi}} e^{-\frac{(x-\mu_x)^2}{2\sigma_x^2}} \tag{10.126}$$

To calculate the mean μ_x and standard deviation σ_x from the output ensemble of your MC runs $x_1, x_2, \ldots, x_i, \ldots, x_n$, you can use the following equation:

$$\mu_x = \frac{\sum_{i=1}^{n} x_i}{n}; \quad \sigma_x = \left(\frac{\sum_{i=1}^{n}(x_i - \mu_x)^2}{n-1} \right)^{\frac{1}{2}} \tag{10.127}$$

With these values the density function $p(x)$ is determined, but it lacks intuitiveness. If we integrate over the function, we get an easier-to-understand probability statement of the random variable x, namely, that the probability that x assumes the values between a and b equals the area under the density function:

$$P(a < x < b) = \int_a^b p(x)\,dx \tag{10.128}$$

You have heard of the so-called one-sigma, two-sigma, and three-sigma accuracy statements. For instance, what does it mean if the three-sigma navigation error is 50 m? With Eq. (10.126) substituted into Eq. (10.128), we can calculate the probabilities. Unfortunately, the value of the integral has to be obtained from tables. Instead of developing tables for every sigma and mean, a normalized table for sigma equals one, and zero mean is provided; it is called the *normal table*. Hence, we normalize the Gaussian density function under the integral by introducing $t = x/\sigma_x$ and null the mean to get the probability areas between $z = x/\sigma_x = \pm 1, \pm 2, \pm 3, \pm 4$:

$$P\left(-z < \frac{x}{\sigma_x} < +z\right) = \frac{1}{\sqrt{2\pi}} \int_{-x/\sigma_x}^{x/\sigma_x} e^{-\frac{1}{2}\left(\frac{x}{\sigma_x}\right)^2} d\left(\frac{x}{\sigma_x}\right) = \frac{1}{\sqrt{2\pi}} \int_{-z}^{z} e^{-\frac{1}{2}t^2}\,dt \tag{10.129}$$

We look up the values of the integral between the z limits and record them in Table 10.2. For instance, the probability that the normalized random variable lies

Table 10.2   **Multiples of standard deviations and associated probabilities**

z	1	2	3	4
P	0.68268	0.95450	0.99730	0.99993
$x = z\sigma$	1σ	2σ	3σ	4σ

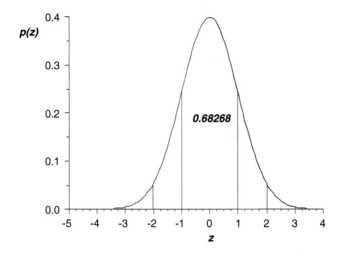

Fig. 10.43 Gaussian normal probability density function.

between ± 1 is 0.68268. With the same probability the random variable x lies between $1 \times \sigma$. Figure 10.43 shows the normalized Gaussian density function. It is useful to visualize this so-called bell-shaped curve when you interpret your output. Applying Table 10.2 to our example, we expect that in 99.73% of occurrences the range error will be within 50 m.

Instead of using the standard deviation to classify the range error, we can also ask the question: what is the 50% probability error bound?[7] Starting with Eq. (10.129), we look for the integration bounds $\pm z$ that contain 50% of the area under the normal density function

$$0.5 = \frac{1}{\sqrt{2\pi}} \int_{-z}^{z} e^{-\frac{1}{2}t^2} \, dt$$

From the normal table we interpolate for the value of 0.5 and receive $z = 0.6745$. For any distribution characterized by σ_x, we obtain, based on $z = x/\sigma_x$, the so-called *range error probable* (REP)

$$\text{REP} = z\sigma_x = 0.6745\sigma_x \tag{10.130}$$

Notice that the REP of 50% embraces smaller bounds than the one-sigma error, which occurs with the probability of 68.268% according to Table 10.2.

As an example, a fighter aircraft prepares to attack a target 3 km downrange. The pilot estimates that, given his one-sigma INS error of 100 m, he has a 50:50 chance to deliver the weapon within ± 67.45 m of the required release point. The bomb itself has a ballistic error of REP = 50 m. Because both errors are statistically independent, they can be root mean squared together to provide the 50% probability that the bomb will land within $\sqrt{67.45^2 + 50^2} = \pm 84$ m down- and up-range of the target coordinates.

The cross-range error or, as it is called, the *deflection error probable* (DEP) is defined in the same fashion. Given the lateral one-sigma error σ_y, then

$$\text{DEP} = 0.6745\sigma_y \tag{10.131}$$

If the range and deflection errors are given by the univariate measures REP and DEP, the implicit assumption is made that the error distributions in the two directions are uncorrelated. You need to judge if this simplistic approach is adequate. When you analyze MC runs, you are more likely to find that the range and deflection errors are correlated, i.e., that they have a bivariate distribution. We now turn to discuss measures of accuracy for these two-dimensional situations.

10.3.1.2   Bivariate Gaussian distribution.
Suppose you developed an air-to-ground guided missile simulation with many error sources and uncertainties. Now you execute MC runs and record the impact points in the x and y directions. From this two-dimensional data set $x_1, y_1; x_2, y_2; \ldots x_i, y_i \ldots; x_n y_n$, you calculate the means μ_x, μ_y and standard deviations σ_x, σ_y. You can adapt Eq. (10.127) to this process

$$\mu_x = \frac{\sum_{i=1}^{n} x_i}{n}; \quad \sigma_x = \left(\frac{\sum_{i=1}^{n}(x_i - \mu_x)^2}{n-1} \right)^{\frac{1}{2}}$$
$$\mu_y = \frac{\sum_{i=1}^{n} y_i}{n}; \quad \sigma_y = \left(\frac{\sum_{i=1}^{n}(y_i - \mu_y)^2}{n-1} \right)^{\frac{1}{2}} \tag{10.132}$$

but you also have to check whether the data are cross correlated by calculating the covariance σ_{xy}^2

$$\sigma_{xy}^2 = \frac{\sum_{i=1}^{n}(x_i - \mu_x)(y_i - \mu_y)}{n-1} \tag{10.133}$$

Take note that the covariance is formulated without the square root. It has the squared units of x and y. Similarly, the *variances* of x and y are their standard deviations squared σ_x^2 and σ_y^2. The *correlation* coefficient can be calculated from the standard deviations and the covariance

$$\rho_{xy} = \frac{\sigma_x \sigma_y}{\sigma_{xy}}$$

It assumes values between zero and one. Your data are uncorrelated if $\rho = 0$. In most cases they will be correlated, and you have to employ the two-dimensional Gaussian probability density function

$$p(x, y) = \frac{1}{2\pi \sigma_x \sigma_y \sqrt{1 - \rho_{xy}^2}} \exp\left[-\frac{1}{2(1 - \rho_{xy}^2)} \right.$$
$$\left. \times \left(\frac{(x - \mu_x)^2}{\sigma_x^2} - 2\rho_{xy} \frac{(x - \mu_x)(y - \mu_y)}{\sigma_x \sigma_y} + \frac{(y - \mu_y)^2}{\sigma_y^2} \right) \right] \tag{10.134}$$

as a model for the distribution of your data. Equation (10.134) portrays a bell-shaped surface over the x, y plane with its peak over the mean coordinates μ_x and μ_y. A plane $p(x, y) = $ const intersects the surface in the form of an ellipse. The volume above the ellipse is the probability that a data pair x, y will appear within the confines of the ellipse. The variances and covariances establish a particular ellipse, which can be represented by the two-dimensional covariance matrix

$$P = \begin{bmatrix} \sigma_x^2 & \sigma_{xy} \\ \sigma_{xy} & \sigma_y^2 \end{bmatrix}$$

It is a symmetric matrix, which can be diagonalized to obtain the major and minor axes of the ellipse. Once diagonalized in the new coordinate system (principal axes), the covariance is zero, and the data set has become uncorrelated.

When you analyze your impact points, it is helpful to draw so-called containment ellipses, centered on the mean values. They indicate to you the stretch of the data and the principal axes. The CADAC utility BIVAR, a subprogram of KPLOT, provides you with this display capability.

Sometimes, your customer may ask, "Could you boil down your misses into a single number?" Then you know that he is asking for the circular error probable CEP or the mean radial error (MRE). "No problem," you say, "as long as you accept the simplifying assumptions."

The definition of the CEP goes back to the times when the accuracy of rifles was determined by picking that circle on a paper target that contained 50% of the impact points. We take the mathematical approach and plot the radial distances r from the center as a histogram. If our simulation has enough noise sources with plenty of MC runs and the x and y dispersions are equal, we should see a density function similar to the Rayleigh distribution, also called the circular normal distribution. Its mathematical form is

$$p(r) = \frac{r}{\sigma_r^2} e^{-\frac{r^2}{2\sigma_r^2}} \tag{10.135}$$

Figure 10.44 presents the Rayleigh for three standard deviations $\sigma_r = 1, 2, 3$. Notice how the curves flatten out as the standard deviation increases. In all cases the area under the curve equals one (100% probability). The standard deviation

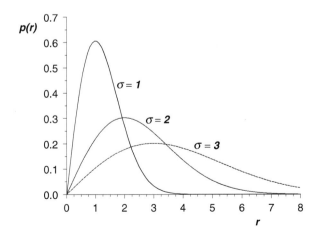

Fig. 10.44 Rayleigh probability density function.

σ_r is calculated from the data set $r_1, r_2, \ldots, r_i, \ldots r_n$ by an equation similar to Eq. (10.127).

$$\sigma_r = \left(\frac{\sum_{i=1}^{n} r_i^2}{n-1} \right)^{\frac{1}{2}} = \left(\frac{\sum_{i=1}^{n} x_i^2}{n-1} + \frac{\sum_{i=1}^{n} y_i^2}{n-1} \right)^{\frac{1}{2}} = \sqrt{\sigma_x^2 + \sigma_y^2} \quad (10.136)$$

Once the probability density function is given, we can calculate the 50% probability from the distribution integral taken from zero to the median m

$$\frac{1}{2} = \int_0^m \frac{r}{\sigma_r^2} e^{-\frac{r^2}{2\sigma_r^2}} \, dr = 1 - e^{-\frac{r^2}{2\sigma_r^2}} \Big|_0^m = e^{-\frac{m^2}{2\sigma_r^2}}$$

Take the natural logarithm on both sides and solve for m:

$$\ln 1 - \ln 2 = -\frac{m^2}{2\sigma_r^2}; \quad m = \sigma_r \sqrt{2 \ln 2} = 1.1774 \sigma_r$$

The CEP is the median m, i.e., the 50% probability that an impact point will occur within a circle of the radius,

$$\text{CEP} = 1.1774 \sigma_r \quad (10.137)$$

I have raised several warning flags that you must recognize in order to properly interpret your results. If you have some insurance that your simulation produces Gaussian distributions, you can apply the Rayleigh function. However, the CEP is only a good single number if the x and y means are near zero and the impact pattern is skewed no more than 2:1. It is always advisable first to plot the mean center and the elliptical distribution before calculating a single CEP number. You can do this with the CADAC/BIVAR program that plots impact points, the mean, ellipses with their major axes, and actually two CEPs, the aimpoint (target)-centered and the mean-centered CEP.

The MRE is not to be confused with the CEP, although their numerical values differ only by 7%. Given a one-dimensional density function $p(r)$, the mean is the expectation integral

$$E(r) = \int_{-\infty}^{\infty} r p(r) \, dr$$

The MRE is the expectation integral of the Rayleigh distribution with r the distance of the impact point from the target. Its solution is given in integral tables

$$E(r) = \int_{-\infty}^{\infty} \frac{r^2}{\sigma_r^2} e^{-\frac{r^2}{2\sigma_r^2}} \, dr = \sigma_r \sqrt{\frac{\pi}{2}} = 1.253 \sigma_r$$

Therefore the MRE of a radial distribution, which conforms to the Rayleigh function, is

$$\text{MRE} = 1.253 \sigma_r \quad (10.138)$$

Compare Eq. (10.137) with Eq. (10.138) and note the slight difference of the two accuracy measures.

We have reached the end of our short tour of Monte Carlo. Although you did not get any help for improving your chances at the roulette table, I hope you are equipped to conduct and analyze your MC simulations. The univariate measures of merit, REP, DEP, and standard deviation portray range and deflection accuracy or one-sigma errors of each INS channel. They are the easiest to analyze. More difficult are the bivariate measures of merit, particularly if the data are highly skewed. In that case we have to work with elliptical distributions. However, the desire to interpret the accuracy by a single number leads to the reduction of the bivariate data to a single radial distribution and the definitions of CEP and MRE. You will have to judge under what circumstances this shortcut is advisable. Some practical applications should help you to sharpen your tools.

10.3.2 Winds

One of the effects studied in six-DoF simulations is the variability of the atmosphere. Air density may deviate from standard conditions, wind and wind shear may dislocate the vehicle, or atmospheric turbulence may shake up the passengers.

In Sec. 8.2.1 I introduced the standard atmosphere and its variations. The incorporation of wind and turbulence requires the fidelity of six-DoF simulations. Three-DoF and pseudo-five-DoF simulations are inadequate because they do not model incidence angles based on the velocity vector but as output from the autopilot.

The air movement over the Earth can be divided into two components. The large-scale horizontal winds and the localized vertical gusts or turbulence. In simulations, tables represent winds as a function of altitude, direction, and magnitude, whereas turbulence, as a random phenomenon, is generated by a stochastic process. We shall discuss both phenomena.

To predict accurately the air movement over the globe is a complex problem. The mathematical model is formulated by partial differential equations and stochastic processes—techniques we would rather dodge. Fortunately, a large database exists of measured wind profiles over the Earth at all seasons, day and night. From this abundance you have to pick the winds that are appropriate for your study. If you have specific locations in mind, the National Oceanographic and Atmospheric Agency (NOAA) will be glad to supply you with archival information.

For sensitivity studies you can pick a particular wind profile that is representative of the winds your vehicle may encounter. A reasonable standard is the winds aloft Wallops Island, Virginia, where NASA's Flight Center is located. Figure 10.45, derived from the *Handbook of Geophysics*,[28] depicts the two extremes, the January and July winds. With little north–south activity it shows only the easterly and westerly components. Taken over many years, the mean winds are plotted against altitude. You can clearly see the jet stream bulges and the severe winds in the stratosphere. I also include a band, one standard deviation wide, to indicate the spread of the data around the mean wind.

You can use Fig. 10.45 or similar plots to define a single wind profile of projected severity, or you can establish an upper bound for winds occurring at a certain probability. For instance, the dashed lines on the right-hand side contain all westerly winds with probability 0.68. If you execute Monte Carlo runs, you can even have the simulation draw the wind profiles from a Gaussian distribution.

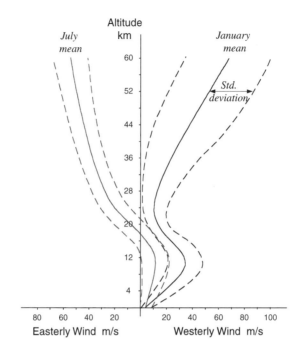

Fig. 10.45   Winds aloft Wallops Island, Virginia.

In six-DoF simulations winds and gusts alter the incidence angles and thus change the aerodynamic forces and moments. The incidence angles are calculated from the velocity vector $[v_B^A]^B$ of the vehicle's c.m. B wrt the air A, expressed in body coordinates $]^B$, [see Eqs. (3.20–3.23)]. To determine v_B^A, we subtract the wind vector v_A^E from the geographic velocity v_B^E

$$v_B^A = v_B^E - v_A^E$$

and introduce the body and geographic coordinates and their transformation matrix $[T]^{BG}$

$$\left[v_B^A\right]^B = [T]^{BG}\left(\left[v_B^E\right]^G - \left[v_A^E\right]^G\right) \qquad (10.139)$$

Ordinarily, the wind, assumed horizontal, is given by its magnitude V_w and direction from north ψ_w. Watch out however for the sign! A north wind blows the air from north to south. In geographic coordinates it has a negative 1^G component. Therefore, the wind vector in geographic coordinates is determined by

$$\left[v_A^E\right]^G = [-V_w \cos \psi_w \quad -V_w \sin \psi_w \quad 0]$$

In CADAC you use the weatherdeck to input winds in tabular form. Go to the G3 Module of the GHAME6 simulation and track down the code that implements Eq. (10.139).

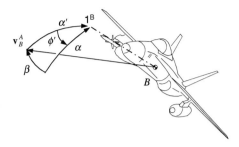

Fig. 10.46   Total incidence angle α' in load factor plane.

10.3.3 Turbulence

Atmospheric turbulence is air movement on a small scale. It is caused by the instabilities of pressure and temperature distributions in clouds, near the ground, and in the wind-shear regions of the jet stream. The fluctuations are time and space dependent. Because we cannot predict with certainty the behavior of the air molecules, we use statistical descriptors like standard deviation and power spectral density to classify the severity and spectral characteristics of turbulence.

For simulation purposes we make four assumptions that will ease the modeling task. The statistical properties of the turbulence are *stationary* (independent of time), *homogeneous* (independent of location in space), *isotropic* (independent of direction), and *larger* than the airplane (all parts of the airplane are equally affected). Two statistical representations are in good agreement with measurements: the von Kármán[29] and Dryden[30] models. Although both yield similar results, the Dryden spectrum is preferable because it is easier to implement in simulations.

I will keep the modeling task simple. We limit ourselves to the one-dimensional Dryden spectrum and superimpose its turbulence on the air mass in the load factor plane of the aircraft or missile. (The load factor plane contains the total incidence angle α'). I pick this plane because our main interest is in the variability of the incidence angle and its effect on the vehicle.

Figure 10.46 shows the total incidence angle α' as it relates to the angle of attack α and sideslip angle β. We already encountered this spherical triangle in Fig. 3.19 with the corresponding angular relationship of Eq. (3.24). The aeroballistic wind axes 1^A and 3^A, which contain also the 1^B body axis are embedded in the load factor plane. The transformation matrix $[T]^{AB}$ is given by Eq. (3.19).

Our goal is to express the movement of the turbulent air particles T wrt to the steady air mass A by the velocity vector v_T^A and coordinate it in geographic axes $[v_T^A]^G$. The turbulence component τ normal to the vehicle's velocity vector v_B^A in the load factor plane is expressed in the aeroballistic wind coordinates

$$\left[\overline{v_T^A}\right]^A = \begin{bmatrix} 0 & 0 & \tau \end{bmatrix}$$

With the two TMs $[T]^{AB}$ and $[T]^{BG}$ we obtain the desired form

$$\left[v_T^A\right]^G = [\bar{T}]^{BG}[\bar{T}]^{AB}\left[v_T^A\right]^A \qquad (10.140)$$

The entire movement of the turbulent air v_T^E wrt Earth consists of the turbulence v_T^A and the steady wind velocity v_A^E. We program this superposition in geographic

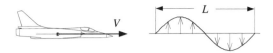

Fig. 10.47   Aircraft entering a sine-shaped gust.

coordinates

$$[v_T^E]^G = [v_T^A]^G + [v_A^E]^G \tag{10.141}$$

In Eq. (10.139) we used $[v_A^E]^G$ as wind vector. If we want to include turbulence, we supplement it by $[v_T^E]^G$ and obtain the expanded expression

$$[v_B^A]^B = [T]^{BG}\{[v_B^E]^G - ([v_T^A]^G + [v_A^E]^G)\} \tag{10.142}$$

The vehicle's velocity wrt air is now dependent on the geographic velocity v_B^E, the turbulent wind v_T^A, and the steady wind v_A^E.

Now we turn to the task of generating the velocity traces τ based on the one-dimensional Dryden turbulence model. First, we need to convert the spatial penetration of turbulent air into a temporal process. Figure 10.47 shows an aircraft with velocity V preparing to fly into a turbulence field with sine-type vertical velocity distribution of length L. The spatial frequency of the wave is $\Omega = 2\pi/L$. It takes an aircraft the time $T = L/V$ to traverse the wave. The aircraft experiences the wave with the temporal frequency $\omega = 2\pi/T = 2\pi/(L/V)$.

Turbulent air consists not just of waves of one frequency but of a whole spectrum. Their spectral density is modeled by Dryden's function, expressed in the nondimensional frequency $\varpi = \omega T = \omega L/V$

$$\Phi_{zz} = \frac{\sigma^2}{2\pi}\frac{L}{V}\frac{1+3\varpi^2}{(1+\varpi^2)^2} \tag{10.143}$$

where σ is the standard deviation of the turbulence in meters/second. The Dryden power spectral density function is shown in Fig. 10.48 for three L/V ratios. They

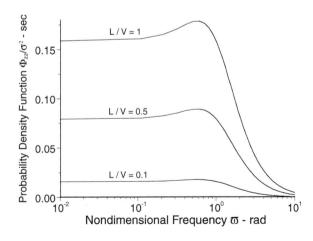

Fig. 10.48   Dryden power spectral density functions.

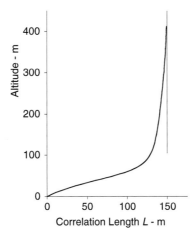

Fig. 10.49   Correlation length of turbulence.

peak near $\varpi = 0.6$. For an aircraft traveling at $V = 300$ m/s and a turbulence correlation length $L = 150$ m, the peak frequency occurs at $\omega = 1.2$ rad/s, not far removed from the pitch frequency of a large aircraft.

Correlation length L and standard deviation σ are obtained from atmospheric measurements. Lumley and Panowsky[31] give a detailed summary of L as a function of altitude. It is only near the ground, below 200 m, that it changes appreciably. At higher altitudes L maintains the constant value of 150 m length, as shown in Fig. 10.49.

The intensity of the turbulence, represented by the standard deviation σ, is a strong function of altitude. We follow Pritchard et al.[32] up to 300 m and use the U.S. MIL Standard MIL-A-8866 for higher altitudes. Figure 10.50 depicts two curves.

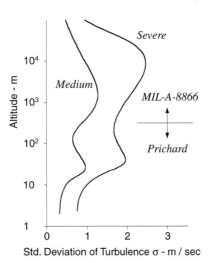

Fig. 10.50   Turbulence intensity.

The medium profile represents the standard deviation at average gust speeds of 4 m/s, whereas the severe trace centers around 8 m/s. Notice the peak gust velocities in the jet stream of 8 ± 2.3 m/s, which is as much as 20 kn.

To generate time traces from the Dryden spectrum, we have to convert the power spectral density into a filter equation. Driving the filter with white Gaussian noise will produce the desired output. Spectral factorization (see Maybeck[33]) of Eq. (10.143) yields the transfer function of the gust variable $\tau(s)$ over the white Gaussian noise $w(s)$

$$\frac{\tau(s)}{w(s)} = \sigma \sqrt{\frac{L/V}{2\pi}} \frac{1 + \sqrt{3}(L/V)s}{(L/V)^2 s^2 + 2(L/V)s + 1}$$

For programming we convert the transfer function into its state-space form:

$$\begin{bmatrix} \dot{x}_1 \\ \dot{x}_2 \end{bmatrix} = \begin{bmatrix} 0 & 1 \\ -\left(\frac{V}{L}\right)^2 & -2\frac{V}{L} \end{bmatrix} \begin{bmatrix} x_1 \\ x_2 \end{bmatrix} + \begin{bmatrix} 0 \\ \left(\frac{V}{L}\right)^2 \end{bmatrix} w(t)$$

$$\tau(t) = \sigma \sqrt{\frac{L/V}{2\pi}} \begin{bmatrix} 1 & \sqrt{3}\frac{L}{V} \end{bmatrix} \begin{bmatrix} x_1 \\ x_2 \end{bmatrix}$$

You can find these equations programmed in Module G2 of the CADAC GHAME6 simulation. A typical trace is shown in Fig. 10.51. This type of trace $\tau(t)$ is inserted into the turbulence vector

$$\left[\overline{v_T^A} \right]^A = \begin{bmatrix} 0 & 0 & \tau \end{bmatrix}$$

converted to geographic coordinates with Eq. (10.140), and combined with the wind velocity to be used for the incidence angle calculations in Module G3.

Winds and gusts are important feature of six-DoF simulations. They help us to investigate their effect on the performance, stability, and ride comfort of airplanes. You should be able to model steady winds and turbulence and use the provided data to conduct exploratory studies. For detailed analysis you should comply with the applicable industry standards and consult with the experts of your project team.

10.3.4 Applications

The applications draw on the two CADAC simulations CRUISE5 and SRAAM6, the cruise missile and the short-range air-to-air missile. We will analyze the output

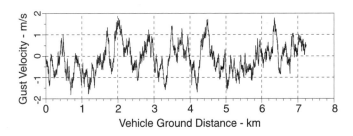

Fig. 10.51   Gust velocity traces for the following conditions: velocity $V = 750$ m/s (Mach $= 2.5$); altitude 13 km; correlation length $L = 150$ m; $L/V = 0.2$; and $\sigma = 1.5$ m/s.

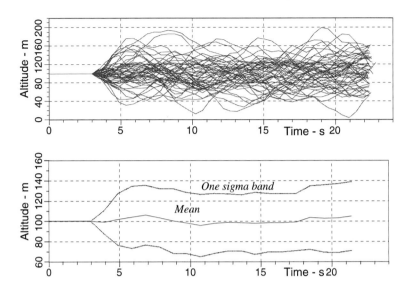

Fig. 10.52 Medium roughness terrain, 30 Monte Carlo runs, mean and standard deviation.

using univariate and bivariate stochastic models. Of particular interest is the modeling of stochastic terrain, INS position error growth, impact accuracy, and launch envelopes. I will present only selected results; you can find more test cases at the CADAC Web site.

In Sec. 9.3.3 I briefly described the random terrain, which models smooth, medium, and rough topography. A second-order power spectral density function is sampled for laying down the terrain ahead of the missile. Figure 10.52 shows the traces of 30 Monte Carlo runs and their univariate analysis, initiated at 3 s into flight. The mean altitude was set at 100 m. The mean of the stochastic traces deviates only slightly from this value. For an infinite number of runs, they would coincide. The spread of the tracks is characterized by the one-sigma band of ± 15 m. In CADAC the Monte Carlo technique is automated, and the univariate analysis is carried out by the utility program MCAP.EXE.

Another interesting study is the propagation of the INS position errors. In Sec. 10.2.4 I derived the error equations for space and terrestrial stabilized INS. Again making use of the CRUISE5 simulation with its terrestrial INS, Fig. 10.53 shows the 30 Monte Carlo replications before and after the start of the GPS updates. At launch the mean is slightly biased, and the one-sigma band is ± 5 m. Starting with the first GPS update at 8 s into flight, the bias is eliminated and the stochastic error reduces to 1 m.

Bivariate distributions are usually related to miss calculations. They arise at aircraft landings, spacecraft docking, and missile intercepts. I use an air-to-air missile to illustrate the application.

If you have solved Problem 4.11, you worked with the intercept plane that is used to record miss distance. Refer to the figure of Problem 4.11 to orient yourself. Fifty Monte Carlo shots were taken and recorded on the intercept plane in

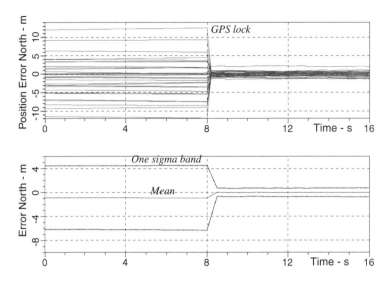

Fig. 10.53   INS position error (north component) before and after GPS update, mean and standard deviation.

Fig. 10.54. The target aimpoint is centered at zero. As the right schematic indicates, miss distance is the vectorial sum of navigation and guidance errors. The crosses mark the point where the onboard processor places the target, and the stars are the actual impact points. On the intercept plane, the crosses cluster around the target, indicating that the navigation error is unbiased and equally distributed in both directions. Its 50% error ellipse is near circular. The impact points however are skewed by the 7-m bias and the 2:1 elliptical distribution. You as analyst will be asked the question whether the miss distance can be represented by the univariate CEP. How would you answer? The CEP is drawn, centered on the target, with a radius of 8 m. Is it an accurate measure of miss? I recommend that you give your customer the complete picture, explain the difference between univariate and bivariate distributions, and let him, according to his needs, make that judgement. The CADAC utility program KPLOT/BIVAR will help you prepare your presentation.

Figure 10.54 is based on a single engagement geometry between launch and target aircraft. It does not portray the performance of the missile under all possible launch conditions. The launch envelope of Fig. 10.55 provides that total picture. The scenario is a close-in engagement with both aircraft maneuvering at 7.5 g. The shooter (launch aircraft) is at the center of the envelope, whereas the target aircraft can be located anywhere, but maneuvering such that both canopies face each other. As long as the target is within the outer contour (at missile launch) it will be intercepted with the CEP indicated by the inner contours. To provide also the bivariate distributions, at selected points the bias and 50% miss ellipse are drawn at a magnified scale, as indicated in the box. For example, pick point 4 as the location of the target aircraft at missile launch. At intercept the CEP is less than 10 m with a bias of 8 m and an elliptical distribution about the mean as

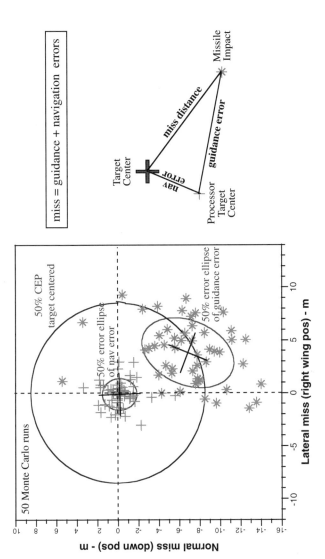

Fig. 10.54   Accuracy analysis in the intercept plane.

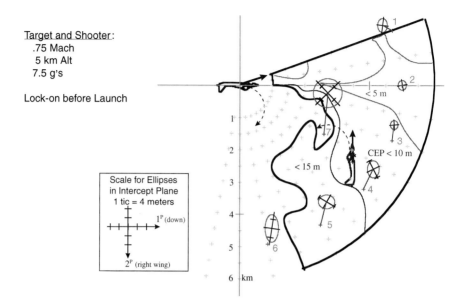

Target and Shooter:
.75 Mach
5 km Alt
7.5 g's

Lock-on before Launch

Scale for Ellipses
in Intercept Plane
1 tic = 4 meters

1^P (down)

2^P (right wing)

Fig. 10.55   Launch envelope, shooter centered.

shown. Surveying the whole envelope, you see how bias and miss ellipse change with the engagement geometry.

Figure 10.55 packs a multitude of information into one graph. It can only be generated in reasonable time if the whole process is automated. The CADAC SWEEP methodology provides the wherewithal. You set up a Monte Carlo sweep run and let your computer chew on it over night. The next morning, you execute the SWEEP utility program with the IMPACT.ASC file from the previous night and produce a graph similar to my example. Your customer will be duly impressed.

Enough said of the elements of six-DoF simulations! I covered a smorgasbord of specialties, which, combined with the information of Chapters 8 and 9, should give you a broad taste of the variety of subsystems. Other specialties that I did not serve up you will have to find in the references. Unfortunately, much of the information lingers in unpublished notebooks of specialists. To round out this chapter, I will briefly describe three typical six-DoF simulations.

10.4   Simulations

The best way to learn the secrets of six-DoF simulations is by taking apart some well-written prototypes. In the early 1970s, when I was stranded at the White Sands Missile Range in New Mexico, I used the time to dissect my first missile six-DoF simulation. It was the spark that ignited the CADAC blaze.

There are a few good six-DoF simulations on the open market. If you can get your hands on one, study it diligently and learn by example. Here, I will describe briefly three prototypes based on the CADAC architecture. They have been mentioned throughout this chapter, and now the time has come to introduce them formally. The code is provided on the CADAC Web site for your scrutiny.

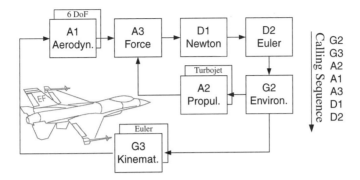

Fig. 10.56   FALCON6 aircraft simulation, six degrees of freedom.

10.4.1 FALCON6

Let us start with a simple model. It is based on the description by Stevens and Lewis[9] and the NASA database.[34] Figure 10.56 depicts the seven essential modules of a six-DoF simulation: Newton's and Euler's equations with the aerodynamic and propulsive forces and the kinematic calculations. I will briefly describe each module.

For the *D1 Newton Module* the translational equations of motions [Eq. (10.3)] are solved with the flat-Earth assumption.

For the *D2 Euler Module* the rotational equations of motion [Eq. (10.18)] are tailored for aircraft with a cross product of inertia term. The angular momentum of the turbojet engine is also included.

For the *G3 Kinematics Module* the incidence angles α and β are calculated. The Euler angle rates are integrated to obtain the direction cosine matrix.

For the *G2 Environment Module* the ISO 62 standard atmosphere is implemented.

The *A1 Aerodynamics Module* is patterned after Eqs. (10.61) and (10.62), but restricted to 0.6 Mach or lower.

For the *A2 Propulsion Module* the turbojet model of Sec. 9.2.2 applies here also. It calculates the thrust as a function of throttle setting, altitude, and Mach. It also includes a lag filter to model the spooling time delay.

The *A3 Force Module* combines the aerodynamic forces and moments with the turbojet thrust and readies them for the Newton and Euler equations.

The simulation does not have any controls. When you make a run, you may encounter an unstable region, and the trajectory will be cut short. I recommend you build your own autopilot, either by following Stevens and Lewis[9] or working Problem 10.2.

10.4.2 GHAME6

Here we meet again the NASA GHAME that we employed in Chapter 8. Now it has fully blossomed into a six-DoF simulation with elliptical rotating Earth and a full suite of autopilot options. Figure 10.57 shows the modules and their special features. For a summary of the essential kinematic and dynamic equations, see Fig. 10.6.

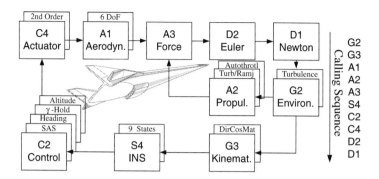

Fig. 10.57   GHAME6 hypersonic vehicle simulation, six DoF.

For the *D1 Newton Module* the translational equations of motion are formulated in inertial coordinates [see Eqs. (10.42) and (10.43)]. Vehicle position is converted to longitude, latitude, and altitude and velocity to geographic coordinates.

For the *D2 Euler Module* the rotational equations of motion (10.44) are expressed in body coordinates.

For the *G3 Kinematics Module* the incidence angles α and β are calculated and the direction cosine matrix integrated from Eq. (10.47).

For the *G2 Environmental Module* the standard atmosphere ARDC1959 is implemented. Options are provided for tabular atmosphere and winds. The turbulence model of Sec. 10.3.3 is also coded.

For the *A1 Aerodynamics Module* hypersonic aerodynamics are modeled according to Eqs. (10.65) and (10.66) and are based on NASA data.[35]

For the *A2 Propulsion Module* the combined cycle engine of Sec. 8.2.4 is modeled, consisting of a turbojet, ramjet, and scramjet phase. An autothrottle controls thrust to maintain constant dynamic pressure.

For the *A3 Force Module* the aerodynamic forces, moments, and propulsive forces are combined and expressed in body coordinates.

For the *S4 INS Module* the space-stabilized INS of Sec. 10.2.4 is implemented. Gyro, accelerometer, and gravitational perturbations are accounted for.

The *C2 Control Module* offers a variety of controllers: roll control, SAS, pitch acceleration controller, altitude hold, and both heading and flight-path-angle hold autopilots. Refer to Sec. 10.2.2 for a detailed description. All autopilot modes have aeroadaptive gains, derived by the pole placement techniques.

For the *C4 Actuator Module* second-order actuators with rate and position limiters control the elevons and rudder. See Sec. 10.2.3 for details.

The CADAC Web site has several test cases that you should exercise. Some results of the input file INCLIMB.ASC are displayed in Figs. 10.58 and 10.59. The run starts 3000 m above Cape Canaveral at a speed of Mach 0.75. Both the heading and flight-path-angle autopilots are engaged, and the autothrottle is set for 50 kPa dynamic pressure. After the climb-out at 20 and then 10 deg, the vehicle cruises at 30 km altitude until after 500 s it starts another shallow limb at 0.6 deg to top out at 40 km.

With the FALCON6 and GHAME6 code you have two simulations that cover the flight regimes of any aircraft over flat and elliptical Earth. Their aerodynamics

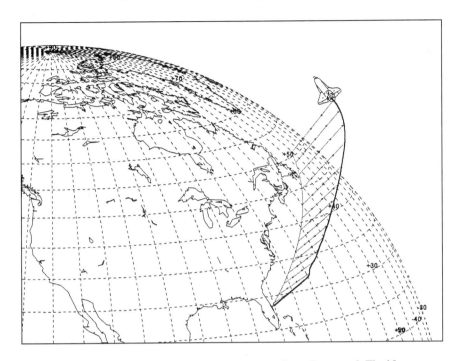

Fig. 10.58   Hypersonic vehicle launch from Cape Canaveral, Florida.

is a function of the angles α and β. We round out our collection by a missile simulation with polar incidence angles α' and ϕ'.

10.4.3 SRAAM6

The SRAAM6 simulation models a generic short-range air-to-air missile. We encountered its five-DoF version already in Chapter 9. Now I present you with the flat-Earth, six-DoF benchmark version. By looking at the modules of Fig. 10.60, you can appreciate the complexity of this simulation. With nearly 6000 lines of FORTRAN code, it is the largest of all CADAC models. Actually, some of the modules are almost identical to the SRAAM5 versions. They are those forming

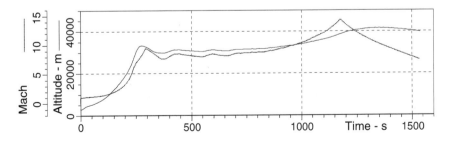

Fig. 10.59   Ascent parameters of Fig. 10.58.

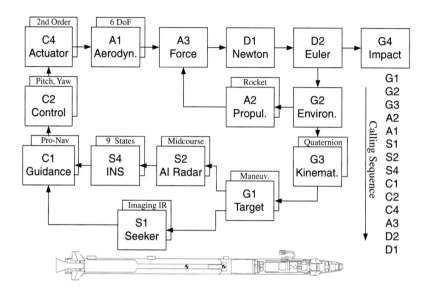

Fig. 10.60 SRAAM6 short-range air-to-air missile, six DoF.

the navigation and guidance loop: G1, S1, S2, S4, and C1. Even the G2 and A2 modules can be adopted with minor modifications. However, the kinematic and dynamic modules are new and reflect the six-DoF implementation. They are summarized in Fig. 10.1. Let us briefly discuss the individual modules.

For the *G1 Target Module* target engagement scenarios are implemented. They are known by special designations, like Pre-Merge (shooter centered), One Circle Fight, Two Circle Fight, Lufbery Circle, Target Centered Engagement, Chase Circle, Head-On Circle, and Twin Circle. You just need to set the flag MINIT and invoke the preprogrammed initial conditions for the target and the shooter aircraft.

For the *G2 Environmental Module* the ISO 62 standard atmosphere provides pressure, density, and temperature.

For the *G3 Kinematics Module* the incidence angles α' and ϕ' are calculated, and the quaternion methodology of Secs. 4.3.3 and 10.1.1 is employed to obtain the direction cosine matrix.

For the *S1 Seeker Module* imaging IR sensors are the current state of the art of short-range air-to-air missiles. Although only generic data are used, the roll/pitch gimbals and the coordinate systems are quite realistic. You can find the description of the seeker in Sec. 10.2.6.

The *S2 Air Intercept Radar* is a simple kinematic model of an acquisition and tracking radar, located in the shooter aircraft. It acquires and tracks the target and transmits that information to the missile at launch and during an optional midcourse phase.

The *S4 INS Module* is used in midcourse only. The nine error state equations (three positions, three velocities, and three tilts) bring realism to the fly-out accuracy. The derivation of the error equations is given in Sec. 10.2.4.

For the *C1 Guidance Module* a midcourse and terminal guidance phase is provided. In midcourse a simple pro-nav law is implemented, whereas the terminal phase relies on the compensated pro-nav law of Sec. 10.2.5.

For the *C2 Control Module*, during separation from the launcher, only rate damping is provided. Thereafter, the proportional/integral autopilot of Sec. 10.2.2.4 provides rapid response to the guidance signals.

For the *C4 Actuator Module* four second-order actuators with rate and position limiters control the four fins. See Sec. 10.2.3.1 for details.

For the *A1 Aerodynamics Module* the aerodynamics are patterned after Eqs. (10.69) and (10.70). The coefficients are expressed in tables as functions of Mach and total angle of attack, for power on/off, and variable c.m. locations. The length of the missile is 2.95 m, and its diameter 0.1524 m.

For the *A2 Propulsion Module* the thrust of a single-pulse rocket motor is given as a table of thrust vs time with backpressure corrections. Vehicle mass, c.m., and MOI are updated. Launch mass is 91.7 kg, and the motor fuel is 35.3 kg.

The *A3 Forces Module* converts the aerodynamic coefficients to force and moment vectors in body coordinates and adjoins the thrust component.

For the *D1 Newton Module* the translational equations of motion are formulated in body coordinates [see Eqs. (10.3) and (10.6)].

For the *D2 Euler Module* the rotational equations of motion (10.16) are solved in body coordinates.

For the *G4 Impact Module* the miss distance is calculated in the intercept plane. Figure 10.54 shows the details of a Monte Carlo run.

A typical engagement is shown in Fig. 10.61. With the shooter and target aircraft at the same altitude, the missile executes primarily a lateral maneuver, expressed by large excursions of the sideslip and heading angles.

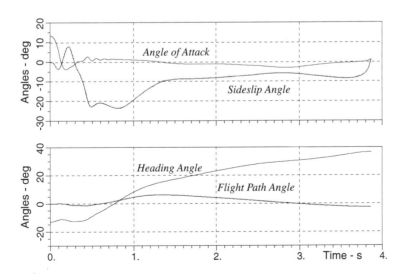

Fig. 10.61 Short-range engagement of the SRAAM6 missile.

My mammoth chapter has come to an end. I did not expect you to study it line by line. Your professional specialization may have drawn you to the aerodynamic models, the autopilot implementations, the guidance laws, or the seeker mechanization. Some of you may have gone straight for the examples. Whatever your interest, you probably griped that your specialty is not adequately represented.

For you novices, unburdened by years of experience, I hope you sensed a certain awe of the multidisciplinary challenge a six-DoF simulation poses. Learn by following in the tracks of others. Study the provided examples or other simulations and work the projects at the end of this chapter. Soon you will be the expert that others will consult in the pursuit of the perfect aerospace vehicle simulation.

References

[1]Rolfe, J. M., and Staples, K. J., *Flight Simulation*, Cambridge Univ. Press, Cambridge, England, U.K., 1986.

[2]Zipfel, Peter H., "On Flight Dynamics of Magnus Rotors," U.S. Army Technical Rept. 117, available at DTIC, ADA 716 345, Nov. 1970.

[3]"Department of Defense World Geodetic System 1984, Its Definition and Relationships with Local Geodetic Systems," 3rd ed., NIMA WGS 84 Update Committee, NIMA TR 8350.2, Bethesda, MD, 4 July 1997.

[4]Torge, W., *Geodesy, An Introduction*, Walter de Gruyter and Co., Berlin, 1980.

[5]Britting, K. R., *Inertial Navigation System Analysis*, Wiley-Interscience, New York, 1971.

[6]AMTEC Corp., "Endo-Atmospheric Non-Nuclear Kill Simulation (ENDOSIM)," U.S. Army Strategic Defense Command, TR1158, TR1163, TR1161, Huntsville, AL, Aug. 1989.

[7]Jordan, W. E., "Simulation Models and Baseline Guidance and Control for Indirect-Fire Missiles with Strap-Down Inertial Guidance," U.S. Army Missile Command, TR GR-76-41, Huntsville, AL, Jan. 1976.

[8]Savage, P. G., "Strapdown System Algorithms," AGARD Lecture Series, No. 133, May 1984, p. 379.

[9]Stevens, B. L., and Lewis, F. L., *Aircraft Control and Simulation*, Wiley, New York, 1992, Eq. (1.3-6).

[10]Lanchester, F. W., *Aerodonetics*, A Constable and Co., London, 1908.

[11]Nielsen, J. N., *Missile Aerodynamics*, McGraw–Hill, New York, 1960.

[12]Gentry, A. E., et al., "The Mark IV Supersonic-Hypersonic Arbitrary-Body Programs," Air Force Flight Dynamics Lab., TR-73-159, Wright–Patterson AFB, OH, Nov. 1973.

[13]Williams, J. E., et al., "Missile Aerodynamic Design Method (MADM)," Air Force Wright Aeronautical Lab., TR-3109, Feb. 1988.

[14]Magnus, A. E., and Epton, M. A., "PAN-AIR—A Computer Program for Predicting Subsonic or Supersonic Linear Potential Flows About Arbitrary Configurations Using a Higher Order Panel Method," NASA CR-3251, Aug. 1982.

[15]White, D. A., and Sofge, D. A. (ed.) *Handbook of Intelligent Control*, Van Nostrand Reinhold, New York, 1992, Chap. 11.

[16]Roskam, J., *Airplane Flight Dynamics and Automatic Flight Controls*, Vol. 1, Roskam Aviation and Engineering Corp., Lawrence, KS, 1979.

[17]Britting, K. R., *Inertial Navigation Systems Analysis*, Wiley-Interscience, New York, 1971.

[18]Widnall, W. S., and Grundy, P. A., "Intertial Navigation System Error Models," Inter-metrics Rept. TR-03-73, May 1973, available from DTIC, AD 912 489L, Alexandria, VA.

[19]Chatfield, A. C., *Fundamentals of High Accuracy Inertial Navigation*, Progress in Astronautics and Aeronautics, Vol. 174, AIAA, Reston, VA, 1997.

[20]Biezad, D. J., *Integrated Navigation and Guidance Systems*, AIAA Education Series, AIAA, Reston, VA, 1999.

[21]Maybeck, P. S., *Stochastic Models, Estimation, and Control*, Vol. 1, Academic Press, New York, 1979.

[22]Stengel, R. F., *Stochastic Optimal Control*, Wiley, New York, 1986.

[23]Hammersley, J. M., and Handscomb, D. C., *Monte Carlo Method*, Methuen and Co., London, 1964, reprint 1975.

[24]Zarchan, P., "Comparison of Statistical Digital Simulation Methods," *Guidance and Control Systems Simulation and Validation Techniques*, AGARDograph No. 273, AGARD, 1988.

[25]Gelb, A., *Applied Optimal Estimation*, MIT Press, Cambridge, MA, 1974.

[26]Maybeck, P. S., *Stochastic Models, Estimation and Control*, Vol. 1, Academic Press, New York, 1979.

[27]Stengel, R. F., *Optimal Control and Estimation*, Dover, New York, 1994.

[28]*Handbook of Geophysics and the Space Environment*, U.S. Air Force Geophysics Lab., National Technical Information Center, ADA 16700, DTIC, Alexandria, VA, 1985.

[29]Von Kármán, T., and Howarth, L., "On the Statistical Theory of Isotropic Turbulence," *Proceedings of the Royal Society of London A*, Vol. 164, 1938, pp. 192–215.

[30]Dryden, H. L., "A Review of the Statistical Theory of Turbulence," *Quarterly of Applied Mathematics*, Vol. 1, 1943, pp. 7–42.

[31]Lumley, J. L., and Panowsky, H. A., *The Structure of Atmospheric Turbulence*, Wiley, New York, 1964.

[32]Pritchard, F. E., et al., "Spectral and Exceedance Probability Models of Atmospheric Turbulence for Use in Aircraft Design and Operation," U.S. Air Force Flight Dynamics Lab., Rept. AFFDL-TR-65F-122, Wright Patterson AFB, OH, Nov. 1965.

[33]Maybeck, P. S., *Stochastic Models, Estimation, and Control*, Academic Press, New York, 1979, p. 188.

[34]Nguyen, J. T., et al., "Simulator Study of Stall/Post-Stall Characteristics of a Fighter Airplane with Relaxed Longitudinal Static Stability," NASA T.P. 1538, Dec. 1979.

[35]White, D., and Sofge, D., *Handbook of Intelligent Control*, Van Nostrand Reinhold, New York, 1992, chap. 11.

Problems

10.1   GHAME6 flight control evaluation.   The hypersonic NASA vehicle GHAME, modeled in GHAME6, has several control modes. You are to check out the dynamics of the open-loop and closed-loop response.

Task 1:   Download from the CADAC Web site the GHAME6 simulation and run the test case INCLIMB.ASC. Try to maximize the terminal energy (altitude and velocity) at 1540 s flight time by modifying the trajectory profile. Plot your final trajectory (altitude, inertial velocity, Mach number, flight-path angle vs time) and record the end conditions of at least five trials in a table.

Task 2:   Run GHAME6 without autopilot. Build the input file INOPEN.ASC with the following modules: G2, ENVIRONMENT; G3, KINEMATICS; A1, AERODYNAMICS; A2, PROPULSION; A3, FORCES; D2, EULER; and D1, NEWTON. With initial BALT = 15 km, DVBE = 720 m/s , and autothrottle set at 50 kPa, trim elevator DELEX to keep the vehicle with ±1000 m for 20 s.

Task 3:   Determine the primary dynamic modes of the pitch, yaw, and roll channels from Mach = 0.5 to Mach = 20 and plot the values.

Task 4:   With the flight conditions of Task 2, execute a pitch doublet without autopilot. Build the input file PITCH.ASC for a doublet of 5 deg and 10-s pulse width. Plot elevator deflection, pitch rate, and angle of attack vs time.

Task 5:   Now, close the control loop by including the following modules: S4, INS; C2, AUTOPILOT; and C4, ACTUATOR. Build input file INSAS.ASC for the yaw damper (SAS). Use ideal INS and maintain altitude with DELECX = 5 deg trim. Check the yaw damper with yaw-rate command pulse of 10 deg/s at 1-s pulse duration. Determine the best value for ZSAS. Why is the yaw rate response so sluggish? Plot yaw-rate command and response vs time.

Task 6:   Check the roll controller. Build input file INROLL.ASC. Keep SAS, and introduce a roll pulse of 10 deg and 4 s duration. Determine the best WRCL and ZRCL in conjunction with the best ZSAS damping. Plot roll command, roll rate, roll position, and aileron deflection vs time.

Task 7:   Check the pitch acceleration controller. Build the input file INACC.ASC. Keep SAS and engage the pitch acceleration controller. Initialize with 1.1 g, then introduce a 0.5-g incremental step command. Select through trial and error the best GAINP, WCLP, ZCLP, and PCLP. Record these values, and plot acceleration command, achieved acceleration, angle of attack, and elevator deflection vs time.

Task 8:   Finally, wrap around the acceleration loop the altitude controller. Build the input file INALT.ASC. Maintain the initial altitude of 15 km, followed after 10 s by a altitude step command of 200 m increase. What are the best values for GAINALT and GAINALTD? Record these values and plot altitude, angle of attack, and elevator deflection vs time.

Task 9:   Summarize your findings in a GHAME flight dynamics report. Show all tables and plots.

10.2   Build a flight controller for the FALCON aircraft.   The Falcon six-DoF simulation, available on the CADAC Web site, comes without autopilot. You are to transfer the flight controller from the GHAME6 simulation, integrate it into the FALCON6 simulation, and optimize the dynamic response.

Task 1:   Familiarize yourself with the FALCON6 simulation, and run the input file INPITCH.ASC. Plot elevator deflection, pitch rate, and angle of attack. Comment on the open-loop stability.

Task 2: Close the control loop with the C2 and C4 modules from GHAME6 and S4 module from SRAAM6. Add subroutine A1DER (derivative calculations) from GHAME6, A1 module to your FALCON6, A1 module. Ensure that all EQUIVALENCEd variables are consistent and that no errors remain after running MKHEAD3.EXE.

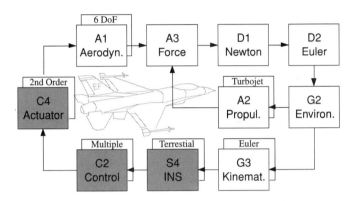

Build input file INYAW.ASC for the yaw damper (SAS). Use ideal INS, and check the yaw damper with a yaw-rate command pulse of 1 deg/s and 2-s pulse duration. Determine the best value for ZSAS.

Task 3: Check the roll controller. Build the input file INROLL.ASC. Keep SAS and introduce a roll pulse of 10 deg and 2-s duration. Determine the best WRCL and ZRCL together with the best ZSAS damping. Plot roll command, roll rate, roll position, and aileron deflection vs time.

Task 4: Check the pitch acceleration controller. Build file input INACC.ASC. Keep SAS, and engage the pitch acceleration controller. Initialize with 1.1 *g*, then introduce a 0.5-*g* incremental step command. Select through trial and error the best GAINP, WCLP, ZCLP, and PCLP. Record these values, and plot acceleration command, achieved acceleration, angle of attack, and elevator deflection vs time.

Task 5: Wrap the acceleration loop around the altitude controller. Build the input file INALT.ASC. Maintain the initial altitude of 1524 m, followed after 10 s by an altitude pulse command of 100 m and 20 s duration. What are the best values for GAINALT and GAINALTD? Record these values, and plot altitude, angle of attack, and elevator deflection vs time.

Task 6: Now build the heading controller, using the input file name IN-HEAD.ASC. Make a 90-deg heading change and limit the roll angle to 70 deg. Determine the best FACTHEAD to reduce GAINPSI. Plot heading angle and roll angle with aileron and elevator deflections vs time.

Task 7: Let us climb with the aircraft. Build INCLIMB.ASC for the gamma-hold controller and climb at 20 deg with a 10-deg heading change. Determine best PGAM, WGAM, and ZGAM. Plot heading and flight-path angles, with angle of attack roll angle, and all three control surfaces.

Task 8:   Wind and nonstandard atmosphere are major disturbance factors. Build the input file INWIND.ASC, based on INCLIMB.ASC for several conditions:
(a) Constant wind of 50 m/s from east.
(b) Shear wind from east: 1000 m, 50 m/s; 10000 m, 100 m/s.
(c) Turbulence and shear wind with correlation length 150 m, sig = 2 m/s.
(d) Input mean wind in January at Wallops Island, Virgina (Fig. 10.45).
(e) Input a test atmosphere by increasing temperature by 10%, pressure by 10%, and reduce density by 10% from the standard atmosphere.
Choose variables that show the effect in each case and plot them vs time.

Task 9:   Summarize your findings in a FALCON flight control report. Show all tables, plots, and input files.

10.3   AGM6 air-to-ground missile.   Convert the SRAAM6 air-to-air missile model into an unpropelled air-to-ground missile simulation AGM6, controlled by an autopilot, optimize the autopilot parameters, and document the results.

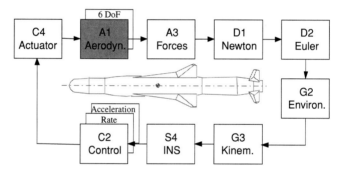

Task 1:   Download the SRAAM6 simulation from the CADAC Web site, run the test cases, and plot selected parameters.

Task 2:   Build the AGM6 simulation with a control loop only (no guidance). Copy the following modules from SRAAM6: A3, G3, S4, C2, C4, D1, D2, and adopt the G2 module from GHAME6. Download from the CADAC Web site the AGM6 aerodeck and program the A1 module. Check carefully the interfaces, and make corrections as necessary. The utility MKHEAD3.EXE should diagnose no errors.

Task 3:   Check the simulation first without control loop. Because the airframe is aerodynamically stable, it should fly smooth ballistic trajectories. Excite the airframe with individual control inputs δp, δq, δr at 1-deg magnitude and pulse width of 1 s. Do the transients damp out?

Task 4:   First close the roll position and the pitch/yaw rate loops. Select optimal values for the roll loop parameters WRCL and ZRCL. Adjust the rate autopilot parameter ZETLAGR for optimum performance.

Task 5:   Now replace the rate loops by the acceleration autopilot. Adjust the acceleration autopilot parameters WACL, ZACL, PACL for optimum performance.

Task 6:   Study the effect of air turbulence on the AGM acceleration autopilot using TRUBL $= 150$ m and TURBSIG $= 3$ m/s.

Task 7:   Summarize the verification of your simulation in an AGM report. Show the open- and closed-loop transient behavior about all axes and of all autopilot modes.

11
Real-Time Applications

The last three chapters gave you the opportunity to execute three-, five-, and six-DoF simulations of aircraft and missiles in the so-called batch mode. You gave the computer an executable form of your simulation and waited until the output came back. As outside agent, you may have yearned to get into the simulation and become a part of its world. Flight simulators, hardware-in-the-loop facilities, and war gaming let you be submerged into their virtual environments. To make the human experience real, the simulations must run in real time or at least be phased by real-time events.

The modeling techniques you have acquired will serve you well in building real-time applications. Preferably, you first construct a batch simulation and then synchronize it with a clock for real-time execution. You will have to balance model fidelity and execution time for best performance on your equipment.

Real-time simulations play a part at all levels of the simulation hierarchy. Recall from Chapter 1 the pyramid of modeling hierarchy. At the foundation lie the engineering simulations, modeling systems, or subsystems with great fidelity. Hardware-in-the-loop HIL simulations are used to check out the performance of the system's hardware and software before flight testing.

At the next higher level are the engagement simulations, which determine the effectiveness of aerospace vehicles. In military applications blue and red platforms may be engaged, firing projectiles or missiles at each other. The testing of equipment and air crew is carried out in *air combat simulators*, a special form of a flight simulator. In commercial applications the *cockpit simulator* is used to test handling qualities and to train airline pilots.

The upper two levels, called *mission* and *campaign* simulations, are tools for strategy or war games. What they lack in engineering detail they make up by a large aggregate of vehicles. As the multitude of entities interact, operational effectiveness is evaluated, and winning strategies are developed.

My purpose is to give you a taste of each of these categories. Drawing from my own experience, I selected topics from piloted simulators, HIL facilities, and war-gaming exercises. The genre however is so vast that you have to consult the specialty literature for particular applications.

11.1   Flight Simulator

In today's world of personal computers, everybody with a joystick has flown some kind of a flight simulator. They are so sophisticated that their flight dynamics and cockpit layout approach the commercial product. However, the visual display confines the experience to a two-dimensional projection of the view out the window.

Workstation simulators are the elder cousins of the gaming simulator. With more computer power they model with greater fidelity the dynamics of the airborne vehicles, the surrounding terrain, and the cockpit layout. Large CRT screens with stick, rudder, and throttle give the pilot a realistic environment, albeit confined to two-dimensional displays and visual sensory feedback only.

The ultimate flying experience, short of air time, is in the six-DoF cockpit simulator mounted on a motion platform with full-scale cockpit and three-dimensional visual cueing. These are the simulators for commercial and military air crew training. They have detailed six-DoF flight dynamics, accurate displays, and tactile feedback.

For military applications, where "blue" engages "red," at least two simulators are needed. In close-in-combat (CIC) training, the two sides must "see" each other to practice evasive maneuvers and missile firing tactics. Dome- or tent-like screens envelop the cockpits and give the pilot a three-dimensional, large field-of-view perspective. One of the issues, which I will address, is the integration of the missiles into the simulation experience. Fidelity of the missile simulation must be balanced with execution speed.

I will define some general terms of flight simulators before treating the individual topics. As reference, I recommend the book by Rolfe and Staples.[1] If you want to spend a delightful week in Cambridge, Massachusetts, attend the Massachusetts Institute of Technology summer course on fundamentals of flight simulation.[2]

Definition: A *flight simulation* is the dynamic representation of the behavior of a vehicle in a manner that allows the human operator to interact with the simulation.

The dynamics are modeled mathematically by Newton's and Euler's laws, whereas the immersion of the operator occurs by sensory stimulants. A flight simulation can mimic any manned vehicle, from single propeller-driven pleasure craft to hypersonic aircraft and spacecraft.

Definition: A *model* is a representation, physical or analytical, of the structure or dynamics of a system or process.

A *physical* model could be a subscale model airplane or a joystick masquerading as the flight controls. We distinguish between tangible and recorded models. *Tangible* models are the cockpit, controls, instruments, and simulator domes. *Recorded* models are airport scenes, terrain models, wind and gust profiles, and sound effects.

Analytical models are based on physical laws expressed in mathematical terms. Some examples are Newton's and Euler's laws, INS error model, aerodynamic forces, actuators, propulsion units, and seekers.

The facility that comprises the flight simulator consists of five major components (see Fig. 11.1): *aircraft model*, the mathematical model of aircraft dynamics; *vision system*, the scene feedback to the pilot's eyes; *motion system*, the vestibular feedback to the pilot's ears and tactile senses; *acoustic system*, the acoustic feedback to the pilots' ears; and *instruments*, cockpit dials and displays. The pilot, receiving visual, acoustic, and vestibular feedback, controls the aircraft with the help of the cockpit instruments. A major part of the facility is devoted to the faithful replication of those stimuli. Deleting the motion system and simplifying the vision system can attain significant cost savings. You can find these shortcuts in some military trainers like the F15 simulator with its emphasis on medium-range intercepts and, of course, in any of the workstation simulators.

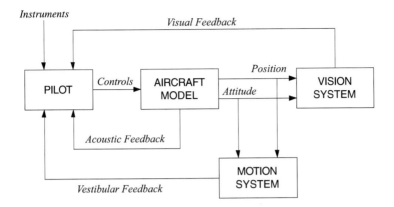

Fig. 11.1 Major elements of a flight simulator.

11.1.1 Workstation Simulator

Workstation simulators are CRT screen-based, man-in-the-loop simulators with processors of high-end workstations. Several processors, linked by Ethernet protocol, support distributed computing. The aircraft and missiles are executed simultaneously with their display utilities. Typically, the pilot sits at a two-screen display with throttle, stick, pedals, and keyboard as input devices. The upper screen displays the background scene, superimposing the heads-up display. It also can contain the radar, tactical display, and an attitude indicator. The lower screen is reserved for the remaining aircraft instrumentation—the so-called lookdown display.

The number of piloted stations depends on the objective of the simulation. Flight training requires just one station, whereas air-to-air combat calls for multiple stations, simulating blue and red aircraft. The maximum number is determined by the outlay of the processors. Workstation simulators make it possible to conduct studies with many piloted aircraft without large investments. Military tactics, like engagement maneuvers, choice of weapons, and egress maneuvers, can be developed and practiced. However, their utility for flight training is restricted because of limited situational awareness: the pilot's view is confined to the flat screen display, and no tactile feedback is provided. In workstation simulators you will find a mixture of models. Highly detailed aircraft dynamics with six-DoF aerodynamics, multimode flight controls, high-bandwidth actuators, navigation aids, gears, and flaps. The missile trajectories are often modeled as simple three-DoF or pseudo-five-DoF simulations. Rarely will you encounter a full six-DoF missile representation. The reason is simple. The pilot's attention is focused on the aircraft's behavior as he delivers and evades missiles. Any obvious discrepancy with his flying experience will distract him from the simulation's objective. The missile fly-out, on the other hand, is autonomous, and only the flight time and the effect on the target are of interest to the pilot. Therefore, simplified missile simulations are acceptable as long as they have been validated previously by full six-DoF models.

Most of the simulation code is programmed in FORTRAN. A long legacy exists of flight dynamics, autopilot and radar models, adapted to the characteristics of new systems, as required. The graphics interface is highly dependent on the simulation hardware and has changed drastically over the past decade. Today, C++ is the preferred language to interface the hardware and the displays. Both FORTRAN and C++ interact harmoniously in workstation simulators, although the purists would like to deal only with a single language.

Let us observe the genesis of a project from planning to execution. Assume the objective is to study the effectiveness of a new air-to-air missile in an air combat environment. Well before the pilots arrive, in some instances 12 month in advance, a planning meeting is held with the customer, the aircraft and missile designers, and the simulator personnel. The objectives are defined, the aircraft and missiles identified, and the scenarios discussed. If new flight systems are introduced, the facility programmer is given their code, which he will integrate into the simulation environment. Here, the first problems can arise. The new missile or aircraft must run as subroutines callable from the main program, usually referred as the man-in-the-loop (MIL) frame. The input and output to the subroutines must be well defined, preferably formalized by an interface document.

You will encounter two approaches to the integration. The older technique, the *distributed integration*, breaks the flight system into its individual components and incorporates only the new subroutines like aerodynamics and flight controls into the MIL frame, while reusing such basic equations as translational and attitude motions. With today's abundance of computer memory, the newer approach, the *compact integration*, inserts the complete missile or aircraft into the MIL frame. It has the advantage that the code, after having been thoroughly checked out in a batch environment, is executed in its entirety in the MIL frame, thus eliminating the time-consuming validation phase.

After the MIL frame programmer has updated his simulation environment, the engineering trials are conducted as final system check. An experienced pilot at this point would be helpful in uncovering any flight anomalies. Once checkout is complete and the air combat simulation has been validated, the test-planning meeting is convened, attended by the customer, the designers, the programmers, and the pilots. They establish the scenarios, concur on the parameter space of the design variables, and lay out a detailed run schedule.

Finally, the time has come to call the pilots for the combat trials. During the first week, they are briefed on the mission objectives and the scenarios and given an opportunity to develop their tactics. Sometimes they question the fidelity of the simulator and the "feel" of the aircraft response. If you are the facility programmer, you have the difficult task to convince the pilots of the adequacy of your simulation for the planned trials. Once they understand that the compromises you had to make do not adversely affect their combat skills, you have won them over, and you can start with the actual test.

The conduct of the trials closely follows the established plan. The engagements are flown, the missiles launched, and the intercepts recorded. Then the parameters are changed, and the cycle repeated. All data are recorded for analysis.

During the post-trial analysis, before the measures of effectiveness are evaluated some of the stressing engagements must be validated. The recorded data are replayed and scrutinized. Miss distance values, which depend on the fidelity of the missile fly-out simulation, must be verified. The analyst may even have to take recourse to the original six degrees of freedom. He or she drives it with the actual target motions and duplicates the engagement based on the high-fidelity simulation. Eventually, the results will be expressed in blue and red fighter exchange rates and conclusions drawn as to the best aircraft and missile designs.

Workstation simulators are particularly effective training tools for managing the aircraft's systems. Radio communication, flight planning, and troubleshooting faulty devices are some of the exercises that can be practiced. The U.S. Air Force uses them as weapon tactics trainers, giving the pilot hands-on experience with fire control radars and weapon release switchology without expending missiles. However, to submerge the trainee into the complete flight experience the more elaborate cockpit simulator is essential.

11.1.2 Cockpit Simulator

The cockpit simulator encompasses all of the elements displayed in Fig. 11.1. The fledgling pilot is subjected to a complete palette of the sensory feedback: visual, vestibular (balance of the inner ear), somatosensory (seat-of-the-pants), and aural (acoustic). If these features are well integrated, the simulator will pass FAA Level D certification, authorizing its use for all flight training without air time. The airlines train their pilots in cockpit simulators, and the military services make use of specialized trainers. For instance, the V-22 Osprey trainer satisfies very demanding requirements and complies with Level D certification.

11.1.2.1 Vision system.

Our eyes are a marvelous creation. Collaborating with the brain, they are able to process information unequaled by any computer. Their specification or I should say their physiology reads as follows: field-of-view for stereoscopic vision, horizontal ±70 deg, vertical ±50 deg; eye movement, ±10 deg at 500 deg/s; head movement, ±100 deg at 10 to 100 deg/s; and resolution in the central region of the retina, 1 arc min (0.3 mrad). These characteristics must be matched by the vision system. It should be stereoscopic, with large field-of-view, high resolution and fast response.

Not all of these requirements can be realized. Fortunately, stereoscopic pictures are not mandatory because the scenes are far removed. However, with close-up CRT screens, the impression of distant scenes must be created by collimating the light that enters the two eyes. Figure 11.2 shows the principle. The beam from the CRT is reflected off the beam splitter, to be returned by the parabolic mirror

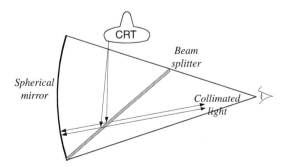

Fig. 11.2 Collimated scene display.

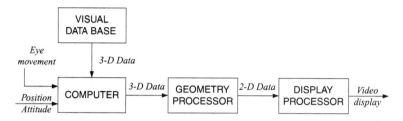

Fig. 11.3   Generation of visual scene.

through the beam splitter into the eye as collimated beams. The eye perceives the scene to be at infinity. A typical value for the horizontal field-of-view is ±48 deg and vertically ±36 deg. The refresh rate should be at least 30 frames per second to be flicker-free.

The scenes themselves derive from video tape or digital imagery. Figure 11.3 depicts the process. The position and attitude of the aircraft and the perspective of the pilot's eyes are correlated with the three-dimensional visual database. A geometry processor projects the selected portion onto the two-dimensional plane and sends the polygons to the video display.

Many of the scenes are computer generated. However, for best effects, photo-graphic pictures are used. You may even find photographic objects superimposed on a digital terrain database. Scene generation is the most computer-intensive part of a simulator. Its realism is a direct function of the invested computer power.

11.1.2.2 Motion system.
Although, according to physiologists, 70% of flying is accomplished with the eye, the ear plays also a very important role. It relates wind and motor noises to airplane speed and status of the engine. But more important than the eardrum is the inner ear with its vestibular set of sensors.

The *vestibular sensors* consist of the inner ear canals, which measure angular acceleration, and the inner ear otoliths (little calcium stones), which sense linear acceleration. Isn't it interesting that an INS senses the same parameters with its gyros and accelerometers? Our brain carries out the integration, although not as accurately as the INS computer. Close your eyes while in the passenger seat of a car. Although you can sense the accelerations, to deduct the velocity or even the traveling distance is a difficult task.

This feature of the human ear is stimulated by the motion platform, which supports the cockpit. It is impossible and unnecessary to duplicate the aircraft's velocity and position, but the accelerations are important stimuli.

The motion platform is supported at three points by a pair of hydraulic cylinders each (see Fig. 11.4). Cylinders from opposing points are paired and attached on the floor. This arrangement gives the platform three translational and three attitude degrees of freedom, a true six-DoF dynamic structure. Although the platform is restrained, the onset of linear and angular accelerations can realistically be simulated. So-called washout filters quickly fade out the signals to restrict travel.

For midsize simulators the linear travel is typically 3 m and the rotational excursions about ±30 deg. Powerful hydraulic lifters support the platform. They deliver

Fig. 11.4 Motion platform.

a motion bandwidth of 2 Hz, commensurate with the upper limit of typical rigid airframe dynamics.

The linear accelerations of civilian airplanes rarely exceed one g, i.e., one times the Earth's gravity. However, fighter pilots, practicing dogfights, routinely experience 3–5 g. They wear pressure suits to prevent blood from accumulating excessively in the lower part of the body, causing blackout. They certainly can attest to the seat-of-the-pants (somatosensory) feeling.

Because the motion system cannot simulate sustained g loads, the pressure suit can be enlisted to mimic g effects. Particularly in military simulators with the pilot willing to suit up, the somatosensory feedback is provided by a specially designed pressure suit. It cannot duplicate the spine-jamming agony, but still conveys the sense of sustained maneuvers.

Simulators penetrate ever-deeper flight training. They are cheaper than air time, allow practicing emergency procedures without endangering life, and are essential for single seat aircraft. The U.S. Air Force is so infatuated with simulators that it broke with tradition and bought only single-seat F-22 fighters. All flight training is done on the ground in two facilities.[3] The Full Mission Trainer (FMT), set inside a geodesic dome with 360-deg view, is used for flight training and emergency egress procedures. The Weapon Tactics Trainer (WTT) is a workstation simulator with up to 21 stations, manned by blue and red pilots. Here the war fighters can hone their skills in acquiring, tracking, and attacking enemy aircraft. The F-22 pilot training lasts about 104 days, 10 days less than for the F-15, but covering a more complex aircraft. Wouldn't you want to be part of the action? Well, you can. Go to your software store, buy an F-22 simulator, and then fly away on your home computer!

11.1.3 Missile Integration for Combat Simulators

As with all engineering endeavors, air combat simulators are benefiting greatly from increased computer prowess. With the latest Silicon Graphics processors, it has become possible to double the number of aircraft and missiles in simulated combat, while even improving the fidelity of their models.

Air combat simulators are piloted flight simulators that engage multiple aircraft and missiles simultaneously. They support all phases of aircraft and missile development. During the definition of requirements, they are used to establish aircraft maneuverability and missile fly-out performance. In the development phase competing designs are evaluated on simulators. Eventually, the airman is trained on them for combat.

Depending on the implementation, we distinguish between dome, workstation, and virtual helmet simulators. At the current state of the art the dome simulators provide the highest situational awareness, but may be replaced by virtual helmets in the future. A poor-man's choice is the tabletop workstation.

The software that drives the air combat simulators has a long history of development. The equations of motion of the vehicles are well understood. However, the fidelity of the models is restricted by the limitations of the computer hardware. Although the aircraft are modeled in six degrees of freedom, the missiles in current simulators are simplified point-mass formulations in three degrees of freedom or pseudo-five degrees of freedom.

With the increased reliance of the developer on prototyping by computer, it has become necessary to improve the fidelity of the missile model. Particularly for close-in combat with its highly dynamic environment, the missile should be modeled with six-DoF fidelity, or at least any simplification should be validated against it. Fortunately, the continued increase in computing power will eventually allow full six-DoF representation of aircraft and missile models in air combat simulators.

To shorten the time of the design and evaluation cycles, the simulation models should be adaptable to both activities: execution in batch mode for design and real-time implementation in simulators. The batch simulation is built first, then converted into a real-time capable subroutine, which is embedded into the air combat simulator.

I will first describe standard air combat maneuvers as they affect modeling choices, discuss the effect of missile model fidelity, and describe a typical conversion process from batch simulation to real-time code. Your knowledge of five- and six-DoF simulations and particularly your familiarity with the SRAAM5 and SRAAM6 models will be an asset.

11.1.3.1 Air combat fundamentals.
Air-to-air missile engagements are categorized by launch range. They are beyond visual range (BVR) and within visual range (WVR). Many variables influence the visual detection threshold, from atmospheric conditions to the choice of aircraft color schemes. It is possible for one combatant to be within visual range while his opponent is still hidden, causing BVR and WVR conditions to overlap. CIC encounters are a special subset of WVR engagements.

In CIC engagements the opponents are maneuvering at high load factors almost continuously. These maneuvers cause loss of speed and altitude. Both aircraft descend in spirals, which looks like two dogs chasing each other; therefore, the expression *dogfight*.

For over 40 years air combat doctrine has called for three types of weapons with overlapping, concentric coverage. Long or medium range air-to-air missiles (MRAAM) are tasked with all BVR engagements. Under BVR conditions the aircraft's fire control radar detects the target, and the MRAAM guidance is matched by a compatible radar frequency seeker. Large warheads compensated for the inaccuracy of the radar seeker. Thus, MRAAMs tend to be heavy vehicles with sluggish flight characteristics.

Short-range air-to-air missiles (SRAAM) are preferred for the outer portion of the WVR engagements. The typical SRAAM uses an IR seeker, because acquisition ranges are consistent with visual detection. Lighter warheads and motors, made possible by smaller miss distances and fly-out ranges, make the SRAAMs lighter and more agile than MRAAMs.

Finally, the gun is dedicated to CIC conditions. Each fighter pilot tries to shoot at his adversary's tail. Only recently, are the tactics of the World War II dogfight beginning to change.

With today's advanced technologies an air-to-air missile can be designed with creditable performance even for CIC conditions. A single missile could be effective over the entire detectable target range, even replacing the gun. In the past, analyses of CIC conditions have generally concentrated on aircraft performance because of the importance of flight quality for effective aerial gunnery. The usual assumption was that the CIC problem had been solved if the shooter could point his aircraft at the target and hold that geometry long enough for his gunfire to take effect.

With the new generations of SRAAMs on the drawing boards, CIC tactics are updated to account for agility and off-axis capability. The missile's high-g maneuvers require greater fidelity simulations in combat simulators. Because of this new challenge, I will concentrate on the close-in engagements and the issues of missile simulation fidelity.

11.1.3.2 Circle fights.

In CIC engagements the opponents are maneuvering at high load factors almost continuously. Historically, the most common form has been a hard lateral turn in a steeply banked attitude. Although this maneuver presumes a body-fixed, forward-firing gun, the same tactics has been used in analyses of short-range missile engagements. Viewed from above, the flight paths of the opposing aircraft resemble a series of circles, leading to the terminology of *circle fight*. Viewed from the side, each successive circle is lower than the previous one, owing to the tendency of aircraft to lose altitude when in a steep bank. Overall, the flight paths of all of the participants in a CIC engagement are descending spirals, as illustrated in Fig. 11.5.

Two stylized types, the single-circle and double-circle fights, are important in CIC analysis. In the single-circle model both opponents hold constant altitude during a circular flight path about a common center of rotation. They could be flying toward each other or chasing each other. In the U.S. the chase version of the single-circle fight is known as a Lufbery circle. The head-on, single-circle fight, however, is considered the more important baseline engagement. In the double-circle fight the opponents fly nonconcentric circular flight paths that are tangent at a single point. Both aircraft turn the same way, either clockwise or counterclockwise.

Before the introduction of air-to-air missiles, these circle fights could easily be assessed. The outcome depended on who could first point the nose of his aircraft at his opponent and keep it pointed there long enough to inflict significant damage with his gun. Aircraft performance and robustness were the deciding factor. These are conflicting design requirements, because an agile aircraft is light, lacking heavy armor and redundant controls. Finding the proper balance shaped the course of fighter development for decades.

Although circle fights were originally used to categorize close-in combat with guns, they have gained new prominence for engagement studies with the new breed

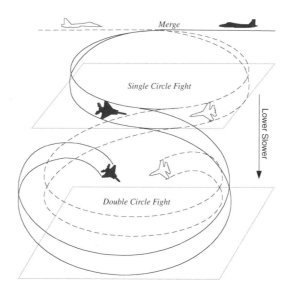

Fig. 11.5   CIC engagement as a sequence of circle fights.

of agile missiles. The format most often seen in the U.S. is the launch acceptable region (LAR) graphic. There are five distinct LAR graphics; those most relevant to circle fights are given the designations LAR-3, LAR-4, and LAR-5, representing the single, double, and Lufbery circles, respectively. The construction of a LAR graphic can best be described by an example.

Figure 11.6 illustrates the LAR-3 scenario. The purpose is to simulate a head-on, single-circle fight, at coaltitude and equal speeds. Its prelude begins at great distance, with both aircraft racing straight for each other and trying to fire off their first shot. If both survive the encounter, they merge and begin turning with facing cockpits, setting up the head-on circle.

The geometry of a single-circle fight can be expressed as the relationship between the velocity vectors of the shooter and the target, as shown in the left panel of Fig. 11.6. The slew angle σ is the angular displacement of the LOS to the shooter velocity vector at launch. If the target velocity vector makes the same angle to the LOS, the result will describe a single-circle fight.

The LAR-3 envelope follows the format illustrated in the right panel of Fig. 11.6. The shooter aircraft is fixed at the center. The slew angle σ is swept incrementally from zero to the largest off-boresight capability of the missile seeker. At each value of σ, the launch range to the target is varied incrementally to identify those initial range conditions that result in hits. The shaded area shown in Fig. 11.6 is the area of successful launch conditions relative to the shooter and is called the launch acceptable region.

Two features of the LAR-3 graphic are noteworthy. First, the full capability of the missile is not normally shown. Because the LAR-3 scenario depicts CIC performance, the envelope is truncated at approximately 6 km. The full performance envelope of a missile could extend to greater distances, but is of little interest.

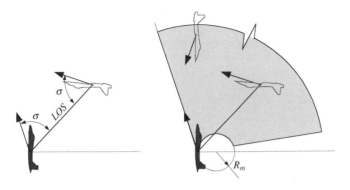

Fig. 11.6   LAR-3 single-circle fight characteristics.

A second feature of the LAR-3 graphic is the inner exclusion circle defined by the turning radius of the shooter. Launch ranges within that circle result in a crossover between the flight paths of the shooter and target, and both opponents must turn through at least 180 deg before a true single circle can be established. The situation is physically realistic, but does not fit the "circle choreography" selected for LAR-3.

We will use the LAR-3 engagement to study the fidelity issues associated with CIC. Although there are several other types of engagements—the twin circle fights and the British combat circle diagrams, the LAR-3 is representative of the stressing engagements of typical CIC encounters.

11.1.3.3 Prototype missile. For our performance studies we use the air-to-air missile, introduced as SRAAM5 in Chapter 9 and SRAAM6 in Chapter 10. It is a 6-in. diam/116 in. long, high fineness ratio, body-tail configuration with a 6:1 ogive, blunted for an IR dome with 1.5-in. radius, and forward strakes (see Fig. 11.7). Its forebody includes an imaging IR seeker, electronics, an INS, a cooling bottle, an active optical target detector, and a 20-lb warhead with an electronic safe and arm system.

The SRAAM is modeled in the CADAC environment. It consists of five- and six-DoF fly-out versions and can be downloaded from the CADAC Web site. The target is modeled as a point-mass vehicle with its attitude aligned in the load factor plane. The maneuvers are defined before missile launch.

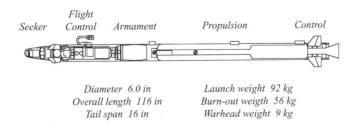

Seeker *Flight Control* *Armament* *Propulsion* *Control*

Diameter 6.0 in   *Launch weight 92 kg*
Overall length 116 in   *Burn-out weigth 56 kg*
Tail span 16 in   *Warhead weight 9 kg*

Fig. 11.7   SRAAM.

SRAAM5 represents the so-called pseudo-five-DoF implementations as presented in Sec. 9.3.2. Three translational DoF are modeled by nonlinear differential equations (Newton's equations) employing tabulated trimmed aerodynamic data. The two attitude degrees of freedom are pitch and yaw (skid-to-turn). They are modeled by linearized differential equations that describe the attitude dynamics of the controlled airframe. In this case Euler's equations are not modeled.

The SRAAM6 version is a full six-DoF simulation. It solves the three translational DoF with Newton's equations and the three attitude degrees of freedom with Euler's equations. All systems of the missile are modeled in detail.

During the missile design phase, CADAC is executed in the batch mode. Its many postprocessing tools support the performance evaluation of the missile concepts. The same missile simulation can be stripped of all unnecessary subroutines using the converter program CONVRT.EXE, which generates a self-contained subroutine suitable for real-time execution. Thus, the tractability of the missile simulation from batch processing to execution in the combat simulator is maintained.

The six-DoF architecture of the batch simulation is shown in Fig. 10.53. Each module represents a major subsystem with its closely controlled interfaces. The calling sequence is important and must be maintained for the simulation. For real-time applications the two modules G1 Target and S2 AI Radar are deleted and are replaced by inputs from the flight simulator.

The five-DoF simulation, presented in Fig. 9.54, was built from the six-DoF simulation by simplifying the aerodynamics and replacing Euler's equations by the response of the attitude autopilots.

11.1.3.4   Fly-out comparison.

Any missile model of a combat simulator should reproduce accurately the sequence of events that affect the pilot's situation awareness. The sequence consists of the missile time of flight, the hit or miss outcome of the firing, and the time required assessing the damage inflicted by hits. This last item is closely related to the missile lethality model. Consideration of lethality however is beyond the scope of this discussion.

The importance of situation awareness cannot be overstated. The purpose for a MIL simulation is to capture the reactions of the pilot and factor them into the assessment of the weapon's effectiveness. If his perception of the overall combat situation is corrupted by simulation errors, he will react differently, and most benefits of a MIL simulation environment will be wasted. This is the principal reason why workstation MIL simulations have only been useful in BVR engagements, but not in CIC battles. The tabletop monitor can provide a good approximation of a radarscope, but cannot reproduce the visual field of view of a human pilot. Therefore, the situation awareness is lost and with it the realistic pilot response, which was the primary goal of the exercise.

Time of flight of the missile and the hit or miss outcome are the foundations of situation awareness. Pilot actions form an unbroken chain of events, based on the pilot's perception. If the simulation time of flight is in error, say by a half-second, the actions of the pilot during that half-second are ambiguous relative to an actual launch. Initial conditions for any subsequent launch have been changed irreversibly, and system effectiveness may have been reduced to speculation. The same can be said of the hit assessment because the actions of a pilot who perceives a miss will normally be quite different from the one who sees a hit.

The fidelity of the missile model must take into account these requirements. From the standpoint of the simulation facility, the missile code should be as compact as possible and executable at the same time step as the aircraft. That is, low-fidelity models would simplify the integration task. However, the simplifications should not jeopardize the pilot's situation awareness. Because CIC is the most demanding engagement, we look into the adequacy of five-DoF models to represent the missile fly-out in LAR-3-type launch zones.

Time of flight is a critical parameter in determining the adequacy of any missile simulation. A comparison between SRAAM5 and SRAAM6 of a fly-out trajectory in the middle of the envelope is shown in Fig. 11.8.

It is instructive to follow the incidence angles from launch to intercept for the two models. The six-DoF simulation replicates accurately the initial roll angle, imparted to the missile by the shooter aircraft, executing a 7.5-g maneuver at an angle of attack of 13 deg. The missile rolls out to a "hooks-up" attitude and develops sideslip for the lateral intercept. On the other hand, five-DoF simplifications cause the missile to be initialized with a large sideslip angle and small angle of attack because the roll DoF has been suppressed.

This different kinematic behavior is also evident in the yaw seeker gimbal angles. The highly banked missile exhibits very little initial yaw gimbal angle, whereas the simplified five-DoF representation starts out with a correspondingly large value.

Naturally, the pilot is not aware of these missile parameters. He is more interested in the intercept time. The endpoints of Fig. 11.8 show that the missile times of flight of the two versions are very close, within about 1%.

For a broader look a typical LAR-3 (see Fig. 11.6) envelope is displayed in Fig. 11.9. Throughout the envelope the flight times agree well, with the trend that the six DoF exhibits slightly longer times. This tendency is caused by the initial transients, which are more accurately duplicated by the six-DoF simulation. The 2.65-s contour is also included, marking the burn-out of the rocket motor.

11.1.3.5   *Miss distance comparison.*

The other important factor for the pilot is the knowledge whether the missile annihilated the target. For our study miss distance serves as criterion, although the damage to a particular aircraft depends also on the missile's warhead and fuse. We assume, somewhat arbitrarily, that a 6-m miss is the cutoff point for a hit. Any larger value constitutes a miss.

The simplifications of the five-DoF model, affecting miss distance, are in aerodynamics, autopilot, and, above all, in the seeker implementation. We replaced the detailed seeker code of the six DoF with an ideal seeker not constrained by gimbals and tracker loop. Multiple runs throughout the LAR-3 envelope with the five-DoF simulation lead, as expected, to hits everywhere. However, if we use the detailed six-DoF simulation with noise-corrupted seeker and sensor models, the envelope changes significantly.

To calculate the LAR-3 envelope for missiles with noise, multiple runs are made against every aimpoint. The miss distances are averaged and displayed as representative terminal miss performance. We use 30 such Monte Carlo runs against each aimpoint, employing the automated SWEEP methodology of CADAC for analysis and plotting.

Figure 11.10 displays the results of an engagement at 5000 m altitude with both aircraft executing 7.5-g maneuvers at 0.75 Mach.

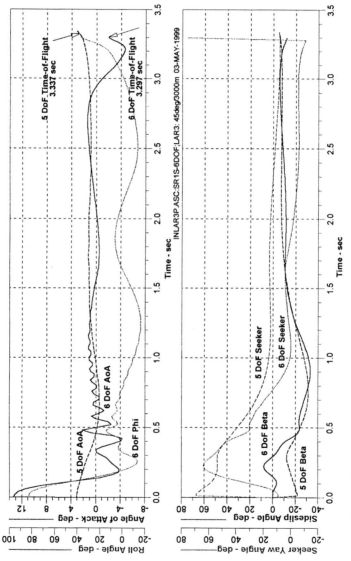

Fig. 11.8   Comparison of six- and five-DoF trajectories. LAR-3: $R = 3000$ m, $\lambda = 45$ deg.

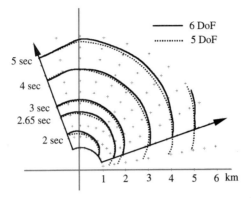

Fig. 11.9   LAR-3 time-of-flight comparison.

Compared to the five-DoF model, which delivers hits everywhere, the envelope is reduced for close-in shots (no hits within 2 km) and at high off-boresight shots beyond 5.5 km. The sweet spot lies at midranges and small look angles. A major reduction in real envelope performance occurs for look angles greater than 70 deg at all launch ranges. It is caused by the seeker gimbal limits. To understand this phenomenon, we need to look at the trajectory more closely.

The maximum look angle of the SRAAM seeker is 90 deg. This allows lock-on before launch throughout the envelope. After launch there is a delay of 0.25 s before guidance is enabled. During this time, the missile tends to align itself with its velocity vector by virtue of its inherent stability. This leads to a horizontal turning away from the target, through an angle perhaps as large as the launch angle of attack and a corresponding increase in seeker pitch-gimbal deflection. Whenever the seeker is locked on before launch at a gimbal angle between 90 and 90-α deg

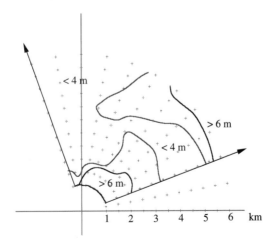

Fig. 11.10   Six-DoF LAR-3 envelope; 30 Monte Carlo runs against each aimpoint.

(α is the angle of attack of the shooter at missile launch), there is a chance that the weathercock stability of the missile will drive the gimbals to their stops. The reduced off-axis performance is shown in Fig. 11.10. Maximum gimbal deflection is exceeded during the first 0.25 s of flight, and the seeker breaks lock on the target whenever the initial look angle exceeds 72 deg. The kinematic seeker simulation ignores the gimbal stops and provides whatever deflection is required to maintain target tracking.

The situation depicted in Fig. 11.10 signifies the physical limitations whenever launches are made at high angle of attack. Such launches can be common in CIC engagements, but are very rare at longer ranges.

Many CIC engagements, particularly those that feature high angles of attack at launch, are initiated from subsonic conditions. As a result, the low-speed maneuverability of the missile plays a dominant role in determining whether or not a gimbal limit violation is likely to occur. The characterization of missile airframes intended for a CIC role should, therefore, include accurate subsonic aerodynamics.

In circle fights maximum look angles determine the all-important first launch opportunity of the shooter. If one aircraft is given an 85-deg capability compared to the baseline 70-deg system, its effectiveness is greatly enhanced. For realism, the look angle boundaries must be accurately modeled. Assuming the six-DoF simulation is the truth model, the five degrees of freedom can be brought into agreement by reducing the limiting gimbal deflection input. A maximum pitch gimbal deflection of 70-deg for the five degrees of freedom should lead to an envelope with a maximum look angle very close to that of the six-DoF simulation.

In summary, five-DoF models, adjusted for six-DoF effects, can be used in CIC—and of course in BVR—engagements. They portray sufficiently accurate the time of flight and the in-range/out-of-range envelopes for the pilot's situation awareness. However, low-speed, close-in, and large off-boresight shots require six-DoF missile models, or, as a minimum, confirmation of such engagements by off-line six-DoF runs.

11.1.3.6   *Real-time conversion.*

Converting the batch simulations of the design phase to real-time code is often a disjointed process. Frequently, the MIL simulator has its own tightly integrated missile fly-out model. In some instances the same aerodynamic and autopilot subroutines are engaged in driving the aircraft and the missile, albeit with different data sets. The batch simulation is broken up and distributed over the subroutines of the combat simulator. You can imagine how difficult it is to validate the converted missile simulation. Many test cases have to be run in both environments, analyzed for inconsistencies and adjusted—only to have to execute the test cases again for revalidation.

Integrating the missile simulation as a complete entity streamlines the process, but sacrifices efficiency. CADAC provides a converter program that takes the batch simulation, strips it of unnecessary code, and translates it into a package that can be "dropped" into the MIL simulation. All missile subroutines remain intact, and only the interface variables have to be adjusted. Although this approach requires more code, execution speed remains essentially the same—a small price to pay considering the abundance of memory in modern computers. The rewards lie in the much simplified validation phase and the assurance that the missile model in the MIL simulator reflects accurately its performance.

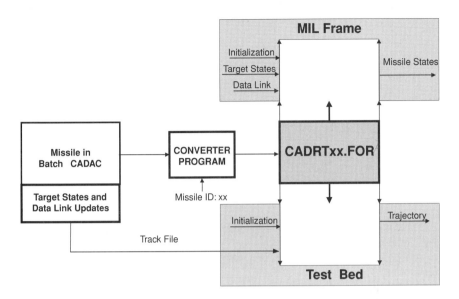

Fig. 11.11 CADAC converter program.

The conversion takes place in several steps and applies equally to five- and six-DoF simulations. After having validated the missile simulation in the batch mode, the programmer runs a test case and records the input to the missile, e.g., target states, on a track file. Then the converter program translates the batch code into a subroutine package, which is validated in the test bed, driven by the track file. Now the code is ready for integration into the MIL frame simulator.

The schematic of Fig. 11.11 depicts this procedure. The missile identification code IDxx, inserted during conversion, attaches itself to every subroutine and labeled common statement. Thus, every missile package becomes unique, and different types of missiles can be flown in the MIL simulator simultaneously.

The communication between the MIL frame and the missile simulation occurs over a common A array. It is the exclusive conduit for missile initialization, target state tracking, and data link transmission, as well as the missile state feedback. Figure 11.12 shows the interface with three missile models, named x1, x2, and x3.

In summary, we reviewed air combat maneuvers and focused on the CIC engagement, stylized by the LAR-3 launch acceptable region, as the most demanding situation. Employing the five- and six-DoF simulations of a generic SRAAM, we assessed the effect of simulation fidelity on the pilot's situation awareness. We concluded that five-DoF models, adjusted for six-DoF effects, could be used in CIC. They portray sufficiently accurate the time of flight and the firing zones for the pilot's situation awareness. However, low-speed, close-in, and large off-boresight shots require six-DoF missile models, or, as a minimum, confirmation of such engagements by six-DoF batch runs. Integration of new missile concepts into combat simulators is greatly simplified by a conversion process that leaves the missile code intact, instead of overlaying it on the aircraft model.

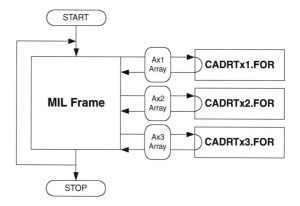

Fig. 11.12   Interfaces between multiple missiles and MIL simulator.

11.2   Hardware-in-the-Loop Facility

Replacing the pilot with hardware turns the flight simulator into a HIL simulation. Of course, I am oversimplifying, but the major difference is the absence of the human touch. Otherwise, both run at real time, consist of a mix of software and hardware, and are expensive to operate.

Autonomous systems, like missiles and spacecraft, operate without a human in the loop and are therefore prime candidates for HIL simulations. The control and guidance loops can be closed electronically, and any of the costly visual and vestibular stimuli are superfluous. So, what is a HIL? A HIL simulation is the dynamic representation of a vehicle in real time with subsystems modeled by a combination of hardware and software.

The main elements are shown in Fig. 11.13. The heart is the computer. Here reside the equations of motion and all of those features that must be mathematically

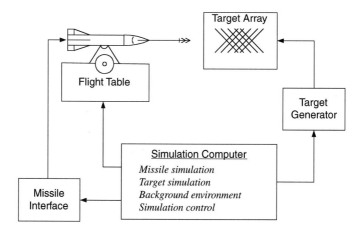

Fig. 11.13   Elements of an HIL simulation.

modeled, like aerodynamics, propulsion, atmosphere, and gravity. The computer tracks the missile and target geometry and controls the event sequence. The missile attitude angles are sent to the flight table to simulate the attitude dynamics, and, through the missile interface, position and velocity are provided to the missile processor. To close the guidance loop, the missile seeker, whether IR or RF, receives the target signature from the target array, which is driven by the target generator.

There are many variations of this basic theme, contingent on the type of vehicle, availability of hardware, and purpose of investigation. The missile in Fig. 11.13 could be replaced by a cruise missile, an unmanned air-vehicle, or the kinetic kill vehicle of a ballistic missile. Depending on the development phase, we may see breadboard, brassboard, or flight-ready hardware in the HIL facility. Key hardware systems are INS, flight controller, actuator, and seeker.

HIL simulations can be used throughout the development cycle of a flight vehicle: from testing of early prototypes to system integration and acceptance testing of flight-ready hardware. Scrutinizing the interfaces between devices from various vendors can prevent much embarrassment during flight testing. (I will never forget the loss of a test vehicle over the Gulf of Mexico because of reversed polarity.)

Not all components of a HIL simulation have to be located at the same facility. For instance, if a missile is to be tested with the armament panel of an aircraft, it is impractical to park the aircraft inside the HIL facility. Instead, the aircraft is linked over fiber optic cable with the wiring harness of the missile.

The degree of distribution can be further expanded. The U.S. Army keeps tank battalions at several locations throughout the world. Each battalion uses tank simulators for training. To practice tactics without burning up fuel, several trainers are linked together to form what is called a *distributed interactive simulation* (DIS). Sometimes several battalions are netted together for a global training exercise. Sending time-position data over disparate telecommunication nets require tight control of the message traffic. The communication protocol that accomplishes this feat carries also the name DIS. Recently, the U.S. Department of Defense substituted DIS by HLA, a higher-order language protocol.

After this brief excursion to distributed simulations, let us discuss the requirements of an HIL facility. The system requirements are determined by the system to be tested. The apparatus should allow the system to be tested to its limits. Therefore, the bandwidth of the flight table should exceed the rigid-body frequency of the vehicle, the target model should expose the seeker to all possible frequencies with realistic noise sources and countermeasures, and the actuator should be loaded by the correct hinge moments.

Mathematical models are used for nonhardware subsystems or to back up hardware. You should by now have a good idea what they contain. Make sure to balance the model fidelity evenly across the HIL. Any or all of the components can be modeled in code. Sometimes the HIL simulation of a new vehicle can start as an all-digital simulation. As the development program progresses, hardware is substituted, as it becomes available. The computer language today is most likely C or C^{++}, but you may have to work with legacy code in Pascal or FORTRAN or even the simulation language ACSL. Above all, it is important to establish and maintain good configuration control.

The credibility of your study depends on the extent of the verification, validation, and accreditation process preceding it. First, verify that the hardware and software performs as specified, then validate your closed-loop simulation against benchmarks and test cases from other sources. Finally, call the user to accredit your simulation by giving their stamp of approval.

The mature HIL simulation is not static. It needs maintenance and must be upgraded. The hardware needs probably more maintenance than the software, but even software in today's volatile upgrade-driven market must be kept up to date. And who can claim a bug-free simulation? As errors are discovered, they must be eradicated and the corrective action documented. Establish a budget for maintenance and realize that upgrades are a way of life.

11.3   Wargaming

Our focus throughout the preceding chapters was on the engineering-level treatment of aerospace simulations. In the last section we made an excursion into the world of real-time simulators, but essentially concentrated on the technical aspects of aerospace systems. However, the treatment of modeling and simulation would be incomplete without an egression to the rapidly growing field of wargaming.

A *war game* is a simulation of a military operation, involving two or more opposing forces, using data and procedures that model a real-life situation. This definition captures the main elements: entities that reflect the real world and are in conflict with one another. Although the emphasis is on military engagements, war-gaming-like techniques are also applied to civilian endeavors. Like nations, the corporate world is subject to competition and conflicts. CEOs, functioning as generals, experiment with winning strategies using war-gaming techniques.

The best known and first war game is the venerable chess game. The chessmen are the entities, their moves model their real-life mobility, and the two opposing players are in conflict for the winning call: checkmate! Generations have enjoyed this game, which was first played in the eighth century. It was much later, in 1811 in the Prussian military, that Herr von Reisswitz formalized the "Kriegspiele" (the German word for war games). In sand tables with elaborate terrain models he confronted blue and red forces and maneuvered them under strict rules of engagement. His fame spread to the United States, and there Major W. R. Livermore, playing catch up, published the "American Kriegspiele" in 1879. The U.S. Navy picked up the idea in earnest with the first fleet maneuvers in 1887. Ever since, the U.S. Navy has led the military services in the art of war gaming.

Many publications have appeared in print. If you are serious about researching this subject, the book by Peter Perla *The Art of Wargaming*[4] is a most quoted text.

I invite you to throw off all engineering inhibitions, abandon high-fidelity six degrees of freedom, and reach out to the wide world of opposing armies, air forces, and navies. Follow my guided tour of building a war game, participate in a demonstration, and learn to critically assess the results.

11.3.1   Building a War Game

Long before the start of a game, possibly a year or so, the responsible agency defines the objectives and the scenarios. The objective is simply the purpose toward which the game is directed. It may be tactical, operational, or strategic. It may serve the evaluation of a weapon concept, the development of doctrine, strategy, or policy, and provide guidance in resource planning. Answers to the following questions can be sought: what is the required number of tactical fighters needed to protect offensive bombers against an air attack, or what are the best tactics of an armored battalion to conquer a heavily defended city? At the operational level the scope is broader, and the objective may be, for instance, the optimum deployment sequence of fighter wings and army corps against an invading adversary. Games with strategic objectives are the favorites of four-star generals. They take on global implications. What strategies should the national command authority pursue to prevent the annihilation of a friendly country? How do space assets enhance global power projection? As you can imagine, the examples are innumerable.

The objectives must be played out in scenarios. The players are immersed in real-life situations that challenge their decision-making capacity. They may be confronted with actual scenarios to prepare for war or with generic situations to explore new weapon concepts. In either case much thought goes into the preparation of scenarios. Only if they stress the players in all aspects of the intended objectives will the game be productive.

Scenarios are categorized by the capability of the opposing forces. *Peer* competitors are nations with similar infrastructure and weapons. A *niche* competitor is inferior to the blue force but may have innovative concepts of deployment or certain political advantages. Among the niche competitors one distinguishes further regional conflicts or small-scale contingencies. These are scenarios of limited scope but can be politically explosive.

A typical peer scenario may consist of two world powers claiming the same oil-rich territory. Their own energy resources have been depleted to such a degree that their economical prosperity depends on the exploitation of these new oil resources. Both competitors have colossal conventional forces and atomic weapons. Their leadership is aware that an all-out war would be catastrophic for both countries. However, the red side believes that through a quick preemptive strike they could occupy the contested territory before the blue side can respond. Faced with the fait accompli, the blue nation may acquiesce to the occupation. Afterwards, over the years, the red nation should prosper economically and would become the dominant world power.

This scenario sketch must of course be expanded in several areas. Detailed maps must be provided with key cities, military installations, and choke points like bridges, mountain passes, and shipping lanes. Question must be addressed like what is the time frame—is it today or 30 years from now? What is the time line of the conflict—build up of tensions, massing of forces, political intervention, and first strike? (Once the first shot is fired, the events are determined by the play.) Given the time frame, the war fighting capability of each side is to be assembled in databases, also called toolboxes. Finally, rounding out the scenario, the political landscape and the infrastructure of both competitors need to be described.

This elaborate process of fleshing out the scenario is called *modeling*. As you may already have suspected, that word has taken on quite a different flavor from that so familiar to us through the preceding 10 chapters. The war gamers like to define a *model* as a "representation of some aspect or attribute of a system," and *modeling* as "the process of constructing a model that represents some aspects of a system." These are quotes from the wargaming literature. As you see, the definition is very broad; it is as broad as you may want to interpret the word "system."

On the heel of modeling follows *simulation*. You will find that the words modeling and simulation are used by war gamers as often as by engineers, but be aware of the difference in meaning. We already dealt with modeling. The definition of simulation is similarly broad: *simulation* is the use or exercise of a model. Another definition reads: Modeling and simulation is the process of designing a model of a system and conducting experiments with this model for the purpose either of understanding the behavior of the system or of evaluating various strategies for the operation of the system. Clearly, a model does not have to be mathematical or be some kind of breadboard, just as simulations do not have to run on computers.

However, I do not want to create the impression that war games do not use computers. To the contrary, you find more computer terminals in a wargaming facility than at a NASA launch pad. As part of the planning process, the game designer must allocate carefully the computer resources. The first question he needs to ask is what computer models will be required. It could be any of the major models, like AWSIM, the U.S. Air Force's warfare simulation; TACWAR, the U.S. Army's tactical warfare model; or ITEM, the U.S. Navy's integrated theater engagement model. In the future it may even be JTLS, the Joint Theater Level Simulation, representing the aggregate wisdom of all three services. Many more examples could be given. Building war-game simulations is big business outside the beltway of Washington, D.C.

If existing computer models do not suit the objectives of the game, the planner must engage a team of experts to develop specialized tools. Prior to usage, the new model must be verified, validated, and accredited. Verification is the responsibility of the developer. He must ensure that the implementation accurately models the conceptual description. Then, jointly with the customer, he validates that the model accurately represents the real world. Finally, the customer or a higher authority certifies that the simulation is acceptable for the intended purpose. Only then is the computer tool accredited for the war game.

Based on the type of war game, the planner specifies the simulation models. If the game is to be played in real time, the simulations must be synchronized to the clock, or for more flexibility the time could also be accelerated or delayed. In either case, as long as the time marches in constant increments, the game is said to be time stepped. However, a conflict that extends over months or years is better simulated by dynamic event scheduling, i.e., the clock jumps from event to event rather than in constant steps. The three games AWSIM, TACWAR, and ITEM are time stepped, whereas the new JTLS supports both phasings.

Another decision the planner has to make is the fidelity of the models. High-fidelity war games require the display of movements of entities and fly-out trajectories of weapons. Low-fidelity engagements are limited to iconic symbols, representing flights of aircraft or tank battalions. Be cautioned however concerning the use of the word *fidelity*. The trajectories of high-fidelity war games are at best three-DoF trajectory models, which we considered low-fidelity trajectory simulations in preceding chapters.

Should our world be deterministic or stochastic? Is the puzzling question often asked during planning meetings. Deterministic models capture only specific events and do not lend themselves to generalizations. Stochastic simulations take the fuzziness of the "fog-of-war" into considerations, but stir up much controversy when the numerical results are to be interpreted. Often, random number generators are enlisted to cover sins of sloppy modeling. Be aware of ignorance hidden in stochastic simulations!

Once the simulation support is defined and the databases established, the rules of engagement must be laid down: who strikes first; will weapons of mass destruction be used and under what conditions; will the combatants abide by the Geneva conventions; when is the game over; and what are the win–lose criteria? At this point in the planning cycle, it is necessary to bring in the umpires who will referee the game. They need to participate in the rule development because they will have to enforce them later.

Although computers have an important part in wargaming, they are dispensable. The further we play in the future, the vaguer the data and the harder it becomes to computerize the moves. We escape to the so-called seminar games. Here, a BOGAG (bunch of guys and gals) sit around tables forming the blue and red forces and making alternate moves. An umpire decides the outcome of each engagement and declares the winner. Seminar games are popular with technologists and are the favorite form played in the corporate boardrooms.

11.3.2 Conducting a War Game

With the planning complete the invitations are sent out to the participants. You receive an order to report to the wargaming institute Monday morning at 0800 sharp. Entering a large briefing room, you find a seat and are bombarded with the in-brief. Depending on your rank or expertise, you may be part of the training audience, the response cell, or the control cell. The *training audience* consists of the decision makers. There you find the commander and staff of the red and blue forces. The whole game is set up for their benefit. The rest of the personnel that surrounds them are just support folk.

The people in the *response cell* are the buffer between the training audience, the computer model, and the control cell. They take the battle orders from the staff, convert them into simulation lingo, and feed them to the beast, while keeping the control cell informed of the moves. The output of the computer is relayed back to the commander and his staff. The response cell is vital to the realism of the game. They organize and operate the joint and combined armed forces; portray the correct levels of allowances, supply, and logistics; and exercise appropriate operational doctrines, tactics, and procedures.

The *control cell* consists of the umpires, analysts, and computer operators. Here you find the experts and most likely the planners who set up the game. Any technical and modeling issues are resolved by them, and they are responsible for debriefs and final reports.

After the prebrief the groups go into their team rooms, and the action begins. You have very little warm-up time. The action is fast paced, combat like, and the training audience is directly immersed in the conflict. The competition is intense. The blue commander, usually the highest-ranking officer, feels that his reputation is on the line. It can happen that a staff member is fired on the spot if his performance is lacking.

Following the intelligence briefing, the first move is made. The red commander decides to invade the oil-rich territory. His staff passes the order of battle to the response cell. There it is checked, fed into the computer, and the control cell is notified. The result is given to the red and blue forces in the appropriate format. Now, the blue commander must counter with the deployment of his forces. He responds with a massive assault, engaging his Air Force, Navy, and Army Corps. This force structure is passed to the response cell, checked, and entered into the computer. Casualties are calculated, loss of war materiel is recorded, and territorial gains are drawn on the maps. One cycle is complete.

These cycles can be time phased or event driven and continue until the exit criteria are reached. Usually time-phased games are more realistic because they portray more faithfully the pressure that limited time can inflict on the decision makers. It is important that the players act out their roles as faithfully as possible and that the losing party does not fall into the temptation of blaming the game setup for their misfortune.

After the play the training audience gets a good night's rest while the control cell burns the midnight oil analyzing the data and composing the debrief. This *hot wash* is given the next day. It summarizes the events from an overall perspective and provides red and blue with the rationale behind all of the important decisions. The sponsor of the games is given an assessment of his objectives. Any discrepancies of the conduct of the game and modeling deficiencies are addressed and hopefully resolved. With the quick-look report in hand, the warriors say goodbye and, despite some ruffled feathers, part as friends until the next encounter.

11.3.3 Assessing a War Game

War is chaotic, unpredictable, and not a good training ground, although senior commanders are selected from battle-hardened soldiers. As new officers advance through the ranks, they must be given the opportunity to practice their decision-making skills in war-like scenarios. War games fill that need. They make people think about war. Players can apply their theoretical knowledge, acquired in command and staff colleges, to specific military conflicts and become battle-hardened without the terrible cost of war.

New strategies and tactics can be explored in war games without risk of life and loss of equipment, except, possibly, the loss of ego. Battle planning, force sizing, and logistic preparations can be conducted. If the commander and his staff make bad decisions, the war is not lost, only the computer has to be reset for replay.

War games are cheaper than command post and field exercises. No actual troops have to be deployed, and no farmer's field is ravaged by tank tracks. They can be played anywhere on the globe or, stretch your imagination, anywhere in space. Because battles are fought over maps and not over actual territory, international treaties are not violated. Furthermore, war games are unaffected by cumbersome safety and environmental restrictions.

Even the technologist clamors for participation in war games. He brings his newest weapon concept to the games and hopes for a rousing support from the training audience. Thus emboldened, he can go home and begin the task of planning a development program.

War games embody many advantages. Ultimately, competing nations may settle their conflicts by simulated war, without firing a single bullet. However, until that age dawns, we have to assess war games in the critical light of real, cruel, and chaotic war. Here lie the shortcomings. War games do not match reality. The movement of troops, aircraft, and space assets can be approximated, but models never can predict what will really occur during their deployment. Human relations, the threat of death, and mechanical failures contribute to the confusion. Without the physical threat players may be more complacent or more aggressive than in actual combat, whereas others may be tempted to play the expected solution, to satisfy the sponsor's preference.

The participants determine the success of a war game. The most common pitfall is the complaint that something about the model is not quite right. Particularly, the losing players are tempted by that response. They may not understand the scenario or the constraints placed on their movements. If the game deals with new doctrines or advanced weapon concepts, their know-how fails them, and they may lose interest in continuing the exercise. Senior experienced commanders have their own ideas how a game should play. To convince them otherwise is often difficult. After all, the model could be wrong!

The sponsoring agency must not prejudge the outcome. Using a game to prove or disprove a point is a travesty. War games are played to raise issues not to settle them. They rarely produce quantitative measures of performance and are unique and do not provide a statistical basis for decision making.

War games do not predict the outcome of a conflict; they only sharpen the decision-making skills of the officers. They can convey a false picture of time passage and combat effects. Particularly, they conceal the reality and difficulty of the command and control functions. The free message flow between the commander, his staff, and the response cell lulls the players into the illusion that the information flow in war is unimpeded.

11.3.4 Summary

I have laid out before you the strength and weaknesses of wargaming. Used with caution, war games become a great tool in the hands of a seasoned leader. Admiral Chester W. Nimitz acknowledged, "The war with Japan has been enacted in the game room here by so many people and in so many different ways that nothing that happened during the war was a surprise—absolutely nothing, except the kamikaze tactics towards the end of the war; we had not visualized those."

Wargaming inhabits the pinnacle of the pyramid of model hierarchy. It is the ultimate campaign simulation. Beneath it are the mission simulations whith single force-on-force conflicts, followed by the engagement simulations of few players. The bedrock foundation is fashioned by the engineering simulations, to which this book is dedicated. All of these tools are necessary to build the pyramid and to undergird the genuineness of war games.

References

[1] Rolfe, J. M., and Staples, K. J. (eds.), *Flight Simulation*, Cambridge Univ. Press, Cambridge, England, U.K., 1986.

[2] Young, L. R., "Fundamentals of Flight Simulation," Summer Course, Man-Vehicle Lab., Dept. of Aeronautics and Astronautics, Massachusetts Inst. of Technology, Cambridge, MA, 1996.

[3] Proctor, P., "Boeing Hones Computer-Based Fighter Training System," *Aviation Week and Space Technology*, 29 Nov. 1999, pp. 50–53.

[4] Perla, P. P., *The Art of Wargaming: A Guide for Professionals and Hobbyists*, U.S. Naval Inst. 1990.

Appendix A
Matrices

Computer modeling of flight dynamics makes extensive use of matrices. You should already be familiar with the basic concepts of matrix algebra. To refresh your memory, the essential facts are summarized here. For practice you can do the exercises at the end of this section.

A.1 Matrix Definitions

An $m \times n$ matrix is a rectangular array of $m \times n$ elements arranged in m rows and n columns. For $m = 3$ and $n = 4$ we have the 3×4 matrix

$$\overset{3 \times 4}{[A]} = \begin{bmatrix} a_{11} & a_{12} & a_{13} & a_{14} \\ a_{21} & a_{22} & a_{23} & a_{24} \\ a_{31} & a_{32} & a_{33} & a_{34} \end{bmatrix}$$

In general, the subscript notation defines an $m \times n$ matrix as

$$a_{ij}, \quad i = 1, 2, 3, \ldots m; \quad j = 1, 2, 3, \ldots n$$

The determinant $|A|$ of a matrix $[A]$ is a scalar, obtained from the determinants of the minors M_{ij} of the matrix $[A]$:

$$|A| = (-1)^{i+j} a_{ij} M_{ij} \quad \text{for } i = 1, \text{ or } 2, \text{ or } 3, \ldots, m$$

The transpose $[B]$ of a matrix $[A]$ is obtained by swapping out the rows and columns $[B] = [\bar{A}]; b_{ij} = a_{ji}$. A vector is always portrayed as a column matrix. For a 3×1 vector

$$[a] = \begin{bmatrix} a_1 \\ a_2 \\ a_3 \end{bmatrix} \quad a_i, i = 1, 2, 3$$

The null matrix consists of zero element $[0]$; $0_{ij} = 0$; for all i and j. The square matrix has the same number of rows and columns

$$\overset{n \times n}{[A]}; \quad a_{ij}, i = 1, \ldots, n; \quad j = 1, \ldots, n$$

A matrix is symmetric if it equals its transposed

$$[\bar{A}] = [A]; \quad a_{ji} = a_{ij}$$

A matrix is skew symmetric if it equals its negative transposed

$$[\bar{A}] = -[A]; \quad a_{ji} = -a_{ij}, i \neq j; \quad a_{ji} = a_{ij} = 0, i = j$$

The skew-symmetric 3×3 matrix is

$$[A] = \begin{bmatrix} 0 & -a_3 & a_2 \\ a_3 & 0 & -a_1 \\ -a_2 & a_1 & 0 \end{bmatrix}$$

The off-diagonal elements of a diagonal matrix are zero. For a 3×3 matrix

$$[D] = \begin{bmatrix} d_{11} & 0 & 0 \\ 0 & d_{22} & 0 \\ 0 & 0 & d_{33} \end{bmatrix}$$

The unit matrix is a diagonal matrix with unit elements. For a 3×3 matrix

$$[E] = \begin{bmatrix} 1 & 0 & 0 \\ 0 & 1 & 0 \\ 0 & 0 & 1 \end{bmatrix}$$

The Kronecker delta is

$$\delta_{ij} = \begin{cases} 1, & i = j \\ 0, & i \neq j \end{cases}$$

The adjoint of a matrix $[A]$ is obtained by $\mathrm{adj}[A] = $ transposed of $\{(-1)^{i+j} M_{ij}\}$, where M_{ij} is the minor determinant of a_{ij}.

The inverse of a matrix $[A]$ is calculated from its adjoint and determinant: $[A]^{-1} = \mathrm{adj}[A]/|A|; |A| \neq 0$.

A matrix $[A]$ is orthogonal if its inverse equals its transposed: $[A]^{-1} = [\bar{A}]$. Other properties are $[\bar{A}][A] = [E]; |A| = \pm 1$.

The rank of a matrix is the largest number r such that at least one rth-order determinant formed from the matrix by deleting rows and/or columns is different from zero. The trace of a matrix is the sum of its diagonal elements.

A matrix $[A]$ can be partitioned into submatrices. For example,

$$\underset{[A]}{\overset{3 \times 4}{}} = \begin{bmatrix} a_{11} & a_{12} & a_{13} & a_{14} \\ a_{21} & a_{22} & a_{23} & a_{24} \\ \hline a_{31} & a_{32} & a_{33} & a_{34} \end{bmatrix} = \begin{bmatrix} \overset{2 \times 1}{A_{11}} & \overset{2 \times 3}{A_{12}} \\ \hline \overset{1 \times 1}{A_{21}} & \overset{1 \times 3}{A_{22}} \end{bmatrix}$$

where A_{11}, A_{12}, A_{21}, and A_{22} are submatrices.

A.2   Matrix Operations

Two matrices $[A]$ and $[B]$ of dimension m, n are equal if all corresponding elements are equal:

$$\overset{m \times n}{[A]} = \overset{m \times n}{[B]}, \quad \text{if} \quad a_{ij} = b_{ij}, \quad \text{for all } i, j$$

The sum of two matrices $[A]$ and $[B]$ with the same dimension m, n is obtained by adding corresponding elements:

$$\overset{m \times n}{[C]} = \overset{m \times n}{[A]} + \overset{m \times n}{[B]}, \quad \text{or} \quad c_{ij} = a_{ij} + b_{ij}, \quad \text{for all } i, j$$

Addition is associative: $([A] + [B]) + [C] = [A] + ([B] + [C+])$. Addition is commutative: $[A] + [B] = [B] + [A]$.

The product of the $m \times n$ matrix $[A]$ and the $n \times r$ matrix $[B]$ is the $m \times r$ matrix $[C]$:

$$\overset{m \times r}{[C]} = \overset{m \times n}{[A]}\overset{n \times r}{[B]}; \quad \text{or} \quad c_{ik} = \sum_{j=1}^{n} a_{ij}b_{jk}; \quad i = 1, \ldots, m; \ k = 1, \ldots, r$$

where $[A]$ and $[B]$ must be conformable: $m \times r = (m \times n)(n \times r)$.

The product of the $m \times n$ matrix $[A]$ by the scalar α is

$$\overset{m \times n}{[C]} = \alpha \overset{m \times n}{[A]}, \quad \text{or} \quad c_{ij} = \alpha a_{ij}, \quad \text{for all } i, j$$

Multiplication is associative: $[A]([B][C]) = ([A][B])[C]$. Multiplication is distributive: $[A]([B] + [C]) = [A][B] + [A][C]$ but not commutative $[A][B] \neq [B][A]$. Important rules are the following:

$$\overline{[A][B]} = [\bar{B}][\bar{A}]$$

$$([A][B])^{-1} = [B]^{-1}[A]^{-1}, \quad \text{if} \quad |A| \neq 0, \ |B| \neq 0$$

$$\overline{([A]^{-1})} = ([\bar{A}])^{-1}, \quad \text{if} \quad |A| \neq 0$$

Differentiation operates on every element of a matrix $[A]$:

$$\frac{d}{dt}[A] \equiv \frac{d}{dt}a_{ij}, \quad \text{for all } i, j \quad \text{and} \quad a_{ij} \text{ must be differentiable}$$

Integration operates on every element of a matrix $[A]$

$$\int [A]\, dt \equiv \int a_{ij}\, dt, \quad \text{for all } i, j$$

Any square matrix $[A]$ can be decomposed as the sum of a symmetric and skew-symmetric matrix:

$$[A] = \frac{1}{2}\left(\underbrace{[A] + [\bar{A}]}_{\text{symmetric}}\right) + \frac{1}{2}\left(\underbrace{[A] - [\bar{A}]}_{\text{skew symmetric}}\right)$$

A.3 Matrix Eigenvalues

The similarity transformation is formed from a square, nonsingular matrix $[T]$ operating on a square matrix $[A]$:

$$[A]^* = [T][A][T]^{-1}$$

Two similar matrices $[A]^*$ and $[A]$ have the same rank, trace, determinant, and eigenvalues; if $[A]$ is symmetric, so is $[A]^*$. If $[T]$ is orthogonal, the similarity transformation is $[A]^* = [T][A][\bar{T}]$.

Orthogonal transformations preserve scalar multiplication of vectors, vector addition, multiplication by scalars, absolute values and distances, orthogonality and orthonormality, and value of trace of a matrix. The eigenvalues of an $n \times n$ matrix $[A]$ are the roots of the characteristic equation

$$|[A] - \lambda[E]| = 0$$

An $n \times n$ matrix has n eigenvalues. A real symmetric matrix has only real eigenvalues. An orthogonal matrix $[T]$ has one real eigenvalue ± 1 and a pair of conjugate complex poles $e^{\pm i\phi}$. Given a square matrix $[A]$ with eigenvalues λ_i, then $\alpha[A]$ has the eigenvalues $\alpha\lambda_i$, and $([A])^p$ has the eigenvalues λ_i^p. A square matrix is nonsingular if and only if all its eigenvalues are nonzero.

A matrix $[A]$ can be diagonalized by an orthogonal matrix $[T]$ if $[A]$ is symmetric. The determinant of a matrix $[A]$ can be calculated from the product of its eigenvalues $|A| = \prod_{i=1}^{n} \lambda_i$.

The trace of a matrix $[A]$ is the sum of all its eigenvalues $Tr(A) = \sum_{i=1}^{n} \lambda_i$.

A real, symmetric quadratic form $[\bar{x}][A][x]$ is symmetric if the matrix $[A]$ is real and symmetric. A real symmetric quadratic form is positive definite, if its eigenvalues are positive.

Problems

A.1   Partitioned matrices obey matrix operations just like regular matrices. Show that the product of two matrices equals the product of their partitioned forms by using the following example:

$$\begin{bmatrix} 1 & 1 & 1 \\ 2 & -1 & 0 \\ -1 & 0 & 2 \end{bmatrix} \begin{bmatrix} 1 & 2 & 3 & -1 \\ 3 & -1 & 1 & 0 \\ 0 & 0 & -2 & 1 \end{bmatrix} = ?$$

A.2   What is the adjoint and the inverse of matrix $[A]$?

$$[A] = \begin{bmatrix} -2 & 1 & 3 \\ 4 & 0 & -1 \\ 3 & 3 & 2 \end{bmatrix}$$

A.3   If $[A]$ and $[B]$ are symmetric matrices of the same dimension, prove that the product $[A][B]$ is symmetric only if $[A][B] = [B][A]$.

A.4   Prove that the square of a skew-symmetric matrix is a symmetric matrix.

A.5   Determine the symmetric and skew-symmetric parts of the matrix $[A]$.

$$[A] = \begin{bmatrix} 1 & 2 & 3 \\ 4 & 5 & 6 \\ 7 & 8 & 9 \end{bmatrix}$$

A.6   Determine the characteristic equation and the eigenvalues of the matrix $[A]$.

$$[A] = \begin{bmatrix} 2 & 0 & 0 \\ 0 & 3 & -\sqrt{3} \\ 0 & -\sqrt{3} & 5 \end{bmatrix}$$

Appendix B
CADAC Primer

Program CADAC, Computer Aided Design of Aerospace Concepts, provides an environment for the development of general purpose, digital computer simulations of time-phased dynamic systems. It manages input and output, generates stochastic noise sources, controls state variable integration, and provides postprocessing analysis and display. CADAC has proven its adaptability to many simulation tasks: air-to-ground and air-to-air missiles, ground-to-space and space-to-ground vehicles, and airplanes. The CADAC environment is suitable for three-, five-, and six-DoF simulations. It supports deterministic and Monte Carlo runs. Output can be listed or plotted.

CADAC is hosted on an IBM-compatible PC with a minimum of 32 MB of RAM, 25 MB of free disk space, and Microsoft's Window operating system. The graphics utilities are best displayed on a 1024×768-resolution screen with font size set to small.

CADAC consists of CADAC Studio and CADAC Simulations. Both can be downloaded from the CADAC Web site (go to www.aiaa.org, under Market Pulse, select Web Links to find CADAC among a list of aerospace-related links). CADAC Studio, written in Visual BASIC, analyzes and plots the output with programs KPLOT and SWEEP and provides utility function for debugging. The vehicle simulations are written in FORTRAN 77 with some common language extensions. The preferred compiler is Compaq's (formerly Digital) Visual FORTRAN for Microsoft's Developer Studio. Other compilers (like Silicon Graphics) may require minor modifications of the CADX3.FOR executive routine.

The CADAC Studio comes with four volumes of documentation:

1) *Quick Start* gets you started with the test case and provides an overview of the input files, executables, and output files.

2) *User Documentation* addresses all capabilities of the CADAC development environment. It should give answers to most questions that come up during the design of a new trajectory simulation. Tables, examples, and matrix utilities provide useful references. The serious CADAC user should read this document in its entirety.

3) *Program Documentation* provides details for many subjects: building the input and the header files, the integration routine, generation of stochastic variables, execution of multiruns, sweep runs, Monte Carlo runs; utilities that aid in building, documenting, and analyzing CADAC simulations. It should be used as reference to answer specific questions.

4) *Real-time CADAC Documentation* addresses the functionality and capabilities of the the real-time CADAC methodology. It explains how to generate the data files for the real-time version of the modules and the validation procedures.

CADAC is essentially a trajectory program. Over the years features were added to make it more useful. Some of these capabilities are the following:

Vector integration   State vectors can be integrated as an entity without breaking them up into scalar components.

Staging   Modes of trajectories, like midcourse or terminal guidance, can be sequenced in sections called stages. They are initiated by IF statements in the INPUT.ASC file.

Multiruns   Several trajectories can be combined. For instance, if the sensitivity of a seeker parameter is to be investigated, new parameter values are scheduled in separate group runs and loaded into the global C-array at the appropriate stage with the keywords LOAD and STAGE.

Reinitializing   To save execution time, the state of the trajectory can be saved at a certain event, say seeker acquisition, and the following group runs are reinitialized at this point. The keyword SAVE will write the global C-array to the file CSAVE.ASC.

Sweep runs   This feature is used to automate the calculation of launch envelopes and footprints. The launch position or target location is swept through a polar grid. The SWEEP utility analyzes and displays the results.

Single Monte Carlo runs   Stochastic runs are generated by the keyword MONTE, followed by the number of desired runs. The noise sources, like wind gusts, INS errors, and seeker noise, are defined by the keywords GAUSS, UNIF, EXPO, or RAYLE. The output file RANVAR.ASC saves the random values, which can be used to rerun a particular trajectory realization. The MCAP program averages the trajectory parameters and calculates means, standard deviations, and correlation coefficients. Histograms and error ellipses are displayed by the KPLOT program.

Sweep Monte Carlo runs   Sweep runs can also be executed as a family of Monte Carlo runs. Just introduce the keyword MONTE with its replication number into a sweep run.

Weather   Atmospheric conditions, like temperature, density, pressure, and wind, can be specified as tabular functions of altitude using the keyword WEATHER in the INPUT.ASC file.

Real-time execution   The CADAC simulation can be converted by the CONVRT.EXE program into a real-time capable code package, suitable for a man-in-the-loop simulation.

The best way to learn CADAC is by trying. So get the code from the Web site and a FORTRAN compiler, then start running the test cases. To help you over the initial hurdles, I include a primer that gets you going.

CADAC PRIMER

... building your own simulation

1. Computer Aided Design of Aerospace Concepts
2. Run GHAME3 test case to get started
3. Look at console display and plot output
4. Modify INPUT.ASC file
5. Define output in HEAD.ASC file
6. Plot output in 3 Dimensions
7. Modify Aerodynamic Module A1
8. Develop your own Module
9. Module assignments and sequencing
10. CADX3.FOR executes integration of state variables
11. Building HEAD.ASC in four steps
12. Building CADIN.ASC
13. Build your own SSTO simulation
14. Your SSTO output should look like this
15. Debugging aids for MODULE.FOR

1. Computer Aided Design of Aerospace Concepts

... simulating the flight dynamics of aerospace vehicles

History

1966 Litton Industry
1978 CADAC-Air Force
1998 CADAC Version 3.0
2000 CADAC Version 3.1

CADAC Family of Simulations

CADAC 2 - 3 DoF, spherical earth, **GHAME3, ROCKET3**
CADAC 3 - 5 DoF, flat earth, air-to-ground, **CRUISE5**
CADAC 4 - 5 DoF, flat earth, air-to-air, **AIM5, SRAAM5**
CADAC 5 - 5 DoF, spherical earth
CADAC 6 - 6 DoF, flat earth, missiles **SRAAM6**
CADAC 7 - 6 DoF, flat earth, aircraft **FALCON6**
CADAC 8 - 6 DoF, elliptical earth, hypersonic **GHAME6**

Compatibility

FORTRAN 77 with extensions
Compact Visual Fortran is preferred compiler
Platforms:
          All Windows platforms
          Adaptable to Silicon Graphics computers

Run-Time Capabilities

Staging
Special functions
Random distributions
Multiple runs
Monte Carlo runs
Re-Initialization of runs
Automated envelope generation

Plotting and Analysis of Output

KPLOT
2-DIM, Strip Charts, 3-DIM, Globe,
Histograms, Bi-variate distributions
SWEEP
Launch envelopes
Footprints
MCAP
Monte Carlo analysis
QPRINT
Listing of variables

2. Run GHAME3 test case to get started

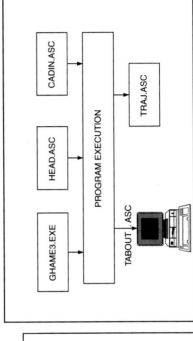

1. Build GHAME3.EXE
2. Place HEAD.ASC in project directory or build from MODULE.FOR
3. Build CADIN.ASC
4. Run GHAME3.EXE
5. Look at Screen output
6. Plot from TRAJ.ASC file

File Names

MODULE.FOR	Vehicle subsystems (modules)
UTL3.FOR	Utility matrix routines (V.3)
CADX3.FOR	CADAC executive (V.3)
INPUT.ASC	Free format input file
CADIN.ASC	Fixed format Fortran input
HEAD.ASC	Defines output to TRAJ.ASC
TRAJ.ASC	Output file for plotting
TABOUT.ASC	Output scrolled to screen
GHAME3.EXE	Compiled Fortran program
MKHEAD3.EXE	Utility program (V.3)
CADIN3.EXE	Utility program (V.3)

3. Look at console display and plot output

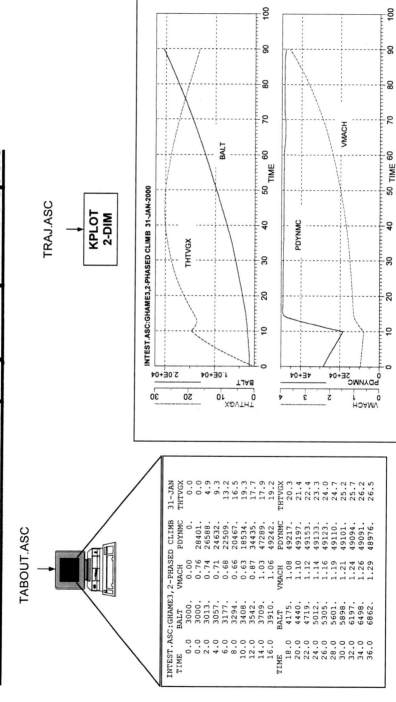

4. Modify INPUT.ASC file

Module calling sequence

Stage at time > 50 s sec

Execute program

Fortran Stop

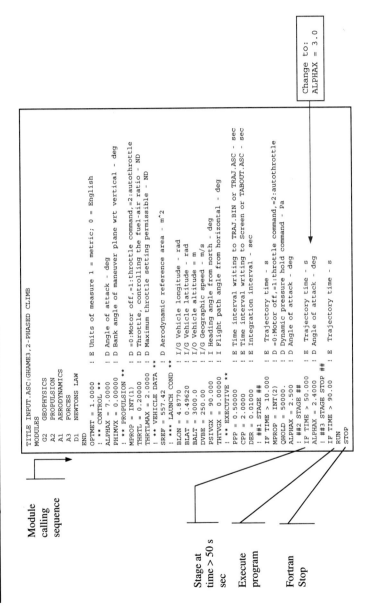

```
TITLE INPUT.ASC:GHAME3,2-PHASED CLIMB
MODULES
  G2    GEOPHYSICS
  A2    PROPULSION
  A1    AERODYNAMICS
  A3    FORCES
  D1    NEWTONS LAW
END
OPTMET = 1.0000          ! B Units of measure 1 = metric; 0 = English
! ** CONTROL **
ALPHAX = 7.0000          ! D Angle of attack - deg
PHIMVX = 0.00000         ! D Bank angle of maneuver plane wrt vertical - deg
! ** PROPULSION **
MPROP = INT(1)           ! D =0:Motor off,=1:throttle command,=2:autothrottle
THRTL = 0.20000          ! D Throttle, controlling the fuel-air ratio - ND
THRTLMAX = 2.0000        ! D Maximum throttle setting permissible - ND
! ** VEHICLE DATA **
SREF = 557.42            ! D Aerodynamic reference area - m^2
! *** LAUNCH COND **
BLON = 4.8770            ! I/G Vehicle longitude - rad
BLAT = 0.49620           ! I/G Vehicle latitude - rad
BALT = 3000.00           ! I/O Vehicle altitude = m
DVBE = 250.00            ! I/G Geographic speed - m/s
PSIVGX = 90.000          ! I Heading angle from north - deg
THTVGX = 0.00000         ! I Flight path angle from horizontal - deg
! ** EXECUTIVE **
PPP = 0.50000            ! E Time interval writing to TRAJ.BIN or TRAJ.ASC - sec
CPP = 2.0000             ! E Time interval writing to Screen or TABOUT.ASC - sec
DER = 0.01000            ! E Integration interval - sec
! ##1 STAGE ##
IF TIME > 10.000         ! B Trajectory time - s
MPROP = INT(2)           ! D =0:Motor off,=1:throttle command,=2:autothrottle
QHOLD = 50000.           ! D Dynamic pressure hold command - Pa
ALPHAX = 2.500           ! D Angle of attack - deg
! ##2 STAGE ##
IF TIME > 50.000         ! E Trajectory time - s
ALPHAX = 2.4000          ! D Angle of attack - deg
! ##3 STAGE STOP ##
IF TIME > 90.00          ! E Trajectory time - s
RUN
STOP
```

Change to:
ALPHAX = 3.0

5. Define output in HEAD.ASC file

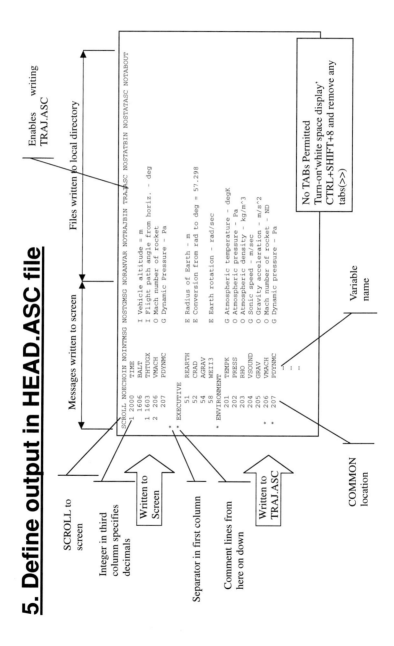

```
SCROLL NOECHO IN NOINTMSG NOSTGMSG NORANVAR NOTRAJBIN TRAJASC NOSTATBIN NOSTATASC NOTABOUT
  1 2000   TIME         I Vehicle altitude = m
  1 1606   BALT         I Flight path angle from horiz. - deg
  1 1603   THTUGX       O Mach number of rocket
  2  206   VMACH        G Dynamic Pressure - Pa
     207   PDYNMC

EXECUTIVE
   51   REARTH         E Radius of Earth - m
   52   CRAD           E Conversion from rad to deg = 57.298
   54   AGRAV          E Earth rotation - rad/sec
   58   WEII3
*
ENVIRONMENT
  201   TEMPK          G Atmospheric temperature - degK
  202   PRESS          O Atmospheric pressure - Pa
  203   RHO            O Atmospheric density - kg/m^3
  204   VSOUND         G Sonic speed - m/sec
  205   GRAV           O Gravity acceleration - m/s^2
  206   VMACH          O Mach number of rocket - ND
  207   PDYNMC         G Dynamic pressure - Pa
*
*                      : :
```

Callouts:

- Enables writing TRAJ.ASC
- Files written to local directory
- Messages written to screen
- SCROLL to screen
- Integer in third column specifies decimals
- Written to Screen
- Separator in first column
- Comment lines from here on down
- Written to TRAJ.ASC
- COMMON location
- Variable name

No TABs Permitted
Turn-on'white space display'
CTRL+SHIFT+8 and remove any
tabst(>>)

6. Plot output in 3 Dimensions

Use **KPLOT-GLOBE** and plot BLON, BLAT, BALT traces

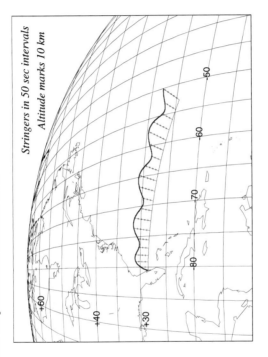

Stringers in 50 sec intervals
Altitude marks 10 km

Challenge: Modify INPUT.ASC by modulating angle of attack to duplicate this trajectory of 1000 sec duration

7. Modify Aerodynamic Module A1

Problem
Increase drag by a factor specified in
INPUT.ASC
Modify A1 Module source code
Include FACTCD in HEAD.ASC
Insert FACTCD into INPUT.ASC
Build CADIN.ASC
Run GHAME3.EXE

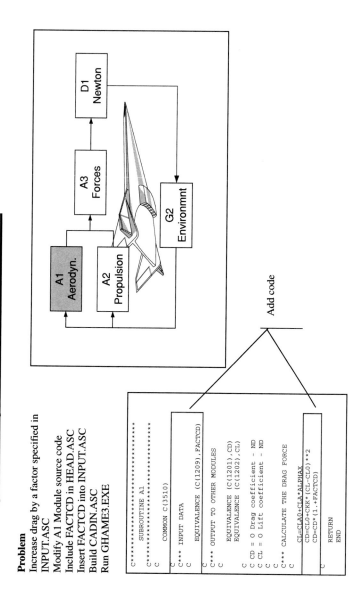

```
C**********************************
C     SUBROUTINE A1
C**********************************
C
      COMMON C(3510)
C
C*** INPUT DATA
C
      EQUIVALENCE (C(1209),FACTCD)
C
C*** OUTPUT TO OTHER MODULES
C
      EQUIVALENCE (C(1201),CD)
      EQUIVALENCE (C(1202),CL)
C
C CD = 0 Drag coefficient - ND
C CL = 0 Lift coefficient - ND
C
C*** CALCULATE THE DRAG FORCE
C
      CL=CLA0+CLA*ALPHAX
      CD=CD0+CKK*(CL-CL0)**2
      CD=CD*(1.+FACTCD)
C
      RETURN
      END
```

Add code

8. Develop your own Module

```
C************************************************
      SUBROUTINE XXI
C************************************************
      COMMON C(3510)
C*** INITIALIZATION
      EQUIVALENCE (C(1210),IX1)
C IX1 = I placeholder for table look-up - ND
      IX1=1
      RETURN
      END
C************************************************
      SUBROUTINE XX
C************************************************
      COMMON C(3510)
C*** INPUT DATA
      EQUIVALENCE (C(1203),ALPHAX)
C ALPHAX = D Angle of attack - deg
C*** INITIALIZATION
      EQUIVALENCE (C(1210),IX1)
C*** INPUT FROM EXECUTIVE
      EQUIVALENCE (C(0052),CRAD)
C*** INPUT FROM OTHER MODULES
      EQUIVALENCE (C(0206),VMACH)
C VMACH= O Mach number of rocket - ND
C*** OUTPUT TO OTHER MODULES
      EQUIVALENCE (C(1201),CD)
      EQUIVALENCE (C(1202),CL)
C CD = O Drag coefficient - ND
C CL = O Lift coefficient - ND
C*** DIAGNOSTICS
      EQUIVALENCE (C(1204),CD0)
      EQUIVALENCE (C(1205),CL0)
      EQUIVALENCE (C(1206),CKK)
      EQUIVALENCE (C(1207),CLA)
>>>>>>> CODE <<<<<<<

      RETURN
      END
```

Initialization module XXI is called once
  Initializes variables
  Identifies state variables to be integrated
Module XX is called twice for every integration step (Euler predictor/corrector)
  Calculates the derivatives of the state variables
  Executes all other computations
  Calls utility subroutines MATyyy, VECyyy, TABLy,TABLPy
  Calls one lower level of subroutines XXyyyy
  Talks to other modules by EQUIVALENCEing to COMMON(3510)

  INPUT DATA: D
  INITIALIZATION: I
  INPUT FROM EXECUTIVE: E
  INPUT FROM OTHER MODULES
  STATE VARIABLES: S
  OUTPUT TO OTHER MODULES: O
  DIAGNOSTICS: G

• Avoid:

  labeled COMMON
  CALLs to subroutines of other modules

9. Module assignments and sequencing

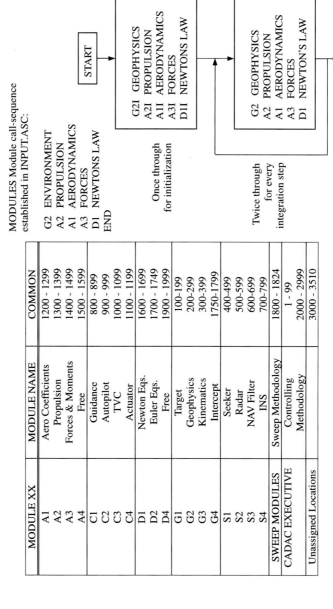

10. CADX3.FOR executes integration

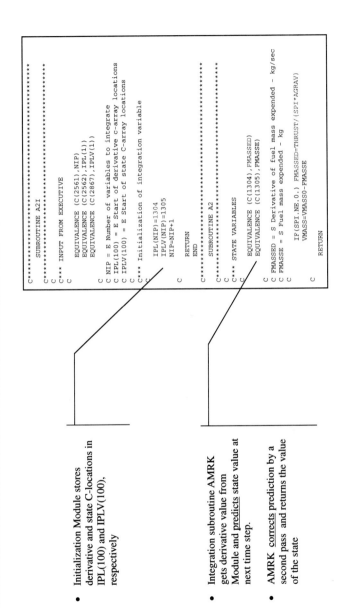

- Initialization Module stores derivative and state C-locations in IPL(100) and IPLV(100), respectively

- Integration subroutine AMRK gets derivative value from Module and predicts state value at next time step.

- AMRK corrects prediction by a second pass and returns the value of the state

```
C***********************************************
      SUBROUTINE A2I
C***********************************************
C*** INPUT FROM EXECUTIVE
C
      EQUIVALENCE (C(2561),NIP)
      EQUIVALENCE (C(2562),IPL(1))
      EQUIVALENCE (C(2867),IPLV(1))
C
C NIP = E Number of variables to integrate
C IPL(100) = E Start of derivative c-array locations
C IPLV(100) = E Start of state C-array locations
C
C*** Initialization of integration variable
C
      IPL(NIP)=1304
      IPLV(NIP)=1305
      NIP=NIP+1
C
      RETURN
      END
C***********************************************
      SUBROUTINE A2
C***********************************************
C*** STATE VARIABLES
C
      EQUIVALENCE (C(1304),FMASSED)
      EQUIVALENCE (C(1305),FMASSE)
C
C FMASSED = S Derivative of fuel mass expended - kg/sec
C FMASSE = S Fuel mass expended - kg
C
      IF(SPI.NE.0.) FMASSED=THRUST/(SPI*AGRAV)
      VMASS=VMASS0-FMASSE
C
      RETURN
```

11. Building HEAD.ASC in four steps

1. Merge Modules into MODULE.FOR
(Window PC use DOS COPY command)

↓

MKHEAD3.EXE

2. Builds columns of C-Locations
Error checking

↓

HEAD.ASC →

DFHEAD3.EXE

3. Inserts variable definitions taken
from MODULE.FOR

↓

HEAD.ASC →

4. Insert variables for scroll list
and asterisks for variables to be
written to TRAJ.ASC

↓

HEAD.ASC

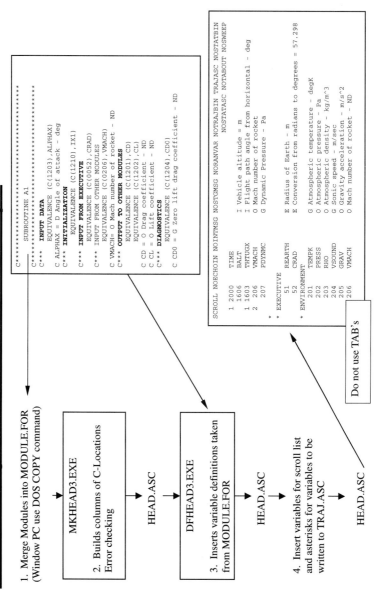

```
C*****************************************
      SUBROUTINE A1
C*****************************************
C***  INPUT DATA
      EQUIVALENCE (C(1203),ALPHAX)
C ALPHAX = D Angle of attack - deg
C***  INITIALIZATION
      EQUIVALENCE (C(1210),IX1)
C***  INPUT FROM EXECUTIVE
      EQUIVALENCE (C(0052),CRAD)
C***  INPUT FROM OTHER MODULES
      EQUIVALENCE (C(0206),VMACH)
C VMACH= O Mach number of rocket - ND
C***  OUTPUT TO OTHER MODULES
      EQUIVALENCE (C(1201),CD)
      EQUIVALENCE (C(1202),CL)
C CD = O Drag coefficient - ND
C CL = O Lift coefficient - ND
C***  DIAGNOSTICS
      EQUIVALENCE (C(1204),CD0)
C CD0 = G Zero lift drag coefficient - ND
```

```
SCROLL NOECHOIN NOINTMSG NOSTGMSG NORANVAR NOTRAJBIN TRAJASC NOSTATBIN
                                           NOSTATASC NOTABOUT NOSWEEP
1 2000    TIME      I Vehicle altitude = m
  1606    BALT      I Vehicle altitude = m
1 1603    THTUGX    I Flight path angle from horizontal - deg
2  206    VMACH     O Mach number of rocket
   207    PDYNMC    G Dynamic Pressure - Pa
*
* EXECUTIVE
    51    REARTH    E Radius of Earth - m
    52    CRAD      E Conversion from radians to degrees = 57.298
* ENVIRONMENT
   201    TEMPK     G Atmospheric temperature - degK
   202    PRESS     O Atmospheric pressure - Pa
   203    RHO       O Atmospheric density - kg/m^3
   204    VSOUND    O Sonic speed - m/sec
   205    GRAV      O Gravity acceleration - m/s^2
   206    VMACH     O Mach number of rocket - ND
```

Do not use TAB's

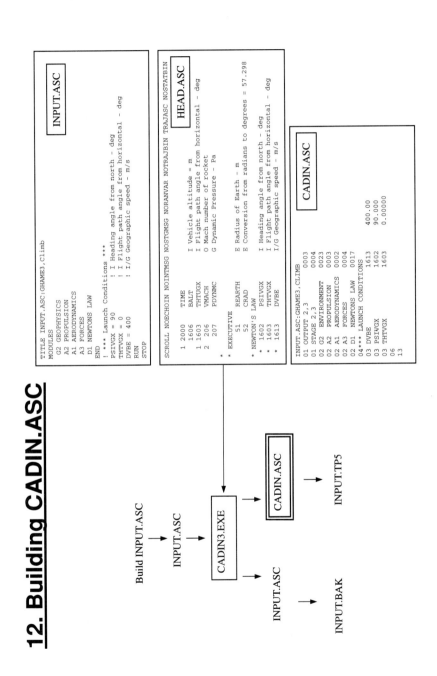

13. Build your own SSTO simulation

A2 MODULE Propulsion

Liquid throttlable rocket motors
Thrust is acting parallel to the body x-axis.

Maximum Thrust = 1.51×10^6 N
Nozzle Exit Velocity = 4860 m/s
Launch Mass = 181,437 kg
Fuel Mass = 156194 kg
Vehicle Mass = 25243 kg (no fuel)

Test Case

Initial Conditions
  Cape Canaveral, Altitude 12 km,
  Geographic Speed 253 m/s, easterly direction
Control Sequence

t sec	α deg	throttle
< 200	22.93	.9
200 –400	5.73	.9
> 400	5.73	.5

A1 MODULE Aerodynamics

The aerodynamics are modeled by a symmetric drag polar of the form

$$C_D = C_{D_0}(M) + k(M) C_{L_\alpha}^2 (M) \alpha^2 = C_{D_0}(M) + \bar{C}_{L_{\alpha^2}}(M) \alpha^2$$

Reference Area = 102 m^2

Mach	C_{D_0}	$C_{L_\alpha}\ (rad^{-1})$	$\bar{C}_{L_{\alpha^2}}\ (rad^{-2})$
0.2	.0417	1.569	0.815
1.2	.0850	1.482	1.185
5.0	.0400	1.115	1.135
10.0	.0290	1.063	1.040
20.0	.0320	1.033	1.022

14. Your SSTO output should look like this

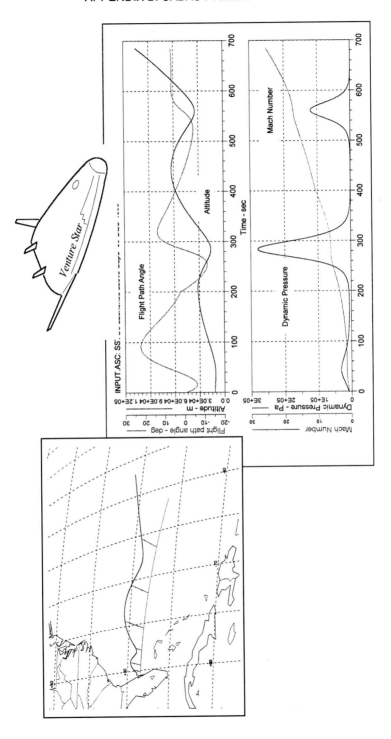

15. Debugging aids for MODULE.FOR

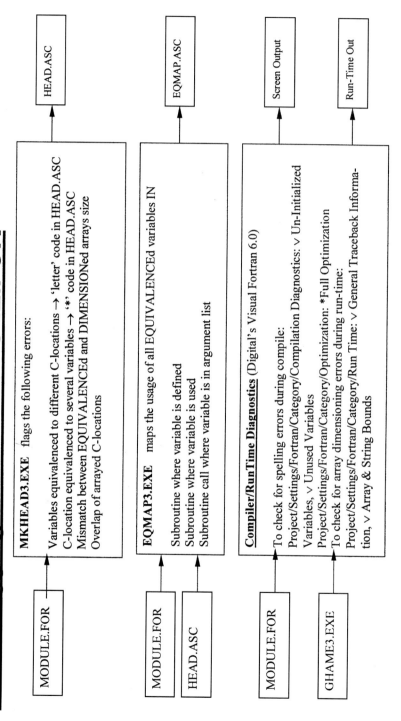

MKHEAD3.EXE   flags the following errors:

Variables equivalenced to different C-locations → 'letter' code in HEAD.ASC
C-location equivalenced to several variables → '*' code in HEAD.ASC
Mismatch between EQUIVALENCEd and DIMENSIONed arrays size
Overlap of arrayed C-locations

MODULE.FOR → HEAD.ASC

EQMAP3.EXE   maps the usage of all EQUIVALENCEd variables IN

Subroutine where variable is defined
Subroutine where variable is used
Subroutine call where variable is in argument list

MODULE.FOR
HEAD.ASC → EQMAP.ASC

Compiler/RunTime Diagnostics (Digital's Visual Fortran 6.0)

To check for spelling errors during compile:
Project/Settings/Fortran/Category/Compilation Diagnostics: ∨ Un-Initialized
Variables, ∨ Unused Variables
Project/Settings/Fortran/Category/Optimization: *Full Optimization
To check for array dimensioning errors during run-time:
Project/Settings/Fortran/Category/Run Time: ∨ General Traceback Informa-
tion, ∨ Array & String Bounds

MODULE.FOR → Screen Output
GHAME3.EXE → Run-Time Out

Appendix C
Trajectory Simulations

Trajectory simulations abound. Government agencies have spearheaded general-purpose simulations, and aerospace companies have developed specialized models. Yet, not all simulations are readily available. Many military simulations are classified, and industry keeps their source code under wraps, away from the competition. Some simulations are available to the public, occasionally at a nominal fee.

To give you an overview, I have assembled the synopsis of four popular simulation environments. If you want to get more information, you should use the contact point cited at the end of each profile.

C.1 Trap Trajectory Analysis Program

C.1.1 Description

TRAP is a general-purpose missile fly-out program with a shooter and aircraft target model.

- Degrees-of-freedom: three-DoF point mass, three-DoF pitch, three-DoF yaw, five DoF, six DoF
- Equations of motion: Reference B. Etkin, *Dynamics of Atmospheric Flight*, Wiley; singularity at vertical climb or dive
- Inertial reference frame: flat Earth
- Aerodynamic model: alpha, beta table look-up
- Propulsion: rocket, turbojet, and ramjet
- Autopilot: many options—acceleration feedback, rate feedback, torque balance, etc.
- Guidance: pro-nav, pursuit, three-point guidance
- Sensors: infrared; passive, semi-active and active radar

C.1.2 Executive Routine

- Integration: Adam Bashforth second-order trapezoidal predictor (new state calculated from current derivative and past state and derivative); transfer function integration by Tustin z transform; integration interval is 0.01 s
- Communication: via common blocks stored in INCLUDE files

C.1.3 Special Features

- Performance reconstruction: determines unknown flight parameters by minimizing the sum of squares of the difference between given and simulated trajectories
- Launch envelop generator: vertical and horizontal planes

C.1.4 Size

30,000 lines of code, 3 Mbytes

C.1.5 Language

FORTRAN 77 with extensions, double precision, IMSL random variable generators

C.1.6 Input/Output

Free-format input, output listing, postprocessor graphics compatibility

C.1.7 Host

Sun, SGI, PC (simple point mass with interactive plotting)

C.1.8 Application

Large selection of missiles both foreign and domestic, air-to-air, surface-to-air, and air-to-ground

C.1.9 Documentation

Coordinate transformations, data dictionary, single parameters data files, tabular data files, common block source code listings, and detailed descriptions

C.1.10 Developer

National Air Intelligence Center (NAIC/TANW), WPAFB, OH, Phone: 937-257-2653, DSN 787-2653

C.1.11 Comment

Mature code, well-documented, configuration controlled by SURVIAC; no Monte Carlo capability, complex code, difficult to modify, rudimentary plotting

C.1.12 Availability

SURVIAC, AFRL/VAVS/SURVIAC Wright Patterson AFB, OH, 45433-6553, DSN 785-4840

C.2 DIMODS—Digital Modular Simulation

C.2.1 Description

DIMODS is a general purpose missile fly-out program for air-to ground missiles.

- Degrees-of-freedom: Six DoF
- Equations of motion: Newton's law solved in Earth axes, Euler's equiations in body axes; integration of direction cosine matrix
- Inertial reference frame: flat Earth

- Aerodynamic model: aeroballistic angles, table look-up
- Propulsion: rocket, turbojet
- Autopilot: acceleration feedback, rate feedback
- Guidance: pro-nav
- Sensors: electro optical, semi-active lasers

C.2.2 Executive Routine

- Integration: fourth-order Runge–Kutta
- Communication: variables equivalenced to blank common communication array C(3615)

C.2.3 Special Features

Store separation model

C.2.4 Size

15,000 lines of code

C.2.5 Language

FORTRAN 77 with extensions

C.2.6 Input/Output

Fixed-format input, output listing

C.2.7 Host

VAX

C.2.8 Application

DoD official simulations for GBU 15, AGM 130, and Hellfire

C.2.9 Documentation

See respective weapon specification documents

C.2.10 Developer

Lockheed/Martin (formerly, Rockwell International)

C.2.11 Comment

Mature code, tailored to special applications, compatible with CADAC

C.2.12 Availability

Company proprietary, except where released in conjunction with a weapon development program

C.3 Endosim Endoatmospheric Kill Simulation

C.3.1 Description

ENDOSIM is a one-on-one missile simulation of a ground-based interceptor engaging a reentry vehicle target in the atmosphere.

- Degrees-of-freedom: Simple missile six DoF with linear rotational dynamics, full missile six DoF, target three-DoF point mass
- Equations of motion: Newton's law solved in inertial axes, Euler's equations in body axes; quaternion integration for body attitudes
- Inertial reference frame: ECI (Earth-Centered inertial, rotating Earth)
- Aerodynamic model: aeroballistic angles, table look-up
- Propulsion: rocket
- Autopilot: adaptive acceleration feedback, rate feedback, attitude, and thrusters
- Guidance: pro-nav, predictive pro-nav, angle only pro-nav
- Sensors: electro optical, MMW radar, INS, ground-based radar

C.3.2 Executive Routine

- Integration: rectangular, second- and fourth-order Runge–Kutta
- Communication: variables equivalenced to blank common communication array $A(5500)$

C.3.3 Special Features

- Monte Carlo runs
- Event-based modeling: discrete and continuous processes, synchronous, asynchronous
- Target kill assessment: engagement statistics for focused and radial warheads

C.3.4 Size

30,000 lines of code

C.3.5 Language

FORTRAN 77 with extensions

C.3.6 Input/Output

Fixed-format input, for VAX an Intelligent Interactive Interface (I3), output listing

C.3.7 Host

CYBER, VAX, PC

C.3.8 Application

ENNK (endo-atmospheric nonnuclear kill), KEW (kinetic energy weapon)

C.3.9   Documentation

User manual, source code listings

C.3.10   Developer

U.S. Army Strategic Defense Command and AMTEC Corporation, Huntsville, Alabama

C.3.11   Comment

Mature code, well documented, compatible with CADAC, no graphics output

C.3.12   Availability

U.S. Government and Industry. Contact: Deputy Commander, U.S. Army Strategic Defense Command, Attn: CSSD-KE-E/Mr. Dan Bradley, P.O. Box 1500, Huntsville, AL 35807-3801.

C.4   MSTARS—Munition Simulation Tools and Resources Simulation System

C.4.1   Description

MSTARS is a unique visual six-DoF munition simulation development and execution environment built around the commercial tool VisSim (from Visual Solutions, Inc.).

- Top-level simulation diagram with easy to understand bitmaps for component access
- Modular organization of component library in Windows directories
- Simulation construction in hierarchical levels
- Uniform structure for component layout
- Model attribute values via external user-defined file
- Model documentation built into the diagrams
- Easy swapping and modification of components
- External code developed DLLs for a number of components
- Sample attribute files for each model component
- Various data file save options for analysis

C.4.2   Special Features

- Aero memo describes how to construct compatible aerodynamic data files from the user's available aero data sources.
- Fast custom code components can be built by the user in virtually any language that supports Windows DLLs.

C.4.3   Size

About 10 MB for the public release version

C.4.4   Language

Visual, based on VisSim from Visual Solutions, Inc., 487 Groton Road, Westford, MA 01886

C.4.5   Input/Output

Visual interaction and text files for input, interactive plots, and text file output

C.4.6   Host

PC-based, runs under MS Windows 95, 98, NT

C.4.7   Application

Can be used to build and execute simple or complex six-DoF simulations for single or Monte Carlo run analysis. Contains sample library of simple munition components and a sample generic guided bomb (GGB) and generic guided missile (GGM) built from these components.

C.4.8   Documentation

Internal object descriptions, help files, and orientation PowerPoint file

C.4.9   Developer

AFRL/MNGG, Eglin AFB, Florida, Phone: 850-882-8195, ext. 3222 (point of contact is Scott Hess).

C.4.10   Comment

Vigorous development process ongoing for new library components

C.4.11   Availability

Public release version available free upon request from AFRL/MNGG (point of contact is Scott Hess); includes free VisSim viewer for viewing and executing simulations.

Index

Modeling and Simulation of Aerospace
Vehicle Dynamics
Peter H. Zipfel 2000
1-56347-456-6

Applied Mathematics in Integrated
Navigation Systems
Robert M. Rogers 2000
1-56347-445-X

Mathematical Methods in Defense
Analyses, Third Edition
J. S. Przemieniecki 2000
1-56347-396-6

Finite Element Multidisciplinary
Analysis
Kajal K. Gupta and John L. Meek 2000
1-56347-393-3

Aircraft Performance: Theory and
Practice
M. E. Eshelby 1999
1-56347-398-4

Space Transporation: A Systems
Approach to Analysis and Design
Walter E. Hammond 1999
1-56347-032-2

Civil Jet Aircraft Design
Lloyd R. Jenkinson, Paul 1999
Simpkin, and Darren Rhodes
ISBN 1-56347-350-X

Structural Dynamics in Aeronautical
Engineering
Maher N. Bismarck–Nasr 1999
ISBN 1-56347-323-2

Practical Intake Aerodynamic Design
E.L. Goldsmith and J. Seddon 1999
ISBN 1-56347-064-0

Integrated Navigation and Guidance
Systems
Daniel J. Biezad 1999
ISBN 1-56347-291-0

Aircraft Handling Qualities
John Hodgkinson 1999
ISBN 1-56347-331-3

Performance, Stability, Dynamics, and
Control of Airplanes
Bandu N. Pamadi 1998
ISBN 1-56347-222-8

Spacecraft Mission Design,
 Second Edition
Charles D. Brown 1998
ISBN 1-56347-262-7

Computational Flight Dynamics
Malcolm J. Abzug 1998
ISBN 1-56347-259-7

Space Vehicle Dynamics and Control
Bong Wie 1998
ISBN 1-56347-261-9

Introduction to Aircraft Flight
Dynamics
Louis V. Schmidt 1998
ISBN 1-56347-226-0

Aerothermodynamics of Gas Turbine
and Rocket Propulsion, Third Edition
Gordon C. Oates 1997
ISBN 1-56347-241-4

Advanced Dynamics
Shuh-Jing Ying 1997
ISBN 1-56347-224-4

Introduction to Aeronautics:
A Design Perspective
Steven A. Brandt, Randall J. Stiles, 1997
John J. Bertin, and Ray Whitford
ISBN 1-56347-250-3

Introductory Aerodynamics and
Hydrodynamics of Wings and Bodies:
A Software-Based Approach
Frederick O. Smetana 1997
ISBN 1-56347-242-2

An Introduction to Aircraft Performance
Mario Asselin 1997
ISBN 1-56347-221-X

Orbital Mechanics, Second Edition
Vladimir A. Chobotov, Editor 1996
ISBN 1-56347-179-5

Thermal Structures for Aerospace
Applications
Earl A. Thornton 1996
ISBN 1-56347-190-6

Structural Loads Analysis for
Commercial Transport Aircraft:
Theory and Practice
Ted L. Lomax 1996
ISBN 1-56347-114-0

Spacecraft Propulsion
Charles D. Brown 1996
ISBN 1-56347-128-0

Helicopter Flight Dynamics:
The Theory and Application of Flying
Qualities and Simulation Modeling
Gareth D. Padfield 1996
ISBN 1-56347-205-8

Flying Qualities and Flight Testing
of the Airplane
Darrol Stinton 1996
ISBN 1-56347-117-5

Published by
American Institute of Aeronautics and Astronautics, Inc.
601 Alexander Bell Drive, Reston, VA 20191